Jean-Yves Girard

The Blind Spot

Lectures on Logic

European Mathematical Society

Author:

Jean-Yves Girard
CNRS, Institut de Mathématiques de Luminy
UMR 6206
163 Avenue de Luminy, Case 907
13288 Marseille cedex 9
France

E-mail: girard@iml.univ-mrs.fr

2010 Mathematics Subject Classification: 03-01, 03A05, 03B40, 03F(03-05-07-10-20-40-52-99), 18A15, 68Q(05-10-12-15-55)

Key words: Logic, proof-theory, incompleteness, sequent calculus, natural deduction, lambda-calculus, Curry–Howard isomorphism, system F, coherent spaces, linear logic, proof-nets, ludics, implicit complexity, geometry of interaction

ISBN 978-3-03719-088-3

The Swiss National Library lists this publication in The Swiss Book, the Swiss national bibliography, and the detailed bibliographic data are available on the Internet at http://www.helveticat.ch.

Contact address:

European Mathematical Society Publishing House
Seminar for Applied Mathematics
ETH-Zentrum FLI C4
CH-8092 Zürich
Switzerland

Phone: +41 (0)44 632 34 36
Email: info@ems-ph.org
Homepage: www.ems-ph.org

Typeset using the author's TEX files: I. Zimmermann, Freiburg
Printing and binding: Beltz Bad Langensalza GmbH, Bad Langensalza, Germany
The cover picture shows the syllogism *Barbara* as a proof-net, see p. 239.

9 8 7 6 5 4 3 2 1

Contents

Foreword xi

Part I The basics 1

1 Existence vs. essence 3
1.1 The opposition existence/essence . . . 3
1.2 Essentialist and existentialist projects . . . 5
1.3 Gödel and after . . . 8
1.A Essentialism vs. platonism . . . 12
1.B Perfect vs. imperfect . . . 13

2 Incompleteness 15
2.1 Technical statement . . . 15
2.2 Hilbert in the face of incompleteness . . . 20
2.3 Incompleteness is not a deficiency . . . 22
2.4 Metaphorical readings . . . 26
2.A More on the classification of predicates . . . 29
2.B Formal arithmetic . . . 31
2.C Techniques of incompleteness . . . 34
2.D Incompleteness and truth . . . 36
2.E Undecidability . . . 38

3 Classical sequents: LK 41
3.1 Generalities . . . 41
3.2 The classical calculus . . . 43
3.3 The cut-free system . . . 48
3.4 Proof of the *Hauptsatz* . . . 51
3.A Around sequent calculus . . . 54
3.B Semantic aspects . . . 59
3.C Infinitary proof-theory . . . 65

4 Intuitionistic logic: LJ, NJ 70
4.1 The intuitionistic sequent calculus LJ . . . 70
4.2 The *Hauptsatz* in LJ . . . 73
4.3 The natural deduction NJ . . . 75
4.4 The signature in natural deduction . . . 81

4.A Existence and disjunction in LJ 83
4.B Natural deduction vs. sequent calculus 86
4.C Around contraction 87
4.D Logic programming 89
4.E Kripke models 91

Part II Around Curry–Howard 95

5 Functional interpretations 97
5.1 Proofs as functions 97
5.2 Pure λ-calculus 103
5.3 The Curry–Howard isomorphism 107
5.A Kreisel and functional interpretation 109
5.B Combinatory logic 110
5.C Other connectives 111
5.D Martin-Löf's type theory 113

6 System F 115
6.1 System F 115
6.2 The normalisation theorem 121
6.A Type theories 126
6.B Heyting's arithmetic 129
6.C System T 129
6.D Expressive power 131
6.E Subtyping 135
6.F Essence, existence and typing 136

7 The category-theoretic interpretation 140
7.1 The three layers 140
7.2 Closed cartesian categories 147
7.3 Examples of CCC 150
7.4 Logic in a CCC 152
7.A Classical logic 154
7.B Various interpretations 157

Part III Linear logic 161

8 Coherent spaces 163
8.1 *Grandeur* and misery of Scott domains 163
8.2 Coherent spaces 164

8.3 Interpretation of system F 171
8.A Asymmetric interpretations 175

9 Linear logic 178
9.1 Linearity in coherent spaces 178
9.2 Perfect linear connectives 182
9.3 Imperfect connectives 185
9.4 The logical system 187
9.A Monoidal categories 194

10 Perfection vs. imperfection 197
10.1 Phase semantics 197
10.2 A perfect world? 202
10.3 The world is imperfect 208
10.A Focalisation 210

11 Proof-nets 216
11.1 ILL 216
11.2 Multiplicative nets 220
11.3 The correctness criterion 227
11.A More on multiplicatives 233
11.B Syllogistic 238
11.C General nets 240

Part IV Polarised interpretations 249

12 A hypothesis: polarisation 251
12.1 Faithfulness of coherent spaces 251
12.2 A prototype 254
12.3 Objections to polarisation 256
12.4 Logic and games 257
12.5 Proofs and tests 260
12.6 Hypersequentialised logic 262
12.A Classical polarity 267
12.B Intuitionistic logic 269
12.C Hypercoherences 270

13 Designs and behaviours 272
13.1 Designs-*dessins* 272
13.2 Designs-*desseins* 277
13.3 Partial designs 282

13.4 Nets and normalisation: *dessins* 283
13.5 Nets and normalisation: *desseins* 288
13.6 Analytical theorems 292
13.7 Introspective vs. extraspective 298
13.8 Behaviours 300
13.9 An example: the shift 304

14 Ludics: the reconstruction 306
14.1 Additives 306
14.2 Multiplicatives 316
14.3 Quantifiers 320
14.A Faithfulness 324
14.B Bihaviours 326
14.C Parsimony 328
14.D Epilogue (?) 329

15 Orthodox exponentials 331
15.1 The perennial perenniality 331
15.2 Exponential nets: normalisation 333
15.3 Categories and classical logic 334
15.4 The system LC 337
15.A Exponentials and analytic functions 340
15.B Exponential ludics 342
15.C Polarised linear logic 343
15.D The $\lambda\mu$-calculus 350

Part V Iconoclasm 355

16 Heterodox exponentials 357
16.1 The quarrel of images 357
16.2 Exponentials 359
16.3 Russell's antinomy 361
16.4 LLL and ELL 364
16.5 Expressive power 366

17 Quantum coherent spaces 369
17.1 Logic vs. quantum 369
17.2 Probabilistic coherent spaces 373
17.3 Quantum coherent spaces 375
17.4 Additives 380
17.5 Multiplicatives 387

17.6 Discussion 392
17.A Initiation to C^*-algebras 393

18 Nets and duality 403
18.1 Duality and correctness 403
18.2 The original criterion 406
18.A Trips and coherent spaces 409
18.B Non-commutative logic 410

Part VI Geometry of interaction 415

19 The feedback equation 417
19.1 Basic examples 417
19.2 Cut-systems 419
19.3 Solving the equation 421
19.4 The normal form 425
19.5 The first GoI 432
19.A Complements on operators 439

20 Babel Tower vs. Great Wall 443
20.1 Idioms 443
20.2 The Babel Tower 446
20.3 A new finitism? 448
20.A Von Neumann algebras 450
20.B Finite algebras 453
20.C Hyperfinite algebras 456
20.D The determinant 459

21 Finite GoI 462
21.1 Projects 462
21.2 Conducts 469
21.3 The social life of conducts 473
21.4 Polarised conducts 475
21.5 Exponentials 477
21.6 Lateralised logic 480
21.7 The social life of behaviours 482
21.A Second-order quantification 485
21.B Truth 488
21.C Truth and intersubjectivity 491

Envoi. The phantom of transparency 497
The transparent world 497
Logics of transparency 498
Semantics 499
From semantics to the cognitive onion 501
Negation 504

Bibliography 509

Index 517

Il faut toujours dire ce que l'on voit.
Surtout il faut toujours, ce qui est
plus difficile, voir ce que l'on voit.

Charles Péguy, Notre Jeunesse, 1910.

Foreword

This series of lectures on proof-theory *a priori* addresses mathematicians and computer-scientists, physicists, philosophers and linguists; and, since we are no longer in the XVIth – not to speak of the XVIIIth – century, it is doomed to failure. Such a prediction is in contrast to a course focusing on subdomains which work quite well (model-theory, set-theory), not that well (temporal or modal logics), or not at all (quantum or epistemic logics) and which would therefore be grounded on a certain technical excellence or, more prosaically, on a well-understood circle of scientific welfare. This being said, plain success is not the only possible goal; mine might simply be the exposition of a disorder in this apparently well-organised universe, in which logic eventually takes its place between two beer mugs and the *Reader's Digest* and no longer disturbs anybody – like a fat cat purring on the carpet.

On the eve of the last century, the cat was rather a wolf-dog, of the strongly barking kind; the XXth century has been a century of totalitarianisms of all possible kinds, in particular the linguistic variety (styled « turn »). This extreme form of *scientism* consisted in the reduction of any mathematical question (therefore, everything being supposedly mathematisable, any question) to a problem of formal, linguistic, bureaucratic, protocols: Kafka was waiting behind the door. Dating back to 1904, the same scientism was involved in improvement of the human species in Namibia, at the hands of the IInd Reich of the blueprint of the *final solution* proper: how many gallows in this treeless country! Modern logic remains basically impregnated with the « 1900 spirit », this sort of pretension at simplifying everything, since one can solve all problems. When, after 1930, incompleteness shook this haughtiness, one hardly observed more than a complexification of the discourse: instead of explaining from the simpler, one started to explain from the « meta ». There began the time of counterfeit coinage. Since that time, logic, unable to effect its own reformation, severed its links with mathematics, physics, etc.

A typical sophism: what is the point of seeking beautiful mathematical structures for logic? Such a thing cannot exist, since, as mathematics, good or bad, can be translated into logic, the logical structure must reflect the worst, i.e., not exist or, at least, remain very bleak. For instance, when looking for a topological, continuous, interpretation of logic, one will head for the worst (e.g., Scott domains) and one will even be proud of it! Among the revealing details is the insistence of logicians on choosing counter-intuitive symbols, in order to make sure that one does not suggest that certain properties – say distributivity – might be more important than others[1]: « More important, really ? How do you define importance ? ». This reminds me of my daughter Isabelle – then very young – « Why not call the door "spoon" and the spoon "door" ? », to which I answered « When one says "Make for the door", it should not be taken as an invitation to supper ». Among the magisterial mistakes of logic, one will first mention quantum logic, whose ridiculousness can only be ascribed to a feeling of superiority of the language – and ideas, even bad, as soon as they take a written form – over the physical world. Quantum logic is indeed a sort of punishment inflicted on nature, guilty of not yielding to the prejudices of logicians… just like Xerxes had the Hellespont – which had destroyed a boat bridge – whipped.

One century ago, very scarce were those daring to oppose scientists' certainties. After one century of slaughter, this is now much easier: even if the same baloney sempiternally comes back, like the intelligent robot, a fantasy of Artificial Intelligence and unlikely prosthesis for those who badly need it, we have won the right to make fun of scientific *Jivaros*. An instance of this is H. Simon, the guy who had his computer « rediscover » Kepler's third law (squares and cubes), forgetting that it is not the law linking the period and the semimajor axis which is hard to find, it is the very idea of such a law, especially for an… astrologer like Kepler.

It would be fair to observe that, in spite of its heavy scientist-created liabilities, the domain of logic, although limited, is not empty. Model-theory and set-theory are doing rather well; even proof-theory has a non-negligible place and, by the way, what would I otherwise start from, since my topic will precisely be *proof-theory*?

At the beginning of the last century, Einstein's relativity and, in a more radical way, quantum physics, called in question our « fundamental intuitions ». Logic, because of its excesses, decided to catapult itself into emptiness; the non-structure, the non-significant « Everything can be coded in everything[2], and also into the sea of the idea of translating images into sound, or rather gurglings ! ». Still, in the « linguistic turn », the idea of pregnancy of the language was deeply inspired and didn't deserve to become this « machine à décerveler[3] » that we just mentioned. With a closer look, the pregnancy of language contains the germ of another form of « relativisation », in fact of *derealisation* of nature. This is the viewpoint I will try to develop.

One has the right to find this project crazy and to prefer a preamble of the style « *A language is a finite alphabet with which one constructs terms, formulas, proofs – syntax; the language is in turn interpreted in a model – semantics; eventually, this is formalised in a meta-system.* ». But then one does not do logic, at least

[1]Witness for instance, on the eve of linear logic, the point of honour taken by those who insisted on writing « par » + and « with » ×, while « par » distributes over « times ».

[2]And conversely, I suppose ! The idea of mutual codings is ancient and universal : think of des Esseintes and his *orgue à liqueurs* (Huysmans, *À rebours* , 1885).

[3]Removal of the brain, according to Alfred Jarry.

not foundations: why not undertake to seal in the Bering strait? The domain, as it ossified during the XXth century is indeed everything but crazy: a cemetery of ideas. In other words, the only excuse in the XXIth century for indulging in « foundations », is a « grain de folie », i.e., a slight madness.

About the title: it was while revising the text (Summer 2005) that I noticed the recurrence of the expression « blind spot ». The blind spot is what one does not see and what one is not even conscious of not seeing[4]. The most trivial blind spot is the cheap modal logic justified by an even cheaper Kripke semantics and *vice versa*; but one finds similar blindings in the most elaborated interpretations. The good news of these lectures is that the *procedural* standpoint seems to be capable of dislodging the unsaid, the unseen. Simply, while the absence of *Hauptsatz* is enough to show that logic **S5** is nonsense, one has to work much more to imagine what could be wrong in the principles justifying – say – the function 2^n.

Acknowledgements. This indeed constitutes the notes of a series of lectures given at the Université Franco-Italienne, from October to December 2004, at Università Roma Tre (Dipartmento di Filosofia). I am very grateful to Michele Abrusci who initiated this course. I particularly thank, for their help, Lorenzo Tortora de Falco and Marco Pedicini; and, – last but not least – the « linchpin » of the whole enterprise, Roberto Maieli, who took care of the organisation of lessons, of the video *streamings*, of the maintenance of the text on the Web.

Three months of lecturing together with the simultaneous composition of lectures notes of 500-odd pages is a very very heavy burden taking six concentrated days per week for the duration of the course. I would never have succeeded without the support of the audience, not very numerous, but fervent. Nor without the constant attention of Louise.

Two additional subsections, 15.C and 15.D have essentially been written by Olivier Laurent. The definitive version takes into account corrections suggested by Thomas Streicher (Chapter 2), Philip Scott (Chapter 8) and the long list of typos found by Akim Demaille.

This English translation owes much to the enthusiasm of Manfred Karbe, who convinced me to translate the French original in English; this is indeed much more than a translation, since the last three chapters have been completely rewritten, thus take care of the latest developments (2011). The book was carefully reread by Edwin Beschler who not only expunged the gallicisms but also clarified the text, so that it is at points superior to the French original. The typesetting is due to Irene Zimmermann; heavy work, but the result is superb!

[4] Kreisel in 1984, speaking of certain Americans: « They have no soul and they don't know that they have none ».

Part I

The basics

Chapter 1
Existence vs. essence

1.1 The opposition existence/essence

The noted Kubrick movie « 2001 » opens on nothing less than the creation of intelligence, bestowed from a monolith falling from outer space over a tribe of dumb monkeys, unable to cope with lions, wild boars, etc. Observe that:

(i) The « great galactics » controlling this intelligence, spreading like a virus, must have been some sort of monkeys as well in their youth and so...

(ii) Intelligence is taken as an absolute attribute, independently of experience, of interaction, a software implemented on some preexisting hardware. Had the same monolith fallen amidst the Galápagos, reptiles would have taken over: the Age of the Tortoise...

In fact, forgetting the afflictive scientism of the author (Arthur C. Clarke), this could be read as a miracle *à la* Fátima.

The reactions to such an « explanation » are very pronounced, instinctive: this is inspired or dumb, whether one is rather *essentialist* or *existentialist*.

Essentialists. Those who think that everything is already there, that one can but repeat archetypes; they believe in flying saucers, especially above Aztec pyramids. In logic, they believe in « inverted foundations »: a system can be explained by a deeper « meta-system »; which in turn can be explained by a meta-meta-system... and this never ends[1]. What could be taken as a faulty construction (to suppose something, nay slightly more) appears as the fascination of the irreducible. The tutelary father of essentialism is Thomas Aquinus; by the way, the medieval Thomist philosophers deeply impregnated our conception of the immutable, of the « necessary » (still present in modal logics) and, above all, of infinity: with the best will, one has difficulties not to « see » the natural numbers as an infinite list, aligned on a wall like hunting trophies. The invention of set-theory, dating back to Cantor, Dedekind, etc. at the end of the XIXth century, is presumably the most unexpected scion of Thomism.

[1] In a most famous fantasy, the World rests on a turtle, which in turn rests on another turtle... « Turtles all the way down ». One can also think of this pyramidal con: multi-level marketing; and more recently, of B. Madoff, whose meta was the interest rate.

Existentialists. This term – of slightly delicate use, after an excessive vogue by the middle of the last century – would rather qualify those who find the ideas of Kubrick infantile (see also the last image of « The Shining », which suggests an eternal return) and do not believe in civilising saucers. Those who do not find « deep » the fact of defining truth as supposedly done by Tarski « $A \wedge B$ is true when A is true **and** B is true » . If truth can be « defined » only through a pleonasm (& := **and**), this means that it has no sense, *even if one can mathematically manipulate it* as in the *reflection schema* (Section 3.B.4). The best possible law has value insofar as one can justify it, i.e., show the effect of non-observance.

By the way, not all movies are as essentialist as those of Kubrick, e.g.:

- Dealing with war prisoners, military principles, social prejudices, *La grande illusion*, or *The Bridge on the River Kwai*, shows in one case how you can follow principles without believing in them, in the other how a strict obedience to the book leads to collaboration with the enemy.

- Classical westerns, for instance those of A. Mann with J. Stewart (e.g., *Bend of the River*) or those of B. Boetticher with R. Scott (e.g., *Ride Lonesome*) deal with too lax or too rigid moral principles.

It should be clear that the viewpoint of these lectures is « rather » that of existence. This choice does not come without certain difficulties:

(i) Thomism, which discretely dictates to Western culture, systematically brings us back to the same rut. For instance, people tried to oppose the official version of infinity (the « actual », ready-made, infinity), another infinite: dynamic, in construction, « potential ». Then, in order to speak of potentiality, one introduced « all » possibilities (Kripke models and other child's play)... and this is how the idea was killed: if the potential is truly potential, how can we pin all possible potentials like ordinary butterflies? This collection cannot make sense[2].

(ii) The existentialist world is a lawless world, in which contestation is total and constant; which is not viable. This viewpoint takes substance only as a reaction against the essentialist haughtiness. In fact, instead of a stiff opposition between two viewpoints (which is also a form of essentialism), one should rather consider their mutual interaction.

The opposition essence/existence is at the very heart of *typing* (Section 6.F).

[2] In modal logics, the dual of necessity, *possibility*, is treated as a poor relative. For instance, in the « possible world interpretation », the interpretation of the formula « if pigs had wings » is ludicrous.

1.2 Essentialist and existentialist projects

1.2.1 Set-theory. The development of « modern » analysis in the second half of the XIXth century prompted a clarification of vague notions: real numbers, continuous functions. Very soon one had to face paradoxes[3]: the curve without tangent, or worse, the Peano curve which « fills » a surface; the worst could be feared, for instance that the distinction curve/surface was only an « optical illusion » ... not to speak of plain inconsistency. In order to fix this problem, set-theory (essentially due to Cantor) tried to reduce everything to a simple common setting. This setting is undoubtedly essentialist, since it rests upon a shameless use of infinity. The first set-theory, styled « naïve », is based on the *comprehension schema*: any property P defines a set X, noted $\{x; P[x]\}$, « the set of x such that P »:

$$\exists X \ \forall x \ (x \in X \iff P[x]) \tag{1.1}$$

Dating back to 1898, a contradiction, an *antinomy*[4] was discovered – not a mere paradox – in set-theory, now remembered through the simplified version given by Russell in 1902: let $a := \{x; x \notin x\}$ (the set of sets not belonging to themselves); hence $a \in a \Leftrightarrow a \notin a$, contradiction[5]. This was not that terrible, since naïve set-theory was still an experimental system; and by the way, it only took a couple of years until Zermelo formulated reasonable restrictions on the schema (1.1), restrictions which are essentially those of the set-theoretic system **ZF**, still in use nowadays, to general satisfaction.

Before leaving set-theory for good, observe that the solution involves the distinction between two sorts of objects: the sets, which are « small » and the classes, which are « huge », typically the $\{x ; x \notin x\}$ of Russell is a class, not a set. Note that the distinction set/class is necessary without being absolute: like the mobile curtain separating business and tourist classes, the divide set/class occurs at the level of no matter which inaccessible cardinal. As to the intended essentialism, this is rather a failure!

1.2.2 The hilbertian project. « Hilbert's Program », essentially formulated in the years around 1920, corresponds to the *formalist* vision of mathematics. This approach has often been simplified into « *Mathematics is pure symbol pushing, with no more significance than the game of chess ; all that matters is the formal consistency of the rule of the game.* ». But this kind of provocation should not be taken literally: behind hilbertian formalism lies a complex thought, even if this thought is a reductionist one. Without questioning for one moment the interest

[3]Not quite formal contradictions, just results contradicting common sense.

[4]Burali-Forti.

[5]This is indeed a contradiction in all « honest » systems, with one remarkable exception: the *light* logics **LLL** and **ELL** (Chapter 16).

of set-theory, Hilbert takes set-theoretic essentialism the wrong way against the fur, especially in its infinite aspects. What follows is an unauthorised version of Hilbert's thought.

(i) Our intuitions concerning infinity are misleading; one speaks about objects and one does not see them. On the contrary, reasoning, which is expressed through logical formalism, is real, tangible. One must therefore understand what is a proof, under its most mathematical aspect, i.e., formal; indeed, understand *what it is good for*.

(ii) A proof produces a theorem A which can be reused (as a lemma) in another proof, through rules like *Modus Ponens*:

$$\frac{A \quad A \Rightarrow B}{B}.$$

B can in turn be used as a lemma to produce (given $B \Rightarrow C$) C and so on; which brings nothing new, indeed, a variant of *Modus Ponens* yields

$$\frac{A \Rightarrow B \quad B \Rightarrow C}{A \Rightarrow C},$$

in other words, one could have made a *Modus Ponens* between A and $A \Rightarrow C$ and directly get C.

(iii) The only hope to get a convincing explanation by this method consists in considering consequences B of an extremely simple form and therefore liable to an immediate analysis.

The most conspicuous choice is $B = \bot$, where $\bot$ stands for absurdity. And remember that, in that case, $A \Rightarrow B$ is equivalent to the negation $\neg A$. What can be said concerning absurdity? Not much, except that one rejects it; therefore it should not be provable. *Modus Ponens* takes the symmetrical form

$$\frac{A \quad \neg A}{\bot},$$

a sort of duality between proofs of A and proofs of $\neg A$, with the peculiarity that its enunciation is self-destructive: this situation can never occur, i.e., the system must be consistent. It cannot prove the absurdity, i.e., a formula and its negation. This is the origin of the idea of a consistency proof.

The inconvenience of this extreme restriction is that it attributes the production of A to its *non*-utilisation (by means of $\neg A$). A less frustrating version consists in admitting that B could be an (in)equation between integers, say $2 + 3 = 5$, $(2 \times 7) + 4 \neq 18$, etc. The constraint is that this (in)equation must be verified, e.g., one cannot get $(2 \times 7) + 4 \neq 18$ that way. This version turns out to be equivalent to consistency: indeed, it is clearly more general (take $0 \neq 0$ as absurdity); but not more if $(2\times 7)+4 \neq 18$ were provable, since one can surely prove $(2\times 7)+4 = 18$ (the axiomatic apparatus is adapted to that task), one would get a contradiction. More generally, B could take the form of a universal (in)equation, $\forall n\ (n+1)^2 = n^2+2n+1$ or $\forall n \forall p \forall q\ n^2+p^2 \neq q^2$, which we call a *recessive* formula, without essential change to the previous discussion. For Hilbert, recessivity is the limit of significance *stricto sensu* and the non-recessive is a sort of dowager which signifies only by delegation; recessive statements are supposed to have an immediate *finitary* meaning and this is why the hilbertian formalism is also called « finitism ». Observe a contradiction, a tension inside finitism: everything is finite, but this everything of finiteness is in turn infinite. Hence, the pattern is slightly unsquare....

A last word to justify the distance between Hilbert and plain essentialism: for him, consistency is a substitute for existence. Instead of proving the existence of an object a such that $A[a]$, one proves the consistency of the formula $\exists x A[x]$. This beautiful idea, which dates back to 1904, turns out to be barren, due to the impossibility[6] of producing consistency proofs. In practice, the consistency of $\exists x A[x]$ comes from the object a and not the other way around.

1.2.3 Brouwer's project. From the same stock as the criticisms – sometimes unfair, but always relevant – of Poincaré against formalists, Brouwer proposed an anti-logicist rereading of infinity. Nowadays, one can discern what is common to Brouwer and Hilbert, but one must admit that, in the 1920s, their relations were rather difficult, a sort of opposition between the priest Don Camillo (Brouwer) and the communist mayor Peppone[7] (Hilbert) – but without the slightest hidden sympathy –, the stumbling-block being scientism. By the way, the school founded by Brouwer, *intuitionism*, claims the primacy of intuition over the language. Brouwer did not reject infinity (contrary to hilbertian finitism which only sees it as a *façon de parler*), but he refused the most Thomist, « actual », aspects of infinity; especially set-theory and the idea that one could define a function of a real variable pointwise, value by value. Some principles, valid in the finite domain, cease to work in the infinite case, typically the excluded middle $A \vee \neg A$. The usual justification of the *tertium non datur* is that a formula A has a truth value (is true or false). However, while one can *compute* a truth value in the finite case, no algorithm can cope with the verification of infinitely many steps: this is why Brouwer called it in question in the infinite case. This approach is very modern, in the sense that truth does not exist independently of the means, the protocols, of verification; one should try to relate to quantum physics and forget the subjectivism, nay the solipsism, in which Brouwer sometimes went astray.

We shall have an opportunity to revisit intuitionism, through linear logic which

[6] Incompleteness theorem.

[7] *Le petit monde de Don Camillo*, Julien Duvivier, 1952.

primarily appears as its « symmetrised » version. Let us conclude this brief encounter with Brouwer by two (imaginary) confrontations with Hilbert.

Modus Ponens: for Hilbert, logical consequence must essentially avoid enormous mistakes, e.g., proving the absurdity; this bestows a second-class citizenship to most logical formulas: they don't make sense by themselves, they should not cause mischief, that's all. *A contrario*, for intuitionists, *Modus Ponens* is more than a legal advisor, it is a door open on a new world: the application of a function ($A \Rightarrow B$) to an argument (A) yielding a result (B). Proofs are no longer those sequences of symbols created by a crazy bureaucrat: they become functions, *morphisms*.

Tertium non datur: Hilbert accepts the excluded middle; not that he believes in a preexisting truth value for A, but because it simplifies matters: « A property which means nothing, but which costs nothing as well ». Technically speaking, this argument is viable. Indeed, one knows (since 1932: Gödel, one more result from him, Section 4.1.3) that classical logic can be faithfully translated in intuitionistic logic: it suffices to put double negations « everywhere ». In particular, the excluded middle, once translated as $\neg\neg(A \vee \neg A)$ becomes intuitionistically provable and it is not more « risky » to add it. However, one can answer back that it is not because one is warranted impunity (consistency) that one is entitled to commit a crime (enunciate an unjustified principle).

1.3 Gödel and after

1.3.1 Failure and decay. With the comfort and easy superiority given by elapsed time, we can see that Hilbert and Brouwer were both saying very interesting things, although astoundingly immature, the first clue to this immatureness being this unbelievable animadversion, typical of the $\mathrm{XX}^{\mathrm{th}}$ century. Dating back to 1930, the failure of Brouwer was conspicuous; excluded from a prestigious institution, the *Mathematische Annalen* (at the hands of Kaiser Hilbert), he had to content himself with presiding over a chapel limited to Holland. His attempt at rewriting analysis encountered general indifference: too complicated and sailing against the stream of a well-established tradition – and satisfactory in its main lines, as one must acknowledge!

The Tarpeian Rock of Hilbert's Program was the incompleteness theorem of 1931, that we shall detail in the next chapter. Without anticipating too much, let us recall that it foredooms any consistency proof. To get the consistency of system $\mathcal{T}$, one needs « more than $\mathcal{T}$ ».

This should have cut short « formalism »; at the least, provoke a salutary reformation… Not at all! One should not overlook the believers, the need-for-believing.

If one cannot give finitistic foundations to mathematics, let us generalise finitism[8]! That this generalisation amounts to supposing the result, forget it! One will eventually find a gimmick to make the pill swallowable. Result: an industry, essentially Germanic, which, during fifty-odd years, produced consistency proofs, of epistemological value close to zero, because of incompleteness. If the secondary parts of this strange adventure are out of place here, one must absolutely mention the figure of the protagonist, Gentzen, whose career was short[9] and paradoxical. Gentzen gave, in 1936 and 1938, two consistency proofs for arithmetic. The first one was very badly received, since it rested on a truth definition: in other words, he gave to the formulas their naïve, infinitary, sense; but, in order to remain « finitary », one cannot step around recessivity. The second one was far more acceptable, being based on a transfinite induction up to the denumerable ordinal ϵ_0: this is the way followed by Gentzen's epigones, especially Schütte. Concerning Gentzen's second consistency proof, André Weil said that « Gentzen proved the consistency of arithmetic, i.e., induction up to the ordinal ω, by means of induction up to ϵ_0 », the venom being that ϵ_0 is much larger than[10] ω; and there is something of the like, even if this is not as bad as it may look at first glance. If the second proof of Gentzen eventually badly aged, the first one, the one without epistemological value, eventually took the lead. Surely Gentzen committed the irrecoverable mistake of showing that a proof establishes a truth, but what he actually produced was not a truism one could be afraid of (axioms are true, rules preserve truth and so...). He gave, in a hesitating terminology, an interactive definition of truth, the proof appearing as a winning strategy in a sort of « game of truth ». Yes, this proof is even less convincing than the second one, which already convinced only the true believers; but, who cares? This was the first *interactive* interpretation of logic.

Creators lack the necessary distance from their own production; witness Kepler who took for his own grandest achievement an alleged correspondence between the thus-known planets (up to Saturn) and regular polyhedra, a law that posterity kindly forgets[11]. In the same way, in order to carry out his consistency Program, Gentzen had to create tools, essentially the « calculus of sequents ». This calculus, a weapon created for a dubious battle, remains one of the main logical achievements of the XXth century. We shall study it in detail, starting with the original formulation and introducing more and more sophisticated *avatars*: natural deduction, proof-nets, geometry of interaction, etc. At the end of the book, Gentzen's theorem, « cut-elimination », will eventually be rephrased in terms of... operator algebras.

---

[8]The first one to say so was... Gödel in his 1931 paper; naïveness of youth, or fear of Zeus' thunder?

[9]His untimely death in 1945 is due to his obstinacy as a little soldier of the Reich: he did not leave a « German town » (Prague, where he was *dozent* since 1943!).

[10]ϵ_0 is the smallest solution to the ordinal equation $\omega^\alpha = \alpha$.

[11]By the way one must wonder why H. Simon did not encourage his computer to study this law: one must also rediscover mistakes.

Among slightly decaying ideas, let us mention (just for the sake of a good laugh) iterations of theories. Starting with a system of arithmetic, say $\mathcal{T}_0$, one « founds » it over a meta-system $\mathcal{T}_1$ ($\mathcal{T}_0$ enhanced with the consistency of $\mathcal{T}_0$). Since $\mathcal{T}_1$ does not prove its own consistency, one can add this principle so as to get a « meta-meta-system » $\mathcal{T}_2$, then $\mathcal{T}_3$... One recognises *matrioshka*-turtles, each of them sustaining the previous one, « Turtles all the way down ». If you think that nobody dared to indulge in such baloney, you are completely wrong; believe it or not, it has even been elaborated! Since the $\mathcal{T}_n$ globally rest on nothing, they must be sustained by a $\mathcal{T}_\omega$, which one must in turn ground on something, say $\mathcal{T}_{\omega+1}$. Transfinite progressions of meta-theories were eventually produced. The question to determine « how many » steps are licit in this *fuite en avant*[12] belongs to the most subtle Byzantine theology. Observe that a bad idea (*matrioshka*-turtles) does not improve through transfinite iteration: it simply becomes a bad transfinite idea.

1.3.2 Second readings and renewal. In the recent history of logic, Kreisel stands as a slightly enigmatic figure. Indeed, he was – especially between 1950 and 1970 – a strongly influential figure, even if his texts are rather illegible and if his technical contribution, although honourable, is not outstanding. One of his main virtues was to « break » post-formalist illusions, especially by attacking the sacrosanct consistency proofs of Gentzen's epigones. He had a tendency to move questions from the ideological arena to a more pragmatic standpoint. Thus, « a consistency proof is not convincing, but it may have mathematical corollaries ». In general, he was insisting on applications of logic to mathematics; which has been by the way implemented by model-theory in algebraic geometry.

His most beautiful achievement might be the *reflexion schema* [68] obtained through the formalisation of truisms of the type « provability preserves truth ». This sophism, conveniently formalised, becomes a powerful metamathematical tool (Section 3.B.4).

A contrario, Kreisel never quite understood intuitionism, which he tried to reduce to its formal aspect, at the very moment the Curry–Howard isomorphism was suggesting, through category theory, a new dimension of logic. His explanation of the functional interpretation of proofs [67] is too formalist to be honest (Section 5.A). His attempted *revival* of Brouwer's analysis eventually turned into a bureaucratic nightmare. The real renewal started in the late 1960s and is due to various people, indeed all in relation with Kreisel, but this movement occurred almost against him.

We shall meet these names, Bill Howard, Dag Prawitz, Per Martin-Löf, Dana Scott, Bill Tait and many others, including myself. We shall have plenty of time to examine these contributions, nay criticise them[13].

---

[12]Forward flight.

[13]Typically Scott domains which are part of this renewal, even if they aged badly.

It takes something like 15 years for a new generation to take control. The renewal of 1970 became conspicuous around 1985, through the « computer revolution ». This year is important in my personal evolution, since it is when I invented *linear logic* which is to some extent the very heart of these lectures.

1.3.3 And tomorrow? One cannot describe in a few lines an evolution that spread over more than 40 years. I would only draw attention to the aspect « Pascalian bet » of these lectures. My hypothesis is the absolute, complete, inadequacy of classical logic and – from the foundational viewpoint – of classical mathematics. To understand the enormity of the statement, remember that Kreisel never departed from a civilised essentialism and that, for him, everything took place in a quite tarskian universe. Intuitionism was reduced to a way of obtaining fine grain information as to the classical « reality », e.g., effective bounds.

My hypothesis is that classical logic, classical truth, are only self-justifying essentialist illusions. For instance, I will explain incompleteness as the non-existence of truth. Similarly, a long familiarity with classical logic shows that its internal structure is far from being satisfactory. Linear logic (and retrospectively, intuitionistic logic) can be seen as a logic that would give up the sacrosanct « reality » to concentrate on its own structure; in this way, it manages to locate the blind spot where essentialism lies to us, or at least refuses any justification other than « it is like that, period ». In 1985, the structuring tool of category theory disclosed, inside logic, a *perfective* layer (those connectives which are linear *stricto sensu*) not obturated by essentialism.

What remains, the *imperfective* part (the exponential connectives) concentrates the essentialist aspects of logic, and categories cannot entangle anything there. To sum up:

$$essence = infinite = exponentials = modalities$$

Note that linear logic reduces essentialism to an opaque modal kernel, especially when one keeps in mind that modalities are a creation of essentialist logicians. By the way, nothing is more arbitrary than a modal logic: « I am done with this logic, may I have another one ? » seems to be the *motto* of modal logicians. Which exposes the deficiency, the lies, of essentialism: when everything comes from the sky, this convinces nobody. It seems that *geometry of interaction*, an interpretation in operator algebras, is capable of « breaking the nutshell ». The idea would be to revisit logic in relation with this phenomenon ignored, despised, by logicians – who treated it with contempt through their calamitous quantum logic – quantum physics. To imagine foundations, if not « quantum », at least in a quantum spirit: proportionately speaking, something of the sort Connes is doing with *non-commutative geometry*. That is the project of the day, enough to be kept busy for a while! Which topsy-turvies the usual relation logic/quantum: instead of interpreting quantum in logic, one tries the opposite.

Obviously, all of this is conjectural: the hypothesis of another regularity, another logic, living its own life, its own geometry, far from any setting « falling from outer space », like the monolith of « 2001 ». As Pascal would say, if it works, that's perfect and if it fails, nothing is lost anyway.

1.A Essentialism vs. platonism

1.A.1 Platonism. These remarks are not those of a professional philosopher; my « philosophical » considerations are more a provocation, a pebble thrown in a pond in which one can no longer see the fish under accumulated mosses, than a systematic – let alone systemic – reflexion. I always followed with the utmost boredom the academistic exposure of logical schools as they were classified at the time of Bernays and I rather looked for my own questions instead. As time elapsed, I had the curiosity of confronting what I had understood concerning logic and what was still mumbled in the Sorbonne and I discovered that this didn't match, didn't match at all.

Let us say that it is dubious that Plato could be called « platonist » in the acceptation of a certain ossified epistemology. This expression seems to basically recover a summary reading of the *Cavern Myth*. Following this classification, all good mathematicians (and all good scientists) should be styled platonists, since they believe in what they are doing. The fact of believing in the *reality* of the objects one is dealing with... without giving too precise a sense to this expression, is primarily the *responsibility* of the scientist: his activity is not arbitrary, what he says corresponds to « something ». The opposite attitude should be called *solipsism*, but should one create an autonomous category for non-solipsism? By the way, the alleged analogy between platonism and essentialism is dubious, since the latter keeps on invoking the sky, thus producing a total derealisation: witness modal logics, the triumph of essentialism and discretionary definitions. One can bet that the specialists of flow-production modal logics do not believe too much in their artifacts, otherwise they would not present a new one every second week... Contrarily to « platonism » which supposes a certain honesty.

1.A.2 Essentialism and morphology. The real debate is not existential, but rather *morphologic*. We observe phenomena, we give shape to them, but what does this form correspond to? I see a snake in my garden; is it God, or is it the Devil, which tries to tempt me (essentialist version), or is there a spring and in this case, is the water drinkable, can it be used to water my flowers (existentialist version)? Take for example a debate between two known logicians, Gödel and Bernays, in the years around 1940. None of them can be called a « platonist » with the slight shade of contempt usually attached to this expression. They speak about the notions of proposition, of proof; according to Gödel, there is first a wild species, the proofs, that one domesticates (by means of logic) to make them accomplish understandable actions: to prove propositions. For Bernays[14], everything is set in advance, as in Thomist theology: first comes the law which defines licit things, the authorised moves. The proof-making phenomenon is no longer primitive, it becomes the perfect activity of a marching army following its general (it is of course not written that way, the debate is much duller than that). By the way, one regrets that Gödel (and this was the case for everybody, up to Kreisel included) could only argue by means of artifacts of the kind « the list of all propositions », « the list of all proofs » and didn't at least integrate the category-theoretic viewpoint, which would impose itself only 30 years later. As to the debate Gödel/Bernays, as late as the mid 1960s (Section 5.A), Kreisel proposed a rather essentialist explanation « all these artifacts belong in a formal system, given in advance »; this shows that he rather stood on the « Bernays side ». This incites to caution in front of too elaborated, too systemic philosophical arguments: they easily turn into sophisms, they show the impossibility, the vanity of no matter which idea. A philosophical discussion must be fed like a fire and the combustible is made of the technical breakthroughs which make conspicuous the door hidden in the labyrinth. This also means that one should not push too much this criticism of essentialism. What we eventually want is no more than a *limited* progression.

Essentialism is indeed as a morphologic *simplism*. For instance, the idea of « naked » sets, which are later dressed like *mannequins* with an algebraic, then a topological structure[15]. Frankly, do the real numbers $\mathbb{R}$, in which everybody believes – independently of any philosophical commitment – make sense without addition, without multiplication, without continuity? Essentialism says « yes » through set-theory, but do we believe in this baloney?

The distinction existence/essence is at the very heart of these lectures. So to speak, we shall resume the debate Gödel/Bernays, trying to justify Gödel's viewpoint, but with more serious mathematical tools. In the role of the essentialist villain, Tarski will star with his truth definition and, more generally, all those logical definitions that presuppose logic.

1.B Perfect vs. imperfect

Another aspect of the opposition existence/essence can be found in the opposition perfect/imperfect, with the linguistic meaning of these terms. Perfection corresponds to unique, well-defined actions, while imperfection is the mode of repetition. It is only with linear logic, (Chapter 9) that this opposition takes a logical sense: the multiplicative/additive *fragment*[16] of linear logic, illustrates the « no reuse »

[14]Logician of rather medium size, especially in comparison to Gödel.

[15]Which is incredibly reminiscent of Genesis and its creation in seven days.

[16]A word indiscriminately used; the most general meaning is that of a sublanguage closed under subformulas. By no means that of a « subsystem » obtained by weakening the axioms, a notion of little interest in logic.

paradigm and one can say that, insofar as one should systemise the viewpoint of existence, it quite fits our anti- (or rather a-) essentialist ambitions.

Linear logic proper must be completed with a modal part, the *exponentials* !, ?, so as to cope with the infinite, which the previous fragment is quite incapable of. $!A$ enunciates the durability, the « perenniality », of A, which therefore becomes imperfect; infinity thus appears as an attribute of perenniality. Imperfection is the mode of generality, like in the titles of James Bond movies: « Diamonds are forever », « You only live twice », ... and the link with essentialism is therefore quite justified. Let us add that the modal character of exponentials reinforces this essentialism.

This being said, some experimental systems (light logics) enable one to consider an infinite that would not be « infinitely infinite » (Chapter 16). It seems that neither categories nor sets are subtle enough to understand these completely atypical systems. The attempted explanation involves *Geometry of Interaction*, an interpretation by means of operator algebras: this is the subject of the ultimate chapters of this book.

Bibliography: [32], [31], [51], [54, 59], [60], [67].

Chapter 2
Incompleteness

2.1 Technical statement

2.1.1 The difficulty of the theorem. For several reasons, it is out of the question to enter into the technical arcana of Gödel's theorem[1]:

(i) This result, like the late paintings of Claude Monet[2], is easy to perceive, but from a certain distance. A close look reveals only fastidious details that one perhaps does not want to know.

(ii) Neither is there a need to know, since this theorem is a scientific *cul-de-sac*: in fact it exposes a way in with no way out. Since it is without exit, there is nothing to seek and it is of no use to be expert in Gödel's theorem.

It is however important to know the general sense and the structure of the proof. Further, since the theorem is a genuine *paradox*[3], one is naturally tempted to get around it – which is indeed the only way to understand it. The examination of various objections which have been raised to the theorem, all of them wrong, requires more than a mere detailed knowledge of the proof. Rather than attempting to tease out those tedious details which « hide the forest », we shall spend time examining objections, from the most ridiculous to the less stupid (none of which, in the long run, prove themselves to be respectable).

2.1.2 The diagonal argument. The argument is as follows: given functions $g(z)$ and $f(x, y)$, we construct $h(x) := g(f(x, x))$; if by any chance h admits the form $h(x) = f(x, a)$, we obtain $h(a) = f(a, a) = g(f(a, a))$; $b := f(a, a)$ is a fixed point of g, which is obviously unexpected. Depending on the context, various consequences will be drawn, most of them paradoxical.

1. **Cantor's paradox:** there is no bijection between $\mathbb{N}$ and its power set. If (X_n) enumerates the subsets of $\mathbb{N}$ and $f(m, n) := 1$ when $m \in X_n$, 0 otherwise and $g(0) = 1, g(1) = 0$, then $g(b) = b$, a contradiction.

2. **Russell's antinomy:** the same story, $\mathbb{N}$ being replaced with the set of all sets. Integers become arbitrary sets so that $f(x, y) = 1$ when $x \in y$ and, with g as above, then $a = \{x; x \notin x\}$ and $b := a \in a$, so that $a \in a \Leftrightarrow a \notin a$.

[1] Nevertheless, one will find detailed technical information in the annexes of this chapter.
[2] Musée Marmottan, Paris.
[3] In the literal sense, « exterior to the dogma », the $\delta \acute{o} \xi \alpha$.

3. Fixed point of programs: if (f_n) enumerates all programs sending $\mathbb{N}$ to $\mathbb{N}$, if g is one among the f_n, then the previous construction yields a fixed point for g. Since most functions admit no fixed point, one concludes that the fixed point often corresponds to a diverging computation. Typically, starting with $g(n) := n + 1$, $a = a + 1 = a + 2 = \cdots$, which shows that we are indeed dealing with *partial* functions.

4. Fixed point of $\boldsymbol{\lambda}$-calculus: if M is a λ-term and $\Omega := \lambda x M(x(x))$, then $\Omega(\Omega)$ is a fixed point of M. 4: is to 3: what 2. (Russell) is to 1: (Cantor).

5. First incompleteness theorem: the fixed point of a program (3:), but replacing the programming language with a formal theory. $f(m, n)$ is the code of $A_n[\bar{m}]$ and g is non-provability: the fixed point is a formula saying « I am not provable ». Note that the theorem also establishes that $g(\cdot)$ is not computable.

To this series, it is correct to add Richard's paradox, which slightly prefigures Gödel's theorem: « the smallest integer not definable in less than 100 symbols », which has just been defined in much less than 100 symbols. One traditionally dispenses with Richard by saying that the word « define » is not well-defined, that the language should be made precise. Gödel's theorem can be seen as a « corrected » version of Richard; incidentally, Gödel explicitly referred to Richard.

2.1.3 Coding. This is traditionally the « difficult » part of the theorem, the one in which some « experts » do their best to mislead the neophyte, perhaps because they themselves do not grasp its general structure. What is this about? Nothing more than the « numerisation » of language, quite revolutionary an idea in 1931, but thoroughly common in the age of computers. Note however that there is a causal link: one should never forget Turing's contribution to computer science, a contribution which mainly rests on a second reading of Gödel's theorem. The fixed point of programs is nothing more than the noted algorithmic undecidability of the halting problem: no program is able to decide whether it or another program will eventually stop; and there is no way to get around this prohibition. This simplified version of the incompleteness theorem loses very little, which is not the case of Tarski's version (Section 2.D.1).

All this tedious poliorcetics[4] can be summarised in a couple of points, easy to understand at the eve of the XXIth century:

Formalism: all operations relevant to a formal system – the creation of a language (terms and formulas), the axioms, the rules, their combination in order to produce proofs, the result of these proofs (theorems), all bureaucratic operations (renaming of variables, substitution of terms for variables) – can be written informatically, provided we don't care about the physical possibilities

[4]The art of besieging towns.

of the computer (no limitation on memory). Such a system is essentially the same as a word-processing software program that checks closing brackets, replaces words, etc. Note that it is a quite formal, bureaucratic, activity: the machine does not tolerate the slightest mistake, e.g., the confusion between "0" and "O".

Numerisation: we all know that language can be coded by numbers, in a binary or a hexadecimal basis; we even encode images and sounds. In 1931, Gödel for the first time associated to any expression a of the language a « Gödel number » $\ulcorner a \urcorner$, which is only a coding, e.g., according to Gödel $\ulcorner (\urcorner = 11$, $\ulcorner) \urcorner = 13$, although our modern ASCII code yields 40, 41 for the two brackets. The coding used by Gödel is at least partially obsolete, its use being justified in the absence of size problems (memory allocation), apart from which it works in the same way as our modern codings. Moreover, the operations (or the properties) of a formal system will immediately be translated into functions (or properties) of the associated codes, which one can represent in formal arithmetic (or in any system in which arithmetic can be translated).

In particular (and this is the « reflexive » aspect of the theorem), a system of arithmetic can represent itself, « speak » about itself, just as a programming language can be represented as a specific program of the same language. Let's have a closer look at this, still avoiding details.

2.1.4 Expansivity vs. recessivity. A property of integers is *expansive* if it can be written $\exists x_1 \ldots \exists x_k A[x_1, \ldots, x_k]$, where A is an arithmetic formula whose quantifiers are of the form $\forall x < p$, $\exists x < p$; these formulas are also called Σ^0_1. The dual class (*recessive* properties, Π^0_1 formulas) is made of all $\forall x_1 \ldots \forall x_k A[x_1, \ldots, x_k]$, where A uses only bounded quantifiers. The terminology reflects the fact that an expansive property can be « approximated » by means of systematic trials: the more we try, the more we have opportunities to verify. It therefore refers to a peculiar style of potentiality. *A contrario*, recessivity means « so far, so good »; in other terms, it is a property that shrinks with trials. Provability is expansive: the more we try, the more theorems we get; while consistency is recessive: the more we try, the more we get chances to find a contradiction. The incompleteness theorem basically says that the recessive and the expansive do not match, do not and cannot match at any price.

A property which uses bounded quantifiers is *algorithmically decidable*, in the sense of an algorithm deciding whether $A[n_1, \ldots, n_k]$ is true or false for every choice of values for the variables $x_1, \ldots, x_k$. For instance, a bounded quantification $\forall x < p\ A[x]$, is checked by successively trying the values $x = 0, \ldots, p-1$. If $\mathcal{T}$ is a « reasonable » system of arithmetic, the same A becomes *decidable in* $\mathcal{T}$; in other terms, for each value $n_1, \ldots, n_k$ of the parameters $x_1, \ldots, x_k$, either the formula or its negation is provable. Why? Simply because a reasonable system of arithmetic must contain enough symbols and axioms to formally reflect computations. Indeed, in a language based upon 0, 1, $+$, $\times$, $=$, $<$, a few axioms are enough for this endeavour. Their list does not matter, what would be the use of a system not containing this vital minimum? Remark that decision in $\mathcal{T}$ is effected in the same sense as the algorithmic decision, simply because decision in $\mathcal{T}$ formally recopies the steps of the algorithm. With a slight shading: if $\mathcal{T}$ is inconsistent, it « overdoes », since it proves both A and its negation.

An expansive property $\exists n_1 \ldots \exists n_k A[n_1, \ldots, n_k]$ is (algorithmically) semi-decidable. This means the existence of a semi-algorithm (algorithm which need not yield an answer) answering « yes » when the property is true and nothing otherwise. This semi-algorithm is easily found: one enumerates the values $n_1, \ldots, n_k$, for instance for $k = 2$,

$$(0,0),\ (0,1),\ (1,0),\ (0,2),\ (1,1),\ (2,0),\ (0,3),\ (1,2),\ (2,1),\ (3,0),\ (0,4),\ \ldots,$$

and one successively tries all possible choices: one will eventually stumble on the right choice, if any: in this case the algorithm for A enables a conclusion. If the property is false, the algorithm yields no answer, which is why it is « semi », partial. From the viewpoint of provability, the same principle holds: in the case when there is a choice of values $n_1, \ldots, n_k$ validating A, then A is provable in $\mathcal{T}$ for this choice and the logical rule for the existential quantifier surely allows one to pass from $A[n_1, \ldots, n_k]$ to $\exists x_1 \ldots \exists x_k A[x_1, \ldots, x_k]$. As usual, with a formal system, there is the danger of overdoing, e.g., if $\mathcal{T}$ is inconsistent; there is even a warped possibility, namely that $\mathcal{T}$ is consistent while being able to prove false expansive formulas, which is for instance the case for $\mathcal{T} + \neg\mathrm{Con}(\mathcal{T})$, which is consistent when $\mathcal{T}$ is consistent (second incompleteness theorem) and which proves a wrong expansive statement: the inconsistency of $\mathcal{T}$[5].

Dually, recessive properties are not conspicuously algorithmic, not even « semi ». There is no longer any reason why a true recessive property should be provable. Indeed incompleteness offers a counterexample: Gödel's formula, both recessive and true, is not provable. At this early stage, the only thing we can assert is that a recessive formula provable in $\mathcal{T}$ must be true, provided $\mathcal{T}$ is consistent. If it were false, its negation, a true expansive formula, would be provable in $\mathcal{T}$ and $\mathcal{T}$ would therefore be contradictory.

2.1.5 The first theorem. Provability in $\mathcal{T}$ is the paragon of expansivity: « there exists a proof of … in $\mathcal{T}$ ». This requires a sort of *travail de fourmi*[6], but clearly understand that, if there is a proof, it is at hand and we will be able to formally translate in $\mathcal{T}$ the fact that this sequence of symbols obeys the bureaucratic constraints.

[5] This example of a consistent theory which is not « 1-consistent », shows that one can be consistent and a liar, as in current life one can be a bandit and escape from justice.

[6] Ant's work.

A contrario, non-provability and in particular the consistency of $\mathcal{T}$ are recessive: remember that consistency is only the fact that absurdity (or $0 \neq 0$) is not provable in $\mathcal{T}$.

The *Gödel formula* G is obtained through a diagonalisation whose principle has been explained and whose precise details are of no interest: it literally means « I am not provable in $\mathcal{T}$ »; it is therefore recessive. If G were provable in $\mathcal{T}$, it would be false, and its negation $\neg G$, both expansive and true, would be provable in $\mathcal{T}$. For $\mathcal{T}$ to prove both G and $\neg G$ would thus be inconsistent.

Theorem 1 (First incompleteness theorem). *If $\mathcal{T}$ is « sufficiently expressive » and consistent, there is a formula G, which is true, but not provable in $\mathcal{T}$.*

Let us mention that Gödel prefers to speak of a formula that is undecidable (neither provable, nor refutable) in $\mathcal{T}$, which has the effect of avoiding the epistemologically suspect notion of truth. For the anecdote, observe that our G could be refutable, while being true: we already mentioned the possibility that $\mathcal{T}$ might prove wrong expansive formulas and remain consistent. This is why the original version of Gödel (undecidable formula) is formulated under an hypothesis on $\mathcal{T}$ stronger than mere consistency[7]. A couple of years later, Rosser found a variant of G, neither provable, nor refutable, under the sole hypothesis of consistency (Section 2.D.3). But, once more, we should not waste too much time with all these details: the subject is dead.

2.1.6 The second theorem. This result is of a biblical simplicity, but its detailed proof is hell. Quite simply, we just established *in a rigorous way* that, if $\mathcal{T}$ is consistent, then G is not provable in $\mathcal{T}$: we proved it by reasoning arithmetically and we can choose $\mathcal{T}$ to be the very system in which we can formalise our proof. So $\mathcal{T}$ formally proves that the consistency of $\mathcal{T}$, noted $\mathrm{Con}(\mathcal{T})$, implies the non-provability of G, i.e., G. In other terms, $\mathcal{T}$ proves the implication $\mathrm{Con}(\mathcal{T}) \Rightarrow G$; not proving G, it cannot prove $\mathrm{Con}(\mathcal{T})$.

Theorem 2 (Second incompleteness theorem). *If $\mathcal{T}$ is « sufficiently expressive » and consistent, then $\mathcal{T}$ does not prove its own consistency.*

A few remarks to conclude this short technical survey:

(i) The fuzziness in « sufficiently expressive » is not quite the same in both cases. For the first theorem, we need very little expressive power in $\mathcal{T}$, basically to be able to mimic a computation step by step. For the second theorem, we need much more, i.e., the possibility to reproduce the (very simple) reasonings of Theorem 1. Essentially $\mathcal{T}$ must allow inductions over formulas of a rather simple structure.

[7] 1-consistency.

(ii) It is the second theorem which is overly weighted with discussion (and advertisement!), although it is presumably less deep than the first one. Indeed, since it is the demise of the most popular version of Hilbert's Program, one sees only it; but the first one refutes the programme too and is more general.

2.1.7 Fast version. We use the term (Kreisel) *provably recursive function* in $\mathcal{T}$ to mean a recursive function (i.e., a function given by an algorithm) from $\mathbb{N}$ to $\mathbb{N}$ such that $\mathcal{T}$ proves the termination of the algorithm. One can enumerate all provably recursive functions: simply, if n encodes a proof of termination, let f_n be the function whose termination has just been established; otherwise f_n is, say, the null function. Diagonalisation constructs $f_n(n) + 1$, an example of a recursive function that is not provable in $\mathcal{T}$[8].

This slightly simplified version shows that $\mathcal{T}$ is unable to recognise all total functions. Those whose algorithmic complexity[9] is too big, cannot be recognised by $\mathcal{T}$ as total functions. In particular, if the solution of an algorithmic question admits a lower bound not provable in $\mathcal{T}$, one gets a concrete form of incompleteness.

2.2 Hilbert in the face of incompleteness

2.2.1 The programme. Even if he denied the fact by invoking in his paper so-called generalised finitistic methods, it is truly capital punishment that Gödel inflicted on Hilbert's Program.

Hilbert's Program aimed at a mathematical justification of mathematics. Bad start, since it is reminiscent of the French Parliament voting itself amnesty: if mathematics is inconsistent and proves its own consistency, what is the point? The objection was overruled by Hilbert in an apparently convincing way: one does not ask the opinion of the Parliament, one asks the Supreme Court. This role is played by a tiny bit of mathematics, *metamathematics*, which is put aside, out of discussion. Metamathematics is supposed to address the properties of formal languages, essentially finite combinatorics: consideration on the length, the parity, of a finite string of symbols... A typical metamathematical argument could have been the following: show, by a simple induction, that all theorems have an even number of symbols (one would have checked it for axioms, then shown that deduction rules respect this property); one would have concluded by remarking that the absurdity $\perp$ has one symbol and is therefore not provable.

In parallel with this justification method, Hilbert develops an ontology in which most mathematical entities have no real room in which to exist, in conformity with the noted sentence of Kronecker: « God created the integers, everything else

[8] Hypothesis: the 1-consistency of $\mathcal{T}$, which excludes the canard of a partial function that $\mathcal{T}$ would consider total.

[9] Warning: we are speaking of huge complexities, not at all of P, NP, etc.

is the deed of man ». Moreover, most properties of integers have no meaning: indeed he recognises as significant only the identities, the universally quantified formulas, that we already met and called *recessive*. Everything else is only a *façon de parler*. Hilbert's Program consists in establishing, by metamathematical means, that general mathematics[10] are *conservative* over recessive formulas. A result proven by « infinite » methods, can, *modulo* transformations – not necessarily very friendly, but possible « in principle » – be established in a strictly finitary setting. This is therefore a principle of *purity of methods*, taken from the practice of number theory (elementary proofs), but which Hilbert tried to mechanise. Note that, although this project failed because of incompleteness, Gentzen's cut-elimination achieved a similar goal, the *subformula property* (Section 3.3.1).

Using some general considerations, we can easily show that conservation is equivalent to formal consistency. Surely it implies consistency, since the absurdity, which can be written $0 \neq 0$, is recessive; if it is provable, it is also « finitarily » provable, but we have no serious doubts as to the « finitary ». Conversely, the argument of Section 2.1.4, which shows that a consistent theory cannot prove a false recessive formula, justifies every recessive theorem on the grounds of the consistency of the theory $\mathcal{T}$ establishing it; a « finitary » consistency proof thus enables one to transform any proof into a finitary one, provided the result is recessive.

From a certain viewpoint, the (immense) accomplishment of Gödel's endeavour is not that surprising. After all, those metamathematics, those finitistic meta-methods, are a significant part of mathematics. Let us be clear: if metamathematics could not have been translated in mathematics, this would have been an incompleteness much more dramatic than Gödel's. In fact, Gödel's work must be seen as the end of the process dating back to Cantor, Dedekind, ... of coding mathematics: the pioneers did encode reals by sets of rationals in turn encodable by integers; Gödel closed the cycle by coding the language.

A last remark: one must be admiring, even if it eventually failed, in the face of the texture of Hilbert's viewpoint. He reduces everything to consistency, which appears, *modulo* Gödel's encoding, as a recessive formula, of the very restricted form on which Hilbert deigns to bestow a significance. Therefore this programme is wrong, reductionist, whatever you like, except stupid. One cannot say the same of the afflictive *remakes* proposed by Artificial Intelligence.

2.2.2 The *coup de grâce*. The second theorem says that the consistency of $\mathcal{T}$ is not provable in $\mathcal{T}$. In other words, even the most brutal version, « auto-amnesty » does not work. It is impossible to prove consistency by finitary methods, since the methods (*a priori* much more powerful) available in $\mathcal{T}$ are not enough. Note that the first theorem suffices for the « conservation » version of the programme:

[10] Nowadays, formalised as set-theory **ZF**; in Hilbert's days, the *Principia* of Whitehead & Russell or even Peano's arithmetic **PA**.

indeed, if $\mathcal{T}$ denotes « finitistic » mathematics, G provides us with an example of a recessive statement provable by « infinite » methods but not provable in $\mathcal{T}$.

While writing this last line, I suddenly began to doubt the rigour of my deductive chain: I found myself skipping too quickly from « true » to « provable by infinite means ». Rather than trying to make sense of this, let us remark that the distinction between truth and provability (inside the current formalisations of mathematics) was not part of the landscape of 1930; so what is the point in a too detailed discussion of Hilbert's failure? Most nuances we can fashion are *posterior* to incompleteness and, even with the utmost good will, there is no way to fix this programme.

A last point: we observed (Section 1.2.2) that the rule of *Modus Ponens* establishes a duality between proofs of A and proofs of $\neg A$. Such a duality suggests an *internal* form of completeness, of the form « A is provable iff $\neg A$ is not provable », without reference to truth. On the other hand, in the presence of such a completeness, one could easily define truth as being provability. Now, what says the incompleteness theorem, the first one? G is not provable, although its negation $\neg G$ is not provable either[11]. In other words, something is rotten in the Kingdom of Formalism.

2.3 Incompleteness is not a deficiency

When something does not work well, we usually try to fix it; an apt example is what was done in the past to people who were deemed to be « crazy »: they were lobotomised, chimicised, ..., although, in most cases, there was simply nothing that could be done to any effect. Incompleteness is, similarly, a disease that cannot be healed and it is ridiculous to look for the « missing piece of the puzzle ». We shall now take a critical glimpse at various proposals, starting with the most indigent ones. This is not so much to indulge in teratology, but rather an amused exploration of the various facets of incompleteness.

2.3.1 Want of rigour? The most radical want is plainly that of rigour. This is why, from time to time, there is a blossoming of refutations of incompleteness; they come in periodic waves (for instance the year 2000 was very productive). There is little to say about them. They are all made from the same mould: one affirms that, since G is true, it must be provable. The problem is not that a few morons keep on repeating over and over the same mistakes, it is the guilty complacency these amateurs find in certain milieus. Artificial Intelligence advertises them insidiously, Emails of the style « we didn't dare to say it, now we know », « hmmm, hmmm, hmmm... », invitations to speak in the name of the « one can exclude nothing »... It is futile to try to repress such baloney, especially when their authors are half-wits.

[11] Rigourously, if $\mathcal{T}$ is 1-consistent, or, better, by replacing G with the Rosser variant, which is neither provable, nor refutable.

On the other hand, the *revisionists* who broadcast such nonsense are anything but stupid; there are even super-talented people in their specialty – which bespeaks a penchant for swimming between two waters. To those people, one should explain the following:

(i) Like it or not, there is no bug in the proof of incompleteness of mathematics, a contention that has even been checked by computer.

(ii) A refutation of the theorem (which cannot be *a priori* excluded, though very improbable) would produce an inconsistency in mathematics.

(iii) But the hypothesis of the theorem is the consistency of mathematics. A refutation would indeed produce even stronger evidence.

In other words, *that which kills me makes me stronger*. What a strange and capricious result, this incompleteness, one of the few results of the human mind to be absolutely irrefragable.

2.3.2 Want of imagination? Here, we contend with the second theorem, tamper with the definition of consistency – be creative or perish! Remember that consistency is the fact that we cannot prove the absurdity $\perp$, but this relies on the fact that absurdity implies no matter what formula: « *ex falsum quod libet* ». A system is consistent when there is something it cannot prove. We start with this equivalent definition and add a formula, typically a « super-absurdity » and we make sure that this thing can *by no means* be provable: for instance, by simply refusing to write it down! Evidently, a system grounded on such a « paraconsistent logic » is trivially consistent. We thus obtain the cheapest of all refutations of the second incompleteness theorem. It is in this way that certain governments fight criminality: thefts of mobile phones are expunged from the record. But ideas should retain their honest meaning and consistency is the most primal form of honesty: if one starts with giving dishonest, inconsistent, definitions of honesty and consistency, everything becomes possible, but what does this mean?[12]

To sum up: the theorem does apply to a *deductive* system, not to a doohickey in which reasoning is made « à la tête du client », i.e., depends on the weather.

2.3.3 Cognitive want? « Epistemic » logics are supposed to illustrate « abductive » principles of reasoning; from the fact that we don't know, one deduces… something. These logics are based upon somewhat infantile metaphors (and are in fact limited to the axiomatisation of these metaphors), like the story of the Baghdad cuckolds who kill their wives because… Rather than reproducing this baloney,

[12]The paraconsistent mob is a bunch of obtuse formalists, not even of the straight sort; they hide their uncouth stratagems behind definitions more difficult to decipher than the map of Shinjuku for a *gaijin*: one seldom gets to the bottom of the first page.

let us give an original, previously unpublished variant of the same: the *Houston cuckolds*. They are only two, V. and W.; each one knows everything concerning the other and the fact that at least one of them has been betrayed; another fact is that there is only one cuckold, W.[13] W. knows that there is one cuckold, he knows that this is not V., but he draws no consequence, because he is a bit slow. On the other hand, V. is very smart and made his PhD on the Baghdad cuckolds, so he thinks: « Gosh, if I were not a cuckold, W. would have concluded that it's him and killed his wife ». Therefore V. slays his innocent spouse. The moral: too much epistemic logic can damage your health.

What is the point of turning this gag topsy-turvy? It illustrates the difference between *constatation* and *deduction*. W. must perform a simple deduction, but he is a moron and is unable to reach a conclusion. From the algorithmic viewpoint, this stresses the difference between old-style computation (to read a result from a table of functions) and modern computation, the execution of a program, which may take an unbelievable, or at least an unpredictable, amount of time.

Epistemic logics are based upon an identification between « not to know » and « to know not ». If such an identification were viable, it would suffice to add the relevant axioms. One sees that this is impossible, since then, G being not provable, this fact would be provable (which is expressed by G), a contradiction. One could object that in certain remote provinces, it is still possible to write a PhD thesis on epistemic logic with results that, while not earth-shaking, are not wrong. Now, these systems should be inconsistent as a corollary of Gödel's theorem. A little enigma, whose solution follows at once: incompleteness only applies to systems with a *minimal* amount of expressive power (not very much indeed). This minimal amount forbids one to predict whether or not a formula will be provable. Epistemic logics have no right to this minimal expressivity and this is why they are confined to being the logic of their own (and ponderous) metaphor.

2.3.4 Want of axioms? The step that epistemic logic did not make, non-monotonic logic took without hesitation. We bluntly add a principle of the sort « if $\neg A$ is not provable, then A is provable ». Granted adequate precautions, these systems are consistent and complete, so what is the complaint? They are simply non-deductive, because there is no way to activate the additional principle. We are no longer dealing with a formal system, since there is no way to know that something is not provable (this is already wrong in a deductive system, so in such a doohickey, good luck!). By the way, let us directly refute the algorithmic analog of « non-monotonicity ». One could *complete* any algorithm as follows: if the algorithm yields the answer « yes », answer « yes », if it says « no », answer « no », if it keeps silent answer no matter what, « yes » or « no », nay « I don't know ». This is impossible, because Turing's undecidability of the halting problem tells us precisely that there is no

[13]In the « official » version, both are cuckolds.

algorithmic way of knowing that one does not know.

Exit non-monotonic logics. It may be amusing to see how these non-systems behave in the face of the second theorem (once more, this is a sterile exercise). Fundamentally, this is only a consistent and complete set of formulas of arithmetic, i.e., essentially a model. Everything depends on the logical complexity of such a set.

(i) There exist consistent and complete sets which can be expressed by an arithmetical formula[14]. In such a case, consistency can be expressed in arithmetic. In spite of the non-computability of deduction, one can reproduce the fixed point argument and obtain a « Gödel formula » G. G is true iff G is not provable, i.e., if $\neg G$ is provable: one sees that the theory makes mistakes and cannot be trusted.

(ii) Most solutions are not definable, typically if one takes true formulas, it is immediate[15] that this set cannot be expressed by a formula. In that case one cannot even express consistency.

2.3.5 Want of truth? In the open Richter scale of mediocrity, what follows has the immense superiority of making no uncouth mistake of logic; but it is an academistic version which is unfair to the earth-shaking originality of incompleteness. This interpretation is simply that incompleteness is « a truth which is not provable ». One can say that I am exaggerating, since I formulated the theorem under this very form!

That's the crux: we must separate a mathematical formula from its interpretation. For instance, there is not the slightest problem in formulating arithmetical truth (even if it cannot be expressed in arithmetic, it can be expressed in set-theoretic terms) and it can be used in mathematical proofs, or even to formulate results like incompleteness. Now, it is not because a concept can be defined in set-theory that the concept makes sense. This is most flagrantly demonstrated for the concept of truth, defined by Tarski by means of a pleonasm, typically:

$$\forall x A[x] \textit{ is true when } A[n] \textit{ is true for any integer } n$$

The truth of A is nothing but A, which is what we called essentialism. One must legitimately doubt a notion that turns out to be so opaque. In place of the academistic interpretation « want of truth », I propose to substitute the more stimulating « truth means nothing » (I didn't say « is not definable in arithmetic », I really meant « no meaning »). Which does not imply that I was wrong in saying that G is true, since we established it. I only say that, in the same way there is no *general* notion of beauty, good, etc., there is no « general » definition of truth. What we logicians

[14]More complex than recessive or expansive, at least Δ^0_2.

[15]Tarski's theorem, a bleak variant of incompleteness (Section 2.D.1).

manipulate under the name « true » is only an empty shell. A last word: one should not forget either that Gödel's formula, this over-ornate artifact, before meaning « I am not provable », says « I mean nothing ». With incompleteness, one gets close to the limbs of signification. *Dixit* René Thom: « la limite du vrai ce n'est pas le faux, c'est l'insignifiant ».

Back to the academistic, « tarskian », reading of incompleteness, once more remembering that there is no « technical » mistake involved, which makes it frankly superior to the previous « readings »: instead of completing bluntly, « non-monotonically », a system into a complete non-deductive one, we can envisage *partial* completions. Typically, we add consistency, then resume. In this way one reaches the myth of nested metas, of transfinite iterations of theories, a barren approach if ever there was one. This matches the essentialist fantasy of multi-layer marketing and we are back to Kubrick and Aztecian flying saucers.

From the theological viewpoint, the academistic approach is rather unbelievable: this unprovable truth that one can mix up with the meta. One could see truth (semantics) as the Father, provability (syntax) as the Son (a.k.a. Verb), the meta playing the go-between, the Holy Ghost. But then incompleteness appears as the non-consubstantiality of the Son, a sort of logical Nestorianism. Maybe it is still dangerous to venture on such grounds in Roma, so let us stop before it's too late!

2.4 Metaphorical readings

2.4.1 Blair's theorem. A recent actual event suggests a « theorem of Blair », not to be confused with Baire, but the Anthony Blair who declared (2004): « weapons of mass destruction (WMD) do exist, but one will never find them ». The analogy with incompleteness is misleading. Indeed, in the field, the occupying army made use of *all* means to « extract information ». To find the enemies hiding places is of an expansive nature; the more one tortures, the more places are found; if nothing has been found this way, it is because there was nothing to find. We are indeed closer to the Liar paradox: « I am lying » than to incompleteness.

A better approximation would be a « theorem of Saddam ». Indeed, he was claiming the absence of WMD, but he was raising various obstacles to the verification of his claim. Which made the absence of WMDs a true recessive statement, but unverifiable.

2.4.2 Anti-mechanism. If we try to broaden the debate, we see that incompleteness opposes the mechanistic view of the world, the view of Hilbert as well as the view of Artificial Intelligence. This is why the results are so cordially hated (under the roses, how many thorns!).

Let us make a bold move and let us say, with the French philosopher Régis Debray, that incompleteness is an anti-totalitarian concept. With a small proviso, the argumentation of Debray (political system = formal system) is plainly dubious. Even so, this argument remains striking. Never forget that XXth century totalitarisms were grounded on scientific illusions – racial theories, historical materialism – and that monstrous crimes were performed in the name of an infallible science[16]. Far from stale totalitarian smells, the debate internal to logic was about the possibility of arriving at « final solutions[17] » into the most abstract scientific debates. These debates had no influence whatsoever on the extermination of such and such groups. Now, if the answer to Hilbert's Program had been « yes », i.e., if all formal questions had – in principle, not necessarily in practice: for instance imagine a « chaotic » mechanisability, impracticable since too expensive –, what an argument for the Hitlers, for the Stalins! Thanks to incompleteness, nobody can contend that every question has an answer.

It should anyway be observed that Gödel's theorem is not an anti-scientistic panacea. For instance, in confrontation with H. Simon (specifically his « Keplerocidal » programme), a limited amount of epistemological common sense is enough: remember that science is primarily seeking questions!

2.4.3 Digression: artificial intelligence. (In order to answer one of the questions: « Clarify your position as to artificial intelligence. »)

The answer is essentially a matter of nuance. If one means the possibility of mechanising the activity of certain zones of the brain, e.g., automatising the recognition of space, building robots capable of landing on their feet, just like cats, or fixing the trajectory of a vehicle whose driver is dozing..., of course yes. But should one call this intelligence, or rather *instinct*? « Artificial instinct » is unproblematic.

If one seeks real intelligence, or creativity, this is more complex and, bluntly, doomed to failure. Gödel's theorem opposes this fantasy, but, more by refuting the totalitarian metaphor of a final and mechanisable science than in quarreling with the details. As to those details, let us exercise some common sense. What constitutes intelligence, in the creative sense, is that one does not expect it, that it places itself in a position of unrest: it is therefore the « grain de folie » already mentioned, the condition *sine qua non* of which is the possibility of making mistakes, the *right to error*. Intelligence follows deviant ways, unexpected, including those of prejudice, ambition, wrath, the seven deadly sins and worse. This remains tolerable in society since, individual power being limited, this « internal whirlpool » has only weakly dramatic effects. When a very smart individual attains supreme power, it is unmistakably the intent that the world will « benefit » from the wanderings of that

[16]By the way, note that the identification between *constatation* and *reflection* that underlies epistemic logic and other *paralogics* is a plausible definition of totalitarianism: the ideal of those *explicit logics* which refuse any hypothetical deduction is the police report.

[17]This is the delicate expression used by Hilbert and his epigones. This term that we now find so shocking was a scientist's commonplace, recycled by the IIIrd Reich.

individual's thought. Imagine that one could delegate to an omnipotent robot the possibility of decreeing, on an impulse, the eradication of mankind, only to find a better idea the next morning! We might as well prefer to be ruled by a G. W. Bush, whose mediocrity limited his ability to do harm.

To sum up: one should first of all avoid simplism or exaggeration. Instead of seeking a software program that would automatically translate Proust in the very style of Proust[18], let us rather try, like Gérard Huet, to make a dictionary of sanskrit, yielding, in the most economical, efficient, possible way, the morphology of expressions, etc., see http://sanskrit.inria.fr. Compared to the fantasy of a robot-Einstein, this may seem limited, but the most *genuine* ambition dresses with the most modest clothes.

2.4.4 The extinction of Popperism. The philosopher Popper[19] proposed an epistemology of « falsifiable » properties that we can as well call « recessive ». It is about general laws, subject to partial verifications, tests, for instance, specific experiments, verification up to a given decimal... We easily recognise Hilbert's influence; it looks like an enhancement of Hilbert's ontology to apply to the complete field of scientific discourse. The pleasant side of the story is an anti-essentialist positioning; laws do not fall from the sky. Now, since Gödel refutes Hilbert, he refutes *a fortiori* Hilbert's epigone Popper.

We can observe that Popper gives no status to Gödel's theorem, since we have already observed that the theorem cannot be refuted or falsified. This is extremely convenient, since, if this formula has no status, we can ignore it as « metaphysical ». It is of interest to look for the flaw in this ontology of the test. It is simply its asymmetry: on one side the test, absolute, on the other side, the law, relative and testable. Which necessarily produces an ontology of the recessive. But we could imagine more intricate situations, where the quality of the test could be questioned; if the law tests the test when the test tests the law, we enter a more open world: « when you gaze long into the abyss, then the abyss also gazes into you » would say Nietzsche. Which will happen in *ludics*, where there is no official referee and everything can be called into question.

2.4.5 Various. As a matter of relaxation after the preceding extreme mental ordeal, let us look at some outstandingly ludicrous interpretations of Gödel's theorem:

(i) The inability to think of oneself: the system $\mathcal{T}$ could not speak of itself. The reduction of thought to a formal activity is not very kind to poets, philosophers and even mathematicians. More, there is an absolute technical misinterpretation, since incompleteness is based precisely on the fact that $\mathcal{T}$ encodes

[18]We are still far from that: only consider that the production of PhD's in epistemic logic is not yet automatised .

[19]« Popperism » is a pun about the book by Napoléon III « L'extinction du paupérisme ».

properties about itself. And consistency, like Achilles' heel, is more or less the unique thing concerning $\mathcal{T}$ « escaping » to $\mathcal{T}$. Between you and me, does the fact that one can do anything while wearing spectacles except fix them deserve endless commentaries on... meta-spectacles?

(ii) Self-reference: what a cliché! The play within the play, the painting displayed in the meta-gallery (the toilets) not to speak of the well-known wedding cake, the cult book Gödel–Escher–Bach! Speaking of cakes, the franco-belgian *entarteur*[20] proposes a cure of his own for pompousness; is he too busy these days?

2.A More on the classification of predicates

2.A.1 First order. In classical logic, every formula can be put in *prenex form*, i.e., be replaced with an equivalent formula of the form $Q_1x_1 \ldots Q_nx_nA$, where the Q_i are quantifiers and A is quantifier-free. In arithmetic, where quantifiers are numerical, bounded quantifiers can be pushed into the propositional part A: only unbounded quantifiers matter. Prenex formulas still make sense outside the classical setting, but they only retain a marginal interest; thus, an intuitionistic combination $\forall\neg\neg\exists$ cannot be made prenex: no way to get rid of « $\neg\neg$ ».

Formulas of classical arithmetic, supposedly in prenex form, can be classified by the number of alternations of (unbounded) quantifiers; thus $\forall\forall\exists\exists\exists$ corresponds to the alternation 2. For $n \geq 1$, the Σ^0_n formulas (Π^0_n) are those with alternation n beginning with an existential (universal) group. Since one can always add « phony » quantifiers, a formula of alternation n can be styled, as we please, Σ^0_m or Π^0_m, as soon as $m > n$. In particular, a formula with bounded quantifiers will be both Σ^0_1 and Π^0_1, i.e., expansive and recessive.

This classification extends to the *sets* defined by such formulas. From this viewpoint appears a third hierarchy, the Δ^0_n, i.e., of the sets admitting both forms, Σ^0_n and Π^0_n. Thus the Δ^0_1 sets are those which are recursive (algorithmically decidable), while the Σ^0_1 are semi-recursive (recognisable by a semi-algorithm). We also call them *recursively enumerable* (r.e.), since a non-empty semi-recursive set is the image of a recursive (computable) function. The Δ^0_n do not form a formal hierarchy, since it supposes a writing under two equivalent forms: this equivalence may be immediate (case of an implication between two Σ^0_1, admitting two prenex forms, $\forall\exists$ and $\exists\forall$, which makes it immediately Δ^0_2) or frankly unprovable (case of a set *a priori* semi-recursive, whose recursive, Δ^0_1, character, depends on a decidability proof).

The fundamental result obtained through coding is that each class Σ^0_n (Π^0_n) contains a *universal* (or complete) element. For instance, for $n = 1$, formal provability

[20]Entarter: to shove a cream pie in the face.

in a system of arithmetic is complete, in the sense that every Σ^0_1 set can be factorised through provability. It suffices to take the definition of the set: indeed, $A[n]$ is true iff $A[\bar{n}]$ is provable, at least when the system is 1-consistent; if it is only consistent, one must perform a modification à la Rosser (Section 2.D.3).

Using diagonalisation, each class is distinct from its dual. Thus, if $f(m,n)$ $(m \in F_n)$ is a Σ^0_1 enumeration of the Σ^0_1 sets, then the Π^0_1 set $\{n; n \notin F_n\}$ is not Σ^0_1. This is Turing's theorem and, if one takes provability as a universal element, the first incompleteness theorem.

2.A.2 Second order. Logic admits a second-order formulation, by means of quantification over predicates. It is important to distinguish between two closely related classifications:

The projective hierarchy: we use first-order quantification on integers, second-order quantifications on sets of integers. The resulting formulas are classified into Σ^1_n and Π^1_n, $(n \geq 1)$, where n counts the alternation of second-order quantifiers. The class $\Delta^1_1 := \Sigma^1_1 \cap \Pi^1_1$ (*hyperarithmetic* sets) is already bigger than all Σ^0_n and Π^0_n. A typical hyperarithmetic set not arithmetical is the set of the (Gödel numbers of) true arithmetical formulas.

The logic hierarchy: same as before, but the first-order quantifiers are not numerical. The notation is – for want of a better one – Σ^n, Π^n, still counting the sole second-order alternation. This hierarchy, finer than the previous one, is essentially one step ahead. It is well-adapted to proof-theory, contrarily to the previous one, rather set-theoretic.

The relation between both hierarchies is obtained by means of the Dedekind definition of natural numbers « the smallest set containing 0 and closed under the successor S ». Which can be expressed by a quantification over a unary predicate X:

$$x \in \mathbb{N} :\Longleftrightarrow \forall X((X(0) \wedge \forall y(X(y) \Rightarrow X(Sy))) \Rightarrow X(x)).$$

A Σ^0_1 formula, say $\exists n\, A[n]$, using a numerical quantification $\exists n$ can be translated in second order as $\exists x(x \in \mathbb{N} \wedge A[x])$, of which we can say that it is Π^1. In the same way, recessive statements are Σ^1. More generally, a Π^1_n formula can be written under a Π^{n+1} form.

2.A.3 Completeness vs. incompleteness. This pair of notions can be discussed at two levels:

Technically: completeness of predicate calculus (Gödel, 1930), is stated: « a closed formula B true in all models is provable ». Say that $B = B[P]$ has a single predicate symbol P; then the second-order formula $A := \forall X B[X]$ is quite closed (no longer any constant) and is Π^1. Moreover $\forall X B[X]$ is true when B is true in all models, i.e., for all choices of an interpretation for P. Completeness is rewritten therefore as « a closed Π^1 formula A which is true is provable ». The relation « true $\Rightarrow$ provable » does not subsist beyond the Π^1 case. Thus, it fails for Σ^1 formulas, since the Π^0_1 $(= \Sigma^1)$ Gödel formula is true without being provable, a result rightly called incompleteness.

Generally: it is important not to reduce notions to their technical expression. Completeness is plenitude, a sort of internal equilibrium. By convenience, one formulates the notion in relation to truth, but I personally prefer the form of completeness coming from cut-elimination (Chapter 3). For the same class of formulas (Π^1), the subformula property (Section 3.3.1) says that « everything is there »: as long as one respects cut-elimination, one can prove nothing more; at most one can give « faster » versions of the « same[21] » proofs. For the dual class $(= \Sigma^1)$, the want of a subformula property causes a complete loss of control. *Ludics* generalises this approach by defining a notion of internal completeness (Section 13.8.4).

2.A.4 Expansivity and polarity. Restricted[22] to reasoning, expansivity corresponds to the deductive mode, while recessivity corresponds to the inductive mode, generalisation from particular cases, perhaps refutable on the basis of a counterexample to come, the « so far, so good ». The (highly personal) terminology reflects more the dynamics than its contents (reasoning, number, image).

As to the relation to *polarity*, we distinguish two groups, on one side positive/answer/explicit/expansive, on the other side negative/question/implicit/recessive. To avoid systemism, it is important to observe that polarity comes from the idea of a « logical onion », of which we peal the successive skins. Thus the opposition answer/question seem to recover the opposition expansive/recessive. On the other hand, positive/negative only relates to the most external skin and, in the same way, explicit/implicit seems to only make sense on the external layer.

2.B Formal arithmetic

2.B.1 The system RR. The first incompleteness theorem is usually formulated for systems based upon a simplistic language, due to Peano. There are four functional symbols, the constant 0 (or $\bar{0}$, as one likes) the successor S (unary) and $+$, $\times$ (binary); thus the integer 5 will be encoded by the term $\bar{5} := SSSSS0$ and in general, n by $\bar{n}$, n symbols S followed by a zero. Observe the regression which often accompanies

[21] They are the same only w.r.t. the category-theoretic reading of proofs.

[22] Anticipating upon the notion of polarity (Chapter 12).

progress: here formalism brings us back to pre-Babylonian numeration. There are two binary predicate symbols, $=$ and $<$. The axioms of system **RR**[23] can be divided in three groups:

Equality: these axioms say that equality is a congruence: $x = x$; $x = y \Rightarrow y = x$; $x = y \wedge y = z \Rightarrow x = z$; $x = y \wedge z = t \wedge x < z \Rightarrow y < t$; $x = y \Rightarrow Sx = Sy$; $x = y \wedge z = t \Rightarrow x + z = y + t$; $x = y \wedge z = t \Rightarrow x \times z = y \times t$. These axioms are enough to prove $x = y \wedge A[x] \Rightarrow A[y]$.

Definitions: these axioms are enough to prove the basic equations and inequations. The group $x+0 = x$; $x+Sy = S(x+y)$; $x\times 0 = 0$; $x\times Sy = (x\times y)+x$ enables one to prove, for any closed term t whose value is n, the equality $t = \bar{n}$, hence all true equalities $t = u$ between closed terms. The group $Sx \neq 0$; $Sx = Sy \Rightarrow x = y$ (third and fourth Peano axioms) enables one to prove all true inequalities $t \neq u$ between closed terms. Observe that these two axioms suppose an infinite domain (otherwise, one could have $\overline{10} = \bar{0}$). Finally, the group $\neg(x < 0)$; $x < Sy \Leftrightarrow x < y \vee x = y$ enables one to get all true inequalities or non-inequalities $t < u$, $\neg(t < u)$ between closed terms. It enables one to prove the formula $x < n \Leftrightarrow x = \bar{0} \vee x = \bar{1} \vee \cdots \vee x = \overline{n-1}$ and thus to handle the crucial step of bounded quantification.

A last axiom: the axiom $x < y \vee x = y \vee y < x$ is of a slightly different nature from the rest, since it is not needed for incompleteness: the representation of expansive properties is handled by the definition axioms. It is used in the representation of recursive functions and therefore in the algorithmic *undecidability* of **RR** and all its consistent extensions. It is also used in the Rosser variant.

These axioms must be manipulated by means of logical rules. We postpone this point to the next chapter. Indeed, there are as many axiomatisations of logic as one wants, all of them without the slightest practical interest: one does not need them for reasoning! So, as to learning rules, one might as well select a formulation with some theoretical interest, e.g., *sequent calculus* (next chapter). Incompleteness usually makes use of classical logic, but there is no inconvenience in using intuitionistic logic, nay linear logic.

Thus, every consistent system containing these few axioms will be incomplete and therefore, undecidable. By this, one means that the system does not decide all properties, but above all that there is no algorithm determining whether or not a formula is provable. To recover decidability, one must use the bludgeon, for instance:

[23] The main virtue of this axiomatics, due to R. Robinson, is to be without quantifiers.

(i) Replace the third Peano axiom with, say, $\overline{10} = \bar{0}$. One is thus in the finite (integers *modulo* 10) and the system becomes decidable. If one simply removes the third axiom without replacing it, one remains in the undecidable.[24]

(ii) Remove a symbol, typically the product (Pressburger arithmetic, decidable). Keeping in mind the crucial role played by the product in the representation of finite sequences, we understand why the system may become decidable.

These systems are decidable because they are (almost) good for nothing.

2.B.2 Peano's arithmetic. Peano originally formulated his arithmetic by means of (a variant of) the previous axioms together with the *induction* schema:

$$A[0] \wedge \forall y(A[y] \Rightarrow A[Sy]) \Rightarrow \forall x A[x]$$

formulated for an arbitrary *property* of integers, with all the fuzziness conveyed by this word. Richard's paradox (1905) shows that this is not tenable: A must be part of a well-defined formal language[25]. What one now calls *Peano arithmetic* is the system **PA** obtained by adding to **RR** (one can skip the last axiom) the induction schema restricted to arithmetical formulas, i.e., to the formulas of the language. While here, observe that there is also a Heyting arithmetic **HA**, the same thing, but on an intuitionistic substrate.

The second incompleteness theorem is essentially a formalisation of the first one. The informal proof of the first theorem involves inductions, therefore this theorem can be established in Peano arithmetic (or its intuitionistic variant). This is indeed too much, since inductions are done on provability statements, thus expansive, Σ^0_1. A good substrate for the second theorem is therefore induction restricted to Σ^0_1 formulas. This is still far from optimality, but that's a good approximation. The « good system » is obtained by adding functional primitives (e.g., the exponential function), so as to be able to replace in some cases unbounded quantifiers with bounded ones. Induction is then restricted to formulas without unbounded quantifiers.

Finally, let us come back to the devastating joke of André Weil concerning Gentzen. Induction can be formulated with two nested variables, three, etc. and these generalised inductions appear as transfinite inductions, e.g., with two variables, up to ω^2, with three variables, up to ω^3 etc. There are therefore two parameters for an induction, on one hand the length of the induction, on the other hand the complexity (alternation of quantifier) of the « inducted » formula. In the case of Gentzen's second proof, the starting induction is of short length (ω), but arbitrary

[24] A is provable in the « basic » system iff $T \Rightarrow A$ is provable in the system without the third axiom, with $T := \forall x\ Sx \neq 0$.

[25] It is the technique of the « mobile curtain », which rather arbitrarily divides tourist class – well-defined things – from the business class which contains the rest.

complexity (any Σ^0_n), while his consistency proof makes use of a long induction (ϵ_0), but of low complexity (without quantifiers). Just to nuance the failure, which is not quite the expected disaster!

2.B.3 More general systems. Peano's arithmetic provides us with a non-finitely axiomatised system, by the way not finitely axiomatisable. We must obviously limit the use of infinite lists of axioms, otherwise there is the pitfall of non-monotonic logics, i.e., the absence of any notion of proof. A natural definition is that of a decidable family of axioms (one can algorithmically tell whether or not an axiom is accepted). A fake generalisation: an expansive axiomatisation, i.e., recursively enumerable, produces nothing new. Indeed, it suffices to replace axiom A_n produced at stage n by an equivalent variant of size at least n (e.g., the conjunction or the disjunction of n copies of A_n) to replace a r.e. axiomatics with a recursive one.

If we dump recursive limitations on provability, we lose expansivity and the relations to computation[26]. On the other hand, all the machinery is intact, we must only take into account the logical complexity of « deduction », remember to use n-consistency... and we easily reach a generalisation colourless, insipid and without savour.

2.C Techniques of incompleteness

2.C.1 The fixed point: Russell. Russell's antinomy, which is just the version of Cantor's theorem adapted to « the set of all sets », produces an immediate antinomy: if $a := \{x; x \notin x\}$, then $y \in a \Leftrightarrow y \notin y$, therefore $a \in a \Leftrightarrow a \notin a$. In « normal » systems, with – unlike paraconsistent logics – a non-contrived deduction, this leads to contradiction. It is important to remark that the logical fixed point (inconsistency) is the same thing as the fixed point of λ-calculus, the latter avoiding contradiction by means of an infinite loop of computations.

This is more than a didactic remark: in his book « Natural Deduction » [87], Prawitz introduced a natural deduction for the « naïve » comprehension schema, hence for an inconsistent logical system. This natural deduction admits a notion of normalisation *which does not converge*. Indeed the normalisation process of Russell's antinomy is precisely the fixed point of λ-calculus. Which establishes an important relation:

$$\textit{inconsistency} \sim \textit{non-termination}$$

[26] Spicy details for non-monotonic logic, which are supposed to solve questions related to computation: the exigency of computability is sacrificed on the altar of efficiency. This is typical of ideological blindness: during WWII, French nationalists worked for the Germans; communists instituted the most inegalitarian system ever, etc.

« Normal » logics have a too brutal (essentialist) vision of infinity, in other terms, they define functions « growing too fast », exponentially or worse; these functions are essentially the normalising time, as we shall see in the chapter on natural deduction. The question is to isolate the principle producing the combinatory « explosion » occurring in normalisation. Now, from the viewpoint of a formal system, there is no difference between an algorithm whose termination cannot be proved and a plain loop in the style of the one induced by Russell's antinomy. The moral is: in order to study huge complexity, let us study divergence; Russell's antinomy: this will lead us to the *light* logics in Chapter 16.

Russell's antinomy splits into two parts, $a \in a \vee a \in a$ and $a \notin a \vee a \notin a$. It becomes plainly contradictory only through the rule of *contraction*, i.e., the possibility to replace a disjunction $A \vee A$ with A. This shows the way to follow, the rule of contraction. This being said, its blunt elimination would be too radical: one should rather bridle it. But this is another story.

Last, a remark on the meta. Dressed with all possible sauces, this doohickey means very little: *grosso modo*, nothing distinguishes a system from a meta-system. However, there is perhaps something to find, provided one allows some geometrical finesse. Thus, in diagonalisation, the function of two arguments $f(x, y)$ is indeed of the form $f_y(x)$, the index y being in a sort of « meta » position w.r.t. x. With an absolute vision, à la Kronecker, of integers, to make $x = y$ looks very natural, but when one witnesses the Pandora's box that this anodyne action opens, should we not have a second thought? From the viewpoint of linear logic, this operation does not respect the depth of boxes (exponential, not of Pandora!). Hence hypothesis, work in progress, etc.: boxes and « meta ».

2.C.2 The fixed point: Gödel. The coding of syntax involves a lot of recursive functions and expansive properties, for instance $\mathrm{Sub}(m, n)$ such that $\mathrm{Sub}(\ulcorner A \urcorner, \ulcorner \bar{q} \urcorner) = \ulcorner A[\bar{q}] \urcorner$ and which therefore represents, at the level of codes, the result of substituting the term coding the integer q in formula A; to be rigorous we should even mention for which variable the substitution occurred (say y_0), not to speak of writing a novelette concerning free and bound variables. One will similarly introduce $\mathrm{Dem}(m, n)$, such that $\mathrm{Dem}(\ulcorner A \urcorner, \ulcorner D \urcorner)$ when D is a proof A, etc. Finally $\mathrm{Con}(\mathcal{T})$, the consistency of $\mathcal{T}$ (recessive property) is expressed as $\forall x \neg \mathrm{Dem}(\overline{\ulcorner 0 \neq 0 \urcorner}, x)$.

To obtain Gödel's formula we thus form $\forall x \neg \mathrm{Dem}(\mathrm{Sub}(y_0, y_0), x)$, which is a formula $A[y_0]$ and we define $G := A[\overline{\ulcorner A \urcorner}]$. G thus expresses its non-provability.

2.C.3 The coding of sequences. The representation of recessive formulas essentially rests upon the coding of sequences of integers by integers. The coding of sequences of a given length can be done by means of polynomials, for instance, in length 2, the pair (m, n) can be coded by $(m + n)(m + n + 1)/2 + n$, which corresponds to the enumeration $(0,0), (0,1), (1,0), (0,2), (1,1), (2,0), (0,3), (1,2), (2,1), (3,0), (0,4), \ldots$; by the way, observe that this coding enables one to reduce to 1 the number of unbounded quantifiers in an expansive (or recessive) formula.

But one should also encode sequences of arbitrary length, typically for the representation of proofs. To do so, Gödel uses a technique which is not quite original, since we know today that he took it from lectures on class field theory that he had just been following... Anyway, the technological transfer was not obvious. Epistemologically, one must remark that, should this coding have not been available, one would have then introduced a specific primitive for the coding of sequences. With a light nuance: the possibility of coding in such a poor language makes the theorem much more spectacular, a dilemma between incompleteness and inexpressivity...which would be less conspicuous if the coding of sequences were more costly, since one would thus have an intermediate zone of decidable systems with average expressive power.

Gödel used the Chinese remainder theorem $\mathbb{Z}/pq\mathbb{Z} \simeq (\mathbb{Z}/p\mathbb{Z}) \times (\mathbb{Z}/q\mathbb{Z})$ when p, q are relatively prime. Concretely, given $a < p, b < q$, we can find a (unique) $c < pq$ such that the respective remainders of the division of c by p and q are a and b. This result immediately generalises to an arbitrary number of mutually prime integers. In order to represent the sequence $(a_1, \ldots, a_n)$, we select an integer N strictly greater than the a_i and the length n of the sequence. We easily check that the numbers $N! + 1, N!.2 + 1, \ldots, N!.n + 1$ are mutually prime. Then there is an integer s whose respective remainders *modulo* $N! + 1, N!.2 + 1, \ldots, N!.n + 1$ are $a_1, a_2, \ldots, a_n$. The sequence can be encoded by means of the three numbers N, n, s. To be mutually prime, the remainder of a division, all this can be expressed in terms of 0, S, $+$, $\times$, $<$, $=$.

The possibility of encoding sequences of an arbitrary length explains why one cannot bridle the dynamics and contrive, as in epistemic logics, proofs to be of length, say 25 (case of the 25 Baghdad cuckolds). The decidability of Pressburger arithmetic is obviously related to the crucial role played by the product in the encoding of sequences of arbitrary length.

2.D Incompleteness and truth

2.D.1 Tarski's theorem. Tarski's theorem says that arithmetical truth cannot be defined in arithmetic. More precisely, there is no arithmetical formula $V[n]$ such that the equivalence $V[\ulcorner A \urcorner] \Leftrightarrow A$ is true for any closed formula A. In other terms, truth is hyperarithmetical without being arithmetical. See also Section 3.B.3.

This result follows from a trivial diagonalisation: in the incompleteness theorem, replace "provable" with "true". One can surmise that Gödel (like anybody, unable to conceive incompleteness *ante literam*) first tried to *define* truth through provability and first thought he had obtained a contradiction – not that far from Russell's antinomy. A second reading made him realise that his definition of truth might be incorrect.

Why is this result attributed to Tarski? It exemplifies his propensity for « dumpster diving »: witness the corny fixed point theorem to which he insisted on attaching his name.

One should not think that the completion of a consistent theory is necessarily hyperarithmetic or worse. With a little know-how, one builds Δ_2^0 models of arithmetic. Of course they are « non-standard », in the sense that they satisfy a lot of false formulas.

2.D.2 1-consistency. To what extent does consistency convey the « honesty » of a system? This surely means something, otherwise there should not be the strong obstruction of incompleteness. This feeling of honesty is reinforced by the contrived character of paraconsistent logics: as we observed, honesty must be honestly defined. But this is true only up to a certain point. Take for instance the idea – surely honest – that consistency is a substitute for existence. It is in fact true solely in the algebraic world, a world of equations: the recessive world of Hilbert. What is problematic is the consistency of manifestly faulty systems, typically $\mathcal{T} + \neg\mathrm{Con}(\mathcal{T})$, when $\mathcal{T}$ is consistent. We can see the limit of the exercise: consistency only deals with the observance of the law by the book, not with its spirit. It is therefore only a *primal* form of honesty.

We introduce Σ_n^0- (or Π_n^0-) *faithfulness*, to mean that any theorem of this class is true. It is immediate that:

(i) Π_1^0-faithfulness is just consistency.

(ii) Σ_n^0-faithfulness is equivalent to Π_{n+1}^0-faithfulness, *a priori* more general.

We thus define n-consistency as Π_{n+1}^0-faithfulness, in particular 0-consistency is just plain consistency. 1-consistency is strictly stronger than consistency; an easy generalisation shows that, more generally, $n+1$-consistency is strictly stronger than n-consistency.

2.D.3 Rosser's variant. Rosser's variant of the first incompleteness theorem solves a small problem left by Gödel, i.e., the hypothesis of 1-consistency[27] necessary to prove that $\neg G$ is not provable. The starting system $\mathcal{T}$ is artificially bridled by the constraint: one can prove A only when $\neg A$ has not been proved (in $\mathcal{T}$) with a proof with a code (Gödel number) smaller. In other terms, $\mathrm{Dem}'(m,n) := \mathrm{Dem}(m,n) \wedge \forall p < n \neg\mathrm{Dem}(\mathrm{Neg}(m),p)$, with $\mathrm{Neg}(\ulcorner A \urcorner) = \ulcorner \neg A \urcorner$. From our viewpoint, $\mathcal{T}$ being supposedly consistent, there is no difference between $\mathcal{T}$ and its bridled version; moreover, the bridled version remains expansive, since it differs from the « normal » one by a bounded quantification (one must verify that the integers smaller than a given n do not code proofs of $\neg A$).

[27] Called ω-consistency in its day, an obsolete terminology.

Formally, $\mathcal{T}$ proves, for each A not provable in $\mathcal{T}$ and each integer n, the formula $\mathrm{Dem}(\overline{\ulcorner\neg A\urcorner}, \bar{n}) \Rightarrow \neg\mathrm{Dem}'(\overline{\ulcorner A\urcorner}, x)$. In fact, the « last axiom » yields $x < \bar{n} \vee x = \bar{n} \vee \bar{n} < x$, in other terms $x = \bar{0} \vee x = \bar{1} \vee \cdots \vee x = \bar{n} \vee \bar{n} < x$ and one then proceeds by cases.

Let R be Rosser's formula, i.e., the Gödel formula of the bridled system. There is nothing to say in case of provability: R is not provable, i.e., is true. If $\neg R$ is provable with a proof of code n, this fact is provable in $\mathcal{T}$; from what precedes, one concludes (in $\mathcal{T}$) $\forall x \neg\mathrm{Dem}'(\ulcorner R\urcorner, x)$, the (bridled) non-provability of R, i.e., R, which is not provable. $\neg R$ is therefore not provable.

Rosser's variant is, you perhaps noticed it, a para-consistent system. This way of choosing, between A and $\neg A$, the formula with the smallest proof is deductively meaningless, since, w.r.t. logical consequence, the smallest proof is most likely to change sides. It would therefore be skulduggery to present this technical trick as meaningful. What makes the value of this « faking » is that it is the formal system $\mathcal{T}$ which is fooled – and not one's fellow man, which the paraconsistent mob try to do.

Rosser's trick can be declined in various techniques to represent functions, etc. in the case of unfaithful theories. We can dispense with the trip, the landscape is without surprise: as everything related to incompleteness, systematic exploration runs into boredom, non-substance. Once more, we have just explored a dead branch of logic and the time is ripe to head for more lively zones.

2.E Undecidability

2.E.1 The undecidable. Incompleteness is deductive undecidability (one cannot decide every question A by proving or refuting it). But it also implies an *algorithmic* undecidability, in the sense that there is no algorithm deciding whether or not a given formula is provable.

The standard way of proving undecidability is to represent all computable (recursive) functions in **RR**: an appropriate diagonalisation enables one (as usual) to such a conclusion. I limit myself to the definition of the notion of representation: a function f is representable by a formula $A[x, y]$ when one can prove, for each integer m:

$$A[\bar{m}, y] \Longleftrightarrow y = \overline{f(m)}.$$

A weaker form of representability would be to require $A[\bar{m}, \bar{n}]$ to be provable when $f(m) = n$, $\neg A[\bar{m}, \bar{n}]$ to be provable when $f(m) \neq n$. This weak form can be converted into the strong form (the quite useful one) by replacing A with $B[x, y] := A[x, y] \wedge \forall z < y \; \neg A[x, z]$. One makes use of the « last axiom » to operate this passage from weak to strong representation.

Finally observe that, since **RR** is finitely axiomatisable, predicate calculus is in turn undecidable.

2.E.2 Inseparability. In the system **RR**, there is not even a *separation* algorithm, i.e., answering « yes » when A is provable, « no » when $\neg A$ is provable, no matter what otherwise, *provided it answers something*; in other terms theorems cannot be *recursively separated* from anti-theorems. This remains trivially true of any consistent extension $\mathcal{T}$ of **RR**, since an algorithm separating $\mathcal{T}$ would also separate **RR**. In other terms, any theory « sufficiently strong » is recursively inseparable.

Which is not the case for predicate calculus, undecidable without being inseparable. For instance one can separate theorems by means of a finite model. My favourite separation method is as follows: forget first order, in other terms, given a formula A of predicate calculus, define A^- as follows:

- Replace predicate $P(\ldots)$, $Q(\ldots)$ with propositional symbols P, Q.
- Ignore quantifiers.

One thus obtains an interpretation of predicate calculus in propositional calculus and it is immediate that this interpretation preserves proofs, which can for instance be verified using sequent calculus. See Section 6.1.4.

2.E.3 Ambiguous functions. Take a system to which incompleteness applies, typically **ZF**; take also two recursive functions f, g and a recessive formula $B = \forall n A[n]$. One defines h to be f as long as one has not refuted B and g « after »:

$$h(n) := f(n) \text{ if } \forall m \leq n A[m], \tag{2.1}$$

$$h(n) =: g(n) \text{ if } \exists m \leq n \neg A[m]. \tag{2.2}$$

Obviously h is either equal to f (if B is true), or equal to g (if B is false), except for a finite number of values. Let us say for simplicity that h is *almost equal to* f or g. Now, if it happens that B is undecidable in **ZF**, one will not be able to decide which of f, g is almost equal to h.

Thus, if f, g are functions constantly equal to respectively 0, 1, h is a Cauchy sequence for which one cannot decide in **ZF** whether it converges to 0 or 1.

This threadbare technique has been used over and over to produce shallow counterexamples to various decidability questions. Typically:

There exists recursive reals a, b, c such that the solvability in $\mathbb{R}$ of the equation $ax^2 + bx + c = 0$ is undecidable in **ZF**.

The proof does not need a deep knowledge of algebraic equations; just take $a = 1$, $b = 0$ and c to be the real defined by the ambiguous Cauchy sequence mentioned above. The equation is either $x^2 = 0$ (one solution) or $x^2 + 1 = 0$ (no solution).

This sort of modest examples being exhausted, the technique has been applied to yield pompous statements, typically:

There exists a recursive function h such that the question $P \stackrel{?}{=} NP$ *relativised to h is undecidable in set-theory* **ZF**.

Which is a trivial corollary of a renowned result of Solovay: there are actually recursive functions f and g such that the answer to $P \stackrel{?}{=} NP$ relativised to f and g is "yes" in one case, "no" in the other. Taking an ambiguous function h, we get the result. In order to conclude, we don't need to know the proof of Solovay's theorem, or anything about complexity theory; we only need the information that relativisation is the same for *almost equal* functions. In particular the claim that this « theorem » shows how hard is the question $P \stackrel{?}{=} NP$ is pure baloney: it is obtained by a mechanical imitation of the argument about the equation of degree 2 and nobody believes that the solution of this equation is « hard ». Here again we may ask the question: « what keeps the *entarteur* ? ».

This sort of bleak slapstick comedy is a perfect illustration of the present state of the area « incompleteness »: a graveyard.

Bibliography: [38], [42].

Chapter 3
Classical sequents: LK

3.1 Generalities

3.1.1 Sequents vs. Hilbert-style systems. *Sequent calculus* is due to Gentzen (1934) [30]. It is a formulation, among others, of *predicate calculus*. For that matter, it is important to clarify an ambiguity as to this « among others »:

- It is therefore legitimate to choose sequent calculus, the sequent « style », as default style. However, one must remark that the system is particularly heavy to handle and that some rules (as in **LJ**, the left rule for implication) are very counter-intuitive.

- The utilisation of this calculus is justified insofar as one makes use of *cut-elimination* and its corollaries. Beware this currently popular con: the writing in sequent style of a system *not enjoying* cut-elimination[1]. Indeed, since sequents are interesting only in relation to this theorem, one always assumes that a system written in this style enjoys cut-elimination. This reminds me of those cheap « Bolex » watches sold in the flea market of Porta Portese, not to be confused with « Rolex ».

Before Gentzen, logic was formulated in « Hilbert-style » formal systems; basically a lot of axioms and a couple of rules, essentially *Modus Ponens*: from A and $A \Rightarrow B$ deduce B and *generalisation*: from $A[x]$, deduce $\forall x A[x]$. These systems have no good structural property, for instance, it is impossible to try the slightest automated deduction in the presence of the *Modus Ponens*: indeed, in order to prove B (supposedly obtained by a *Modus Ponens*), one must guess the premise A, which can be any formula, including B. The only setting in which Hilbert-style systems are justified is the narrow context of the incompleteness theorem: indeed, one does not seek positive properties of formalism, but rather its limits. In this very negative perspective, Hilbert-style systems do no worse than other systems and some of them have the advantage of being immediately understandable[2].

3.1.2 Sequents. Every « old style » introduction to logic, i.e., based upon Hilbert-style systems, begins with a « deduction theorem »; remember that a formula without a free variable is said to be *closed*.

[1] This is what usually happens with « tableaux », the poor man's sequents.

[2] The proposition of an autodidact (Dijkstra): a Hilbert style system based upon logical equivalence, shows that one can cumulate the absence of structure and illegibility.

Theorem 3 (Deduction theorem). *If B is provable in the formal system $\mathcal{T} + A$, with A closed, then $A \Rightarrow B$ is provable in $\mathcal{T}$.*

Proof. It depends on the system and also works for intuitionistic and linear[3] logics. The restriction « A closed » is made necessary by the rules like « generalisation »: one wants to avoid $A[x] \Rightarrow \forall x A[x]$. □

Indeed, even Hilbert-style systems are compelled to introduce the notion of « hypothetical » proof, at least under the form of a lemma. Under hypothesis A, one gets B; one concludes that $A \Rightarrow B$. The novelty of Gentzen is the introduction of hypothetical deduction as a primitive; besides the implication $A \Rightarrow B$, there coexists the sequent $A \vdash B$: « B under hypothesis A » . One will never insist enough: from a brutal standpoint (what can be, cannot be proved: let us say the viewpoint of completeness, i.e., at layer -1, see Section 7.1.1.), this creation makes no sense: it is a pure duplicate, since the deduction theorem equates the two notions. Sequent calculus makes sense only when one steps beyond mere provability, when one works *en finesse*.

In practice, we are led to iterate the deduction theorem, i.e., to make several hypotheses, which is reflected by the creation of sequents $\Gamma \vdash A$, where Γ is a finite sequence of hypotheses, (which is called an *intuitionistic* sequent); it is less immediate that hypotheses can be transformed into auxiliary conclusions, i.e., that one can write $\Gamma \vdash A, \Delta$. This counter-intuitive (and non-intuitionistic) move corresponds to the excluded middle, or, if one prefers, to contraposition, typical of classical logic[4]. One eventually reaches the following definition:

Definition 1 (Sequents). A *sequent* is any expression $\Gamma \vdash \Delta$, where Γ and Δ are finite sequences of formulas.

Let us mention a few variants:

Sets: instead of sequences, one can take sets. The advantage is that the structural group (*infra*) becomes much simpler (reduced to weakening). By taking « weakened » identity axioms $\Gamma, A \vdash A, \Delta$ one can even dispense with weakening. This being said, this simplification is a false cognate, which impedes any fine grain discussion of structural rules.

Right formulation: one can formulate sequent calculus « on the right », i.e., with sequents $\vdash \Delta$. This is possible in the classical and linear cases, thanks to the involutive negation[5]. This formulation aims at economy: it divides the number of rules by two. In the linear case where there are twice more connectives, this keeps the size of the system reasonable.

---

[3] In the linear case, don't take linear implication, choose instead $A \Rightarrow B := !A \multimap B$.

[4] The left/right symmetry, *infra*.

[5] Negation is basically the exchange of the two zones determined by the symbol $\vdash$.

In order to introduce sequents, we started with $\Gamma \vdash A$, nay $\Gamma \vdash A, \Delta$, where Γ and Δ play the role of a *context* for A (e.g., the hypotheses made to get A). Sequents relativise the notion of context. Thus, in $A, B \vdash C, D$, one might as well declare that the context of C is $A, B \vdash -, D$, or that the context of B is $A, - \vdash C, D$. A sequent is therefore a mixture of several formulas and one will, as we please, *focus* (act) on one of them, the others (the context) remaining passive.

3.1.3 Signification. A sequent is written with formulas, the symbol « turnstile », $\vdash$, as well as commas. The intuitive significance of the sequent $\Gamma \vdash \Delta$ is « if all elements of Γ are true, then one of the elements of Δ is true ». One therefore sees that:

- The left comma means « and ».
- The right comma means « or »; not used in the intuitionistic setting (Chapter 4).
- The sign $\vdash$ means « implies ».

Observe that I did not use the logical symbols for conjunction, disjunction, implication. Indeed, this explanation is not limited to classical or intuitionistic logics; it also works for linear logic, in case one is being precise about which conjunction, disjunction, implication is at stake (in fact their multiplicative versions).

Some important particular cases:

- If $\Delta = A$, the sequent means that A is true under the hypothesis Γ. Moreover, if $\Gamma = \emptyset$, the sequent simply means A. In other words, when one wants to prove A, one proves the sequent $\vdash A$.
- If $\Delta = \emptyset$, the sequent means that Γ is contradictory, in particular, $A \vdash$ means $\neg A$. As to the sequent $\vdash$, it is a way to express contradiction[6]. This would remain valid intuitionistically if we had allowed empty Δ's. On the other hand, in linear logic, the empty sequent is not contradictory.

3.2 The classical calculus

The classical sequent calculus **LK** is a system of deductive rules divided in three groups: identity, structure and logic.

[6]Thus, the consistency proofs of Gentzen « establish » that $\vdash$ is not provable in such and such system.

3.2.1 Identity

$$\frac{}{A \vdash A}\ (identity) \qquad \frac{\Gamma \vdash A, \Delta \quad \Lambda, A \vdash \Pi}{\Gamma, \Lambda \vdash \Delta, \Pi}\ (cut)$$

This group means (no joke!) « A is A, *and conversely* ». It relates two *occurrences* of the same formula A, one to the left, one to the right. The identity rule is 0-ary (no premise), in other terms it is an axiom[7]. The cut-rule is binary. Keeping in mind that $\vdash$ has the same value as implication, one sees that *Modus Ponens* and transitivity of implication are particular cases of a cut:

$$\frac{\vdash A \quad A \vdash B}{\vdash B}\ (modus\ ponens) \qquad \frac{A \vdash B \quad B \vdash C}{A \vdash C}\ (transitivity)$$

Indeed, a cut is only a contextual version of *Modus Ponens*.

Anticipating the geometrical interpretations issued from linear logic (proof-nets, geometry of interaction), it is possible to give a very concrete version of this group of rules[8]. A denotes a type of electric plug, which occurs as female (on the left, rewritten as $\neg A$ on the right) or male (on the right). The identity axiom corresponds to an extension cord,

while the cut

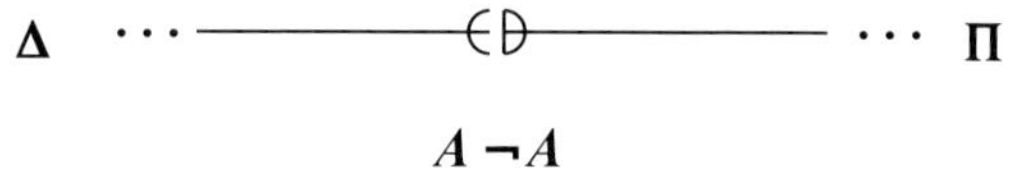

corresponds to the *connexion* of two complementary plugs. We shall not push this analogy further for this moment. We simply remark that, in principle, one can dispense from connexions: typically, connecting an extension cord is doing nothing but *delocating* a plug:

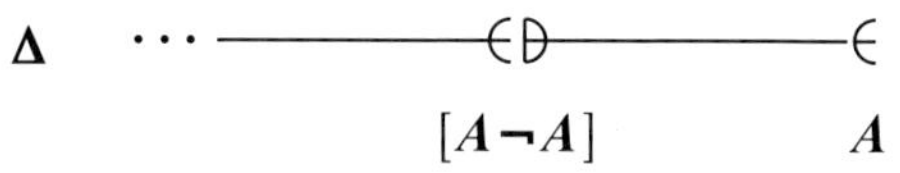

The expression « cut » comes from the fact that A disappears during the rule; but, if this rule is the rule of communication, to call it « cut » is disquieting! Cut is immediately placed under the sign of schizophrenia, which will be confirmed by the fact that one spends one's time *eliminating* this rule – the essential rule of reasoning.

[7]Contrary to Hilbert-style systems, sequent calculi have very few axioms.

[8]What follows refers to the « right » version of the calculus (*infra*, Section 3.2.5).

Sequent calculus is organised around the following theorem, due to Gentzen[9]:

Theorem 4 (Hauptsatz). *In the sequent calculus* **LK**, *the cut-rule is* ***redundant****, in other terms, any sequent provable with cut is also provable without cut.*

This result *of cut-elimination* is hard to prove: we shall it admit for a while... especially since we don't yet know the full system. It will persist for intuitionistic and linear logics; this last system is indeed built around the *Hauptsatz*.

It serves to emphasise the paradoxical character of this result. In Hilbert-style systems, based on common sense, the only rule is (more or less) *Modus Ponens*: reasoning proceeds by linking together lemmas and consequences. We just said that one can get rid of that: it is like crossing the English Channel with fists and feet bound!

3.2.2 Structural group

$$\frac{\Gamma \vdash \Delta}{\sigma(\Gamma) \vdash \tau(\Delta)}\ (X)$$

$$\frac{\Gamma \vdash \Delta}{\Gamma \vdash A, \Delta}\ (\vdash W) \qquad \frac{\Gamma \vdash \Delta}{\Gamma, A \vdash \Delta}\ (W \vdash)$$

$$\frac{\Gamma \vdash A, A, \Delta}{\Gamma \vdash A, \Delta}\ (\vdash C) \qquad \frac{\Gamma, A, A \vdash \Delta}{\Gamma, A \vdash \Delta}\ (C \vdash)$$

This group is devoted to an (apparently) insignificant task, the maintenance of sequences:

Exchange: this group enables permutations on each side of the sequent. It therefore enunciates the *commutativity* of conjunction and disjunction. There exist non-commutative systems, e.g., [4], (refusing exchange), but they remain rather experimental. This is why we shall not take exchange into account (otherwise this turns into a bureaucratic nightmare) and shall consider sequents up to order. This remains legitimate as long as we don't focus on non-commutative logic, which is the case of (the essential of) these lectures. For instance, the conclusion of the rule ($\vdash$ W) might as well be written $\Gamma \vdash \Delta, A$.

Weakening: this rule (at least its left version) enunciates the possibility of adding hypotheses. It is a monotonicity property, the more hypotheses, the more conclusions. This rule is called into question by linear logic, but this questioning is in turn questionable – which cannot be said about the similar questioning of contraction. There exists a completely obsolete literature (and much anterior

[9]The word *Hauptsatz* means « main result »; one uses it by habit or by snobbery.

to sequent calculus) dedicated to « material implication », i.e., the effect without a cause. And also a philosophical tradition of « relevant » logics, a rather barren topic, dedicated to systems accepting contraction, but not weakening. These systems do not usually enjoy cut-elimination and therefore their expression in sequent style is a perfidy[10]: the decomposition identity/structure/logic loses its functionality in the absence of *Hauptsatz*.

Contraction: this rule (at least its left version) permits the reuse of hypotheses. It states the *perenniality* of logical properties, in contrast to, say, physical properties (the state of a system). This durability is essential to the expression of infinity: without being too formal, one understands that the hypothesis $\forall x \exists y\ x < y$ can, in an appropriate context, produce as many distinct objects as we want, provided one reuses it: for instance, 1000 uses will produce 1001 distinct elements. Contraction enables one to produce these 1001 elements from a single hypothesis. Not only is this rule necessary for the proof of properties linked to infinity; but it is also responsible for the complexity of the computation linked to normalisation, as we shall later see, including the contradiction in Russell's antinomy (Section 2.C.1). Eventually, the subformula property (*infra*) almost produces a decision algorithm which fails because of contraction.

To sum up, this rule is everything, except anodyne: it concentrates in it, much more than the logical rules of quantification, the infinite aspects of logic, in one word, essentialism. This is why its study is, implicitly, at the very heart of these lectures.

3.2.3 Logical group. We must now fix a language, more precisely logical connectives. In classical logic, one can take as basic connectives $\wedge, \vee, \neg, \forall, \exists$. This list is not limitative, for instance one could include implication, or constants for truth and falsity. Implication is usually handled as a defined connective $A \Rightarrow B := \neg A \vee B$; if one insists on having constants, say $\mathbf{v}$, $\mathbf{f}$ (*verum, falsum*), one should add the axioms $\vdash \mathbf{v}$ and $\mathbf{f} \vdash$.

$$\frac{\Gamma, A \vdash \Delta}{\Gamma \vdash \neg A, \Delta}\ (\vdash \neg) \qquad \frac{\Gamma \vdash A, \Delta}{\Gamma, \neg A \vdash \Delta}\ (\neg \vdash)$$

$$\frac{\Gamma \vdash A, \Delta \quad \Gamma \vdash B, \Delta}{\Gamma \vdash A \wedge B, \Delta}\ (\vdash \wedge) \qquad \begin{array}{c} \dfrac{\Gamma, A \vdash \Delta}{\Gamma, A \wedge B \vdash \Delta}\ (l \wedge \vdash) \\ \dfrac{\Gamma, B \vdash \Delta}{\Gamma, A \wedge B \vdash \Delta}\ (r \wedge \vdash) \end{array}$$

[10] The *intuitionistic* version is slightly better off, since it even admits a functional interpretation (λI-calculus).

$$\frac{\Gamma \vdash A, \Delta}{\Gamma \vdash A \vee B, \Delta}\ (l \vdash \vee) \qquad \frac{\Gamma, A \vdash \Delta \quad \Gamma, B \vdash \Delta}{\Gamma, A \vee B \vdash \Delta}\ (\vee \vdash)$$

$$\frac{\Gamma \vdash B, \Delta}{\Gamma \vdash A \vee B, \Delta}\ (r \vdash \vee)$$

$$\frac{\Gamma \vdash A, \Delta}{\Gamma \vdash \forall x A, \Delta}\ (\vdash \forall) \qquad \frac{\Gamma, A[t/x] \vdash \Delta}{\Gamma, \forall x A \vdash \Delta}\ (\forall \vdash)$$

$$\frac{\Gamma \vdash A[t/x], \Delta}{\Gamma \vdash \exists x A, \Delta}\ (\vdash \exists) \qquad \frac{\Gamma, A \vdash \Delta}{\Gamma, \exists x A \vdash \Delta}\ (\exists \vdash)$$

The rules of quantification require some clarification. The rules $(\forall \vdash)$ and $(\vdash \exists)$ involve the substitution of a term t for a variable. Remember that this notion requires a few precautions, namely that t does not use bound variables of A. The $(\vdash \forall)$ and $(\exists \vdash)$ are subject to a more serious restriction: the variable x must not occur in the context $\Gamma \vdash \Delta$. In more formalist presentations, the premise of these two rules makes use of a special variable, called an *eigenvariable*, say v, which replaces x, so that the premise of $(\vdash \forall)$ is rigourously $\Gamma \vdash A[v/x], \Delta$. I chose a laxist tradition consisting in being rather fuzzy as to the name of bound variables, that one changes when needed. There are enough important things so that we can dispense with wasting time on one of the few insignificant details of logic – at least at that early stage.

3.2.4 The left/right symmetry. The calculus enjoys a left/right symmetry[11]. The rules are unchanged if one swaps left and right, provided we perform the permutations $\vee/\wedge$ and $\exists/\forall$. One knows that negation exchanges these connectives: one concludes that negation is nothing but the exchange left/right. This is, modest but essential, our second procedural remark (after the interpretation of the identity group):

Negation is the exchange left/right.

In natural deduction, negation will correspond to the exchange up/down, hypothese(s)/conclusion.

This is an obvious change of paradigm: instead of concentrating on the intended meaning of negation, one works on its geometry, which does not care about our intentions. Compare with the tarskian truism:

$\neg A$ *is true when A is* ***not*** *true.*

[11]Which persists *mutatis mutandis* in linear logic, including its non-commutative version.

3.2.5 Right-hand calculus. Symmetry enables one to drastically reduce the size of sequent calculus. Forbidding the expression $\neg A$ when A is not atomic, we are led to *define* $\neg A$, by means of the De Morgan laws:

$$\begin{array}{cc} \multicolumn{2}{c}{\neg(\neg p) := p,} \\ \neg(A \wedge B) := \neg A \vee \neg B, & \neg(A \vee B) := \neg A \wedge \neg B, \\ \neg\forall x A := \exists x \neg A, & \neg\exists x A := \forall x \neg A. \end{array} \tag{3.1}$$

In this simplified syntax, one can restrict to sequents $\vdash \Delta$ without a left-hand side. The idea is to translate $\Gamma \vdash \Delta$ by $\vdash \neg\Gamma, \Delta$, where $\neg\Gamma$ is the sequence of the negations of the elements of the sequence Γ. This idea will be found again in linear logic, including its non-commutative version, in which case Γ must not only be negated, but also order-reversed.

$$\frac{}{\vdash \neg A, A}\ (identity) \qquad \frac{\vdash \Delta, A \quad \vdash \neg A, \Pi}{\vdash \Delta, \Pi}\ (cut)$$

$$\frac{\vdash \Delta}{\vdash \tau(\Delta)}\ (X) \qquad \frac{\vdash \Delta}{\vdash A, \Delta}\ (\vdash W) \qquad \frac{\vdash A, A, \Delta}{\vdash A, \Delta}\ (\vdash C)$$

$$\frac{\vdash A, \Delta}{\vdash A \vee B, \Delta}\ (l \vdash \vee) \qquad \frac{\vdash B, \Delta}{\vdash A \vee B, \Delta}\ (r \vdash \vee) \qquad \frac{\vdash A, \Delta \quad \vdash B, \Delta}{\vdash A \wedge B, \Delta}\ (\vdash \wedge)$$

$$\frac{\vdash A, \Delta}{\vdash \forall x A, \Delta}\ (\vdash \forall) \qquad \frac{\vdash A[t/x], \Delta}{\vdash \exists x A, \Delta}\ (\vdash \exists)$$

3.3 The cut-free system

The *Hauptsatz* establishes the redundancy of the cut-rule. What is so special about cut-free systems?

3.3.1 The subformula property. Let us compare premises and conclusions in the rules of sequent calculus. More precisely, given the conclusion $\Gamma \vdash \Delta$ of a rule, can I say something concerning its premises? It is obvious that nothing can be said when the last rule is a cut, since the effect of this rule is to make a formula disappear (the origin of the expression « cut »). On the other hand, in all other cases, we will observe that we can limit possible sequents to those formed with the formulas of Γ, Δ and « simpler » ones; for instance, if Δ contains a conjunction $A \wedge B$, the premises may use A or B. It is not that perfect in the case of quantification, but there is still a simplification.

Definition 2 (Subformulas). The *subformula* relation (in the sense of Gentzen) is the reflexive/transitive closure of the following particular cases (immediate subformulas):

$$A \preccurlyeq \neg A,$$

$$A, B \preccurlyeq A \wedge B,\ A \vee B,\ A \Rightarrow B,$$

$$A[t/x] \preccurlyeq \forall x A,\ \exists x A.$$

Since an immediate subformula has strictly less connectives, the notion of subformula is well-founded.

Theorem 5 (Subformula property). *In a cut-free proof of* $\Gamma \vdash \Delta$, *all sequents are made of subformulas of formulas occurring in the list* Γ, Δ.

Proof. Immediate by inspection of the rules other than cut. □

In particular, this result makes us realise the paradoxical character of cut-elimination. Reasoning consists in establishing generalities, which one eventually particularises. Now, say that I established $\exists x A$ and $\forall x(A \Rightarrow B)$ and that I want to conclude $\exists x B$; in the presence of cut, this is easy, it suffices to establish the sequent $\exists x A, \forall x(A \Rightarrow B) \vdash \exists x B$ and to perform two cuts. On the other hand, one cannot see any obvious cut-free method, since neither $\exists x A$ nor $\forall x(A \Rightarrow B)$ are subformulas of $\exists x B$. In other terms, even if these two premises have been established without a cut, there is no simple, immediate, way to obtain a cut-free proof of $\exists x B$. There is indeed a powerful algorithm[12] behind the *Hauptsatz.*

3.3.2 Subformula and decision. Let us pause our argument for a while and look at automated deduction. It is immediate that the cut-free calculus effects a drastic reduction of the search space for proofs. The reduction is so spectacular that it is not far from yielding a decision procedure, which *logic programming* tried to exploit in its day. But predicate calculus is undecidable (corollary of Gödel's theorem) and it is instructive to seek the reason for this undecidability. We shall see that it can essentially be ascribed to the contraction rule.

A decision algorithm will be obtained if one can restrict to a finite search space. What are therefore the reasons impeding such a restriction?

[12] Not easy and of enormous algorithmic complexity: towers of exponentials.

Quantification: it is the usual suspect. We must, in certain cases, « guess » a *witness* t; there are infinitely many possibilities. But this is too quickly said; for instance predicate calculus without functional symbols is undecidable and it is plain that the witness can, in this case, be chosen from a finite list, basically the present variables. This is indeed the general case: the technique of *unification* (Section 3.A.4) due to Herbrand[13] enables one to restrict the choice of possible t in the presence of functional symbols.

Contraction: in fact, quantification poses problems only in the presence of contraction. Because of this rule, one must consider sequents with repetitions of the same formula: typically, in order to prove $\vdash A$, one can try to prove $\vdash A, A$. This becomes perverse only in the presence of quantification, since a contraction creates in fact new *eigenvariables* and therefore increases the stock of different subformulas (see Herbrand's theorem, Section 3.A.3).

3.3.3 The signature. When a formula is (unambiguously[14], i.e., as a specific *occurrence*), subformula of another formula, one can introduce its *signature*, i.e., distinguish *positive* from *negative* occurrences. Essentially, if A occurs in B, then it occurs in $\neg B$ (or $B \Rightarrow C$) with the opposite signature. In the same way, one can speak of the occurrence of a subformula in a sequent $\Gamma \vdash \Delta$: then, if A occurs in one formula of Γ, it will occur in the sequent with the opposite signature, etc.

If one traces back the history of an occurrence in a sequent calculus proof[15], one observes that the rules preserve the global signature: for instance, the rules of negation perform two changes of signature, introduction of a negation and change of side. The case of identity is special, since this group relates – while simultaneously introducing or eliminating them – two occurrences of the same formula with opposite signatures.

This *signature property* suggests the definition of schizophrenic models, where the predicates of the system (and more generally the formulas) receive distinct interpretations according to their signature, say A^- and A^+. In fact, this schizophrenia is not tenable because of the identity group which forces the equality between A^- and A^+. But, in a cut-free system, where the identity group is restricted to the axiom, the only constraint is that A^- be stronger than A^+. In fact, there are three possibilities, depending on the truth values $(\mathbf{f}, \mathbf{f})$, $(\mathbf{f}, \mathbf{v})$, $(\mathbf{v}, \mathbf{v})$ taken by the pair (A^-, A^+). These three cases induce a three-valued logic, the three values being respectively denoted **f**, **i**, **v** (**i** for « indeterminate »). The truth tables, due to Kleene, are more or less obvious: a conjunction $A \wedge B$ is true iff A and B are true, false if

[13]Carrier is even more fulgurant than Gentzen! Herbrand (1908–1931), died in a mountain hike, leaving a noted theorem, which is a sort of prefiguration of the *Hauptsatz*; by the way he called his result « théorème fondamental », what an imagination, my God! Herbrand's theorem is limited to the sole classical logic.

[14]From the locative viewpoint, this proviso is useless (Section 5.1.5).

[15]This is very general, by no means restricted to the classical case.

one of them is false, indeterminate in the other cases, etc. Indeed, the values taken on the argument **i** correspond to the two possible « decisions », into **v** or into **f**: if both cases yield the same value, take it, otherwise answer **i**.

It is immediate that cut-free rules are validated by the interpretation « if all Γ are true, then one Δ is not false ». The cut-rule, which requires that « not false » implies « true » is not validated by this interpretation. Indeed, completeness holds, i.e., a sequent is cut-free provable iff it is validated by all three-valued interpretations.

This remark, due to Schütte [91], enables us to give a non-effective proof of the *Hauptsatz* at a low price. Indeed, suppose that $\vdash A$ is not cut free provable; then, by three-valued completeness, A is false in some three-valued model $\mathcal{M}$. Let us refine $\mathcal{M}$ into $\mathcal{N}$ by deciding (in an arbitrary way) the indeterminate part of atomic formulas: for instance, **i** in $\mathcal{M}$ becomes **v** in $\mathcal{N}$. It is obvious that, for non-atomic formulas, this « decision » induces new values which are compatible with the values already defined in $\mathcal{M}$: if B takes a value distinct from **i** in $\mathcal{M}$, it retains it in $\mathcal{N}$. Let us come back to our argument: we refined $\mathcal{M}$ into $\mathcal{N}$, a « real » two-valued model, so as not to contradict previously defined values. A therefore remains false and cannot be provable, even with cuts.

Like all many-valued logics, the one we just encountered is a bleak system. Typically, $A \Rightarrow A$ will take the value **i** when A is indeterminate; in general, in a *real* three-valued model, the indeterminate value phagocytoses the others: everything tends to become indeterminate. The triviality of these many-valued logics exposes the limitations of the non-algorithmic version of the *Hauptsatz*[16]. One will find again signature and asymmetric interpretations in an intuitionistic functional setting in Section 8.A and a nice semantic proof of the *Hauptsatz* – for linear logic – in Section 10.1.6.

3.4 Proof of the *Hauptsatz*

It is necessary to master at least one *algorithmic* proof of the *Hauptsatz*. The simplest one is obtained through a translation of classical logic into intuitionistic logic, followed by a normalisation and a backwards translation. But the original proof of Gentzen, whose main lines follow, is a direct one.

3.4.1 The key cases. The very heart of cut-elimination consists in replacing a cut with a cut on a subformula. This exceptional situation occurs when the cut-formula is introduced, in each of the two premises, by a logical rule. For instance:

$$\dfrac{\dfrac{\Gamma \vdash A, \Delta \quad \Gamma \vdash B, \Delta}{\Gamma \vdash A \wedge B, \Delta}\,(\vdash \wedge) \qquad \dfrac{\Lambda, B \vdash \Pi}{\Lambda, A \wedge B \vdash \Pi}\,(r \vdash \wedge)}{\Gamma, \Lambda \vdash \Delta, \Pi}\,(cut)$$

[16]See [38], chapter 3, for a detailed discussion.

In which case the cut will be replaced with a « simpler » cut:

$$\frac{\Gamma \vdash B, \Delta \quad \Lambda, B \vdash \Pi}{\Gamma, \Lambda \vdash \Delta, \Pi}\ (cut)$$

The same can be done in all other « key cases »; a cut which both premises come through logical rules introducing the cut-formula is replaced with cuts on subformulas, on A or on B when the cut-formula is $A \wedge B$, $A \vee B$, on $A[t/x]$ when it is $\forall x A$, $\exists x A$. This method remains valid in the intuitionistic case; the key case for implication replaces a cut on $A \Rightarrow B$ with *two* cuts, one on A and one on B.

This outline of an algorithm indicates how to « simplify »a cut, but we don't see termination *stricto sensu*. It occurs when one of the premises is obtained through an identity axiom:

$$\frac{\dfrac{}{A \vdash A}\ (identity) \qquad \Lambda, A \vdash \Pi}{A, \Lambda \vdash \Pi}\ (cut)$$

It is plain that we didn't quite need the cut to get the conclusion, which already occurs as a premise.

3.4.2 Commutations. Key cases unfortunately very seldom occur: in general the premises are established by rules other than the logical rules associated with the cut-formulas. The *leitmotive* here is « please, commute ». In other words, if the last rule does not act on the cut-formula, perform the cut before the rule. For instance:

$$\frac{\dfrac{\Gamma, B \vdash A, \Delta \quad \Gamma, C \vdash A, \Delta}{\Gamma, B \vee C \vdash A, \Delta}\ (\vee \vdash) \qquad \Lambda, A \vdash \Pi}{\Gamma, B \vee C, \Lambda \vdash \Delta, \Pi}\ (cut)$$

is replaced with

$$\frac{\dfrac{\Gamma, B \vdash A, \Delta \quad \Lambda, A \vdash \Pi}{\Gamma, B, \Lambda \vdash \Delta, \Pi}\ (cut) \qquad \dfrac{\Gamma, C \vdash A, \Delta \quad \Lambda, A \vdash \Pi}{\Gamma, C, \Lambda \vdash \Delta, \Pi}\ (cut)}{\Gamma, B \vee C, \Lambda \vdash \Delta, \Pi}\ (\vee \vdash)$$

3.4.3 Structural rules. Unfortunately, the worst is still to come. Indeed, we must also take care of the case where the cut-formula is obtained through a structural rule, weakening or contraction[17]. The case of weakening is easy:

$$\frac{\dfrac{\Gamma \vdash \Delta}{\Gamma \vdash A, \Delta}\ (\vdash W) \qquad \Lambda, A \vdash \Pi}{\Gamma, \Lambda \vdash \Delta, \Pi}\ (cut)$$

[17] Remember that we ignore exchange.

can be replaced with a sequence of weakenings (a sequence of structural rules is traditionally signaled by a double line):

$$\frac{\Gamma \vdash \Delta}{\Gamma, \Lambda \vdash \Delta, \Pi}\,(W)$$

In the case of a contraction,

$$\dfrac{\dfrac{\Gamma \vdash A, A, \Delta}{\Gamma \vdash A, \Delta}\,(\vdash C) \qquad \Lambda, A \vdash \Pi}{\Gamma, \Lambda \vdash \Delta, \Pi}\,(cut)$$

the idea would be to operate a replacement with:

$$\dfrac{\dfrac{\Gamma \vdash A, A, \Delta \quad \Lambda, A \vdash \Pi}{\Gamma, \Lambda \vdash A, \Delta, \Pi}\,(cut) \qquad \Lambda, A \vdash \Pi}{\Gamma, \Lambda \vdash \Delta, \Pi}\,(cut)$$

and the double line signals that a certain number of contractions have been performed on the context $\Lambda, \Lambda \vdash \Pi, \Pi$. This idea works well in the case of the intuitionistic system **LJ**, but fails in the case of **LK**. Indeed, in case both premises are obtained through a contraction of the cut-formula:

$$\dfrac{\dfrac{\Gamma \vdash A, A, \Delta}{\Gamma \vdash A, \Delta}\,(\vdash C) \qquad \dfrac{\Lambda, A, A \vdash \Pi}{\Lambda, A \vdash \Pi}\,(C \vdash)}{\Gamma, \Lambda \vdash \Delta, \Pi}\,(cut)$$

The previous procedure does not converge. Indeed contraction is eliminated by means of a duplication: in the case at stake, each side will indefinitely duplicate the other. Thus, starting with the previous configuration, one will try to proceed by symmetrically eliminating the topmost cut between $\Gamma \vdash A, A, \Delta$ and $\Lambda, A \vdash \Pi$, which induces a duplication of that A not used as a cut-formula and reintroduces a contraction $(\vdash C)$ on A.

To get out of this vicious circle, two methods are known:

Cross-cuts: this is Gentzen's original technique. It is a combination of key cases and duplication, with the consequence that we no longer reintroduce the same cuts. Frankly speaking, it is completely illegible; and nobody has ever been able to find the slightest structuring reading of this approach.

Polarisation: this amounts to forbidding structural rules (weakening and especially contraction) on the left or on the right, depending on the first logical symbol of A. If A is negative (begins with $\wedge$, $\forall$), one forbids right structural rules, if A is positive (begins with $\vee, \exists$), one forbids left structural rules[18]. This is because the missing structural rules can be proved. Typically, using two cuts, one deduces $\Gamma \vdash A, A, \Delta$ from $\Gamma \vdash \forall x A, \forall x A, \Delta$ and one concludes by means of a contraction on A:

$$\dfrac{\dfrac{\Gamma \vdash A, A, \Delta}{\Gamma \vdash A, \Delta}}{\Gamma \vdash \forall x A, \Delta}$$

In this way one is back to a locally intuitionistic version, which will be analysed in Section 7.A.6. One will object that this does not fix the atomic case. But once cuts on non-atomic formulas have been eliminated, one is reduced to eliminating cuts in a logic-free fragment: one verifies that contraction is useless in this setting.

3.4.4 Finalisation. Putting everything together, we get the following proof:

- The possibility of eliminating, in a proof ending with a cut, this final cut, at the price of the introduction of cuts on strict subformulas.
- The possibility of eliminating all cuts which are maximal *modulo* $\preccurlyeq$ at the price of the creation of new cuts on strict subformulas of the maximal cuts. This is only an iteration of the previous process.
- The possibility of eliminating all cuts. First, maximal cuts; next, the maximal cuts of the new proof, etc. Something diminishes, i.e., the maximum number of logical symbols occurring in a cut (the *degree*).

The bounds on the size of the cut-free proof are given by towers of exponentials. We shall detail this in the setting of *natural deduction* which produces an equivalent algorithm, but which is much more limpid and efficient.

Note that our proof works perfectly (more simply: there are much fewer commutations) in the *intuitionistic* setting.

3.A Around sequent calculus

3.A.1 Exercise: deduction theorem for sequents. Add the axiom $\vdash A$ to sequent calculus; if the modified system proves $\vdash B$, show that $A \vdash B$ is provable. The idea is simple: translate the modified system in the « normal » calculus by systematically adding A on the left of sequents. The translation of the new axiom is provable and, with a heavy use of contraction and weakening, this extends to arbitrary proofs. In order to cope with the quantification rules ($\vdash \forall$) and ($\exists \vdash$), A must be closed.

[18]The polarity in case of a negation is determined by an implicit use of De Morgan.

3.A.2 Achilles and the Tortoise. The name of Lewis Carroll is related to everything but his professorship of logic in Oxford. However, one of his « nonsenses » will interest us: it is a dialogue between Achilles and the Tortoise, something like:

Tortoise: I propose you another infinite race. You have A and $A \Rightarrow B$, but you will never reach B.

Achilles: Come on, it is well-known that $(A \wedge (A \Rightarrow B)) \Rightarrow B$: from A and $A \Rightarrow B$, one gets B.

Tortoise: I accept the formula, but not the inference.

Achilles: But $((A \wedge (A \Rightarrow B)) \wedge ((A \wedge (A \Rightarrow B)) \Rightarrow B)) \Rightarrow B$ is true: from A and $A \Rightarrow B$ and $(A \wedge (A \Rightarrow B)) \Rightarrow B$ one gets B.

Tortoise: Your new formula is correct, but not your inference.

Achilles: But...

And so on, indefinitely. The analogy with Zeno is slightly forced, in particular, the iteration is genuine nonsense. But there is a point here, the difference between implication and *Modus Ponens*: the Tortoise is « cut-free ».

Let us give a cheaper version of the same thing. One can replace a cut between $\Gamma \vdash A$ and $\Lambda, A \vdash B$ by a left rule for implication[19]:

$$\frac{\Gamma \vdash A \quad \Lambda, A \vdash B}{\Gamma, \Lambda, A \Rightarrow A \vdash B} \ (\rightarrow \vdash)$$

It's all we can do without cuts. We can recover the result of the original cut by means of a cut with the provable sequent $\vdash A \Rightarrow A$:

$$\frac{\dfrac{}{A \vdash A} \ (\textit{identity})}{\vdash A \Rightarrow A} \ (\vdash \Rightarrow)$$

This cut on a more complex, but provable, formula, is essentially what Achilles does in his logical forward flying. This replacement is technically very interesting, since structurally simpler: one of the two premises of the new cut is known. However, the iteration of this process, leading to a cut with $(A \Rightarrow A) \Rightarrow (A \Rightarrow A)$, is pointless.

3.A.3 Herbrand's theorem. Herbrand's theorem is, in the classical case, a rather explicit version of the *Hauptsatz*. In a prenex formula A, let us remove quantifiers and express the universals as *formal* functions of the previous existentials. Typically, $\exists x \forall y \exists z \forall w R[x, y, z, w]$ becomes $R[x, f(x), z, g(x, z)]$. Herbrand's theorem says

[19]For convenience, I use the intuitionistic **LJ**, *infra*; but this works the same in **LK**.

that A is provable iff there exists $n > 0$ and terms $t_1, \ldots, t_n, u_1, \ldots, u_n$ such that the disjunction

$$R[t_1, f(t_1), u_1, g(t_1, u_1)] \vee \cdots \vee R[t_n, f(t_n), u_n, g(t_n, u_n)] \tag{3.2}$$

is propositionally valid. For instance, with $A = \exists x \forall y (P(x) \Rightarrow P(y))$,

$$(P(t_1) \Rightarrow P(f(t_1))) \vee (P(t_2) \Rightarrow P(f(t_2))) \tag{3.3}$$

validates A, with $t_1 = x$, $t_2 = f(x)$; here, $n = 2$.

Let us try to understand this result in relation to the *Hauptsatz*. A cut-free proof of A will use contractions, which are made explicit by the Herbrand disjunction. Now, each contraction creates additional copies of A and its subformulas. For each rule $(\vdash \forall)$ it is necessary to create a distinct *eigenvariable*, in fact a « fresh » variable. How to do this? $\forall y \exists z \forall w R[t, y, z, w]$ asks indeed for a variable not already used in t: Herbrand proposes to write it $f(t)$, because $f(t)$ is surely not contained in t. Hence, the functional expression $f(t)$ means « fresh variable » w.r.t. t; similarly, $g(t, u)$ constructs a variable that is fresh w.r.t. t, u. Since one cannot give an *a priori* bound on the length of the disjunction, one thus generates an infinite search space.

3.A.4 Unification. However, for a given disjunctive length, the search for a solution is decidable. Indeed, to get a propositional tautology, we must equate certain atoms. Decision thus amounts to searching values $t_1, \ldots, t_n, u_1, \ldots, u_n$ realising certain literal identities. This problem is solved by *unification*, a technique introduced by Herbrand:

Theorem 6 (Principal unifier). *If there exists a substitution θ of terms for the free variables of t, u identifying the terms $t\theta$ and $u\theta$, then there exists a* ***principal*** *θ_0 from which they all derive.*

Proof. In other terms, there is a « mother of all substitutions » equating t and u: any other such substitution factorises as $\theta_0\theta$. This is relatively easy: if t is a variable x, either x is strictly contained in u, in which case unification fails, or $\theta_0(x) := u$; if t, u are not variables, one looks at their respective first symbols: if they differ, failure; if, say, $t = f(t_1, t_2)$, $u = f(u_1, u_2)$, one is led to unify t_1, u_1 and t_2, u_2, then to unify the unifiers. □

Unification yields « the » solution to the substitution if any. It is advisable to keep the principal unifier under the form of a product of basic substitutions rather than trying to write it down: since a substitution may induce, say, a doubling of the size in case of repetition of variables, the result of n substitutions can easily be of exponential size.

The unification algorithm therefore almost induces a decidability of predicate calculus. Therefore it is really contraction, which, by creating disjunctions of arbitrary lengths, eventually impedes decision. See also logic programming in Section 4.D.

3.A.5 More about Herbrand. Herbrand's theorem is equivalent to the *Hauptsatz*, provided one restricts to classical logic and concentrates on quantifiers – its propositional contents being indeed empty. When only the quantifier structure matters, one can prefer Herbrand to Gentzen; for instance, it enables one, in the classical case, to give another proof of the reflexion schema of Section 3.B.4.

The theorem can be obtained as a corollary of the *Hauptsatz*, but this is rather tedious. We will favour a proof in two steps:

(i) The purely existential case, for instance: if $\exists x R$, R quantifier-free, is provable, then one can find $t_1, \dots, t_n$ such that $R[t_1] \vee \cdots \vee R[t_n]$ is a propositional tautology.

(ii) The reduction of the prenex case, e.g., $A = \exists x \forall y \exists z \forall w R[x, y, z, w]$ to the existential case $B := \exists x \exists z R[x, f(x), z, g(x, z)]$: A is provable iff B is provable. In one way, the implication $A \Rightarrow B$ is obviously provable. Conversely, if A is not provable, this means that $\forall x \exists y \forall z \exists w \neg R[x, y, z, w]$ is true in some model; we express y as a function of x and w as a function of x, z[20]: $y = \varphi(x)$, $w = \psi(x, z)$, which, with $f = \varphi$, $g = \psi$, yields a model in which B is false.

A implies B, but the reverse implication holds only at the level of *provability*; for instance in case $A := \forall x R$, it is plain that the implication $R[a] \Rightarrow \forall x R$ is wrong. In particular, the ludicrous idea to dispense one from the notion of proof and replace it by truth in a professed « Herbrand model »– an idea that was fashionable in the day of logic programming – finds its limits in the fact that B does not imply A.

3.A.6 ε-substitution. This is perhaps the most noted contribution of Hilbert to proof theory; but let us say it bluntly, a total failure too! The idea is to « eliminate » quantifiers in a purely formal way: one introduces the term $\varepsilon x A$ to mean « the test case for $\forall x A$ ». Naively, this means that $\varepsilon x A$ is some a such that $A[a/x]$ fails, if such an a can be found, anything otherwise. Thus, the axiom $A[\varepsilon x A/x] \Rightarrow A[t/x]$ enables one to get rid of universal quantification (now the propositional expression $A[\varepsilon x A/x]$). It was of course perfectly legitimate in the 1930s to explore such a direction. But it turned out to be completely sterile:

(i) Contrary to Herbrand's theorem, this formulation has no explicit contents: it is pure bureaucracy.

[20]This is called « skolemisation ».

(ii) It creates completely new formulas: the $B[\varepsilon x A/y]$ with which one must now cope without any reasonable interpretation for them.

(iii) The implication $A[\varepsilon x A/x] \Rightarrow \forall x A$ on which the method rests is basically the formula studied *supra*: $\exists x \forall y (P(x) \Rightarrow P(y))$. Its Herbrand expansion (equation (3.3)) is of length 2: in other terms, anticipating linear logic, quadratic; while, as we shall see, quantification can be handled linearily.

In other words, instead of a clarification, ε-substitution induces an obfuscation. These limitations have been exploited in a dubious way by Bourbaki, whose hatred of foundational issues is well-known; for that reason, his official syntax for mathematics is based upon ε-substitution, indeed the dual form $\tau x A := \varepsilon x \neg A$. This doohickey is occasionally used as a universal choice function: $\tau x \; x \in a$ picks up an element of a, if any. But rather than saying « we are using Zermelo–Fraenkel set-theory with a universal choice function », which would raise the various foundational issues related to choice, Bourbaki simply writes the usual axioms of **ZF** over the ε-substitution setting, to the effect that universal choice becomes provable. Worse: with this artificial language, it becomes impossible to discuss the axiom of choice: separating real formulas from the artificial ones created by ε-substitution becomes an impossible task. Long ago, George Orwell introduced *Newspeak*, a language designed in such a way that certain matters could not even be discussed.

3.A.7 Equality. The big shame of proof-theory: equality, which is a logical primitive, does not *work* well. No good solution is known. Some authors will take as axioms the closure of the equality axioms under a cut, i.e., all sequents $\Gamma \vdash \Delta$ made of atoms and provable: not very glorious! One might as well work with a partial elimination, which produces almost the same effect, but without hypocrisy, see *infra*.

In the case of equality in a free system – for instance, the arithmetic terms built from $0, S$ – for which equality corresponds to identity, there is a beautiful solution:

$$\frac{}{\vdash t = t}\;(\vdash =) \qquad\qquad \frac{}{t = u \vdash}\;(\emptyset = \vdash) \qquad \frac{\Gamma\theta \vdash \Delta\theta}{\Gamma, t = u \vdash \Delta}\;(\theta = \vdash)$$

The left-hand introduction splits into two cases: if t, u are not unifiable, the rule $(\emptyset = \vdash)$, which basically says that $t \neq u$; if t, u admit the principal unifier θ, the rule $(\theta = \vdash)$.

The interest of this system is that it proves the third and fourth Peano axioms. That's all we need for *second-order* arithmetic. On the other hand, the idea fails in the presence of sum and product, since equality does no longer correspond to the free system.

3.A.8 Partial eliminations. In presence of proper axioms, of which the most general form is that of sequents $\Gamma \vdash \Delta$, one loses the *Hauptsatz*. One can however state a « partial » version: it is possible to eliminate all cuts, except those the premise of which is a proper axiom. In the case where the proper axioms contain free variables, one must close their list under substitution.

Partial elimination is of interest only when the proper axioms are of a very simple form, i.e., have « few » subformulas. For instance, one can formulate equality by means of the axioms $\vdash t = t$ and $t = u, P[t/x] \vdash P[u/x]$, with P atomic and everything works well.

On the other hand, the formulation of – say – Peano's arithmetic in this setting is without interest: every formula A is a subformula of an induction axiom (on any B containing A). But a system in which induction has been restricted to « simple » formulas can be studied by means of partial elimination.

3.B Semantic aspects

3.B.1 The completeness theorem. The traditional link between logic and models is organised around two results: *soundness*: « what is provable is true » and *completeness*: « what is true is provable ». Completeness can be established by means of sequent calculus. We outline the construction: as expected, the very details are tedious.

Search tree. It is necessary to use the variant « set », free from structural rules, of sequents. We rewrite the rules so as to make premises bigger than conclusions: if $\Gamma' \vdash \Delta'$ is a premise and $\Gamma \vdash \Delta$ is the conclusion of a rule, then $\Gamma \subset \Gamma'$ and $\Delta \subset \Delta'$ (it suffices to modify the rule with the help of contraction). We must also « weaken » the identity axioms into $\Gamma, B \vdash B, \Delta$.

Given a closed formula A, we write a sort of universal proof-tree for $\vdash A$: we get it by systematically trying all possible rules. How to do this? The procedure is not very elegant: starting from the conclusion $\vdash A$, we progressively build larger and larger finite proof-trees by stacking rules one upon another. If $\Gamma \vdash \Delta$ has yet nothing above, we choose a rule with that conclusion, which makes the tree grow. But this choice is not arbitrary, it must be done in such a way that, along an arbitrary infinite branch starting with $\Gamma \vdash \Delta$, « all » possible rules with conclusion $\Gamma \vdash \Delta$ have been performed, of course, with some delay. This is the tedious part of the proof: the choice of the appropriate rules, so as to forget nobody. But this offers no difficulty, although the details are painful. For instance, if $B \wedge C \in \Gamma$, we decide that a $(l \wedge \vdash)$ introducing $B \wedge C$ will be performed on all branches two notches above and that a $(r \wedge \vdash)$ will be performed on all branches eleven notches above; if $\exists x B \in \Delta$, we decide that a $(\vdash \exists)$ on term t will be performed on each branch $\ulcorner t \urcorner$ notches above. The coordination of those various constraints makes the thing completely illegible.

When $\Gamma \cap \Delta \neq \emptyset$, we halt (identity axiom). In particular, if all branches are finite, since all branchings are binary, König's lemma tells us that the tree is finite; it is therefore a proof of $\vdash A$.

This construction, which rests upon contraction, would be impracticable for the intuitionistic system **LJ**: indeed, we are forced to « commit » ourselves on the right, for instance we can make only one choice of a term t in $(\vdash \exists)$ and in case of a mistake, no second chance! Same remark for linear logic.

Infinite branch. Otherwise, there is an infinite branch. We build a model for $\neg A$ by taking as domain the terms (with free variables) occurring in this branch. And we give the truth value **v** to those formulas appearing on the left-hand of a sequent of the branch, the value **f** to those occurring on the right-hand; by the halting clause a formula cannot be both true and false. Now, let us see how these truth values socialise. For instance, if $\Gamma \vdash B \wedge C, \Delta$ is in my branch (so that $B \wedge C$ is false), a right rule introducing $B \wedge C$ is necessarily performed upstairs, with two premises, one containing B, one containing C; the branch proceeds through one of them and the corresponding formula is false. If $\Gamma, B \wedge C \vdash \Delta$ is in my branch (so that $B \wedge C$ is true), the two left conjunction rules are performed somewhere above in the branch and we conclude that B and C are true. If a cut has not been used, we have constructed a three-valued structure, which does not valuate all formulas, but which is compatible with classical truth. This is the starting point of the semantic version of the *Hauptsatz*.

But one can also impose the systematic use of all cuts. The effect is that, in the branch, at some moment a cut on B is performed and this one of the two premises which remains in the branch induces a truth value for B. We thus obtain without contradicting truth tables a structure valuating all formulas of the branch. This is therefore a model in which A is false. To sum up, either A is provable or it is false in a model.

3.B.2 Takeuti's conjecture

Second-order logic. Second-order logic is obtained by adding predicate variables as well as quantifications on these variables. Indeed the arity (the number of arguments) of these predicates is of little interest. We will consider only two particular cases:

Arity 1: unary predicates; instead of $X(t)$, we shall write $t \in X$.

Arity 0: propositions.

For each arity, one introduces *abstraction terms*, namely:

Arity 1: the term $\{x; A\}$.

Arity 0: the proposition A.

One must then define the substitution of an abstraction term for a variable of the same arity. In arity 0, nothing special; in arity 1, the formal expression $t \in \{x; A\}$ must be replaced with $A[t/x]$. Then one can formulate the rules of *second-order sequent calculus*:

$$\frac{\Gamma \vdash A, \Delta}{\Gamma \vdash \forall X A, \Delta}\ (\vdash \forall_2) \qquad \frac{\Gamma, A[T/X] \vdash \Delta}{\Gamma, \forall X A \vdash \Delta}\ (\forall_2 \vdash)$$

the abstraction term T being of the same arity as the variable X; and symmetrical rules for existence.

The system proves a very strong principle, the *comprehension schema*. Indeed, starting with $\vdash \forall x(A \Leftrightarrow A)$, a rule $(\vdash \exists_2)$ on the abstraction term $T := \{x; A\}$ yields $\vdash \exists X \forall x(x \in X \Leftrightarrow A)$.

Second order arithmetic (or *analysis*[21]) $\mathbf{PA_2}$, denotes the above system based upon the primitives $=$, 0, S, to which the third and fourth Peano axioms and the principle of induction on all formulas have been added: $\mathbf{PA_2}$ is considerably *stronger* than Peano's arithmetic **PA**. $\mathbf{PA_2}$ nevertheless reduces to « pure » second-order logic, by *relativising* first-order quantifications to Dedekind integers

$$\mathbb{N} := \{x; \forall X(0 \in X \wedge \forall z(z \in X \Rightarrow Sz \in X) \Rightarrow x \in X)\};$$

one can prove $x \in \mathbb{N} \vdash A[0] \wedge \forall z(A[z] \Rightarrow A[Sz]) \Rightarrow A$ by means of a left rule $(\forall_2 \vdash)$ on the term $T := \{x; A\}$. There remain only a couple of small axioms which admit a cut-free formulation (Section 3.A.7).

The conjecture. Takeuti's conjecture [97] states cut-elimination in this second-order system. Schütte [91] reduced the conjecture into a relation between two-valued and three-valued models. This was 90% of the solution, but, overestimating the difficulty, he gave up, leaving to Tait the task of concluding [95].

This is a very easy result: starting with a three-valued model $\mathcal{M}$, one builds a binary model $\mathcal{N}$ whose second-order part is made of all binary predicates which « refine » a predicate of $\mathcal{M}$; this construction respects the truth values already defined and it therefore suffices to show that $\mathcal{N}$ enjoys comprehension. But, if $A[\cdot]$ is interpreted by a three-valued predicate of $\mathcal{M}$, the interpretation of $A[\cdot]$ in $\mathcal{N}$ is

[21] Since real numbers can be formalised in it.

a refinement of this predicate and therefore belongs to the predicates of $\mathcal{N}$ by construction. Incidentally, we see that comprehension justifies comprehension: this is completely circular and rather vain, an illustration of the blind spot of logic.

Takeuti and subformulas. Note that Takeuti's conjecture is a strange animal. Indeed, the rules $(\forall_2 \vdash)$ and $(\vdash \exists_2)$ do not follow the subformula property. To say that $A[T]$ is a subformula of $\forall XA$ is pure baloney: T may be much more complex than A. The cut-elimination theorem however yields a genuine subformula property when the proven formula has no positive second-order existential quantifier (nor negative universal). Concretely, for Π^1, i.e., Σ^0_1, formulas.

The absence of a subformula property excludes any direct combinatorial proof of the *Hauptsatz*, including ordinal methods, see *infra*. Indeed, one would like to replace a cut on $\forall XA$ by a cut on $A[T/X]$, but one hardly sees what could decrease in this replacement. This is all the difficulty of the effective version, (Section 6.A.1).

A Π^0_1 formula $A = \forall x \in \mathbb{N}\ B$ is such that any formula is a subformula of A, which means that cut-elimination tells us nothing as to A. Which can be put side by side with the following semantic remark: the formula defining integers, $x \in \mathbb{N}$, takes the value **i** in any real three-valued model; thus A can only be false in a binary model. In other words, the *Hauptsatz* is trivial for A.

An empty shell? In fact, by using the principle of the Tortoise, one can replace a proof with cuts of $\Gamma \vdash \Delta$ with a proof without cuts of $\Gamma, x \in \mathbb{N} \vdash \Delta$. The idea is that $x \in \mathbb{N}$ « phagocytoses » the cuts. For instance, starting, as in Section 3.A.2, with a cut between $\Gamma \vdash A$ and $\Lambda, A \vdash B$:

$$\dfrac{\dfrac{\dfrac{\Gamma \vdash A \qquad \dfrac{\dfrac{\dfrac{}{A \vdash A}\,(identity)}{\vdash A \Rightarrow A}\,(\vdash \Rightarrow)}{\vdash \forall z(A \Rightarrow A)}\,(\vdash \forall)}{\vdash A \wedge \forall z(A \Rightarrow A)}\,(\vdash \wedge) \qquad \Lambda, A \vdash B}{\Gamma, \Lambda, (A \wedge \forall z(A \Rightarrow A)) \Rightarrow A \vdash B}\,(\Rightarrow \vdash)}{\Gamma, \Lambda, x \in \mathbb{N} \vdash B}\,(\forall_2 \vdash)$$

The last rule uses the abstraction term $\{y; A\}$, where y does not occur in A. Thus, a proof of $\Gamma \vdash \Delta$ is recursively transformed into a cut-free one; we make use of contraction to keep only one occurrence of $x \in \mathbb{N}$ on the left.

Starting with a proof of $\vdash \forall x(x \in \mathbb{N} \Rightarrow B)$, we deduce $x \in \mathbb{N} \vdash B$; our technique enables us to get a cut-free proof of $x \in \mathbb{N}, x \in \mathbb{N} \vdash B$. Then the three rules $(C \vdash)$, $(\vdash \Rightarrow)$ and $(\vdash \forall)$ yield a cut-free proof of $\vdash \forall x(x \in \mathbb{N} \Rightarrow B)$.

We eventually see that Takeuti's conjecture is often an empty shell; this will no longer be the case with its effective, algorithmic, version (Chapter 6).

3.B.3 Bounded truth predicates. Peano arithmetic **PA** is a very expressive system in which we can formalise almost everything we want; *grosso modo*, every correct reasoning will be formalisable, with obvious exceptions:

- Analysis and, in logic, infinite set-theoretic arguments. This being said, transfinite inductions on ordinals strictly smaller than ϵ_0 can be formalised: it is the underside of Gentzen's work, see *infra*.
- Recurrences dealing with properties involving a variable (i.e., unbounded) alternation of quantifiers, typically truth.

For instance, Peano arithmetic almost succeeds in giving an internal model of itself, i.e., an interpretation of its closed expressions. For instance it can define the value of (the Gödel number of) a closed term:

$$\begin{aligned} V(\ulcorner 0 \urcorner) &:= 0, \\ V(\ulcorner St \urcorner) &:= SV(\ulcorner t \urcorner), \\ V(\ulcorner t + u \urcorner) &:= V(\ulcorner t \urcorner) + V(\ulcorner u \urcorner), \\ V(\ulcorner t \times u \urcorner) &:= V(\ulcorner t \urcorner) \times V(\ulcorner u \urcorner). \end{aligned} \tag{3.4}$$

with the result that $V(\ulcorner t \urcorner) = t$ is provable. The formula $V(x) = y$ is enunciated as the existence of a finite sequence – corresponding to the values of the subterms of the term coded by x – and obeying (3.4).

The same can be tried with closed formulas, which amounts to formalising the tarskian truism:

$$\begin{aligned} V(\ulcorner t = u \urcorner) &\iff V(\ulcorner t \urcorner) = V(\ulcorner u \urcorner), \\ V(\ulcorner t < u \urcorner) &\iff V(\ulcorner t \urcorner) < V(\ulcorner u \urcorner), \\ V(\ulcorner A \wedge B \urcorner) &\iff V(\ulcorner A \urcorner) \wedge V(\ulcorner B \urcorner), \\ V(\ulcorner A \vee B \urcorner) &\iff V(\ulcorner A \urcorner) \vee V(\ulcorner B \urcorner), \\ V(\ulcorner \neg A \urcorner) &\iff \neg V(\ulcorner A \urcorner), \\ V(\ulcorner \forall x A \urcorner) &\iff \forall x V(\ulcorner A[\bar{x}] \urcorner), \\ V(\ulcorner \exists x A \urcorner) &\iff \exists x V(\ulcorner A[\bar{x}] \urcorner). \end{aligned} \tag{3.5}$$

But this does not work, in accordance with Tarski's theorem. This is linked to quantifiers; for instance one cannot express truth by means of a finite sequence – as we did for the value of terms – because truth requires infinitely many steps. Indeed a truth definition for Π^0_n formulas necessarily requires n alternations of quantifiers: there cannot be any global solution.

This being said, there are solutions when the alternation of quantifiers can be bounded. The most typical example is – in relation to sequent calculus – the truth of the closed subformulas (or sequents made of closed subformulas) of a given closed formula A. Indeed, strictly speaking – i.e., by not allowing term substitutions –, A admits only a finite, concrete, number of subformulas, say $B_1, \dots, B_{981}$ and any closed subformula of A in the sense of Gentzen is obtained from one of these 981 formulas by substitution of closed terms. One can therefore define the truth of a subformula of A by a disjunction of 981 cases, for instance case number 7:

$$V(\ulcorner B_7[t] \urcorner) \iff B_7[V(t)]. \tag{3.6}$$

This definition extends to sequents by

$$V(\ulcorner C_0, \dots, C_{m-1} \vdash D_0, \dots, D_{n-1} \urcorner) \iff \forall i < m V(\ulcorner C_i \urcorner) \Rightarrow \exists j < n V(\ulcorner D_j \urcorner).$$

Observe the bounded quantifiers: we cannot write conjunctions or disjunctions, since m, n are variables.

3.B.4 The reflexion schema. Let $\mathcal{T}$ be a finitely axiomatised subsystem of **PA** and let $\mathrm{Thm}_{\mathcal{T}}(x)$ be the formula expressing the provability in $\mathcal{T}$ of the formula with Gödel number x.

Theorem 7 (Reflexion schema, [68]). *If A is a formula with one free variable,* **PA** *proves*

$$\mathrm{Thm}_{\mathcal{T}}(\ulcorner A[\bar{x}] \urcorner) \Rightarrow A[x]. \tag{3.7}$$

Proof. The theorem illustrates the extraordinary expressive power of **PA**. In a first step, we pack the axioms of $\mathcal{T}$ in a single closed formula B, so that the provability of C in $\mathcal{T}$ reduces to the provability of $B \Rightarrow C$ in **LK**; this, provably in **PA**. From $\mathrm{Thm}_{\mathbf{LK}}(\ulcorner B \Rightarrow A[\bar{x}] \urcorner) \Rightarrow (B \Rightarrow A[x])$ we easily get (3.7). We are thus led back to the case of **LK**. We use then the fact that the *Hauptsatz* is provable in **PA** and we are led back to showing that $\mathrm{Thm}^{-}(\ulcorner A[\bar{x}] \urcorner \Rightarrow A[x]$, where $\mathrm{Thm}^{-}(\cdot)$ refers to cut-free provability: we show, by induction on cut-free proofs, that « provable implies true », which does not pose the slightest problem: it is a succession of truisms. Of course, we make an essential use of a bounded truth predicate for the sequents made of subformulas of $\forall x A$. □

The reflexion schema is an important result, a power tool of formalisation, that we shall later use (see Sections 6.D.2 and 6.D.3). More anecdotally, it shows that arithmetic is not finitely axiomatisable: the reflexion schema applied to $0 \neq 0$ shows that **PA** proves the consistency of $\mathcal{T}$.

It illustrates the best of Kreisel's contribution to logic. « Provable implies true » is a nameless truism, the native tarskianism: axioms are true and rules preserve truth. To get something from that is quite acrobatic. As usual, Kreisel formalises and, for once, it works. This is not always the case, as we shall see with his dubious incursion into the functional interpretation of proofs (Section 5.A).

3.C Infinitary proof-theory

3.C.1 Gentzen's second consistency proof. The proof-theory of arithmetic stumbles on two problems of unequal importance:

- The axioms of **RR**[22] do not naturally enjoy a cut-free presentation. This being said, they can be written as sequents made of atomic formulas, closing this list under substitution, so as to be in front of a *partial* elimination – an efficient technique, if not an elegant one.
- Induction axioms do pose a more serious problem. Indeed, we cannot hope for a cut-elimination with a genuine subformula property: this is one of the possible meanings of the reflexion schema. There are two imperfect remedies to this situation:
  - ► Restrict cut-elimination to proofs of closed Σ^0_1 formulas: which Gentzen did.
  - ► Introduce infinitary deduction rules, so as to recover a genuine cut-elimination: which Schütte did.

The induction rule is written:

$$\frac{\Gamma \vdash A[\overline{0}], \Delta \quad \Lambda, A[z] \vdash A[Sz], \Pi}{\Gamma, \Lambda \vdash A[t], \Delta, \Pi}\ (I)$$

Gentzen gave an algorithm to eliminate this rule in case the conclusion is closed and quantifier-free; in particular when the conclusion is the empty sequent (consistency proof):

(i) Eliminate all cuts one can eliminate.

(ii) Take a downmost induction rule; then verify that the term t can be chosen to be closed: this is because there is no rule $(\forall)$ below, so that one can replace any variable in t with a closed term, say $\overline{0}$.

(iii) Replace t with its value, say $t = \overline{3}$, in which case the rule can be eliminated in favour of three cuts:

$$\frac{\dfrac{\dfrac{\Gamma \vdash A[\overline{0}], \Delta \quad \Lambda, A[\overline{0}] \vdash A[\overline{1}], \Pi}{\Gamma, \Lambda \vdash A[\overline{1}], \Delta, \Pi} \qquad \Lambda, A[\overline{1}] \vdash A[\overline{2}], \Pi}{\Gamma, \Lambda \vdash A[\overline{2}], \Delta, \Pi} \qquad \Lambda, A[\overline{2}] \vdash A[\overline{3}], \Pi}{\Gamma, \Lambda \vdash A[\overline{3}], \Delta, \Pi}$$

[22]One can forget the last one, provable by induction.

This can be generalised to the case where the conclusion is a closed Σ_1^0 formula. One must be careful and pay some attention to the bounded universal quantifiers of the conclusion.

One must still show the convergence of this algorithm; to do so, Gentzen introduces a notion of *size* for the proof in the form of an ordinal strictly bounded by ϵ_0, the smallest ordinal α such that $\omega^\alpha = \alpha$ and shows that the size strictly decreases with the reduction of proofs. Cut elimination eventually follows from a *transfinite* induction up to ϵ_0.

3.C.2 The ω-rule. Revisited by Schütte [90], Gentzen's proof becomes more limpid. One restricts to closed formulas; first-order quantification is thus treated like a sort of denumerable conjunction:

$$\frac{\dots \quad \Gamma \vdash A[\bar{n}/x], \Delta \quad \dots}{\Gamma \vdash \forall x A, \Delta}\ (\vdash \forall) \qquad \frac{\Gamma, A[\bar{n}/x] \vdash \Delta}{\Gamma, \forall x A \vdash \Delta}\ (\forall \vdash)$$

The right rule, also called ω-rule, has infinitely many premises, one for each integer n. It is no wonder that induction can be proven using these uneven objects: sequent calculus easily establishes $A[\bar{0}], \forall z (A[z] \Rightarrow A[Sz]) \vdash A[\bar{n}/x]$ for each integer n, typically for $n = 2$:

$$\dfrac{\dfrac{\dfrac{\dfrac{\overline{A[\bar{0}] \vdash A[\bar{0}]} \quad \overline{A[\bar{1}] \vdash A[\bar{1}]}}{A[\bar{0}], A[\bar{0}] \Rightarrow A[\bar{1}] \vdash A[\bar{1}]}\ (\Rightarrow \vdash)}{A[\bar{0}], \forall z (A[z] \Rightarrow A[Sz]) \vdash A[\bar{1}]}\ (\forall \vdash) \quad \overline{A[\bar{2}] \vdash A[\bar{2}]}}{A[\bar{0}], \forall z (A[z] \Rightarrow A[Sz]), A[\bar{1}] \Rightarrow A[\bar{2}] \vdash A[\bar{2}]}\ (\Rightarrow \vdash)}{A[\bar{0}], \forall z (A[z] \Rightarrow A[Sz]) \vdash A[\bar{2}]}\ (\forall \vdash)$$

These particular cases put together enable one to prove induction:

$$\frac{\dots \quad A[\bar{0}], \forall z (A[z] \Rightarrow A[Sz]) \vdash A[\bar{n}/x] \quad \dots}{\Gamma \vdash \forall x A, \Delta}\ (\vdash \forall)$$

by means of a proof whose ordinal height is slightly bigger than ω, with the consequence that any proof in **PA** translates into an infinite proof of height strictly bounded by $\omega + \omega$.

Cut-elimination in such a *semi-formal* system can be done without problem: the key case of a cut between a $(\vdash \forall)$ and $(\forall \vdash)$ « is reduced » by means of a cut on a subformula $A[\bar{N}]$, where $\bar{N}$ is the term involved in the rule $(\forall \vdash)$. Indeed, the increase in size is still bounded by towers of exponentials. Now the starting heights are no longer finite. The ordinal numbers

$$2^{2^{2^{\cdot^{\cdot^{2^h}}}}} \tag{3.8}$$

with $h = \omega + \omega$ approximate ϵ_0. Cuts are therefore eliminated at the price of an increase of size up to ϵ_0 (excluded). One can use this measure of size to understand the ordinal numbers used by Gentzen in his own proof.

Restricted to closed Σ^0_1 conclusions, infinite rules disappear in the cut-free case and we are eventually left with a *finite* cut-free proof. Unfortunately this is not enough, since one must express the function which yields the cut-elimination in that case: starting with a finite proof, one translates it in an infinite system and one finally recovers a finite proof after cut-elimination. One must introduce an auxiliary system of recursive coding for ω-proofs. Which can be very easily done, but it remains the vague impression that this finite recoding hides something: this is the blind spot of the method.

The real cut-elimination therefore never goes beyond the closed Σ^0_1 case, which corresponds to formulas for which completeness holds; which illustrates one of the theses of this textbook:

completeness = cut-elimination

Which *ludics* will make explicit, (Section 13.8.4). By the way, what could be the status of infinitary logic in this perspective? This is simple, the extension to ω-proofs defines a logic which is complete w.r.t. Π^1_1 formulas, a class encompassing all arithmetical formulas. Since completeness has the same value as cut-elimination, one therefore obtains cut-elimination for all arithmetic formulas.

3.C.3 The Munich school. The infinitary version of Schütte has been generalised to various extensions of Peano's arithmetic; indeed to systems intermediate between **PA** and Takeuti's second-order logic: roughly speaking, comprehension is restricted to logical complexities close to Π^1_1. This requires the building of systems of *ordinal notations* for ordinals bigger than ϵ_0, systems which are heavier and heavier and less and less understandable.

On one hand a list of formal systems, on the other hand a list, a sort of *Panzerdivision*, of ordinals; on either side, this is not very exciting. More precisely, because one hardly understands the nature of this relation: the strategic failure of this connection lies in the absence of any structuring view.

The starting correspondence was that of consistency proofs. But, after the – justified – sarcasms of Kreisel, the Munich School preferred to present its results in a different way:

Provably recursive functions: one defines a hierarchy of recursive functions φ_α, indexed by ordinals – say – up to ϵ_0. One reformulates Schütte's result by means of one of its corollaries:

Theorem 8 (Provably recursive functions). *If the recursive function f is such that one can prove its termination in* **PA**, *then $f \leq \varphi_\alpha$ for an appropriate $\alpha < \epsilon_0$,*

which is obtained by use of recoding techniques.

Provably recursive ordinals: since well-foundedness is a Π^1_1 formula which cannot be expressed in arithmetic, this version does not quite concern **PA**; but it applies to second-order systems. If $\preceq$ is a recursive relation defining a total order on integers, the formula

$$\mathbf{WO}_\preceq := \forall X \forall x (\forall y (\forall z (z \prec y \Rightarrow z \in X) \Rightarrow y \in X) \Rightarrow x \in X)$$

expresses that $\preceq$ is a well-ordering. One verifies that, if such a formula is cut-free provable by means of the ω-rule, then the height of the proof is greater than $\preceq$. Hence, if one proves in $\mathcal{T}$ that the ordinal α is well-founded, then $\alpha \leq \alpha_0$, where α_0 is the ordinal associated to $\mathcal{T}$.

3.C.4 Reverse mathematics. The corollaries (provable functions or ordinals) are quite stereotyped, the theory being often as obscure as « its » ordinal. This being said, we seek applications under the form of concrete incompleteness results. Thus, in a combinatory result of the form $A = \forall m \exists n B[m,n]$, if I show that the recursive function yielding n as a function of m increases faster than the $\varphi_\alpha(\alpha \leq \epsilon_0)$, I will have established that A is not provable in **PA**. The same with well-foundedness; let us mention this exceptional lucky find of H. Friedman as to *Kruskal's theorem*: this theorem, which deals with the combinatorics of finite trees, implies the well-foundedness of an ordinal slightly bigger than ϵ_0, the Veblen ordinal Γ_0 and therefore the mathematical theory *corresponding* to Γ_0 does not prove Kruskal.

The only weak point is that nobody cares about this theory. On the other hand, the result of Friedman does show the necessity to go beyond arithmetical induction. The programme of *reverse mathematics* is devoted to the demonstration of the need of axioms beyond the elementary ones in order to prove certain results. This is therefore about concrete incompleteness. In spite of a couple of spectacular successes, such as the result concerning Kruskal's theorem, one must say that it is often slightly specious. Moreover, all these results belong to combinatorics, i.e., to the backyard of logic: they are neither about algebraic geometry nor about the distribution of prime numbers.

Reverse mathematics establish a distinction between formal systems, depending on whether they prove more or less « concrete » theorems. But *quid* of the *nature* of this distinction? This is the question one should not ask: *reverse mathematics* is a sort of essentialist chapel dedicated to the infinite; instead of establishing the existence of God, one proves the necessity of more and more infinite variants of infinity. In order to do so, one seeks simple questions whose solution involves functions of extremely violent growth. Let us remark that there is here a distortion: indeed it is not clear that it is the right way to approach infinity. Thus, an essential question such as $\mathrm{P} \stackrel{?}{=} \mathrm{NP}$ deals with functions taking values in a two-element set, thus trivially bounded functions, for which the previous techniques are inoperative.

I have a doubt as to this « brute force » approach, as to this identification between mathematical complexity and « big axioms ». I rather believe in finesse, resting on the evidence that the best, the deepest mathematics use in fact in their formal expression only a minuscule part of Peano'a arithmetic. By the way, proving R by means of the « big axiom » A, isn't this the same as proving $A \Rightarrow R$? Formally, there is a big gap between the two, but, intellectually speaking, they are strictly the same. The status of the « big axiom » is a marginal question that is posed *a posteriori*: must one admit it or must one place oneself in conditions in which it holds? Real creativity is no longer there.

Chapter 4
Intuitionistic logic: LJ, NJ

Yet two more systems! I hope that this bureaucratic spree is justified by the interest of the notions introduced.

4.1 The intuitionistic sequent calculus LJ

4.1.1 Sequents.

Definition 3 (Intuitionistic sequents). An *intuitionistic sequent* is an expression $\Gamma \vdash A$, where Γ is a finite sequence of formulas and A is a formula.

As in the classical case, there are variants; e.g., replace sequences with sets (this only makes sense for the left-hand side Γ), and also a variant allowing empty right-hand members:

Empty right-hand member: an *intuitionistic sequent* is any expression $\Gamma \vdash \Delta$, where Γ, Δ are finite sequences of formulas and Δ consists of at most one formula. This is technically practicable, but unsuited for natural deduction (Section 4.3) or functional interpretations (Chapter 5). The more restrictive version we adopted uses a constant $\mathbf{0}$ for absurdity[1]; negation is no longer primitive, it is *defined* by $\neg A := A \Rightarrow \mathbf{0}$. One « fills » the empty right-hand member with $\mathbf{0}$, that can therefore be seen as a substitute for emptiness.

4.1.2 The intuitionistic calculus. The language is based on the connectives $\wedge$, $\vee$, $\Rightarrow$, $\forall$, $\exists$, $\mathbf{0}$.

Identity group

$$\frac{}{A \vdash A}\ (identity) \qquad\qquad \frac{\Gamma \vdash A \quad \Lambda, A \vdash B}{\Gamma, \Lambda \vdash B}\ (cut)$$

[1] Usually noted $\perp$; we use the linear notation for the sake of internal coherence.

Structural group

$$\frac{\Gamma \vdash A}{\sigma(\Gamma) \vdash A}$$

$$\frac{\Gamma \vdash B}{\Gamma, A \vdash B}$$

$$\frac{\Gamma, A, A \vdash B}{\Gamma, A \vdash B}$$

Logical group

$$\frac{\Gamma \vdash A \quad \Gamma \vdash B}{\Gamma \vdash A \wedge B}\ (\vdash \wedge) \qquad \frac{\Gamma, A \vdash C}{\Gamma, A \wedge B \vdash C}\ (l \wedge \vdash) \qquad \frac{\Gamma, B \vdash C}{\Gamma, A \wedge B \vdash C}\ (r \wedge \vdash)$$

$$\frac{\Gamma \vdash A}{\Gamma \vdash A \vee B}\ (l \vdash \vee) \qquad \frac{\Gamma \vdash B}{\Gamma \vdash A \vee B}\ (r \vdash \vee) \qquad \frac{\Gamma, A \vdash C \quad \Gamma, B \vdash C}{\Gamma, A \vee B \vdash C}\ (\vee \vdash)$$

$$\frac{\Gamma, A \vdash B}{\Gamma \vdash A \Rightarrow B}\ (\vdash \Rightarrow) \qquad \frac{\Gamma \vdash A \quad \Lambda, B \vdash C}{\Gamma, \Lambda, A \Rightarrow B \vdash C}\ (\Rightarrow \vdash)$$

$$\frac{}{\Gamma, \mathbf{0} \vdash A}\ (\mathbf{0} \vdash)$$

$$\frac{\Gamma \vdash A}{\Gamma \vdash \forall x A}\ (\vdash \forall) \qquad \frac{\Gamma, A[t/x] \vdash B}{\Gamma, \forall x A \vdash B}\ (\forall \vdash)$$

$$\frac{\Gamma \vdash A[t/x]}{\Gamma \vdash \exists x A}\ (\vdash \exists) \qquad \frac{\Gamma, A \vdash B}{\Gamma, \exists x A \vdash B}\ (\exists \vdash)$$

4.1.3 Gödel's translation. Intuitionistic negation is not involutive; it is not, contrary to what one could imagine, « bad will »: we indeed observed that negation is the operation exchanging left and right; now left and right are non-isomorphic zones that no operation can therefore swap. It is however clear that, from $A \vdash B$, one can deduce $\neg B \vdash \neg A$:

$$\dfrac{\dfrac{\begin{matrix}\vdots\\ A \vdash B\end{matrix} \qquad \dfrac{}{\mathbf{0} \vdash \mathbf{0}}\,(\mathbf{0}\vdash)}{A, \neg B \vdash \mathbf{0}}\,(\Rightarrow\vdash)}{\neg B \vdash \neg A}\,(\vdash\Rightarrow)$$

This example illustrates the ability, for intuitionistic negation:

(i) To migrate a formula (here, B), from right to left, at the cost of a negation. The right-hand member is then occupied by the « hole » **0**.

(ii) To migrate a formula (here, A), from left to right, to the cost of a negation, when the right-hand member is « empty » (i.e., occupied by **0**).

Still in this « double migration »[2]:

$$\dfrac{\dfrac{\dfrac{}{A \vdash A}\,(\mathit{identity}) \qquad \dfrac{}{\mathbf{0} \vdash \mathbf{0}}\,(\mathbf{0}\vdash)}{A, \neg A \vdash \mathbf{0}}\,(\Rightarrow\vdash)}{A \vdash \neg\neg A}\,(\vdash\Rightarrow)$$

Classical logic admits the inverse possibility, i.e., *contraposition* a.k.a. reduction to absurdity. The purest form of contraposition is the passage from $\neg\neg A$ to A. In intuitionistic logic, $\neg\neg A$ is established through the sequent $\neg A \vdash \mathbf{0}$, the crucial point being that contraction is available on the left. Due to the left/right symmetry, the same reasoning could be done without negations in classical logic; one would instead use contraction on the right, in other terms:

contraposition = *right contraction*

Let us come back to the left/right symmetry. It is refused by intuitionism, since the zones are handled in a different way. The right zone is a sort of *convento* where contraction is forbidden, since this rule replaces *two* occurrences of A with one, while there cannot be two formulas on the right. On the contrary, the left zone is a sort of *casino* where one can freely contract. In other words, imagine that I want to perform right contractions on A, the only way will be to « jump the wall » to the left, by using negation; later on, the same negation will bring back A to the right. But not scot-free: A will bear the stigmas of its escapade, a double negation[3].

[2]Combined with the previous principle, this establishes the equivalence $\neg\neg\neg A \Leftrightarrow \neg A$ in intuitionistic logic: three negations can be reduced to one.

[3]Double negation produces genuine effects only if one uses left contraction; in other terms, contraposition is useful only if the absurd hypothesis is used several times.

This underlines Gödel's translation (1932), which *faithfully* embeds classical logic into intuitionistic logic – which must not have pleased the intuitionists. If A is a classical formula, A^g is obtained by inserting double negations before atoms and connectives[4] of A: thus $(\forall x \exists y (p(x,y) \wedge (q(x,x) \vee p(x,x))))^g :=$ $\neg\neg\forall x \neg\neg\exists y \neg\neg(\neg\neg p(x,y) \wedge \neg\neg(\neg\neg q(x,x) \vee \neg\neg p(x,x)))$.

Theorem 9 (Gödel's translation). *A is classically provable iff A^g is intuitionistically provable. This remains true for cut-free provability.*

Proof. Classically, $\neg\neg B \Leftrightarrow B$ and it is easily seen that A is formally equivalent to its translation A^g; moreover, intuitionistic logic « proves less ». In other words, if A^g is intuitionistically provable, it is also classically provable and so is A.

Conversely, one interprets the classical sequent $\Gamma \vdash \Delta$ by the intuitionistic sequent $\Gamma^g, \neg^{-1}\Delta^g \vdash \mathbf{0}$, where $\neg^{-1}\Delta^g$ is obtained from Δ^g by removing the first negations of its formulas, which anyway begin with double negations. All the classical rules (in particular the right contraction) are translated by means of left rules, except the right logical rules: in that case, one moves to the right, one performs the *ad hoc* rule, then one comes back to the left. □

4.2 The *Hauptsatz* in LJ

Theorem 10 (Hauptsatz). *In the sequent calculus* **LJ**, *the cut rule is* ***redundant***, *in other terms, every formula provable with the cut rule is also cut-free provable.*

4.2.1 Proof. The proof sketched in Section 3.4, in the classical case, still works. One can prefer a justification through natural deduction, see *infra*; if we want the result for all connectives, including existence and disjunction, one needs the full version (commutative reductions, etc.).

4.2.2 Subformula property. It is obvious that the cut-free part of **LJ** enjoys the subformula property, just like **LK**. With an interesting corollary:

Theorem 11 (Intuitionistic propositional decidability). *Intuitionistic propositional calculus is decidable.*

Proof. In the absence of quantifiers, there is only a finite number of subformulas. Moreover, if one contracts as soon as possible, one can never get more than three occurrences of the same formula on the left of the same sequent $\Gamma \vdash A$; one can thus restrict the search to a finite set of sequents[5]. Finally it is of no use to consider proofs in which a branch contains the same sequent twice; in other words, one can restrict the search to a finite set of proofs. □

[4] It is enough to do so in front of atoms, disjunctions and existences.

[5] Variant: use the alternative set-theoretic version of sequents.

Remark 1. This applies to the classical case as well, but as a hammer crushing a fly. Remember that, in terms of complexity theory, classical propositional calculus is coNP-complete; the intuitionistic propositional calculus is PSPACE-complete.

4.2.3 Intuitionistic existence and disjunction. The cut-free calculus enjoys two remarkable properties:

Theorem 12 (Disjunction property). *If* $\vdash A \vee B$ *is provable, then either* $\vdash A$*, or* $\vdash B$*, is provable.*

Theorem 13 (Existence property). *If* $\vdash \exists x A$ *is provable, then* $\vdash A[t/x]$ *is provable for an appropriate* t*.*

Proof. Immediate in both cases: restricting to cut-free proofs, the only last possible rule is a right-hand rule $(l \vdash \vee)$, $(r \vdash \vee)$ or $(\vdash \exists)$. □

A faulty reading of these twin (and fundamental) properties would be: « a proof of $A \vee B$ is either a proof of A or a proof of B ». Almost total misinterpretation: indeed, who would care to enunciate $A \vee B$ if he has gotten B? In other terms, the intuitionistic calculus will naturally prove properties $A \vee B$, without telling us which of the disjuncts is valid. It is only after cut-elimination, a process which is long and complex and to the antipodes of the deductive spirit, that one can « know which one ». This is natural from the viewpoint of computer science (Section 6.F.3).

In other words, cut-free systems are *explicit*, since non-deductive. In particular, one should not style intuitionism an « explicit logic » – this would be an *oxymoron* – but rather an *explicitable*[6] logic. Which is not the case of classical logic, because of contraction on the right:

$$
\dfrac{\dfrac{\dfrac{\dfrac{\dfrac{}{A \vdash A}\,(identity)}{A \vdash A \vee \neg A}\,(l \vdash \vee)}{\vdash A \vee \neg A, \neg A}\,(\vdash \neg)}{\vdash A \vee \neg A, A \vee \neg A}\,(r \vdash \vee)}{\vdash A \vee \neg A}\,(\vdash C)
$$

This classical proof of the excluded middle (the antithesis of the disjunction property) illustrates the crucial role played by contraction: in a cut-free classical proof, the last rule is most likely a contraction. Herbrand's theorem makes explicit the classical/intuitionistic difference: in the case of $\exists x A$, A quantifier-free, one gets an almost property of existence, under the form of a finite disjunction of cases

[6] Pushing the analogy with computer science, an explicit logic or an explicit mathematics, would be like a table of logarithms.

$A[t_1] \vee \cdots \vee A[t_n]$. On the other hand, if A becomes more complicated, there remain only complex cross-dependencies, existential/universal with no immediate, i.e., no explicit, sense.

If one considers the absurdity $\mathbf{0}$ as a 0-ary disjunction, then we have consistency:

$\mathbf{0}$ *is not provable*

provides us with an « absurdity property »; on the other hand, one can propose nothing for the other connectives $\wedge$, $\Rightarrow$, $\forall$. This is our first encounter with *polarity*: $\vee$, $\exists$ are *positive* in other terms active, explicit(able); in the cut-free regime, they convey an information (a left/right *bit* for disjunction, a witness t for existence). On the other hand the other connectives are *negative*, in other terms passive, implicit, incomplete, since they are waiting; thus $\forall x A$ refers to an implicit x « give me a value v for x and I will show you $A[v]$ » and the implication $A \Rightarrow B$ means « give me A and I will give you B back », which especially makes sense in a functional setting (Section 6.F.3).

4.3 The natural deduction NJ

Gentzen also set the basis of *natural deduction*; but this approach should mainly be associated with the name of Prawitz [87].

4.3.1 The system. Of a much easier access than the sequent calculus **LJ**, **NJ** is the pet system of beginners which one can understand! But one must know that, under its most pleasant aspects, it hides unsuspected difficulties. Because it has the defects of its qualities – few rules, hence less limpid; in particular everything related to *signature* is delicate.

The system consists in posing hypotheses from which we may draw a conclusion, so that the demonstrative structure looks like a tree:

$$\begin{array}{c}\Gamma\\ \vdots\\ A\end{array}$$

We speak of a *deduction* of *conclusion* A and *hypotheses* Γ; when Γ is empty, we speak of a *proof* of A.

Hypothesis. Any formula A can be posed as a hypothesis:

$$A$$

is a *deduction* of A under hypothesis A. Beware, during the proof process, some hypotheses will be *discharged*. This rule quite corresponds to the identity axiom of **LJ**.

It remains to write logical rules, which split into two groups: on one hand, *introductions*, corresponding to the right-hand rules of **LJ**, enable one to create formulas; on the other hand, *eliminations*, corresponding to the left-hand rules of **LJ**, enable one to use, i.e., to destroy, formulas.

Conjunction. Conjunction admits the following rules (introduction, eliminations):

$$\frac{\begin{matrix}\vdots & \vdots\\ A & B\end{matrix}}{A \wedge B}\,(\wedge I) \qquad \frac{\begin{matrix}\vdots\\ A \wedge B\end{matrix}}{A}\,(l \wedge E) \qquad \frac{\begin{matrix}\vdots\\ A \wedge B\end{matrix}}{B}\,(r \wedge E)$$

Implication. Implication admits the following rules (introduction, elimination):

$$\frac{\begin{matrix}[A]\\ \vdots\\ B\end{matrix}}{A \Rightarrow B}\,(\Rightarrow I) \qquad \frac{\begin{matrix}\vdots & \vdots\\ A & A \Rightarrow B\end{matrix}}{B}\,(\Rightarrow E)$$

We should recognise here the deduction theorem and *Modus Ponens*. The introduction rule is very peculiar: we start with a deduction of B under hypothesis A, which means that we have selected a certain number of occurrences of A not yet discharged; the application of the rule discharges them, which we indicate by square brackets. For instance:

$$\frac{[A]}{A \Rightarrow A}\,(\Rightarrow I)$$

is a deduction of $A \Rightarrow A$ without hypotheses, i.e., a *proof*.

Since we can discharge as many hypotheses A as we want, the rule hides contraction and weakening. However, the correspondence between contraction and multiple occurrences of A is approximative (Section 4.C.3).

Universal quantification. Universal quantification admits the following rules (introduction, elimination):

$$\frac{\begin{matrix}\vdots\\ A\end{matrix}}{\forall x A}\,(\forall I) \qquad \frac{\begin{matrix}\vdots\\ \forall x A\end{matrix}}{A[t/x]}\,(\forall E)$$

The introduction rule is subject to the restriction that the *eigenvariable* x be not free in the hypotheses. For instance:

$$\dfrac{\dfrac{[A]}{\forall x A}\,(\forall I)}{A \Rightarrow \forall x A}\,(\Rightarrow I)$$

is faulty. On the other hand:

$$\dfrac{\dfrac{[A]}{A \Rightarrow A}\,(\Rightarrow I)}{\forall x(A \Rightarrow A)}\,(\forall I)$$

where the hypothesis has been discharged before the quantification, is correct.

Absurdity. This constant admits only one rule, an elimination:

$$\dfrac{\begin{matrix}\vdots\\ \mathbf{0}\end{matrix}}{A}\,(\mathbf{0}\,E)$$

Disjunction. Disjunction admits the following rules (introductions, elimination):

$$\dfrac{\begin{matrix}\vdots\\ A\end{matrix}}{A \vee B}\,(l \vee I) \qquad \dfrac{\begin{matrix}\vdots\\ B\end{matrix}}{A \vee B}\,(r \vee I) \qquad \dfrac{\begin{matrix}\\ \vdots\\ A \vee B\end{matrix} \quad \begin{matrix}[A]\\ \vdots\\ C\end{matrix} \quad \begin{matrix}[B]\\ \vdots\\ C\end{matrix}}{C}\,(\vee E)$$

The elimination rule is natural (reasoning by cases); this being said, its graphical expression is rather fabricated. Same remark for existence, see *infra*.

Existence. Existential quantification admits the rules (introduction, elimination):

$$\dfrac{\begin{matrix}\vdots\\ A[t/x]\end{matrix}}{\exists x A}\,(\exists I) \qquad \dfrac{\begin{matrix}\\ \vdots\\ \exists x A\end{matrix} \quad \begin{matrix}[A]\\ \vdots\\ B\end{matrix}}{B}\,(\exists E)$$

In the elimination rule, the variable x cannot be free, neither in the conclusion B, nor in the hypotheses (other than the discharged occurrences of A).

Some terminology: in the elimination rules, one premise (and exactly one) bears the eliminated symbol. It is called the *main premise*; the other premises, if any, are called *minor premises*.

4.3.2 Cuts and normalisation. This looks like a sectarian drift: you enter the Party for – say – freedom, but once you have pushed open the door, it is nothing of the like. In the same way, I advertised intuitionism in the name of existence and disjunction and now I dispense with them! The remainder of the chapter is concerned with the sole fragment $\wedge$, $\Rightarrow$, $\forall$ of intuitionistic logic. For the missing connectives, $\vee$, $\exists$, $\mathbf{0}$, see annex 4.A.

If one looks at a rule in the sense premise $\rightarrowtail$ conclusion, one observes that:

- The premises of an introduction are subformulas of the conclusion.
- The main premise of an elimination (except for $\vee$, $\exists$, $\mathbf{0}$, this is why they have just been excluded), is a super-formula of the conclusion.

If one circulates in a deduction from premise (main if this makes sense) to conclusion, one observes a zigzag-like behaviour: there are therefore maximal and minimal formulas. One calls a *cut* the succession of an introduction and an elimination such that the conclusion of the introduction is the main premise of the elimination: which corresponds to a local maximum[7].

The *normalisation* procedure aims at eliminating cuts so as to « reduce the bumps » of the deduction: once completed, only remain minimal formulas, i.e., from top to bottom the succession decrease (eliminations) then increase (introductions) and that's all.

4.3.3 Immediate reductions. For each case of cut, we define an *immediate reduction*, which replaces the incriminated configuration with a simpler one. This reduction can modify neither the conclusion nor the (active) hypotheses. In what follows, the left-hand member (before reduction) is called *redex*, the right-hand member (after reduction) is called *contractum*. Note that the *contractum* can be bigger than the *redex*.

Conjunction, left case

$$\begin{array}{c}\vdots \qquad \vdots\\ \dfrac{\dfrac{A \qquad B}{A \wedge B}\,(\wedge I)}{A}\,(l \wedge E)\\ \vdots\end{array} \qquad \rightsquigarrow \qquad \begin{array}{c}\vdots\\ A\\ \vdots\end{array}$$

[7] Minimal formulas correspond to identity axioms; since it is out of question to eliminate axioms, one does not care much about minimal formulas. See however Section 7.4.2.

Conjunction, right case

$$\begin{matrix}\dfrac{\dfrac{\begin{matrix}\vdots & \vdots\\ A & B\end{matrix}}{A\wedge B}\,(\wedge I)}{B}\,(r\wedge E)\\ \vdots\end{matrix} \qquad\leadsto\qquad \begin{matrix}\vdots\\ B\\ \vdots\end{matrix}$$

Implication

$$\begin{matrix}\dfrac{\begin{matrix}\vdots\\ A\end{matrix}\qquad \dfrac{\begin{matrix}[A]\\ \vdots\\ B\end{matrix}}{A\Rightarrow B}\,(\Rightarrow I)}{B}\,(\Rightarrow E)\\ \vdots\end{matrix} \qquad\leadsto\qquad \begin{matrix}\vdots\\ A\\ \vdots\\ B\\ \vdots\end{matrix}$$

Observe that reduction substitutes a deduction of A for *each* discharged occurrence of A: the size of the deduction is therefore roughly multiplied by a factor corresponding to the number of these occurrences. The fact of discharging several occurrences corresponds to the rule of contraction[8], of which we once more see that it is everything but an innocent rule. In particular the consecrated expression « reduction » is unwelcome.

Universal quantification

$$\begin{matrix}\dfrac{\dfrac{\begin{matrix}\vdots\\ A\end{matrix}}{\forall x A}\,(\forall I)}{A[t/x]}\,(\forall E)\\ \vdots\end{matrix} \qquad\leadsto\qquad \begin{matrix}\vdots\\ A[t/x]\\ \vdots\end{matrix}$$

Here one substitutes in the deduction with conclusion A, the term t for the variable x. This steps makes a crucial use of the fact that x is not free in the hypotheses: otherwise, reduction would modify them.

[8]Weakening corresponds to the fact of discharging nothing.

4.3.4 The Church–Rosser theorem. *Reduction*, still noted $\rightsquigarrow$, is defined as the reflexive/transitive of the immediate reductions of the previous section.

Theorem 14 (Church–Rosser). *Reduction is* ***confluent****: if* $\mathfrak{d} \rightsquigarrow \mathfrak{d}', \mathfrak{d}''$*, one can find* $\mathfrak{d}'''$ *such that* $\mathfrak{d}', \mathfrak{d}'' \rightsquigarrow \mathfrak{d}'''$.

Proof. The proof is an adaptation, *mutatis mutandis*, of a noted result of λ-calculus, that we shall prove in Section 5.2.3. □

Definition 4 (Normal form). A *normal form* for a deduction $\mathfrak{d}$ is any *normal*, i.e., without cut, deduction $\mathfrak{e}$ such that $\mathfrak{d} \rightsquigarrow \mathfrak{e}$.

If $\mathfrak{d}$ has two normal forms $\mathfrak{d}', \mathfrak{d}''$, one can find $\mathfrak{d}'''$ with $\mathfrak{d}', \mathfrak{d}'' \rightsquigarrow \mathfrak{d}'''$; but a normal deduction can reduce only in itself, hence $\mathfrak{d}' = \mathfrak{d}''$:

Corollary 14.1. *The normal form, when it exists, is unique.*

4.3.5 The weak normalisation theorem

Theorem 15 (Weak normalisation for NJ). *Any deduction admits a normal form.*

Proof. The *degree* of a formula is defined as the number of its logical connectives; the degree of a cut is defined as the degree of the formula which is both a conclusion of an introduction and a premise of an elimination; this formula has a first connective, hence this number is non-zero. The degree of a deduction is defined as the greatest cut-degree to be found in it, so that a deduction of degree 0 is normal. To each deduction one associates two numbers (d, n), where d is the cut-degree and n is the number of cuts of degree d. These numbers are lexicographically ordered: one first compares the degrees; if this comparison turns to equality, one compares the second components. The theorem is established by a double induction on the pairs (d, n)[9].

Let us look at the four immediate reductions; they all have the effect of eliminating one cut. This being said, they can reintroduce new ones, since they put into contact occurrences once distant, respectively of A, of B, of A and B[10], of $A[t/x]$. Any new cut created in this process will be of strictly lesser degree than the original one. In particular, if one reduces a cut of maximal degree, it seems most likely that one succeeds in lowering the degree, or at least the number of cuts of maximum degree.

Unfortunately, this is not quite the case: we forgot that, besides the creation of new cuts, one can witness the proliferation of previous cuts by duplication. This is exactly what may happen for the cuts located above A in the implicative reduction. Thus, the cut we reduce is not simply chosen of maximum degree, it is also chosen

[9] This is a very elementary case of transfinite induction, here up to the ordinal ω^2.
[10] Case of implication: each occurrence of A can yield a new cut.

« topmost », i.e., without a cut of the same degree in the tree above it. Duplication is still likely to increase the number of cuts, but not those of maximum degree. One has therefore succeeded in decreasing the pair (d, n). □

The choice of the reductions to be applied to obtain the normal form is very discretionary. But, by Church–Rosser, this does not affect the result. Observe however that the theorem does not exclude the following unpleasant situation: an infinite sequence of reductions which do not lead to a normal form, since it systematically chooses the wrong redex: this is why one speaks of *weak* normalisation. The destruction of whole pieces of deduction, for instance in the left conjunctive case, suggests the (theoretical) possibility of an infinite sequence of reductions. Indeed one can surmise that this infinite sequence is wholly located in the sub-deduction above B, a premise eventually destroyed during normalisation and that one can anyway destroy at any moment. But if one forgets to do so…

Fortunately, this does not happen: one can prove *strong* normalisation, i.e., the existence of a finite bound on the length of all sequences of immediate reductions starting with a given deduction. See Sections 4.B.3 and 6.D.4.

4.4 The signature in natural deduction

4.4.1 The main hypothesis. Sequent calculus is downwards oriented, in particular, the cut-free rules are monotonic (w.r.t. the subformula relation). Nothing of the like exists with natural deduction: introductions grow downward while eliminations grow upward, in fact main-premiseward; one must introduce two ideas:

- Consider (active) hypotheses as sorts of hidden conclusions.
- Think that the « real » last rule may not be the conspicuous one displayed at the bottom of the tree, but the one eliminating the *main hypothesis*.

Definition 5 (Main hypothesis). If the deduction $\mathfrak{d}$ of A is normal and does not end with an introduction, one defines the *main hypothesis* to be the unique hypothesis obtained by recursively climbing up, starting with A, from conclusion to main premise: this is a real (undischarged) hypothesis. Alternative terminology: head premise, by analogy with « head variable » (Section 6.1.7).

Reasoning by induction on *normal* deductions involves three cases:

- Hypothesis.
- Introduction.
- Elimination: the main hypothesis is considered as the real conclusion.

The writing of linear logic by means of *proof-nets* helps us to understand (Chapter 11). Indeed the hypotheses are written as negated conclusions; in the presence of several conclusions, there is an ambiguity as to the last rule. A main hypothesis, which appears in a deduction inside a piece:

$$\dfrac{\begin{matrix}\vdots\\ A\end{matrix} \quad A \Rightarrow B}{\begin{matrix}B\\ \vdots\end{matrix}}\,(\Rightarrow E)$$

becomes, provided one « topsy-turvies » B:

$$\dfrac{\begin{matrix}\vdots\\ \sim B\end{matrix} \quad \begin{matrix}\vdots\\ A\end{matrix}}{\sim B \otimes A}\,(\otimes)$$

and observe that $\sim B \otimes A$ is the linear negation of the (linear) implication $A \multimap B$, which, for the present discussion, behaves like usual implication.

4.4.2 The subformula in natural deduction. The subformula property can be stated in the setting of natural deduction.

Theorem 16 (Subformula property). *If $\mathfrak{d}$ is normal, then any formula A occurring in $\mathfrak{d}$ is a subformula of either the conclusion or of one of the* (*active*) *hypotheses.*

Proof. Almost immediate induction, provided one does not mistreat the case where the last rule is an elimination. In that case one looks for the main hypothesis, say $B \Rightarrow C$. If A is not $B \Rightarrow C$, either it occurs in the sub-deduction « above » B, or in the sub-deduction « below » C; in both cases the induction hypothesis enables one to conclude. □

4.4.3 Relation with LJ. The main question is the relation between **LJ** and **NJ**. The fact that the two systems prove the same theorems is really immediate. We obtain something much more interesting by relating the cut-free proofs of $\Gamma \vdash A$ in **LJ** to the normal deductions of A under the hypotheses Γ in **NJ**. *Grosso modo*, there is an obvious correspondence at the level of logical rules: right rules ($\vdash *$) correspond to introductions ($*I$), left rules ($* \vdash$) correspond to eliminations ($*E$). Natural deduction has no structural rule (neither identity nor cut rule). As a consequence, the same deduction admits several writings in sequent calculus.

Translation NJ $\mapsto$ LJ. By induction on the size of the normal deduction to be translated, starting with a hypothesis, which becomes an identity axiom. The case of implication concentrates all the difficulties of the induction step:

- If the last rule is a $(\Rightarrow I)$ and if the deduction of B under hypothesis A has been translated by a proof of $\Gamma, A, \ldots, A \vdash B$, one can obtain $\Gamma, A \vdash B$: if there are more than two occurrences of A, by contractions, if there are none, by weakening. One proceeds with a $(\vdash \Rightarrow)$

- If the last rule is an elimination, one climbs up to the main hypothesis, which is the main premise of an elimination. If it is a $(\Rightarrow E)$, with premises A and $A \Rightarrow B$, one applies the induction hypothesis for the sub-deduction starting with B, which yields a proof of some $\Lambda, B \vdash C$ and to the sub-deduction of conclusion A, which yields a proof of $\Gamma \vdash A$. One proceeds with a $(\Rightarrow \vdash)$.

Translation LJ $\mapsto$ NJ. In the other direction, it is important to precisely determine, when A occurs in the left part of $\Gamma \vdash$, a « packet » of hypotheses of the associated deduction corresponding to this occurrence of A. Thus, in the presence of a contraction on A, the packets associated to two distinct occurrences will merge. This being said, a rule introducing A on the left will require several eliminations, one for each hypothesis of the packet associated with A. In case of a $(\vdash \Rightarrow)$ introducing $A \Rightarrow B$, the full packet associated with A will be discharged.

4.A Existence and disjunction in LJ

4.A.1 « Direct » eliminations. We already spoke of *polarity*, i.e., of the distinction between negative connectives – to which our normalisation has so far been restricted – and positive connectives. From the geometrical standpoint, the difference is that the two groups *do not evolve in the same sense*. The treelike structure of natural deduction is badly suited for positives: for instance one should write the elimination of disjunction as:

$$\frac{A \vee B}{A \quad B}$$

which is not *a priori* shocking. Except that one cannot « branch » both upwards and downwards. The drama of the positives is that the deductive connectives, e.g., implication, are negative: they obviously take priority over positive connectives. This is why positive eliminations are so badly behaved.

Let us write immediate reductions for I/E cuts in the positive case.

Disjunction, left case

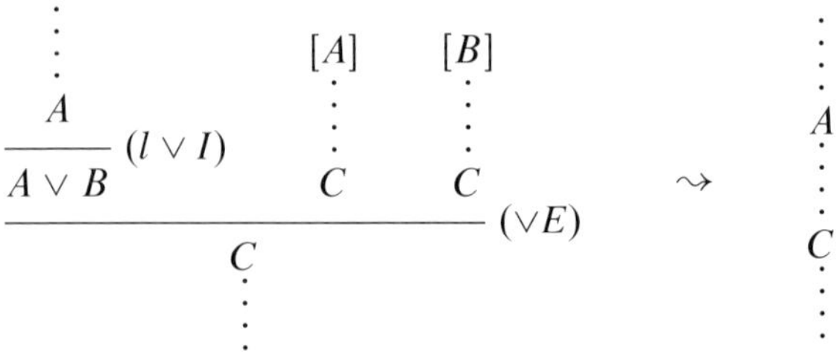

Disjunction, right case

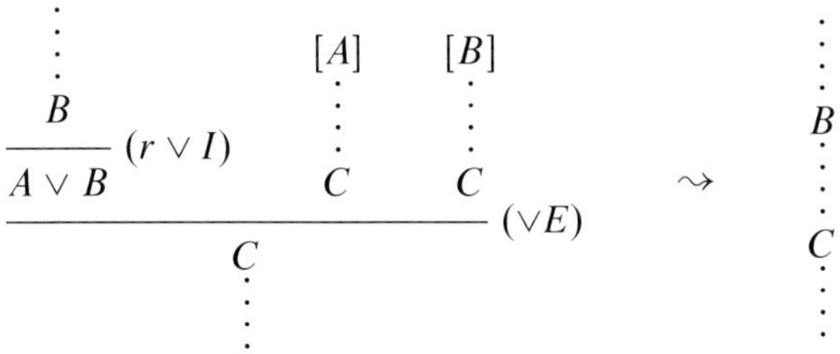

Existence

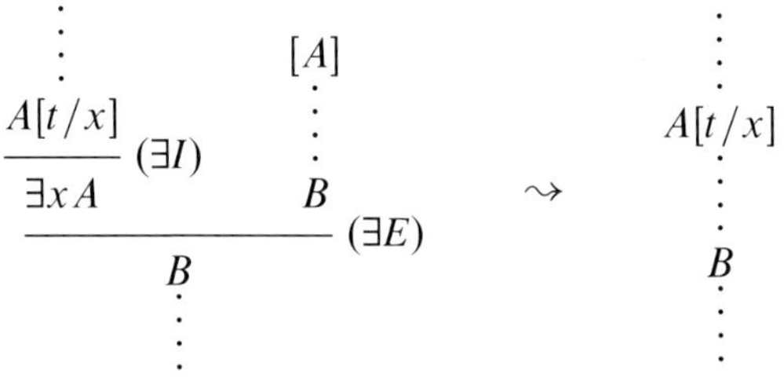

The normalisation theorem can be extended without difficulty so as to include these additional immediate reductions. This extension yields the existence and disjunction properties.

4.A.2 Commutative cuts. But we are far from finished, since we do not get any subformula property in this way. This is due to a certain type of configuration, called « commutative cut »: it is a matter of an elimination whose main premise is the conclusion of one of the rules $(\vee E)$, $(\exists E)$, $(\mathbf{0})$. Which induces $3 \times 7 = 21$ cases. For each of these commutative cuts one defines immediate reductions. The subject not being that hot, I will content myself with three examples:

Commutation $\mathbf{0}E/\Rightarrow E$

$$\begin{array}{c}\dfrac{\begin{array}{c}\vdots\\ A\end{array}\qquad \dfrac{\begin{array}{c}\vdots\\ \mathbf{0}\end{array}}{A\Rightarrow B}\,(\mathbf{0}E)}{B}\\ \vdots\end{array}\quad\rightsquigarrow\quad\begin{array}{c}\vdots\\ \dfrac{\mathbf{0}}{B}\,(\mathbf{0}E)\\ \vdots\end{array}$$

All the other eliminations with $(\mathbf{0}E)$ in « introduction » are built on the same mould: the *contractum* is simply a more direct application of the same $(\mathbf{0}E)$.

Commutation $\vee E/\Rightarrow E$

$$\dfrac{\begin{array}{c}\vdots\\ C\end{array}\qquad \dfrac{\begin{array}{c}\vdots\\ A\vee B\end{array}\quad \begin{array}{c}[A]\\ \vdots\\ C\Rightarrow D\end{array}\quad \begin{array}{c}[B]\\ \vdots\\ C\Rightarrow D\end{array}}{C\Rightarrow D}}{D}\quad\rightsquigarrow\quad \dfrac{\begin{array}{c}\vdots\\ A\vee B\end{array}\quad \dfrac{\begin{array}{c}\vdots\\ C\end{array}\quad\begin{array}{c}[A]\\ \vdots\\ C\Rightarrow D\end{array}}{D}\quad \dfrac{\begin{array}{c}\vdots\\ C\end{array}\quad\begin{array}{c}[B]\\ \vdots\\ C\Rightarrow D\end{array}}{D}}{D}$$

In other word, one makes the rule $(\Rightarrow E)$ commute and one executes it first. The remaining cases are treated in the same way. The most complicated being the following.

Commutation $\vee E/\vee E$

$$\dfrac{\dfrac{\begin{array}{c}\vdots\\ A\vee B\end{array}\quad \begin{array}{c}[A]\\ \vdots\\ C\vee D\end{array}\quad \begin{array}{c}[B]\\ \vdots\\ C\vee D\end{array}}{C\vee D}\qquad \begin{array}{c}[C]\\ \vdots\\ E\end{array}\quad \begin{array}{c}[D]\\ \vdots\\ E\end{array}}{E}$$

$$\rightsquigarrow\quad \dfrac{\begin{array}{c}\vdots\\ A\vee B\end{array}\quad \dfrac{\begin{array}{c}[A]\\ \vdots\\ C\vee D\end{array}\quad \begin{array}{c}[C]\\ \vdots\\ E\end{array}\quad \begin{array}{c}[D]\\ \vdots\\ E\end{array}}{E}\quad \dfrac{\begin{array}{c}[B]\\ \vdots\\ C\vee D\end{array}\quad \begin{array}{c}[C]\\ \vdots\\ E\end{array}\quad \begin{array}{c}[D]\\ \vdots\\ E\end{array}}{E}}{E}$$

4.A.3 Weak normalisation. The system enlarged with commutative reductions is still confluent and enjoys normalisation (weak or strong). This is rather technical, being neither difficult nor innovative. In fact, for a formula C, one introduces a *commutative degree*: it is null, except when C is a conclusion of a positive elimination rule; in this case it equals the supremum of the commutative degrees of the minor premises plus 1. This quantity has a propensity for decreasing with commutative reductions. One reasons by induction on a 3-tuple (d, c, n), where d is the cut-degree[11], c is the greatest commutative degree of a cut of degree d and n is the number of cuts corresponding to the data d and c.

Normal proofs now enjoy the subformula property. Indeed, one can still define the notion of main hypothesis, which enables one to resume the proof of Theorem 16: one must consider the case where the deduction admits a positive main hypothesis, say $A \vee B$, which is the (main) premise of an elimination; in this case the conclusion C of the elimination is also the conclusion of the deduction, since otherwise it would be the premise of another elimination, which would yield a commutative cut; one easily concludes.

4.B Natural deduction vs. sequent calculus

4.B.1 Small comparison. The comparison between natural deduction and sequent calculus is rather informative. Sequent calculus is *grosso modo* an analytical system, which decomposes a proof in useful parts, without looking for economy. It is naturally redundant, with no claim as to unicity. In particular, the translation of **NJ** in **LJ** is anything but univocal. On the contrary, **NJ** is a synthetic system presenting a « primitive » artefact: a natural deduction can be written in only one way. This becomes manifest when one compares the *Hauptsatz* and normalisation of the negative fragment. In the latter case, one only needs the immediate reductions which are the exact analogue of the key cases: no need for endless commutations, nor for a specific handling of structural rules.

The problem of commutative reductions shows that natural deduction justifies its synthetic ideal, but only *in extremis*. Indeed, what are these rules, if not the expression of a non-unicity? They indeed correspond to the commutations in the proof of the « Hauptsatz ». Although everything eventually works, we felt the wind of the cannonball.

4.B.2 Proofs of the *Hauptsatz*. The *Hauptsatz* is unpleasant to prove; and slightly frustrating, since it accumulates *ad hoc* operations. The simplest method therefore consists in transiting through natural deduction: a proof with cuts is translated in

[11]A small detail: the formula **0** has degree 1: this is because **0** has been considered as a 0-ary connective..

natural deduction, then normalised and eventually translated back into a cut-free proof.

Still in the « let's not overwork ourselves » style, one can prove the classical *Hauptsatz* by means of Gödel's translation. One first translates **LK** inside **LJ**, next one eliminates cuts: which yields a cut-free proof of the Gödel translation. It must then be converted back into a classical cut-free proof: let us paint in red the negations introduced by the translation; then erase these red negations, at the cost of the removal of some left/right obstacle race. For instance, a left formula starting with a single red negation is put back to the right, period.

4.B.3 The Gandy method. An elegant idea due to Gandy enables us to reduce strong normalisation to weak normalisation; we shall see it later (Section 6.D.4) in full detail. The idea is that strong normalisation is combinatorially too complex; the problem is therefore handled by a moron, the system itself. We can modify the rules of the system so as to make them size-increasing and then check for Church–Rosser and weak normalisation. In a weakly normalising system where size strictly grows with reduction, there cannot be any infinite sequence of reductions, since the normal form must be of size greater than all elements of the sequence.

The properties of Gandy's system, weak normalisation and Church–Rosser, are established by means of a banal translation in the « ordinary » system.

4.C Around contraction

4.C.1 Classical natural deduction. Some authors insist upon a classical natural deduction **NK**, with the rule

$$\dfrac{\begin{matrix}[\neg A]\\ \vdots\\ \mathbf{0}\end{matrix}}{A}\,(\mathbf{0}E)$$

We could write reduction rules for this rule, but the result is rather ugly, much below commutative reductions: in this case one would not avoid the cannonball. Indeed, it amounts, without saying it, to using Gödel's translation. Since there is no miracle, this system **NK** inherits the defects of Gödel's translation. We must use a *polarised* version to obtain a satisfactory system (Section 7.A.6).

4.C.2 Prawitz and naïve comprehension. Prawitz proposed a version of naïve comprehension based upon a logical interpretation of the symbol $\in$. In a language based on the sole predicate symbol $\in$ (binary), one introduces abstraction terms $\{x; A\}$ and one writes the two rules, which are reminiscent of the substitution in second-order logic (Section 3.B.2):

$$\dfrac{\begin{matrix}\vdots\\ A[T/x]\end{matrix}}{T \in \{x; A\}}\,(\in I) \qquad\qquad \dfrac{\begin{matrix}\vdots\\ T \in \{x; A\}\end{matrix}}{A[T/x]}\,(\in E)$$

This is enough to prove Russell's antinomy: with $T := \{x; x \notin x\}$,

$$\dfrac{\dfrac{\dfrac{\dfrac{[T \in T] \quad \dfrac{[T \in T]}{T \notin T}\,(\in E)}{\mathbf{0}}\,(\Rightarrow E)}{T \notin T}\,(\Rightarrow I)}{T \in T}\,(\in I) \qquad \dfrac{\dfrac{[T \in T] \quad \dfrac{[T \in T]}{T \notin T}\,(\in E)}{\mathbf{0}}\,(\Rightarrow E)}{T \notin T}\,(\Rightarrow I)}{\mathbf{0}}\,(\Rightarrow E)$$

Since $\mathbf{0}$ has no introduction this antinomy *cannot* be normalised; there is however an obvious reduction procedure for the cuts $\in I/\in E$. The process must therefore diverge, i.e., never halt. The final cut is reduced by putting a copy of the left-hand member on each discharged hypothesis $T \in T$ of the right-hand member; one meets up with a version of the original proof, augmented with a section $\in I/\in E$, which once simplified, yields the original proof.

The reduction process thus loops. Contraction plays here an essential role; indeed, without repetition of discharged hypotheses, the size would decrease.

4.C.3 Contraction vs. repetition of hypotheses. There is no strict relation between contraction and repetition of discharged hypotheses. Indeed, let us have a close look at the system **LJ**: one will observe the difference between the right rule for conjunction, treated « additively », the same context on both premises and the right rule for implication, treated « multiplicatively », disjoint contexts[12]. Thus, the deduction

$$\dfrac{A \quad A}{A \wedge A}\,(\wedge I)$$

corresponds to a proof of $A \vdash A \wedge A$ which does not use contraction.

4.C.4 Effective bounds. Let us try to find bounds on the size (number of rules) of the normal form; we should be happy with the negative case. If I start with a proof of size h, I have at most h cuts of maximal degree; each introduction of

[12]The terminology comes from linear logic (Section 9.2).

implication discharges at most h premises: the reduction of one cut increases the size by a multiplying coefficient at most h. I can refine my choice of *redex*: all I need is that there is no cut of maximal degree above the minor premise; among all choices of this type, I choose a downmost cut. The fact of reducing it will not affect the multiplicative coefficients of the other cuts of maximal degree, which remain at most h. Reducing at most h cuts, each of them of multiplicative coefficient at most h, induces a bound of $h^h \leq 2^{h^2}$: the effect is to decrease by 1 the maximal degree. The full process receives as bound a tower of exponentials whose height is the degree.

I didn't quite refine my bounds. This bound cannot be fundamentally improved: it can be shown it by coding iterated exponentials (Section 5.3.2). All that can be done is to lower this height, but one cannot avoid a shape:

$$2^{2^{2^{\cdot^{\cdot^{2^h}}}}} \tag{4.1}$$

with $2_0(h) := h$, $2_{k+1}(h) := 2^{2_k(h)}$, whose height increases with the degree. The actual height only depends on the implicative nesting, a more refined notion than the degree: $\delta(A \Rightarrow B) := \sup(\delta(A) + 1, \delta(B))$. Indeed we see that only the left side of an implication has a real duplicative power.

4.D Logic programming

4.D.1 The resolution method. *Logic programming* is based on the idea of seeking the atomic theorems of a theory whose axioms are *Horn clauses*. Those are sequents $\mathcal{S}$ of the form $P_1, \ldots, P_n \vdash Q$ where $P_1, \ldots, P_n, Q$ are atomic. To each axiom $\mathcal{S}$ one can naturally associate introduction rules:

$$\frac{\begin{array}{c}\vdots\\ P_1\theta \quad \ldots \quad P_n\theta\end{array}}{Q\theta}\ (\mathcal{S}\theta\ I)$$

where θ is a substitution. One easily shows:

- Cut-elimination (immediate, there is no $\mathcal{S}$-elimination).
- For formulas of the shape $\exists x_1 \ldots \exists x_p(R_1 \wedge \cdots \wedge R_q)$, $R_1, \ldots, R_q$ atomic, classical provability matches intuitionistic provability.

Thus the following idea: in such an axiomatic system, try to prove formulas of the form $\exists x_1 \ldots \exists x_p(R_1 \wedge \cdots \wedge R_q)$. A cut-free proof will provide one with explicit values (existence property). The search for cut-free proofs is done by means of an algorithm, the *resolution method*, based on unification.

4.D.2 PROLOG, its *grandeur*. The idea is to consider a Horn theory as a program. One will make queries of the form « find the solutions $x_1, \ldots, x_p$ to the conjunction $R_1 \wedge \cdots \wedge R_q$ », which amounts to finding proofs of $\exists x_1 \ldots \exists x_p (R_1 \wedge \cdots \wedge R_q)$, indeed cut-free. The search is done by unification: one seeks to prove R by finding a clause $\Gamma \vdash S$ (axiom written as a sequent) whose *head* (right formula) unifies with R, by means of θ; then one is led back to similar problems for the formulas $\Gamma\theta$ of the *tail* where θ has been performed. One stops in case one steps on empty tails (success); on the other hand, when no unifier can be found, this is failure. But by far the most likely possibility is that of a search that neither succeeds nor fails, since it does not terminate. The culture of incompleteness is here to inform us as to the pregnancy of this unpleasant eventuality.

Whatsoever, what we just described is the paradigm of *logic programming*, which was so glamourous in the years around 1980, mainly because of the Japanese enthusiasm for « the fifth generation ».

4.D.3 PROLOG, its misery. Logic programming was bound to failure, not because of a want of quality, but because of its exaggerations. Indeed, the slogan was something like « pose the question, PROLOG will do the rest ». This paradigm of *declarative* programming, based on a « generic » algorithmics, is a sort of all-terrain vehicle, capable of doing everything and therefore doing everything badly. It would have been more reasonable to confine PROLOG to tasks for which it is well-adapted, e.g., the maintenance of data bases.

On the contrary, attempts were made to improve its efficiency. Thus, as systematic search was too costly, « control » primitives, of the style « don't try this possibility if... » were introduced. And this slogan « logic + control[13] », which forgets that the starting point was the logical soundness of the deduction. What can be said of this control which plays *against* logic[14]? One recognises the sectarian attitude that we exposed several times: the logic of the idea kills the idea.

The result is the most inefficient language ever designed; thus, PROLOG is very sensitive to the order in which the clauses (axioms) have been written.

4.D.4 Negation in PROLOG. Negation in PROLOG hardly deserves inclusion in a serious textbook. Nevertheless, it is not that stupid: « a query that fails is false » is a procedural idea.

It is unfortunately the only positive thing that can be said. The amateur logicians recruited by hundreds around the « fifth generation » wrote pure gibberish about this. For instance that a query that does not succeed is false. Since non-success

[13] This « control » is something like: disable the automatic pilot and navigate at sight.

[14] This is not that far from *labeled deductive systems*, a doohickey in the paraconsistent style, where one can disable the logical rules on request: these things are not deductive in any honest sense.

in PROLOG is recessive, one is faced with a non-axiomatisable notion, the *closed world assumption* or *CWA*.

How come that such a horror could ever have been conceived? In the day of logic programming, resolution was presented as the refutation of an attempt at building a model of $\neg A$, the refutation of a refutation[15]. Too much joggling with negations and the model slipped out of our hands: provability was mistaken with truth in a professed « smallest Herbrand model ». Which is factually true for formulas $\exists x_1 \ldots \exists x_p (R_1 \wedge \cdots \wedge R_q)$, but which is a conceptual aberration. Such an error of perspective mechanically entailed a conceptual and technical catastrophe: the truth of a universal formula in a bizarre model, what is it, exactly? It is the *CWA*, a pretext for ruining a lot of paper and for discrediting an approach, on the whole, interesting and original.

It is however possible to axiomatise *negation as failure*, i.e., the fact that the algorithm halts for want of unifiers. It is « less worse » than *CWA*, but the result is nevertheless not very exciting. Indeed, it is natural to optimise proof-search by saying that, if one searches for $A \wedge B$, one should start with A: since one must search for both, which one to begin with is irrelevant, at least in case of success. But, at the level of failure, everything goes badly: in fact, if A fails and B neither succeeds nor fails, one will observe either a failure, or nothing, depending on whether one has written $A \wedge B$ or $B \wedge A$. A procedural negation should work in relation to *all* possible searches and not w.r.t. a specific optimisation.

4.E Kripke models

Definition 6 (Kripke models). A *Kripke model* is a non-empty partially ordered set $(I, \preccurlyeq)$, equipped with a relation $i \Vdash P$ between elements of I and propositional atoms. A *monotonicity* condition must be satisfied:

$$i \Vdash P \wedge i \preccurlyeq j \Rightarrow j \Vdash P \tag{4.2}$$

For each A and each $i \in I$ we define $i \Vdash A$:

$$\begin{array}{lll} i \nVdash \mathbf{0} & & \\ i \Vdash A \wedge B & \Longleftrightarrow i \Vdash A \wedge i \Vdash B & \\ i \Vdash A \vee B & \Longleftrightarrow i \Vdash A \vee i \Vdash B & \\ i \Vdash A \Rightarrow B & \Longleftrightarrow \forall j \succcurlyeq i (j \Vdash A \Rightarrow j \Vdash B), & \text{hence} \\ i \Vdash \neg A & \Longleftrightarrow \forall j \succcurlyeq i \ (j \nVdash A) & \end{array} \tag{4.3}$$

We can verify that, if A is provable and $i \in I$, then $i \Vdash A$. Conversely, let I_0 be the set of sets i of formulas which are *saturated*, i.e.:

[15]How to avoid mistakes with two negations? You are not without ignoring that...

- Consistent: $\mathbf{0} \notin i$.
- Stable: i is closed under logical consequence.
- Satisfying the disjunction property: if $A \vee B \in i$, then $A \in i$ or $B \in i$.

We define $i \Vdash P$ by $P \in i$; I_0 is ordered by inclusion, then:

Proposition 1. $i \Vdash A \Leftrightarrow A \in I$

which yields completeness, *modulo* a lemma (which uses König's lemma):

Lemma 1.1. *$A \Rightarrow B$ is provable iff B belongs to all saturated sets containing A.*

Predicate calculus requires domains $|\mathcal{M}_i|$ such that $i \preccurlyeq j \Rightarrow |\mathcal{M}_i| \subset |\mathcal{M}_j|$ and then:

$$\begin{aligned} i \Vdash \exists x A &\iff \exists a \in |\mathcal{M}_i| \; i \Vdash A[a] \\ i \Vdash \forall x A &\iff \forall j \succcurlyeq i \; \forall a \in |\mathcal{M}_j| \; j \Vdash A[a] \end{aligned} \tag{4.4}$$

We can give a complete interpretation of modal logic **S4** (Section 10.3.3) in Kripke models: monotonicity (4.2) is relinquished and classical connectives are interpreted « pointwise », e.g., $i \Vdash \neg A \quad \Leftrightarrow \quad i \nVdash A$; as to modalities:

$$\begin{aligned} i \Vdash \Box A &\iff \forall j \succcurlyeq i \;\; j \Vdash A \\ i \Vdash \Diamond A &\iff \exists j \succcurlyeq i \;\; j \Vdash A \end{aligned} \tag{4.5}$$

We lose monotonicity (4.2), which becomes an attribute of *necessity* « $\Box$ ».

« Algebraic semantics », of which Kripke models are the finial, are usually mediocre. Phase semantics is an exception (Section 10.1.7): to sum up, phase spaces enable one to distinguish between principles that can be stated at first order (as an algebraic constraint) and those that need a second-order wallop: they thus detect certain abuses, which is not the case of Kripke models.

Kripke models are incredibly compliant: they are broken watches, of which one can freely move the hands, so as to display the time one wants to see. Among all modal logics, one of the worst is without contest **S5**, based upon the axiom $\forall x \Box A \Rightarrow \Box \forall x A$, a pure waste of paper without *Hauptsatz*[16]. Believe it or not, this system is complete w.r.t. Kripke models with constant domains[17].

We might as well consider topological interpretations: each formula is interpreted by an open set in a topological space. $\mathbf{0}$, $\wedge$, $\vee$ respectively become $\emptyset$, $\cap$, $\cup$; implication is interpreted by the interior of $A^c \cup B$. Completeness is interpreted by the same structure and the same lemma: one topologises I by taking as basic

[16]This corresponds, *modulo* the translation of intuitionistic logic in **S4** (Section 10.3.3) to the faulty principle $\forall x \neg\neg A \Rightarrow \neg\neg \forall x A$.

[17]This is no longer the duality syntax/semantics, it's a criminal association!

open sets the $\mathcal{O}_A := \{i; i \in A\}$. **S4** corresponds to the abandonment of openness: $\Box A$ then becomes the interior and $\Diamond A$ the closure.

Beginners usually like those semantic diversions: it is helpful to see that the provability of $\neg\neg A$ means that A is a dense open set; on the compact set $[0, 1]$, the negation of the open set $[0, 1/2[$ is the open set $]1/2, 1]$: $A \vee \neg A$ is the whole space except for one point. Topological models are superior to Kripke models: thus, they balk in front of **S5** (a denumerable intersection of open sets is hardly open!). This is because they use a constant domain: this is obviously an interesting constraint.

But, contrary to natural deduction of which one never gets tired, one soon discovers the limits, the shallowness, of this approach. Indeed, it is in the functional, category-theoretic, world, that one should look for the real sense of intuitionism; we are moving precisely there.

Part II

Around Curry–Howard

Chapter 5
Functional interpretations

5.1 Proofs as functions

5.1.1 The « semantics » of proofs. Around 1930 appeared several explanations of logic. Tarski and his truth, but also an interpretation due to Kolmogorov, and independently Heyting[1]. One often uses for that matter the expression *semantics of proofs*.

The expression « semantics » conveys a large scale of significance: its semantics is confusing! Normally, this expression applies to the sense of an expression; a well-established use links it to « syntax » in an opposition syntax/semantics whose poles are *completeness* and *soundness*. *Semantics* has been dressed in all possible sauces, for instance the opposition between *denotational semantics* (the category-theoretic interpretation of logic) and *operational semantics* (the tiresome and unimaginative paraphrases of a programming language). Not to speak of *algebraic* semantics (the interpretation of a system in itself by changing the character style: syntax in *italics*, semantics in **boldface**) and more generally of a very creative activity trying to *obfuscate* the meaning, so as to produce one more useless paper (or PhD). This is why I prefer to confine the use of this dubious word to its original sense and use for the rest « interpretation », « explanation », which have the advantage of being clear and honest, without any subliminal background.

As to the importance of the choice of words, the fact that some of them are purposely chosen to ossify the discussion, see the appendix of Orwell's « 1984 »: the *newspeak*. An example of Orwellian misunderstanding is the expression « game semantics »: one (correctly) interprets logic, even if this is hardly more than a good metaphor, as a game between a player (who tries to prove) and an opponent (who tries to refute); the player therefore replaces syntax and the opponent replaces semantics. But to style this a semantics amounts to introducing a second interpretative layer, a methodological confusion which forgets that it is at the level of the opponent – and not at that of the game – that one should seek the *semantics*. In other words game semantics would be a semantics of « semantics », a « meta-semantics »: the expression carries a whole conception of the world.

I know that my insistence on *giving a purer sense to the words of the tribe*[2] can sometimes be irritating. But it is one of the lessons I took from Kreisel: try to use the right word. This is why, in this text, I am using « antinomy », an expression

[1]Most likely Brouwer, of whom Heyting was the assistant and who cordially hated logic.

[2]Donner un sens plus pur aux mots de la tribu (Mallarmé).

stronger than « paradox », to speak of Russell; which, by the way, restores the paradoxical status of Gödel's theorem. It is why, in this chapter at least, I try to respect a distinction between « application » and « function ». This is also why I banned « meta » from my vocabulary because of the essentialism it conveys. On the other hand, I made no attempts at modifying honest expressions; thus « syntax » which means *language* and nothing more.

5.1.2 The functional interpretation. What follows is indeed a *procedural* interpretation: one defines the expression « θ is a *proof* of A », for, say, the language of Heyting's arithmetic **HA**. In what follows, A is a closed formula and θ is an object whose formal status must be left as vague as possible. Any premature attempt at a formalisation[3] could only weaken the idea: it is much more than that.

Atom: if A is an atom $\mathbf{0}$, $t = u$, $t < u$, then θ is a proof of A iff A is true.

Conjunction: θ is a proof of $A \wedge B$ iff θ is a *pair* (θ_1, θ_2), where θ_1 is a proof of A and θ_2 is a proof of B.

Disjunction: θ is a proof of $A \vee B$ iff θ is a *pair* (i, θ_1), where, either $i = 1$ and θ_1 is a proof of A, or $i = 2$ and θ_1 is a proof of B.

Implication: θ is a proof of $A \Rightarrow B$ iff θ is an application associating to each proof θ' of A a proof $\theta(\theta')$ of B.

Universal quantification: θ is a proof of $\forall x A$ iff θ is an application which associates to each integer n a proof $\theta(n)$ of $A[\bar{n}/x]$.

Existential quantification: θ is a proof of $\exists x A$ iff θ is a pair (n, θ_1), where n is an integer and θ_1 is a proof of $A[\bar{n}/x]$.

A few remarks:

- It is not a matter of formal proofs. A formal proof is a sequence of symbols, by no way an application; it is rather an *interpretation* of formal proofs, or again the attempt at *explaining* logic out of a primitive material external to formalism.

- However this approach could, in disguise, be an alternative definition of formal proofs. This is tenable for all operations, except implication and universal quantification which refer to applications whose domain is not finite (neither definite in the case of implication). Kreisel's attempt to overcome this mismatch foundered into sectarianism (Section 5.A).

[3] E.g., reducing it to a recursive or category-theoretic interpretation.

- The cases of existence and disjunction (which are reminiscent of the well-known properties of system **LJ**) show that one has in mind cut-free, *explicit* proofs: one is quite far from the *deductive* world.

- The disjunctive clause does not only mean « a proof of A or a proof of B », it also says *which* one. This immediately induces, even in the finite case, an immense difference with semantics. Indeed, anything is a proof of $0=0$; but a proof of $0 = 0 \vee 0 = 0$ is not a proof of one or the other (in this case it would not matter); it is a pair (i, θ) where θ does not matter, but where i is a *bit* making a left/right choice. This is a radical novelty w.r.t. semantics; for instance, the not quite exciting Kripke models. The functional interpretation is not concerned with the raw fact of knowing that A is true, it says *how*: here, *leftwise* or *rightwise*.

It is out of the question to check this interpretation, which is not even *formalised*. Note however that the identity map $\mathrm{id}(\theta) := \theta$, which sends any proof of A to itself, is a proof of $A \Rightarrow A$. Similarly, one can justify the induction schema: if θ is a proof of $A[0]$ and ψ is a proof of $\forall z(A[z] \Rightarrow A[Sz])$, then φ, defined by recurrence by $\varphi(0) := \theta$, $\varphi(n+1) := \psi(n)(\varphi(n))$ is a proof of $\forall x A[x]$.

5.1.3 Blind spots. This definition leaves some blind spots. For instance, since $\mathbf{0}$ is false, θ is a proof of $\neg A := A \Rightarrow \mathbf{0}$ iff A has no proof at all; indeed, θ must send the proofs of A into the proofs of $\mathbf{0}$, i.e., into the empty set, which is possible only in case A has no proof at all; in which case θ does not matter. In particular, doubly negated formulas are quite mistreated by this interpretation. Either $\neg A$ has a proof (and no matter what is a proof of $\neg A$) or it is $\neg\neg A$ who has a proof (and no matter what is a proof of $\neg\neg A$). All one can object to my remark is poor: I am using the excluded middle, so my « meta » is inappropriate! Witness how this Mister Meta comes in time to tangle up the cards!

In the same way *recessive* formulas are mistreated. Indeed, if A has only bounded quantifiers, a truth computation for A induces a proof of A, so A has a « proof » iff it is true, *idem* for $\neg\exists x A$; in which case it does not matter if there is a proof of $\neg\exists x A$, but how do we know this? Only since it is true... In that case, the functional interpretation of proofs does not do better than Tarski: it is a pure tautology.

The functional interpretation is a matrix from which one can build interesting things: category-theoretic, ludic... interpretations, but that one should not take literally. One can by the way make it say almost anything: for instance, by taking a corny set-theoretic standpoint, one can associate a proof to any classical truth computation. The functional object thus obtained is, on the other hand, non-effective, non-computable and of no interest.

5.1.4 NJ functionally revisited. For want of something better[4], we shall keep the set-theoretic viewpoint and follow the rules of **NJ** stepwise. Here, we forget the arithmetic substrate; in other words, we assume that the interpretation of non-logical atoms has been given to us. To each closed formula A we associate a set $|A|$, as follows:

$$\begin{aligned} |A \wedge B| &:= |A| \times |B|, \\ |A \vee B| &:= |A| + |B|, \\ |A \Rightarrow B| &:= |B|^{|A|}, \\ |\mathbf{0}| &:= \emptyset, \\ |\forall x A| &:= \prod_{d \in \mathbb{D}} |A[d/x]|, \\ |\exists x A| &:= \sum_{d \in \mathbb{D}} |A[d/x]|. \end{aligned} \tag{5.1}$$

We have used: the cartesian product, the disjoint sum $X + Y := \{1\} \times X \cup \{2\} \times Y$ and the space Y^X of all functions from X to Y. Quantifiers make use of sums and products indexed by a domain $\mathbb{D}$ supposedly given: we shall lose almost all interest in them[5]. Although the interpretation works well in this case, it is *second-order* (propositional) quantification which is of interest (Chapter 6).

To each deduction of a formula A in **NJ**, under the hypotheses Γ, one associates a *function* $\varphi[\cdot]$ sending arguments $\vec{x}^\Gamma$ chosen in the $|\Gamma|$ to the result $\varphi[\vec{x}^\Gamma] \in |A|$. We carefully distinguish, at least in this chapter, the notion of function, $\varphi[\vec{x}^\Gamma]$, from that of *application* (or map) $\varphi(\vec{x}^\Gamma)$. By a function, we quite mean a function, by an application, we mean – for want of anything better – its set-theoretic graph. Please, do not think that this is a matter of hairsplitting, of fly wings weighted in scales made of spider web thread: the distinction function/application is reminiscent of the distinction property/set, of which we know, by Russell's antinomy that it is explosive.

Hypothesis: the function associating to the argument x^A the result x^A.

Conjunction: this case makes use of the *pairing function* $(\cdot,\cdot)$, as well as the two *projections* π_l, π_r.

> **$\wedge$-introduction:** if the two premises of the rule have been interpreted by functions $\varphi_1[\vec{x}^\Gamma]$ and $\varphi_2[\vec{x}^\Gamma]$, then $\varphi[\vec{x}^\Gamma] := (\varphi_1[\vec{x}^\Gamma], \varphi_2[\vec{x}^\Gamma])$.
>
> **Left $\wedge$-elimination:** if the premise of the rule has been interpreted by a function $\varphi_1[\vec{x}^\Gamma]$, then $\varphi[\vec{x}^\Gamma] := \pi_l \varphi_1[\vec{x}^\Gamma]$.

[4] Categories, Section 7.2.

[5] However, see Section 6.A.2.

Right $\wedge$-elimination: if the premise of the rule has been interpreted by a function $\varphi_1[\vec{x}^\Gamma]$, then $\varphi[\vec{x}^\Gamma] := \pi_r\varphi_1[\vec{x}^\Gamma]$.

Disjunction: this case makes use of the *injections* ι_l, ι_r and the *conditional*.

Left $\vee$-introduction: if the premise of the rule has been interpreted by a function $\varphi_1[\vec{x}^\Gamma]$, then $\varphi[\vec{x}^\Gamma] := \iota_l\varphi_1[\vec{x}^\Gamma]$.

Right $\vee$-introduction: if the premise of the rule has been interpreted by a function $\varphi_1[\vec{x}^\Gamma]$, then $\varphi[\vec{x}^\Gamma] := \iota_r\varphi_1[\vec{x}^\Gamma]$.

$\vee$-elimination: if the three premises of the rule have been interpreted by functions $\varphi_1[\vec{x}^\Gamma]$, $\varphi_2[\vec{x}^\Gamma, y^A]$ and $\varphi_3[\vec{x}^\Gamma, z^B]$ then

$$\begin{aligned}\varphi[\vec{x}^\Gamma] &:= \varphi_2[\vec{x}^\Gamma, y^A] \quad \text{if } \varphi_1[\vec{x}^\Gamma] = \iota_l y^A,\\ &:= \varphi_3[\vec{x}^\Gamma, z^B] \quad \text{if } \varphi_1[\vec{x}^\Gamma] = \iota_r z^B.\end{aligned} \tag{5.2}$$

Implication: this case makes use of two operations which seem to go without saying, on one hand the association to a function $\psi[x]$ of an application, its *graph* $\lambda x\psi[x]$ (which is a set); on the other hand the possibility to *apply* a functional graph g (element of Y^X) to an element $x \in X$, yielding an element $g(x) \in B$.

$\Rightarrow$-introduction: if the premise of the rule has been interpreted by a function $\varphi_1[\vec{x}^\Gamma, y^A]$, then $\varphi[\vec{x}^\Gamma] := \lambda y^A\varphi_1[\vec{x}^\Gamma, y^A]$.

$\Rightarrow$-elimination: if the two premises of the rule have been interpreted by functions $\varphi_1[\vec{x}^\Gamma]$ and $\varphi_2[\vec{x}^\Gamma]$, then $\varphi[\vec{x}^\Gamma] := \varphi_2[\vec{x}^\Gamma](\varphi_1[\vec{x}^\Gamma])$.

Absurdity: this case uses the canonical inclusion $\emptyset$ of the empty set into anything.

0-elimination: if the premise of the rule has been interpreted by a function $\varphi_1[\vec{x}^\Gamma]$, then $\varphi[\vec{x}^\Gamma] := \emptyset\varphi_1[\vec{x}^\Gamma]$.

5.1.5 Occurrences, locative aspects. It is time to pause for an embrarrassing ambiguity: the two deductions

$$\dfrac{\dfrac{\dfrac{[A] \quad [B]}{A \wedge B}\,\wedge I}{B \Rightarrow A \wedge B}\,\Rightarrow I}{A \Rightarrow (B \Rightarrow A \wedge B)}\,\Rightarrow I \qquad\qquad \dfrac{\dfrac{\dfrac{[B] \quad [A]}{B \wedge A}\,\wedge I}{B \Rightarrow B \wedge A}\,\Rightarrow I}{A \Rightarrow (B \Rightarrow B \wedge A)}\,\Rightarrow I$$

correspond to $f(x)(y) := (x, y)$ and $g(x)(y) := (y, x)$. Now, make $A = B$:

$$\dfrac{\dfrac{\dfrac{[A] \quad [A]}{A \wedge A}\,\wedge I}{A \Rightarrow A \wedge A}\,\Rightarrow I}{A \Rightarrow (A \Rightarrow A \wedge A)}\,\Rightarrow I$$

The notation is ambiguous, since one cannot tell which *occurrence* of A has been discharged at the occasion of a $\Rightarrow$-introduction[6]. The best solution is due to De Bruijn: it consists in indexing each discharged hypothesis by means of an integer yielding the *relative* location of the discharging rule, i.e., how many steps below. The two previous cases will be distinguished by:

$$\dfrac{\dfrac{\dfrac{[A]_3 \quad [A]_2}{A \wedge A}\,\wedge I}{A \Rightarrow A \wedge A}\,\Rightarrow I}{A \Rightarrow (A \Rightarrow A \wedge A)}\,\Rightarrow I \qquad\qquad \dfrac{\dfrac{\dfrac{[A]_2 \quad [A]_3}{A \wedge A}\,\wedge I}{A \Rightarrow A \wedge A}\,\Rightarrow I}{A \Rightarrow (A \Rightarrow A \wedge A)}\,\Rightarrow I$$

This solution is excellent, especially in practice. I propose an explanation based on *locativity*, a recent idea, see [51]. Let us not forget that the goal of any deductive system is to prove formulas without context (hypotheses); in order to prove Φ, one uses subformulas of Φ: $A, B \dots$ and one must avoid mixing up the various *occurrences* of the same subformula A of Φ. I propose a radical solution: to *abolish the notion of occurrence*. Thus, in $A \Rightarrow (A \Rightarrow A \wedge A)$ one no longer deals with four *occurrences* of the *same*[7] subformula A: one has four *distinct* isomorphic subformulas, which one should note $A_1 \Rightarrow (A_2 \Rightarrow A_3 \wedge A_4)$. In what sense are they distinct? They are distinct since they occupy distinct *locations* of the arborescence of Φ:

$$\dfrac{1 \qquad \dfrac{2 \qquad \dfrac{3 \quad 4}{5}\,\wedge}{6}\,\Rightarrow}{0}\,\Rightarrow$$

We have indicated the other locations of subformulas by $0, 5, 6$: thus Φ is located in 0. The ambiguity as to occurrences disappears: one replaces equality with isomorphism. This poses a notational problem with the identity axiom, since it

[6] There are indeed four possibilities, since one can discharge both hypotheses at once.

[7] In the same way, twins are not occurrences of the same person, they are distinct, although similar, individuals.

has not the same location whether it is a hypothesis (1 or 2) or a conclusion (3 or 4). One should normally note A_4^2 the *delocation* in conclusion, in location 4 of the hypothesis located in 2. This is the most satisfactory *theoretical* solution. De Bruijn's notation fundamentally does the same thing: since it locates the discharged hypotheses in a formula located below.

This being said, in practice, we shall take no account of it. This sort of solution is akin to the bracket-free, styled *Polish*, notation: $\Rightarrow AB$ instead of $A \Rightarrow B$. Theoretically perfect, very good for machines, but we, poor human beings, lose our milestones. Our occurrence-free approach could be useful in relation to *subtyping*, an interesting, but sometimes confused, idea (Section 6.E).

5.2 Pure λ-calculus

5.2.1 A naïve function theory. The set-theoretic functional interpretation is quite brutal. One has difficulties in believing that logical operations may live in something as little constructive as set-theory. λ-calculus will provide us with another space of reference[8].

We can never repeat too much that nothing is more fecund than a mistake, provided one gets out of it. Originally, the inventors of this system sought a sort of « naïve function theory ». The correspondence sets/properties is declined into applications/functions – a nuance introduced here for the sake of pedagogy: by « application » I mean the functional object, while the function is rather the passage argument $\mapsto$ result. To the comprehension schema (1.1) « every property defines a set », corresponds a functional analogue « every function defines an application ». Here, it serves to clearly differentiate the functional dependency $f[a]$ (the value of f on a) from the applicative dependency, $f(a)$ (f applied to a). In the same way a property $P[a]$ induces a set $\{x; P[x]\}$, a function $f[a]$ induces an application $\lambda x f[x]$. The application corresponds to the function just like the set corresponds to the property

$$\lambda x f[x](a) = f[a], \tag{5.3}$$

and equality replaces logical equivalence.

This is obviously a bad start, since one knows that sets can be encoded by means of their characteristic functions: we shall be able to import Russell's antinomy. We have exact analogues of all the notions of naïve set-theory, except negation, which will come later; let us call it N. The literal translation of $\{x; \neg(x \in x)\}$ will be, once N has been determined, $A := \lambda x N(x(x))$. An immediate application of equation (5.3) yields

$$A(A) = N(A(A)). \tag{5.4}$$

[8]Speaking of references: Barendregt's book [9], the encyclopedia of λ-calculus.

Exit λ-calculus? No; however, if though recovered, there was, for a while, a definite hesitation among the inventors, Church, Kleene, Rosser... Eventually came a second thought: the logical setting was replaced with a *computational* setting, which amounts to a passage from essence to existence. If it is no longer a matter of total functions, but of *partial* functions, then « l'honneur est sauf »: one gets a fixed point for negation under the form of a *diverging* algorithm. More generally, since I was cautious in not making N precise, one sees that *any application admits a fixed point*. This is the fundamental principle behind *recursive* functions, i.e., functions defined in terms of themselves. A definition of the form $\Phi(x) := \ldots \Phi \ldots x \ldots$ will be expressed in λ-calculus as a fixed point of the expression $\lambda y \lambda x \ldots y \ldots x \ldots$

Originally a variant of set-theory, λ-calculus developed into a (theoretical) programming language. In general all relations with set-theory are very shallow and produce the wrong intuition. There is an incompatibility of temperament between set-theory and λ-calculus, between unbridled infinity and computability, between essence and existence.

5.2.2 Examples. Before proceeding further, let us indicate how to represent a certain number of data and functions.

Booleans. Truth values are traditionally represented by $\boldsymbol{v} := \lambda x \lambda y x$ and $\boldsymbol{f} := \lambda x \lambda y y$. This is because one can thus represent the *conditional* « if t then a else b » by $(t(a))(b)$. Indeed, $((\lambda x \lambda y x)(a))(b) = a$, while $((\lambda x \lambda y y)(a))(b) = b$. Hence one can define the boolean connectives; thus, negation by $N := \lambda z(z(\boldsymbol{f}))(\boldsymbol{v})$.

Note that these are not quite the logical truth values, but their close algorithmic analogues. It is indeed a matter of the answers one can give to a binary question (yes/no, left/right, *spin* up/down).

Pairs. It is possible to encode ordered pairs by means of $(a, b) := \lambda x(x(a))(b)$. This coding is operational, since it enables one to represent both projections by $\pi_l c := c(\boldsymbol{v})$, $\pi_r c := c(\boldsymbol{f})$: indeed, one verifies that $\pi_l(a, b) = a$, $\pi_r(a, b) = b$.

Natural numbers. Natural numbers are encoded by the *iterators*

$$\bar{n} := \lambda x \lambda y x(x(\ldots(x(y))\ldots)),$$

a.k.a. « Church integers », with n occurrences of x after the prefix $\lambda x \lambda y$: $\bar{n}$ sends x on $x \circ \cdots \circ x$. This coding is operational, since it enables one to represent definitions by *recurrence*, of the type $\varphi[0] := a$, $\varphi[n+1] := f[\varphi[n]]$: take $\varphi[z] := (z(\lambda w f[w]))(a)$. Note that the « successor » function is represented by $S[z] := \lambda x \lambda y(x(z(x)))(y)$. Since the fixed point enables the possibility of recursive (i.e., self-referential) definitions, it will be possible to represent any partial recursive function in λ-calculus. And nothing more, since the *expansive* nature of the equation of λ-calculus forces any partial numerical function representable in it to be recursive.

5.2.3 The Church–Rosser theorem

The syntax of λ-calculus. A change of standpoint enabled us to avoid a *logical* contradiction. But the system could still be *algorithmically* inconsistent. This would be the case if the use of equation (5.3) lead to absurdities, e.g., $\boldsymbol{v} = \boldsymbol{f}$. Observe that there is no walloping argument in favour of algorithmic consistency. The only one we could think of would be set-theoretic, but a set-theoretic interpretation of the type « application = graph » immediately requires naïve comprehension (1.1), which is contradictory.

We will eventually produce an *elementary* consistency proof, a real, a good one. Gödel has nothing to object to, since all potentially explosive logical artifacts have been carefully expelled from the λ-calculus.

We must first carefully define a *syntax*. *λ-terms* are constructed by means of three operations:

Variable: a variable $x, y, z, \ldots$ is a λ-term.

λ-abstraction: if t is a λ-term, then $\lambda x t$ is a λ-term. Observe that the variable x is bound. We should therefore not hesitate to rename it in case of ambiguity[9].

Application: if t, u are λ-terms, $(t)u$ is a λ-term.

Observe the notation: $(t)u$ instead of $t(u)$, while we really mean « the function t applied to the argument u ». This notation – due to Krivine – avoids avalanches of parentheses. W.r.t. the codings of the last section, this yields:

- $\lambda x((x)a)b$ for the pair (a, b), $(a)\lambda x \lambda y x$ for the left projection $\pi_l a$.
- $\lambda x \lambda y(x)(x)(x)y$ for the integer 3, $\lambda x \lambda y(x)((z)x)y$ for $S[z]$, and $\varphi[z] := ((z)\lambda w f[w])a$ for recurrence.

We proceed by defining an *immediate reduction*, a.k.a. *β-reduction*:

$$(\lambda x t)u \rightsquigarrow t[u/x]; \tag{5.5}$$

more precisely, we can replace in an arbitrary term, any part of the form $(\lambda x t)u$ (*redex*) with its *contractum* $t[u/x]$ (respecting the freshness of bound variables): this is immediate reduction. *Reduction* $\rightsquigarrow$ is defined as the reflexive and symmetric closure of immediate reduction.

[9] This bureaucratic baloney is the pretext to a specific rule, α-conversion; better to ignore it, this leads nowhere.

One-step reduction. The Church–Rosser theorem establishes computational consistency by means of confluence (Theorem 14, Section 4.3.4). The idea is to express reduction as the reflexive/transitive closure of a confluent *one step reduction* $\leadsto_1$.

Lemma 17.1. *If* $\leadsto_1$ *is confluent, the same is true of* $\leadsto$.

Proof. If $t = t_{00} \leadsto_1 t_{10} \leadsto_1 \dots t_{m0} = t'$ and $t = t_{00} \leadsto_1 t_{01} \leadsto_1 \dots t_{0n} = t''$, add new pieces t_{ij} such that $t_{i+1j}, t_{ij+1} \leadsto_1 t_{i+1j+1}$, as in a construction set. Eventually, the ultimate piece $t_{mn} = t'''$ is such that $t', t'' \leadsto t'''$. □

One-step confluence. The obvious candidate for one-step reduction is immediate reduction. But this does not work. Take for instance $(\lambda x(x)x)(\lambda xx)\lambda xx$, which contains two *redexes*, call them r, s. By reducing r, we get $((\lambda xx)\lambda xx)(\lambda xx)\lambda xx$, i.e., $(s)s$; by reducing s we get $((\lambda(x)x)\lambda xx)$. These two terms are indeed confluent on $(\lambda xx)\lambda xx$, but respectively in two and one steps. The idea of Tait – to whom this proof is due – consists in allowing the simultaneous reduction of a *family* of *redexes*. Which is not the same as the unlimited iteration of immediate reduction: in the previous example, we can at most simultaneously reduce r, s, which will lead to $(\lambda xx)\lambda xx$.

Definition 7 (Reduction of a family). If $\mathcal{F}$ is a family (set) of *redexes* of t, one denotes by $t \leadsto_{\mathcal{F}} u$ the result of the *simultaneous* reduction of all *redexes* of $\mathcal{F}$.

Although not defined pedantically, this informal definition hides no ambiguity. Now, if I reduced a family $\mathcal{F}$ of t, $t \leadsto_{\mathcal{F}} u$ and if $\mathcal{G}$ is another family of t, I consider $\mathcal{G} \setminus \mathcal{F}$ in u. The notation $\mathcal{G} \setminus \mathcal{F}$, after the reductions have been performed, is of course abusive; indeed these *redexes* have « images » in u, which are sometimes duplicated, sometimes bluntly erased: thus, if $\mathcal{F} = \{r\}, \mathcal{G} = \{s\}$, $\mathcal{G} \setminus \mathcal{F}$ is made of two distinct *redexes* of u, i.e., the two *occurrences* of s in $s(s)$.

Lemma 17.2. *If* $t \leadsto_{\mathcal{F}} u \leadsto_{\mathcal{G} \setminus \mathcal{F}} v$, *then* $t \leadsto_{\mathcal{G} \cup \mathcal{F}} v$.

Proof. Immediate. □

In particular, if one defines one-step reduction as the reduction of an arbitrary family, the notion is confluent.

Computational consistency

Theorem 17 (Church–Rosser theorem). *The reduction of* λ*-calculus is confluent.*

Proof. Immediate from the previous lemmas. □

Corollary 17.1. *The relation defined by* $t = u \ :\Leftrightarrow\ \exists v\ t, u \leadsto v$ *is an equivalence.*

Proof. Transitivity is a consequence of Church–Rosser. □

Corollary 17.2. *λ-calculus is algorithmically consistent.*

Proof. Simply because equality (equivalence) between *normal* terms, i.e., without *redex*, is the identity. For instance $\boldsymbol{v}$ and $\boldsymbol{f}$ are not equal. □

Note that equality is an expansive notion: the more we reduce, the more equalities we get. This is why the functions representable in λ-calculus are partial recursive.

5.3 The Curry–Howard isomorphism

5.3.1 The simply typed λ-calculus. A set-theoretic analogy, for once not too catastrophic: after the discovery of the antinomies in the naïve theory, one proposed *typing*. It is a sort of *superego*[10] forbidding certain forms of logical *incest* of the kind $x \in x$. For instance, in a typed set-theoretic system like the *Principia Mathematica* of Whitehead & Russell, a set of type n can only belong to a set of type $n+1$. We observed that λ-calculus is algorithmically consistent; the superego is thus not supposed to guarantee consistency, it indeed ensures the *termination* of computations. It is still based on the prohibition of incest: a function cannot take itself as argument, apply to itself, like in $(f)f$. The discussion essence/existence will bounce into a discussion typed/pure λ-calculus. Therefore, do not say that the philosophical preamble of Chapter 1 was useless, or that it is a remake of oldies of the kind platonism/formalism...

The simplest typed λ-calculus comprises only one primitive, implication: in other terms, starting with atomic types, one can form new types only by use of implication. The type $A \Rightarrow B$ therefore designates the functions sending type A into type B. The rules of term formation are as follows:

Variable: a variable $x^A, y^A, z^A, \ldots$ is a term of type A.

λ-abstraction: if t is a term of type B, then $\lambda x^A t$ is a term of type $A \Rightarrow B$.

Application: if t, u are terms of respective types $A \Rightarrow B$ and A, $(t)u$ is a term of type B.

This system of terms is governed by a rewriting (typed β-conversion)

$$(\lambda x^A t)u \rightsquigarrow t[u/x] \tag{5.6}$$

of which we can predict without risk that it is confluent.

[10]Therefore subject to essentialist deviances.

5.3.2 Sum, product, exponential. Let us make this system work. Starting with a constant $\boldsymbol{o}$[11], I can define types $\boldsymbol{n}$ by $\boldsymbol{n+1} := \boldsymbol{n} \Rightarrow \mathbf{n}$. One can then write a typed version of Church integers in all types $\boldsymbol{n+2}$: one defines $\bar{p}^{\boldsymbol{n+2}} := \lambda x^{\boldsymbol{n+1}} \lambda y^{\boldsymbol{n}}(x)(x)\ldots(x)y$. Then we introduce binary functions: $+, \times, \cdot^{\cdot}$, yielding a result of type $\boldsymbol{n+2}$ when their first argument is of type $\boldsymbol{n+2}$ and their second argument is of respective type $\boldsymbol{n+2}, \boldsymbol{n+2}, \boldsymbol{n+3}$:

$$\begin{aligned} X + Y &:= \lambda x^{\boldsymbol{n+1}} \lambda y^{\boldsymbol{n}}((X)x)((Y)x)y, \\ X \times Y &:= \lambda x^{\boldsymbol{n+1}}(X)(Y)x, \\ X^Y &:= (Y)X. \end{aligned} \tag{5.7}$$

The sum is the composition of a function iterated p times with a function iterated q times, it is therefore an iteration $p+q$ times. Their product consists in iterating p times a function iterated q times, it is therefore an iteration $p \cdot q$ times. Finally, the exponential consists in iterating q times the p times iteration, it is therefore an iteration p^q times. Which is confirmed by computation, e.g. $(\bar{q}^{\boldsymbol{n+3}})\bar{p}^{\boldsymbol{n+2}} \rightsquigarrow \overline{p^q}^{\boldsymbol{n+2}}$.

In particular, we see that the function $2^{2^{\cdot^{\cdot^{2^x}}}}$ can be represented by a term of type $\mathbf{2}$ depending on a variable x of type $\boldsymbol{k+2}$, where k is the height of the tower of exponentials. This shows, *modulo* Curry–Howard, that the bounds found in Section 4.C.4 for natural deduction cannot essentially be improved.

5.3.3 The isomorphism. The Curry–Howard isomorphism enunciates the equivalence, total, complete, between two viewpoints:

Natural deduction: formulas A, deductions of A, normalisation in natural deduction.

Typed calculus: types A, terms of type A, normalisation in typed λ-calculus.

This presents no difficulty. We must only be careful with matters of *occurrences*. One can use a notation in the style of De Bruijn[12].

Conceptually speaking, the locative viewpoint of Section 5.1.5, transposed in the world of λ-calculus, would yield something like:

For each type, there is only one variable.

On the other hand, the basic primitive, the variable, becomes a *delocation*, i.e., the operation replacing x^A (argument of type A) with its isomorphic image of type A'. As before, it is a splendid theoretical idea that one will ignore in practice for questions of legibility.

[11]Not to be confused with $\mathbf{0}$.

[12]Originally introduced for λ-calculus.

Curry–Howard is not a mere correspondence, it is really a matter of isomorphic *structures*. It is surprising if one keeps in mind that the two domains developed in parallel, on one side the typed calculus, on the other side natural deduction and sequents. With system **F** and the forgetful functor, we shall discover in Section 6.1.2 the unexpected (and unpremeditated) correspondence between the logical translations à la Dedekind and the codings in λ-calculus

The main idea of the isomorphism, i.e., the correspondence between functional terms and proofs, in the limited, but essential, context of implication, is due to Curry. Howard extended it to all logical connectives and, above all, replaced the obsolete Hilbert-style systems used by Curry (in relation with *combinators*, Section 5.B) with Gentzen-style formulations.

5.3.4 Strong normalisation

Definition 8 (Strong normalisation). A term t is *strongly normalisable* (sN) if the supremum $|t|$ of the lengths of all sequences of immediate reductions starting from t is finite.

In particular, t is normal when $|t| = 0$. A weaker definition is that any sequence of immediate reductions is finite; since one has at each step the choice between a finite number of *redexes*, this weaker version is – *modulo* König's lemma « a well-founded tree with finite branchings is finite » – equivalent to the stronger version we have given. This strong version turns out to be much more manageable, typically for matters of formalisation.

We postpone to the next chapter, for questions of thematic coherence, the proof of strong normalisation. Due to the Curry–Howard isomorphism, this result implies the strong normalisation of **NJ**, for which we so far only have *weak* normalisation (Theorem 15, Section 4.3.5).

5.A Kreisel and functional interpretation

We already noticed the infinite (and indefinite) character of the explanation of implication and universal quantification: we may have a proof without being able to *verify* its status (which would at least require an infinite process): this is typically the case for recessive formulas (Section 5.1.3). This is why we tried to « improve » these two cases by adding a second component to the application θ: a proof that it « actually does what it is supposed to do ».

That's pure baloney; indeed, let us take a true recessive formula $\neg\exists x A$. As soon as this formula is true, the actual choice of a proof hardly matters! Thus a proof $\neg\exists x A$ becomes the pair of an object θ devoid of any interest and... of a proof of $\neg\exists x A$.

Kreisel therefore proposed in [67] to remove this circularity by requiring the auxiliary proofs to be formal ones, moreover in a formal system $\mathcal{T}$ *given in advance*. In return for which, he was able to establish a sound correspondence between the functional notion and formal proofs in $\mathcal{T}$. Nothing astonishing in that, since it is more or less *ad hoc*.

This is the way the best ideas are killed. This sort of Kreiselic regression[13] unleashed passions, especially fanatic adhesions. This is regrettable and especially ludicrous if one keeps in mind that all this stays at the level zero of reflexion: who can *seriously* believe in such a two-bit explanation? This is presumably because it was indefensible that the Kreiselians (Troelstra, etc.) made a case of excommunication out of this matter..

5.B Combinatory logic

This variant of λ-calculus[14] is a calculus without variables, based on the sole application and two primitive *combinators* K and S for which one writes the immediate reductions:

$$\begin{aligned} ((K)a)b &\rightsquigarrow a, \\ (((S)a)b)c &\rightsquigarrow ((a)c)(b)c. \end{aligned}$$

They can be encoded in λ-calculus by $K := \lambda x \lambda y x$ and $S := \lambda x \lambda y \lambda z((x)z)(y)z$. One can conversely code λ-calculus in combinators, by representing λ-abstraction. For instance the identity $\lambda x x$ becomes $((S)K)K$.

Obviously this works as well in the typed case. With Curry–Howard in mind, we immediately see that this coding of λ-abstraction is the analogue of the *deduction theorem* of Section 3.1.2 which enables one to translate hypothetical reasoning (i.e., terms with variables) into proofs « à la Hilbert » (i.e., combinators). This intuition is confirmed by the fact that the usual axioms for implication in a Hilbert-style system are

$$A \Rightarrow (B \Rightarrow A), \tag{5.8}$$

$$(A \Rightarrow (B \Rightarrow C)) \Rightarrow ((A \Rightarrow B) \Rightarrow (A \Rightarrow C)), \tag{5.9}$$

both inhabited by typed variants of the combinators K and S, $\lambda x^A \lambda y^B x$ and $\lambda x^{A \Rightarrow (B \Rightarrow C)} \lambda y^{A \Rightarrow B} \lambda z^A((x)z)(y)z$. By the way, the proof of the deduction theorem begins with deducing $A \Rightarrow A$ by means of two *Modus Ponens* by making $B := A$, next $B := A \Rightarrow A$ in (5.8), $B := A \Rightarrow A$, $C := A$ in (5.9), which yields, *modulo* Curry–Howard, a typed variant of $((S)K)K$.

[13] In the same style, he required applications to be recursive.

[14] Misnamed, since λ-calculus carefully expelled any logical content.

Combinatory logic remains globally inferior to λ-calculus. We can understand it by comparing Curry to Howard in « Curry–Howard »: the ancient version puts in relation combinators and Hilbert-style systems, the modern one λ-terms and natural deduction. There is the same nuance between combinators and λ-terms as between Hilbert-style proofs and natural deduction.

5.C Other connectives

The Curry–Howard isomorphism extends to the other logical connectives. One should use $\times$ for conjunction, $+$ for disjunction, but, keeping in mind the isomorphism, I prefer to stick to the logical notation.

5.C.1 Conjunction

Pairing: if t and u are terms of respective types A and B, then (t, u) is of type $A \wedge B$.

Left projection: if t is of type $A \wedge B$, then $\pi_l t$ is of type A.

Right projection: if t is of type $A \wedge B$, then $\pi_r t$ is of type B.

We introduce the immediate reductions:

$$\begin{aligned} \pi_l(t, u) &\rightsquigarrow t, \\ \pi_r(t, u) &\rightsquigarrow u. \end{aligned} \tag{5.10}$$

Curry–Howard is immediately extended; Church–Rosser still holds.

5.C.2 Disjunction

Left inclusion: if t is of type A, then $\iota_l t$ is of type $A \vee B$.

Right inclusion: if t is of type B, then $\iota_r t$ is of type $A \vee B$.

Conditional: if t, u, v are of respective types $A \vee B$, C and C, if x^A, y^B are variables of respective types A, B, then $\delta(x^A u)(y^B v)t$ is of type C.

The reduction rules are manifold, because of the commutations[15]

$$
\begin{aligned}
\delta(x^A u)(y^B v)\iota_l t &\rightsquigarrow u[t/x^A],\\
\delta(x^A u)(y^B v)\iota_r t &\rightsquigarrow v[t/y^B],\\
(\delta(x^A u)(y^B v)t)w &\rightsquigarrow \delta(x^A(u)w)(y^B(v)w)t,\\
\{\delta(x^A u)(y^B v)t\}C &\rightsquigarrow \delta(x^A\{u\}C)(y^B\{v\}C)t,\\
\pi_l\delta(x^A u)(y^B v)t &\rightsquigarrow \delta(x^A\pi_l u)(y^B\pi_l v)t,\\
\pi_r\delta(x^A u)(y^B v)t &\rightsquigarrow \delta(x^A\pi_r u)(y^B\pi_r v)t,\\
\delta(z^C r)(w^D s)\delta(x^A u)(y^B v)t &\rightsquigarrow \delta(x^A\delta(z^C r)(w^D s)u)(y^B\delta(z^C r)(w^D s)v)t.
\end{aligned}
\tag{5.11}
$$

5.C.3 Absurdity

Empty: if t is of type $\mathbf{0}$, then $\emptyset^A t$ is of type A.

There are only commutative reductions:

$$
\begin{aligned}
(\emptyset^{A\Rightarrow B}t)u &\rightsquigarrow \emptyset^B t,\\
\{\emptyset^{\forall XA}t\}B &\rightsquigarrow \emptyset^{A[B/X]}t,\\
\pi_l\emptyset^{A\wedge B}t &\rightsquigarrow \emptyset^A t,\\
\pi_r\emptyset^{A\wedge B}t &\rightsquigarrow \emptyset^B t,\\
\delta(x^A u)(y^B v)\emptyset^{A\vee B}t &\rightsquigarrow \emptyset^C t,\\
\emptyset^A\emptyset^{\mathbf{0}}t &\rightsquigarrow \emptyset^A t.
\end{aligned}
\tag{5.12}
$$

Between us, this is not of the utmost interest.

5.C.4 Quantifiers. The same works for quantifiers: this amounts to writing in a functional setting things we already encountered in natural deduction and which resembles the cases conjunction/disjunction… So, should we dispense with it?

5.C.5 Normalisation. Anticipating the next chapter (Section 6.2.1) let us indicate how to adapt the proof of strong normalisation to the conjunctive case. One first extends simplicity.

[15]The fourth rule anticipates the next chapter: it a commutation corresponding to system **F**.

Definition 9 (Simplicity). t is simple when it does not begin with λ, Λ or $(\cdot,\cdot)$.

The definition of reducibility is adapted too:

Conjunction: a term t of type $A \wedge B$ is reducible iff $\pi_l t$ and $\pi_r t$ are reducible.

Everything adapts, *mutatis mutandis*: this is the sign of a well-lubricated machinery.

As to disjunction (and marginally, absurdity), everything becomes complicated because of commutative reductions. I did not find a truly nice proof taking little space. Since strong normalisation for disjunction is an extremely marginal and outdated question, a technique closed on itself, I propose to ignore the question.

5.D Martin-Löf's type theory

Martin-Löf's system [80] is the only logical system integrating the Curry–Howard isomorphism in its syntax. It is based upon *dependent types* (coming from the system *Automath* of De Bruijn), $\Pi x \in A\, B[x]$ and $\Sigma x \in A\, B[x]$. There is an *intended* ambiguity between two readings:

Logical: $\forall x \in A\, B[x]$, $\exists x \in A\, B[x]$, two most welcome combinations $\forall/\Rightarrow$, $\exists/\wedge$. The formula $t \in A$ reads « t is a proof of A ».

Functional: a matter of dependent products and sums, therefore subject to the usual functional interpretations.

The main originality of the system is therefore its handling of first-order terms. They don't refer to an external domain, but to the proofs themselves.

$$\frac{\begin{array}{c}[x \in A]\\ \vdots\\ t \in B[x]\end{array}}{\lambda x t \in \Pi x \in A\, B[x]}\ (\Pi I) \qquad\qquad \frac{\begin{array}{c}\vdots\\ u \in A\end{array}\qquad \begin{array}{c}\vdots\\ t \in \Pi x \in A\, B[x]\end{array}}{(t)u \in B[u]}\ (\Pi E)$$

which yields the immediate reduction

$$\begin{array}{c}\dfrac{\begin{array}{c}\vdots\\ u \in A\end{array}\qquad \dfrac{\begin{array}{c}[x \in A]\\ \vdots\\ t \in B[x]\end{array}}{\lambda x t \in \Pi x \in A\, B[x]}\ (\Pi I)}{(\lambda x t)u \in B[u]}\ (\Pi E)\\ \vdots\end{array} \quad\rightsquigarrow\quad \begin{array}{c}\vdots\\ u \in A\\ \vdots\\ t[u/x] \in B[u]\\ \vdots\end{array}$$

The same for the dependent sum:

$$\dfrac{\begin{matrix}\vdots & \vdots\\ t\in A & u\in B[t]\end{matrix}}{(t,u)\in \Sigma x\in A\; B[x]}\,(\Sigma I) \qquad \dfrac{\begin{matrix} & [x\in A]\quad [y\in B[x]] \\ \vdots & \vdots \\ t\in\Sigma x\in A\; B[x] & v\in C\end{matrix}}{\delta(xyv)t\in C}\,(\Sigma E)$$

which yields the immediate reduction

$$\begin{matrix}\dfrac{\dfrac{\begin{matrix}\vdots & \vdots\\ t\in A & u\in B[t]\end{matrix}}{(t,u)\in \Sigma x\in A\; B[x]}\,(\Sigma I) \quad \begin{matrix}[x\in A]\quad [y\in B[x]] \\ \vdots \\ v\in C\end{matrix}}{\delta(xyv)(t,u)\in C}\,(\Sigma E) \\ \vdots\end{matrix} \quad\rightsquigarrow\quad \begin{matrix}\vdots & \vdots \\ t\in A & u\in B[t] \\ & \vdots \\ & v[t/x,u/y]\in C \\ & \vdots\end{matrix}$$

As well as, if one insists, the appropriate commutative reductions.

The original Martin-Löf system harboured primitives coming from system **F**. After I found an antinomy in the system (1971), the second-order features were removed. In order to compensate for this want of expressivity, Martin-Löf added other primitives (universes, etc.), however, much less original and exciting than dependent products and sums.

In 1985, Coquand proposed a *theory of constructions* [16] more in conformity with the original intentions, e.g., encompassing system **F** (next chapter). This system is well-adapted to tasks of formalisation and verification and served as a basis for project *Coq* launched by Huet at *INRIA*.

Chapter 6
System F

My first work in logic [34], [35], freely inspired from Gödel's **T** (annex 6.C).

6.1 System F

6.1.1 Generalities. System **F**, contrary to simply typed λ-calculus, is constructed *around* Curry–Howard, as the isomorphic image of intuitionistic second-order *propositional* calculus. Where we say system « **F** », there is an ambiguity as to the basic connectives: one can choose either a minimal system (based on $\Rightarrow$, $\forall$) or richer systems, involving the connectives $\wedge$, $\vee$, $\mathbf{0}$, $\exists$. We shall opt for the minimal choice, this for two reasons:

- As usual, these other connectives induce bureaucratic complications.
- System **F** is so expressive that the missing connectives can be translated in the « basic » version, provided one relinquishes commutative conversions[1]; remember that the existence and disjunction properties hold without commutative conversions.

The *types* of system **F** are built from *type variables* $X, Y, Z, \ldots$ by means of implication and universal quantification: thus, $\forall X(X \Rightarrow X)$. The rules of term formation are those of the simply typed calculus (Section 5.3.1) to which have been added:

Generalisation: if t is a term of type A and if the type variable X is not free in the type of a free variable of t, then $\Lambda X t$ is a term of type $\forall X A$.

Extraction: if t is a term of type $\forall X A$ and if B is a type, then $\{t\}B$ is a term of type $A[B/X]$.

The restriction on generalisation is a direct consequence of Curry–Howard. Indeed, one can pass from $\Gamma \vdash A$ to $\Gamma \vdash \forall X A$ only if X is not free in Γ: remember that the left part of the sequent corresponds to active hypotheses, in other terms to free variables. By the way, what could be the meaning of $\Lambda X x^X$ where x is free, but without type (since X is a bound variable)? On the other hand, one can form $\Lambda X \lambda x^X x$, the « universal identity » of type $\forall X(X \Rightarrow X)$.

This system is equipped with an immediate reduction, given by (5.6) and

$$\{\Lambda X t\}B \rightsquigarrow t[B/X] \tag{6.1}$$

[1] This is not the case at first order where all connectives are independent.

which satisfies the Church–Rosser property, thus the unicity of the normal forms; existence is clearly less obvious.

6.1.2 The forgetful functor. To each term of system **F** one can associate a pure λ-term. For this, it suffices to erase anything connected to typing. In other terms, erase the types of variables: x^A, y^B, z^C becoming x, y, z, erase generalisations and extractions. Thus the term $(\lambda x^{\forall X(X \Rightarrow X)}(\{x\}\forall X(X \Rightarrow X))x)\Lambda X \lambda x^X x$ of type $\forall X(X \Rightarrow X)$, which reduces in three steps in $\Lambda X \lambda x^X x$, becomes $(\lambda x(x)x)\lambda xx$, which reduces in two steps in λxx. One sees in this example that oblivion preserves normalisation: the only thing that happens is that the steps (6.1) disappear.

A contrario, one can see a term of **F** as a *typing* of its « underlying λ-term ». If one considers the typing as a « superego » designed to avoid non-termination, the example we just gave shows that it surely does not prevent incest, since it briskly « types » $(x)x$: the worst can be feared.

6.1.3 Translation of connectives. We shall translate the logical connectives of certain data types. As to this matter, a very amazing phenomenon occurs: for all the cases under consideration, there is a logical tradition (dating back to Dedekind or Russell, nay Prawitz for the most recent), enabling one to translate proofs in second order, here in system **F**; we have, on the other hand, an independent tradition of coding in pure λ-calculus. *Modulo* Curry–Howard, we can see the translation in **F** as a *typing* of the codings of λ-calculus.

Conjunction. Define $A \wedge B := \forall X((A \Rightarrow (B \Rightarrow X)) \Rightarrow X)$. The corresponding operations become

$$\begin{aligned} (t,u) &:= \Lambda X \lambda x^{(A \Rightarrow (B \Rightarrow X))}((x)t)u, \\ \pi_l t &:= (\{t\}A)\lambda x^A \lambda y^B x, \\ \pi_r t &:= (\{t\}B)\lambda x^A \lambda y^B y. \end{aligned} \tag{6.2}$$

This translation respects the rules of reduction of conjunction (5.10). This is also a typing of the pairs of pure λ-calculus (Section 5.2.2).

Disjunction. Define $A \vee B := \forall X((A \Rightarrow X) \Rightarrow ((B \Rightarrow X) \Rightarrow X))$. The corresponding operations become

$$\begin{aligned} \iota_l t &:= \Lambda X \lambda x^{A \Rightarrow X} \lambda y^{B \Rightarrow X}(x)t, \\ \iota_r t &:= \Lambda X \lambda x^{A \Rightarrow X} \lambda y^{B \Rightarrow X}(y)t, \\ \delta(x^A u)(y^B v)t &:= ((\{t\}C)\lambda x^A u)\lambda y^B v. \end{aligned} \tag{6.3}$$

This translation respects the first two reduction rules of disjunction (5.11), but not commutative reductions. It is also a typing of the sum of pure λ-calculus, which we did not introduce, but which would have been

$$\begin{aligned} \iota_l t &:= \lambda x \lambda y(x)t, \\ \iota_r t &:= \lambda x \lambda y(y)t, \\ \delta(xu)(yv)t &:= ((t)\lambda xu)\lambda yv. \end{aligned} \tag{6.4}$$

Absurdity. Define $\mathbf{0} := \forall X X$. And

$$\emptyset^A t := \{t\}A. \tag{6.5}$$

This translation verifies nothing (there are only commutative rules); its only virtue is to exist!

Existence. Define $\exists X A := \forall Y(\forall X(A \Rightarrow Y) \Rightarrow Y)$. Existential types are not attractive enough to spend much time with them; one can toy with writing schemas of term construction corresponding to the rules

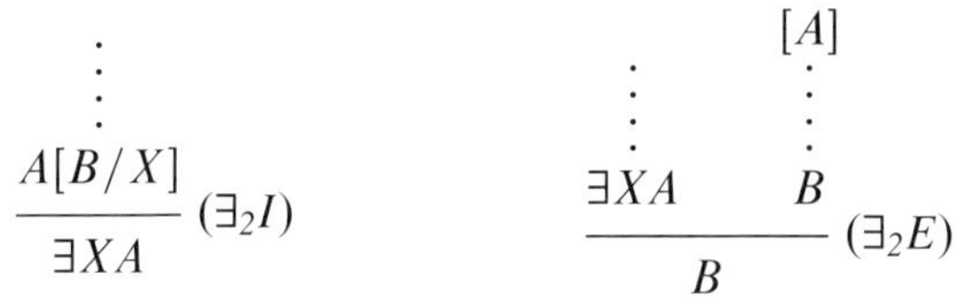

and the reduction corresponding to

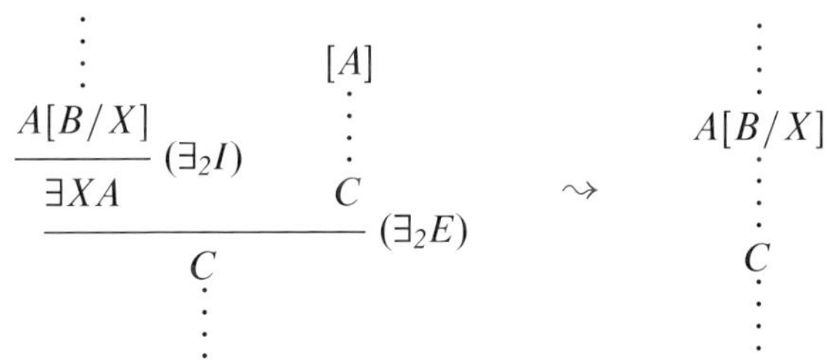

6.1.4 Another forgetful functor. Instead of resting upon a propositional system, we could have taken Takeuti's second-order predicate calculus (Sections 3.B.2 and 6.A.1) whose atoms are of the form $t \in X$ and which therefore uses two universal quantifications, a first-order one, another on predicates. This variant possesses a forgetful functor towards **F** which erases the first order: the atoms become type variables, first-order quantification disappears. In other words, one can successively forget the first order (which brings us in **F**), then the typing for good (one arrives in pure λ-calculus).

6.1.5 Translation of free structures. The second-order definition of integers: « the smallest set containing 0 and closed under S » is due to Dedekind; it generalises to the definition of the free[2] structure generated by functions $f_1, \dots, f_k$ of arbitrary arities and possibly taking their arguments in auxiliary types. Let us take an example, the free structure S generated by a constant $a \in S$, a binary function $f : S \times S \mapsto S$ and a binary function $g : S \times T \mapsto S$, where T is an auxiliary type. Following Dedekind, one would write

$$\begin{aligned} x \in S :\iff \forall X(a \in X \Rightarrow (\forall y \forall z(y \in X \Rightarrow (z \in X \Rightarrow f(y,z) \in X)) \\ \Rightarrow (\forall y \forall z(y \in X \Rightarrow (z \in T \Rightarrow g(y,z) \in X)) \Rightarrow x \in X))). \end{aligned} \tag{6.6}$$

In fact, to translate this free structure, it is enough to take the « underlying proposition »:

$$\forall X(X \Rightarrow ((X \Rightarrow (X \Rightarrow X)) \Rightarrow ((X \Rightarrow (T \Rightarrow X)) \Rightarrow X))). \tag{6.7}$$

In fact, it is what we already did for first-order connectives:

Conjunction: the free structure S generated by a binary function $g : T \times U \mapsto S$, where T, U are given.

Disjunction: the free structure generated by two unary functions $f : T \mapsto S$ and $g : U \mapsto S$.

Absurdity: the free structure generated by... nothing at all.

6.1.6 Translation of data types. The current data types (lists, trees, etc.) can be translated in system **F**, with, in each case, several *constructors* and one *destructor*.

Booleans. The free structure S generated by two constants $a, b \in S$:

$$\mathbf{bool} := \forall X(X \Rightarrow (X \Rightarrow X)).$$

The constructors are typings of the booleans of λ-calculus, $\Lambda X \lambda x^X \lambda y^X x$ and $\Lambda X \lambda x^X \lambda y^X y$. The destructor types the conditional: if v, a, b are of types **bool**, C, C, then « if v then a else b » is expressed by $(\{v\}C)a)b$, of type C.

By the way, observe that the type $\forall X(X \Rightarrow X)$ corresponds to a space with one point.

[2]System **F**, so comfortable with free structures, is rather helpless in front of quotients.

Natural numbers. Dedekind's definition, once first-order has been forgotten, yields: $\mathbf{nat} := \forall X(X \Rightarrow ((X \Rightarrow X) \Rightarrow X))$. The constructors are typings of the integers of λ-calculus: $\bar{0} := \Lambda X \lambda x^{X \Rightarrow X} \lambda y^X y$ of type **nat** and, given t of type **nat**, the successor $St := \Lambda X \lambda x^{X \Rightarrow X} \lambda y^X (x)((\{t\}X)x)y$ of type **nat**. The destructor enables one, given t, a, b of respective types $\mathbf{nat}, C, C \Rightarrow C$ to form $((\{t\}C)b)a$ of type C: this is the *iterator*.

We will remark the close relationship of this typing with our coding of integers in the simply typed calculus (Section 5.3.2). Indeed, our type could be written[3] $\forall X X + \mathbf{2}$; but while the simply typed calculus cannot type exponential as a binary function sending a type in itself, it is child's play for system **F**: for instance, if x, y are of type **nat**, $x^y := \Lambda X(\{y\}(X \Rightarrow X))\{x\}X$ types the exponential at the same type **nat**.

Binary integers. The previous integers are « Cro-Magnon integers », anterior to Babylonian numeration. A more modern version of integers requires finite sequences of zeros and ones. They can be represented by means of the type

$$\mathbf{bin} := \forall X(X \Rightarrow ((X \Rightarrow X) \Rightarrow ((X \Rightarrow X) \Rightarrow X))).$$

This case is very close to the previous one, since one has two successors: « add a 0 », « add a 1 », instead of the sole: « add a stroke ».

Lists. Finite lists of objects of type A can be represented by

$$\mathbf{list}(A) := \forall X(X \Rightarrow ((X \Rightarrow (A \Rightarrow X)) \Rightarrow X)).$$

This type will be studied in detail in the next section.

Binary trees. A last example, binary trees with leaves of type A:

$$\mathbf{bintree}(A) := \forall X(A \Rightarrow ((X \Rightarrow (X \Rightarrow X)) \Rightarrow X)).$$

6.1.7 Properties of the translations. We shall study the translations of connectives (existence excepted) and free structures. We will do it on one example, the most complex considered: $\mathbf{list}(A)$.

Constructors and destructors. Lists admit the following operations:

Empty list: encoded by the term $<> := \Lambda X \lambda x^X \lambda y^{X \Rightarrow (A \Rightarrow X)} x$.

Concatenation: if t is of type $\mathbf{list}(A)$ and a is of type A, one can form $t \frown a$ of type $\mathbf{list}(A)$: $t \frown a := \Lambda X \lambda x^X \lambda y^{X \Rightarrow (A \Rightarrow X)}((y)((\{t\}X)x)y)a$.

Destruction: given C and t, u, v of respective types $\mathbf{list}(A)$, C, $C \Rightarrow (A \Rightarrow C)$, one can form $((\{t\}C)u)v$ of type C by *destruction* of t.

[3] Almost: X and $X \Rightarrow X$ have been swapped for technical reasons (Section 7.4.2).

Constructors encode, *modulo* oblivion and Curry–Howard, the fact that the definition à la Dedekind of the structure S is actually a structure of the desired kind. The destructor codes the fact that this structure is free: if C is equipped with two constructors of the kind « empty list » and « concatenation », the free structure embeds (uniquely) in C. Less pedantically, one will see in the destructor a process of *iteration on lists*. It is important to remark that destruction involves a use of extraction.

One verifies the following reductions linked to the destructor:

$$\begin{aligned} ((\{<>\}C)u)v &\rightsquigarrow u, \\ ((\{t^\frown a\}C)u)v &\rightsquigarrow ((v)((\{t\}C)u)v)a. \end{aligned} \tag{6.8}$$

Representable objects. The list $< a_1, \ldots, a_k >$ of terms of type A will naturally be encoded by the term: $\Lambda X \lambda x^X \lambda y^{X \Rightarrow (A \Rightarrow X)}((y)((y) \ldots ((y)x)a_1 \ldots)a_{k-1})a_k$.

Conversely, assume that t is a term of type **list**(A); is it the case that t is of the form $< a_1, \ldots, a_k >$? One cannot say much, unless t is normal and closed. It will also be necessary to require A to be Π^1: remember that A is Π^1 when universal quantifiers occur only positively in it.

Let us introduce the notion of *head variable*, analogous to that of main hypothesis: when a normal term does not begin with a λ or a Λ, either it is a variable and we are done, or it is of the form $(t)u$ or $\{t\}B$ and we proceed with t; graphically, the head variable is the leftmost occurrence of a variable in the term.

Suppose that the type A is Π^1 (and closed) and that t, of type **list**(A), is normal and closed. Since it has no head variable, t begins with an introduction (i.e., with a Λ), say $t = \Lambda X u$ and the same for u, say $t = \Lambda X \lambda x^X v$. The variable x^X being of atomic type, it cannot play the role of head variable for v: we therefore get $t = \Lambda X \lambda x^X \lambda y^{X \Rightarrow (A \Rightarrow X)} w$. w is of atomic type X and therefore has necessarily a head variable. There are two possibilities, either $w = x^X$ and $t = <>$, and we are done. Or the head variable is $y^{X \Rightarrow (A \Rightarrow X)}$ in which case $w = ((y)w')a$ for a certain a of type A. Since A is Π^1, a cannot contain[4] the variables x, y:

Lemma 18.1. *Let s be a normal term whose type is Π^1 and depending on variables whose type is Σ^1 and maybe of « other variables » of respective types $X, A \Rightarrow X$ or $X \Rightarrow (A \Rightarrow X)$ where X is a variable occurring neither in the type of s, nor in the other variables. Then s does not depend on the « other variables ».*

Proof. Recurrence on s. The case where s begins with a λ or a Λ is immediate. If s admits a head variable z, then its type cannot be $X, A \Rightarrow X, X \Rightarrow (A \Rightarrow X)$, since otherwise the type of the conclusion would contain X. It is therefore one of the « non-other » variables, whose Σ^1 type is of the form $B_1 \Rightarrow (\ldots (B_n \Rightarrow Y))$, with

[4] On the other hand, $\lambda z^{\forall Y Y}(\{z\}(X \Rightarrow \forall Y Y))x$ yields an example of a normal term of (non-Π^1) type $\forall Y Y \Rightarrow \forall Y Y$ containing x.

$Y \neq X$: s therefore writes $(((z)s_1)\ldots)s_l$, with $l \leq n$. By induction hypothesis, no variable of type $X, A \Rightarrow X$ or $X \Rightarrow (A \Rightarrow X)$ occur in the terms $s_1, \ldots, s_l$, whose Π^1 types do not contain X. □

Therefore a is closed and $t = \Lambda X \lambda x^X w'^\frown a$. We proceed with w', etc. We eventually prove:

Theorem 18 (Representability). *If A is Π^1 and closed, any closed normal term of type* **list**(A) *can be written as a list $< a_1, \ldots, a_k >$ of (closed normal) terms of type A.*

This result is made more interesting by the remark that, if A is Π^1, **list**(A) is Π^1 too. Representability results hold for our other codings; this works in the same way, except that we must be cautious with a tiny, but irritating, technicality (Section 7.4.2).

Remark about conjunction. Applying literally our machinery to conjunction, one would indeed get the following destructor: given t, a of respective types $A \wedge B$, C, form $(\{t\}C)\lambda x^A \lambda y^B a$. The two projections are much more sympathetic. This « exception » to the rule can be understood through linear logic: conjunction is translated in the spirit of the tensor product $\otimes$, but since we are in a non-linear mode – intuitionism –, one can « cheat » and formulate destruction with the direct product &.

6.2 The normalisation theorem

Normalisation for system **F** is a big cake to swallow; I therefore begin with the simply typed case.

6.2.1 Simply typed case. We define, for each type A, the notion of a *reducible* term of type A (notion due to Tait, see [96]):

Atoms: for A atomic, a term t of type A is *reducible* iff it is sN (strongly normalisable).

Implication: a term t of type $A \Rightarrow B$ is *reducible* iff for all reducible terms u of type A, $(t)u$ (which is of type B) is reducible.

We will prove properties (R1)–(R3) of reducibility. (R1) and (R2) are sorts of eliminations; (R3) is rather an introduction, the idea being that if all immediate *contractums* of t are reducible, the same is true of t. Which is not tenable: this, combined with (R1), would yield an immediate equivalence between « reducible » and sN[5]. This is why (R3) is restricted to *simple* terms.

[5] Indeed all terms are reducible, so that the equivalence actually holds; but it is not provable. See also Section 6.E.2.

Definition 10 (Simplicity). A term t is *simple* when it does not begin with a λ.

Proposition 2. *Reducibility satisfies the following properties:*

(R1): *every reducible term is sN.*

(R2): *if t is reducible and $t \rightsquigarrow t'$, then t' is reducible.*

(R3): *if t is simple and all immediate* contractums *of t are reducible, then t is reducible.*

Proof. By induction on the type; the atomic case being obvious, one proceeds with the case of an implication $A \Rightarrow B$, assuming (R1)–(R3) to hold for the types A and B. When t is sN, remember that $|t|$ denotes the maximum length of a reduction sequence starting with t (Definition 8).

(R1): condition (R3) applied to type A shows that a variable x^A is reducible. If t of type $A \Rightarrow B$ is reducible, then $(t)x^A$ is reducible too and by (R1) sN. But $|t| \leq |(t)x^A|$, hence t is sN.

(R2): if t is reducible and $t \rightsquigarrow t'$ and if u of type A is reducible, then $(t)u$ is reducible and, since $(t)u \rightsquigarrow (t')u$, (R2) yields the reducibility of $(t')u$. t' is therefore reducible.

(R3): if t does not begin with a λ and all immediate *contractums* of t are reducible, take u reducible of type A; u is sN by (R1). By induction on $|u|$, one shows that $(t)u$ is reducible, which will establish that t is reducible. Since t does not begin with a λ, the immediate *contractums* of $(t)u$ are of the form $(t')u$, where t' is an immediate *contractum* of t or $(t)u'$, where u' is an immediate *contractum* of u. $(t')u$ is reducible by hypothesis and $(t)u'$ is reducible by induction hypothesis, since $|u'| < |u|$. By (R3), one concludes that $(t)u$ is reducible. □

In particular, one gets:

Proposition 3. *If t is a term of type B and if for all reducible u of type A, $t[u/x^A]$ is reducible, then $\lambda x^A t$ is reducible.*

Proof. Let u be reducible of type A; we must show that $(\lambda x^A t)u$ is reducible. Observe that t is reducible (take $u := x^A$), hence sN, just like u: one can work by induction on $|t| + |u|$. Since $(\lambda x^A t)u$ does not begin with a λ, one can try (R3): the immediate *contractums* of this term are of the form $(\lambda x^A t')u$ with $|t'| < |t|$, $(\lambda x^A t)u'$ with $|u'| < |u|$, reducible by induction hypothesis, or $t[u/x^A]$, reducible by hypothesis. We conclude by means of (R3). □

Theorem 19 (Reducibility). *All simply typed terms are reducible.*

Proof. One indeed shows, by induction on the term t, that, if one substitutes *reducible* terms of appropriate types for the free variables of t, then the result is reducible.

Variable: immediate.

λ-abstraction: by Proposition 3.

Application: by definition of reducibility.

This property really establishes the theorem, since, variables being reducible by (R3), the identical substitution yields a reducible term, t itself. □

Corollary 19.1 (Strong normalisation). *All simply typed terms are normalisable.*

Proof. Apply (R1). □

6.2.2 A faulty generalisation. We shall now « extend » the previous proof to system **F**; this faulty generalisation is indeed my first version of the proof.

The idea is to define reducibility for universal types by:

Universal quantification: t of type $\forall X A$ is *reducible* iff for all B $\{t\}B$ is reducible.

We prove the same properties as in the simply typed case. One must adapt *simplicity*: a simple term can begin neither with a λ nor with a Λ.

Indeed, the structure of the proof seems to work well:

Proposition 4. *Reducibility enjoys* (R1)–(R3).

For instance (R3) is rather simplified, since there are less immediate reducts for a term $\{t\}B$ than for a term $(t)u$. One would proceed with an analogue of Proposition 3:

Proposition 5. *If t is a term of type A whose free variables do not contain X and if $t[B/X]$ is reducible for all B, then $\Lambda X t$ is reducible.*

This is easily established. We would end with a proof in the style of Theorem 19. If t is a term of **F**, one substitutes types for its type variables, which yields $u := t[B_1, \dots, B_k/X_1, \dots, X_k]$. Then, in u, one substitutes reducible terms for typed variables. And everything works well, one must only consider two new cases:

Λ-abstraction: by Proposition 5.

Extraction: by definition of reducibility.

6.2.3 The fault. This definition is faulty. Indeed we are applying inductions on non-well-founded structures. More precisely we stumble on the fact that, at second order, there is no subformula property: the « definition » makes no sense, since, if t is of type $\forall X(X \Rightarrow X)$, $\{t\}B$ is of type $B \Rightarrow B$. We tried to say:

*t is **reducible** iff for any type B and any reducible term u of type B, $(\{t\}B)u$ is reducible.*

But we end in a circle: the reducibility of *a single* universal presupposes the reducibility of *all* types.

Even if we could define reducibility, the proof of (CR3) would stumble on the fact that we use an induction hypothesis on the types $A[B/X]$, more complicated than $\forall XA$. In other terms, the deductive chaining would be locally correct, but still an erroneous induction.

6.2.4 Reducibility candidates. There is a problem with substitution; the idea will be to define reducibility of type $A[B/X]$, not as the real reducibility (that we don't yet know), but from an « arbitrary definition » of reducibility of type B. Later on, once we have defined the « real » reducibility of type B, we will be able to use this notion and everything will fall into line (substitution lemma).

Definition 11 (Candidates). Let A be a type; a *reducibility candidate* (CR) of type A is a set $\mathcal{C}$ of terms of type A enjoying the following conditions:

(CR1): every term of type $\mathcal{C}$ is sN.

(CR2): if $t \in \mathcal{C}$ and $t \rightsquigarrow t'$, then $t' \in \mathcal{C}$.

(CR3): if t is simple and all immediate reducts of t are in $\mathcal{C}$, then t is in $\mathcal{C}$.

Observe that, for any type A, including type variables, there is a reducibility candidate, precisely sN_A, the set of strongly normalising terms of type A.

Let A be a type and let, for each free variable X_i of A, C_i be a type and $\mathcal{C}_i$ be a CR of type C_i. One defines the *parametric* reducibility $\mathrm{red}_A[\ldots \mathcal{C}_i/X_i \ldots]$, a property of terms of type $A[\ldots C_i/X_i \ldots]$. For reasons of legibility, the following definition is restricted to the case of a single variable X:

Variable: t of type C satisfies $\mathrm{red}_X[\mathcal{C}/X]$ iff $t \in \mathcal{C}$.

Implication: t of type $A[C/X] \Rightarrow B[C/X]$ satisfies $\mathrm{red}_{A\Rightarrow B}[\mathcal{C}/X]$ iff for all u such that $\mathrm{red}_A[\mathcal{C}/X]$, $(t)u$ satisfies $\mathrm{red}_B[\mathcal{C}/X]$.

Quantification: t of type $\forall YA[C/X]$ satisfies $\mathrm{red}_{\forall YA}[\mathcal{C}/X]$ iff for all D and all CR $\mathcal{D}$ of type D, $\{t\}D$ satisfies $\mathrm{red}_A[\mathcal{C}/X, \mathcal{D}/Y]$.

Let us see what our definition says in the case of $\forall X(X \Rightarrow X)$:

t is **reducible** *iff for all types B and all CR $\mathcal{B}$ of type B,*
if $u \in \mathcal{B}$, then $(\{t\}B)u \in \mathcal{B}$,

a mathematically correct definition.

Proposition 6. *Parametric reducibility enjoys* (R1)–(R3).

Proof. Let us establish it on one example, the type $\forall X(X \Rightarrow X)$:

X: reducibility $\mathrm{red}_X[\mathcal{C}/X]$ is exactly the appartenance to the CR $\mathcal{C}$ and (R1)–(R3) are then satisfied.

$X \Rightarrow X$: t satisfies $\mathrm{red}_{X\Rightarrow X}[\mathcal{C}/X]$ iff for all $u \in \mathcal{C}$, $(t)u \in \mathcal{C}$. (R1)–(R3) result from the previous case, just as the proof of Proposition 2.

$\forall X(X \Rightarrow X)$: properties (R1)–(R3) result from the previous case, as in « proposition » 4, the faulty generalisation of Proposition 2. □

6.2.5 The « real » reducibility. When A is closed, the parametric definition becomes absolute. Now, observe that, by the comprehension schema[6], reducibility (a property) will become a set (indeed a CR by Proposition 6). If $A[B/X]$ is closed, one has a choice between the direct definition of reducibility and the definition parametrised by the *real* reducibility of type B. These two definitions coincide, as a consequence of a more general property:

Theorem 20 (Substitution lemma). *The reducibilities* $\mathrm{red}_{A[B/Y]}[\mathcal{C}/X]$ *and* $\mathrm{red}_A[\mathcal{C}/X, \mathrm{red}_B[\mathcal{C}/X]/Y]$ *are equivalent.*

Proof. The lemma is more complex to write correctly than to prove. Here I also restricted to a single parameter X. This is trivial[7], but it formally uses the comprehension schema, which ensures that $\mathrm{red}_B[\mathcal{C}/X]$ defines a set of terms. □

The monstrous algorithmic complexity of system **F** is located in extraction and normalisability of extraction comes from the substitution lemma and therefore from comprehension.

6.2.6 The proof. An imitation of Proposition 3 yields:

Proposition 7. *If t is a term of type $A[\mathcal{C}/X]$ and if for all type D and all CR $\mathcal{D}$ of type D, $t[D/Y]$ enjoys* $\mathrm{red}_A[\mathcal{C}/X, \mathcal{D}/Y]$, *then $\Lambda Y t$ enjoys* $\mathrm{red}_{\forall YA}[\mathcal{C}/X]$.

[6] I pinpoint its formal use, it is the sort of thing one does not see otherwise.

[7] The proof of normalisation has been formally checked by Berardi. Against all expectations, the « trivial » substitution lemma posed difficulties to the machine… These animals definitely don't see the world like us!

One can conclude:

Theorem 21 (Strong normalisation for **F**). *All terms of* **F** *are strongly normalisable.*

Proof. The theorem is a consequence of a general result of parametric reducibility. Suppose that t is of type A, with free variables of types $B_1, \ldots, B_k$ and for each type variable, X_i, let us choose a type C_i and a CR $\mathcal{C}_i$ of this type. One performs the substitution C_i/X_i in t; let $b_1, \ldots, b_k$ be terms of respective types $B_1[\ldots C_i/X_i \ldots], \ldots, B_k[\ldots C_i/X_i \ldots]$, reducible in the sense of

$$\text{red}_{B_1}[\ldots \mathcal{C}_i/X_i \ldots], \ldots, \text{red}_{B_1}[\ldots \mathcal{C}_i/X_i \ldots].$$

Then $t[\ldots C_i/X_i \ldots, b_1/x^{B_1[\ldots C_i/X_i \ldots]}, \ldots, b_k/x^{B_k[\ldots C_i/X_i \ldots]}]$ is reducible in the sense of $\text{red}_A[\ldots \mathcal{C}_i/X_i \ldots]$.

This formulation is extremely painful, but it is nothing but what we did in the simply typed case: we only had to use a parametric reducibility. The proof splits without problem into five cases:

Variable: trivial.

λ: uses Proposition 3.

Application: uses the definition of parametric reducibility for implication.

Λ: uses Proposition 7.

Extraction: uses the definition of parametric reducibility for quantification as well as the substitution lemma. Formally, this part makes use of the comprehension axiom, since we must replace a property with the set of terms satisfying it. □

6.A Type theories

6.A.1 Takeuti's conjecture. The effective version of Takeuti's conjecture (Section 3.B.2) consists in showing that the replacement of a cut on $\forall XA$ with a cut on $A[T/X]$, in the key case corresponding to second-order quantification, produces a converging algorithm. *Modulo* appropriate translations à la Gödel, one can concentrate on second-order intuitionistic logic. And, *modulo* Curry–Howard, on an obvious analogue of system **F**, based on second-order *predicate calculus* instead of *propositional calculus*. In this variant, there is a first-order *generalisation* Λxt and a first-order *extraction* du $\{t\}d$ where x and d respectively stand for a first-order variable and term as well as the immediate reduction $\{\Lambda xt\}d \rightsquigarrow t[d/x]$.

One can bluntly imitate the proof of strong normalisation of **F** in this (slightly) more general setting. One can also protest against this sort of unimaginative generalisation and reduce Takeuti's conjecture to normalisation of system **F**. Starting with t of type A in the system « with $\forall_1$ », one forms t^- of type A^- in **F** by forgetting first order. If $t = t_0 \rightsquigarrow t_1 \rightsquigarrow \cdots \rightsquigarrow t_n \rightsquigarrow \cdots$ is a succession of immediate reductions, one sees that $t^- = t_0^- \rightsquigarrow t_1^- \rightsquigarrow \cdots \rightsquigarrow t_n^- \rightsquigarrow \cdots$: a succession of immediate reductions or « non-reductions » in **F**. But the clusters of non-reductions come from finite configurations $\{\ldots\{\{\Lambda x_n \ldots \Lambda x_2 \Lambda x_1 \Lambda\}t_1\}t_2 \ldots\}t_n$ and are therefore of bounded length. One concludes that $t = t_0 \rightsquigarrow t_1 \rightsquigarrow \cdots \rightsquigarrow t_n \rightsquigarrow \cdots$ is of finite length, hence that t is sN.

6.A.2 More on first-order quantification. I did not try to extend Curry–Howard to this case, since this leads to a faulty reading, in the style « quantification = infinite sum/product ». Indeed, whereas one actually proves a conjunction by proving both components, a quantification is not proved by establishing each particular case: otherwise we would never have witnessed the slightest proof! In other terms, $\forall x$ does not generalise $\wedge$; although there is a strong analogy.

I propose[8] to use a *variable of a domain*, say $\mathbb{D}$ and to consider that a first-order quantification is implicitly restricted to $\mathbb{D}$. Thus $\forall x \exists y A$ would indeed mean $\forall \mathbb{D} \forall x{\in}\mathbb{D} \exists y{\in}\mathbb{D}\, A$. This enables one to accept without reticence the interpretation by sum and product indexed by $\mathbb{D}$: the universal quantification on $\mathbb{D}$ will eventually, because of the uniformity it supposes, make disappear the infinite and unpleasant aspects of sum and product.

Of course, nothing opposes the consideration of several domain variables corresponding to several types of objects. As to this matter, one must be prepared for empty domains, which is usually forbidden; but this is only a matter of syntactical adjustments.

One can also step out of implicit quantification and quantify explicitly, for instance existentially, on domains.

6.A.3 Type theory à la Russell. Whitehead & Russell's type theory, the *Principia Mathematica* [101], is completely forgotten nowadays; not by historians, but by mathematicians, including those specialised in logic. We only remember the idea of avoiding antinomies in the style of... Russell by *typing*, which can be expressed by a predicate calculus of finite type. For instance, besides predicate variables, we could have variables for « predicates on predicates », etc. By writing the appropriate abstraction terms, for instance $\{X; a \in X\}$, we get a predicate calculus, which essentially corresponds to the theory of finite types à la Russell [89]. The method of reducibility candidates allows us to extend, without the least difficulty – nor the least creativity – Takeuti's conjecture to this setting.

Modulo propositional oblivion, we obtain an extension – not quite earth-shaking – of system **F**: besides propositional variables, we shall have variables of connectives, etc.

[8] At least in the absence of functional symbols: I didn't push this idea.

6.A.4 Gentzen at his worst. We just mentioned Russell's type theory. The basic type $\boldsymbol{o}$ is the type of natural numbers. This being said, if one does not specify the basic type, it becomes *obvious* that the theory is consistent: indeed, the type $\boldsymbol{o}$ can be interpreted by a one-element set $\{a\}$, the type $\mathbf{1}$ by its power set $\wp\{a\}$, the type $\mathbf{2}$ by the powerset $\wp\wp\{a\}$ of the previous, etc. Which yields a finite model, at least if one restricts to finitely many types.

To say that type theory – without axioms at type $\boldsymbol{o}$ – is consistent is therefore a triviality of little interest. This is however the object of a paper by Gentzen, [33], obviously a very bad one. This paper has nevertheless its place in our reflexion. If Gentzen had written: « the finite iteration of the power set of $\{a\}$ yields a model », the paper would not have been published. He instead associated – in a sort of cabalistic way – numbers to proofs, so as to show that the empty sequent cannot be proved. However, it is not too difficult to see that he actually enumerates the set $\wp\wp\wp\ldots\wp\{a\}$, performs substitutions: all this amounting to a non-avowed truth computation.

From a formal viewpoint, this « proof » respects a formalist ideal, which is to avoid the notion of truth[9], considered as suspect. But this is however what he constructs – without using the word: this is styled hypocrisy. This is especially ridiculous since it is hard to see what is dubious with truth in a finite model. This is no longer ideas that govern us, this is the words one uses to speak of them: call the dog a cat and she will start to meow.

6.A.5 From Martin-Löf to constructions. Note that abstraction terms in the theory of types à la Russell look very close to the terms of the simply typed calculus: compare $t \in \{x; A\} := A[t/x]$ and $(\lambda xu)t = t[u/x]$. One could therefore try to make type theory à la Russell « benefit » from a typing in the style of $\mathbf{F}$. In other terms, one would type sets like in system $\mathbf{F}$ and not uniquely by finite types. This is what I tried in 1971 in an unpublished work – this for an excellent reason: I found there an antinomy in the style of Burali-Forti.

This is what Martin-Löf was simultaneously doing with the first version of his system, in which I had no problem to import the contradiction I had found in my prototype.

The theory of constructions [16] appears as a reasonable compromise: it is a variant of Russell à la Martin-Löf, in which sets are typed by means of dependent sums and products which are quite simple, i.e., fundamentally of the same nature as usual simple types, but more flexible.

---

[9]The same hypocrisy is at work in the paper of Schütte [91]: the three-valued models are called *valuations*; in this way the Devil is not named.

6.B Heyting's arithmetic

6.B.1 Second-order translation. Heyting's arithmetic translates into second-order logic by relativising first-order quantifiers to $\mathbb{N}$, i.e., to Dedekind's definition of integers: which freely yields the induction schema. Note that only a marginal fringe of the abstraction terms is used, namely those translating arithmetical formulas.

Let us add to the system the axioms of **RR**, except what deals with the inequality $<$, that one prefers to translate:

- One can prove $x \in \mathbb{N}, x \in X \vdash \forall y \in \mathbb{N}(x = y \Rightarrow y \in X)$: induction on x, with a subordinate induction on y.
- Inequality is definable by $x < y \;:\Leftrightarrow\; \exists z\; y = S(x + z)$. One proves $x \in \mathbb{N}, y \in \mathbb{N} \vdash x < y \vee x = y \vee y < x$.
- In the style « useful exercise », one will toy with proving $x \in \mathbb{N}, y \in \mathbb{N} \vdash x = y \vee x \neq y$.

6.B.2 Existence and disjunction in HA

Theorem 22 (Existence and disjunction). *If a* closed *formula $A \vee B$ is provable in* **HA**, *then either A, or B is provable in* **HA**.

If a closed *formula $\exists x A$ is provable in* **HA**, *then $A[\bar{n}/x]$ is provable in* **HA** *for a certain n.*

Proof. Let R be the conjunction of the axioms of **RR** (without $\vee$), universally quantified[10]. If a closed formula $A \vee B$ is provable in **HA**, then $R \vdash A \vee B$ is second-order provable, hence cut-free provable. An easy induction, linked to the fact that R is first order, without existence nor disjunction, shows that $R \vdash A$ or $R \vdash B$ is second-order provable. □

6.C System T

System **T** is used to give functional interpretations of **HA**.

6.C.1 The *Dialectica* interpretation. Published in 1958 in a rather obscure journal, *Dialectica* [56], Gödel's functional interpretation dates back to the year 1943. This is an interpretation of interactive, « game », style, but the result hardly matches the ambitions.

The idea is to associate to a formula A of **HA** an interpretation of the form $\exists x^S \forall y^T a = 0$, where the quantifiers $\exists x^S$, $\forall y^T$ refer to finite-type functionals

[10]The quantifiers should not be relativised to $\mathbb{N}$.

above the integers. The types increase with the connectives, thus, if B has been interpreted by $\exists z^U \forall w^V b = 0$, the implication $A \Rightarrow B$ becomes

$$\exists Z^{S \Rightarrow U} \exists Y^{S \Rightarrow (V \Rightarrow T)} \forall x^S \forall w^V (a[((Y)x)w/y] = 0 \Rightarrow b[(Z)w/x] = 0).$$

One can, if one wants, regroup quantifiers of the same nature with the help of a product type; one can also bring back the implication $\cdots = 0 \Rightarrow \cdots = 0$ to an equality $\cdots = 0$.

The main drawback of this interpretation is a bad global structure. The existential quantifier proposes something against the refutations that could be brought by the universal quantifier. It is therefore a *positive* game, since the existential comes first; this is different from a *negative* game where the universal would come first, $\forall y^T \exists x^S a = 0$. This writing is well-adapted to the positive connectives $\exists$, $\vee$. But it works not that well in the negative case: indeed, an expression $\forall y^T \exists x^S a = 0$ is replaced with $\exists X^{S \Rightarrow T} \forall y^T a[(X)y/x] = 0$.

If polarity were taken into account, the implication between $\exists x^S \forall y^T a = 0$ (positive) and $\forall z^U \exists w^V b = 0$ (negative) could more modestly be written $\forall x^S \forall z^U \exists y^T \exists w^V (a = 0 \Rightarrow b = 0)$ without introducing those illegible functional dependencies which render *Dialectica* unmanageable. But should one try to fix half-baked ideas?

6.C.2 System T. System **T**, originally introduced for the sake of the *Dialectica* interpretation survived, while Gödel's interpretation no longer interests anybody. It is a simply typed calculus, based on a primitive type **nat**. The type **nat** has two constructors, corresponding to zero and successor; there is also a destructor, the recursor *Rta*, for each type A, see *infra*.

This system almost immediately translates into **F**. By the way, let us translate the induction schema of **HA** in second order: the induction step $\forall z(A[z] \Rightarrow A[Sz])$ translates into $\forall z \in \mathbb{N}(A[z] \Rightarrow A[Sz])$, which one can in turn transform, using $B := z \in \mathbb{N} \wedge A$, into $\forall z(B[z] \Rightarrow B[Sz])$. In other words, induction on A is established by means of a comprehension on B.

When one forgets first order, induction becomes *recurrence* : from a, t of types A, $\mathbf{nat} \Rightarrow (A \Rightarrow A)$, construct $u := Rta$ of type $\mathbf{nat} \Rightarrow A$ such that

$$\begin{aligned} (u)0 &\rightsquigarrow a, \\ (u)Sv &\rightsquigarrow ((t)v)(u)v. \end{aligned} \tag{6.9}$$

Which does not quite correspond to the iterator, the official destructor of **nat** in system **F**. This being said, the recursor of type A, which corresponds to induction via oblivion, can be reduced to the iterator of type $\mathbf{nat} \wedge A$... Only roughly, since there is however a small leak: typically, the recursor enables one to define a predecessor function satisfying $(p)Sx \rightsquigarrow x$ « one step predecessor », while the coding by the iterator only yields $(p)\overline{Sn} \rightsquigarrow \bar{n}$, a computation in n steps.

But, apart from this nuance between iterator and recursor, it is fair to say that **T** is translatable in **F**. In particular, **T** enjoys strong normalisation. Remember that this system was the prototype of **F**.

6.C.3 Realisability. We can try to make precise the functional interpretation of proofs of Section 5.1.2 by means of a notion of *realisability*, a technique mainly due to Kleene [66]. One simply takes the definition « θ is a proof of A » given in Section 5.1.2 and one replaces the notions with their precise definitions in λ-calculus: one knows how to encode integers, ordered pairs. One thus defines θ Ⓡ A when θ is a λ-term, for instance:

Implication: $\theta \circledR A \Rightarrow B :\Leftrightarrow \forall\theta' \, (\theta' \circledR A \Rightarrow (\theta)\theta' \circledR B)$.

Universal quantification: $\theta \circledR \forall x A :\Leftrightarrow \forall n \, (\theta)\bar{n} \circledR A[\bar{n}/x]$.

One can also used a typed, « modified », realisability; the types A^r of the realisers are as follow:

$$\begin{aligned} (A \wedge B)^r &= A^r \wedge B^r, \\ (A \vee B)^r &= A^r \vee B^r, \\ (A \Rightarrow B)^r &= A^r \Rightarrow B^r, \\ (\forall x A)^r &= \mathbf{nat} \Rightarrow A^r, \\ (\exists x A)^r &= \mathbf{nat} \wedge A^r. \end{aligned} \tag{6.10}$$

Observe that the type of the realiser is nothing but, once translated in **F**, the term obtained by forgetting first-order; and, given a proof, the associated realiser is the forgetful term corresponding to the proof. At least if we don't look too carefully at what happens at the level of atoms. To sum up, neglecting a few technicalities linked to atomic formulas:

- The typed realisability of **HA** in **T** is first-order oblivion.
- The realisability of **HA** in λ-calculus is type oblivion.

Would there be a propensity of logic for running in circles?

6.D Expressive power

6.D.1 Provably recursive functions. Suppose that one can prove in **HA** that a recursive function is total; from the proof of the Π^0_2 formula « f is total », one extracts[11] a definition of the function in system **T**.

[11]The methods are manifold, including those coming from the second proof of Gentzen [31].

Theorem 23 (Provably recursive functions of HA). *The provably recursive functions (Section 2.1.7) of* **HA** *are exactly the functions defined by a closed term of type* **nat** $\Rightarrow$ **nat** *of* **T**.

Proof. In one sense, by realisability. Conversely, given a closed term of type **nat** $\Rightarrow$ **nat**, one can formalise the reducibility proof given in the setting of **F**; due to the peculiarity of the situation, this formalisation can indeed be made in **HA**. □

Heyting's arithmetic[12] is a very expressive system; as to its second-order extension, little mathematics cannot be formalised in them – if one excepts results confined to set-theory. It is indeed extremely difficult to find a computable function which cannot be expressed in **T**; indeed the functions vertiginously increase when one toys with iteration:

(i) The iteration of function $n + 2$ yields $n.2$.

(ii) The iteration of function $n.2$ yields 2^n.

(iii) The iteration of function 2^n yields a tower of exponentials of variable height n. This function is already beyond the expressive power of the simply typed calculus.

(iv) The iteration of the previous function yields a monster which no longer makes much sense.

But we could do much worse. Instead of iterating a function, we could iterate the functional of iteration, this is the *Ackermann function*:

$$\begin{aligned} \varphi(x, 0) &:= x + 2, \\ \varphi(0, y + 1) &:= 1, \\ \varphi(x + 1, y + 1) &:= \varphi(\varphi(x, y + 1), y). \end{aligned} \tag{6.11}$$

This monster[13] uses iteration at type **nat** $\Rightarrow$ **nat**, a minuscule part of the possibilities! One understands that it is very difficult to find total recursive functions non-representable in **T**, not to speak of **F**. The only way to do so is based on an *abstract nonsense*: the function, which, to the code $\ulcorner t \urcorner$ of a closed term of type **nat** of **F** associates n such that $t \rightsquigarrow \bar{n}$, cannot be represented in **F** (diagonalisation exercise).

6.D.2 Peano vs. Heyting. One could think that **HA** is much weaker than **PA**. This is true in a certain sense: for instance, a Σ^0_2 formula provable in **PA** – of the kind « the equation $t = u$ has only finitely many solutions » – has little

[12] Or Peano's, *infra*.

[13] In the etymological sense of a thing that one shows: one does not use it, one *shows* it as an illustration of the expressive power of system **T**.

chance to be provable in **HA**: one would indeed need an *effective* proof, yielding – or rather inducing – a bound on the number N of solutions. But this is not a genuine weakness of **HA**: it is rather that formulas take a subtler sense.

Such a thing does not happen with simpler complexities, i.e., Π^0_1, Σ^0_1 and even Π^0_2: the two systems prove the same theorems of these shapes:

$\boldsymbol{\Pi^0_1}$**:** it is easy to verify that, if A uses only bounded quantifiers, then $A \vee \neg A$ is provable in **HA**; therefore A is provably equivalent to its Gödel translation A^g. Remembering that the Gödel translation does not require $\neg\neg$ before the negative connectives $\wedge$, $\multimap$, $\forall$, one concludes that any Π^0_1 formula is provably equivalent (in **HA**) to its Gödel translation. Since the Gödel translations of the induction axioms are still induction axioms, it follows that any Π^0_1 theorem of **PA** is provable in **HA**.

$\boldsymbol{\Sigma^0_1}$**:** in the same way, if the closed formula $\exists y A$ is provable in **PA**, one deduces that $\neg\neg\exists n A$ is provable in **HA**. Here an interesting remark of H. Friedman: one can replace, in Gödel's translation, the negation $\neg B$ with $\neg_0 B := B \Rightarrow A_0$, where A_0 is an arbitrary formula, in particular, $A_0 := \exists y A$. One deduces that $\neg_0\neg_0\exists y A$, i.e., $(\exists y A \Rightarrow \exists y A) \Rightarrow \exists y A$ is provable in **HA** and therefore $\exists y A$ as well.

$\boldsymbol{\Pi^0_2}$**:** this extends to Π^0_2 formulas: if $\forall x \neg\neg\exists y A[x, y]$ is provable in **HA** – and therefore in a finitely axiomatised subsystem $\mathcal{T}$ –, then for each integer n, one can prove, in the same $\mathcal{T}$, $\neg\neg\exists y A[\bar{n}, y]$; hence, by what precedes, $\neg\neg\exists y A[\bar{n}, y]$. This is formally provable in **HA**, hence, by reflexion of $\mathcal{T}$ in **HA** (Section 3.B.4, indeed, its intuitionistic version), one formally deduces that $\exists y A[n, y]$ is true. Which is a formal proof in **HA** of $\forall x \exists y A[x, y]$.

In particular, the provably recursive functions of **PA** are the same as those of **HA**, i.e., they are the terms of type **nat** $\Rightarrow$ **nat** of **T**. And, obviously, there is a similar correspondence at second order between $\mathbf{PA_2}$ (or $\mathbf{HA_2}$) and system **F**.

6.D.3 Formalisation. Let us come back to the proof of strong normalisation in the simply typed case (Section 6.2.1). This proof is not combinatoric since it uses a *logically complex* notion, reducibility. One even sees that this notion is *a priori* as complicated as a truth predicate, since one cannot write it with a fixed alternation of quantifiers: if reducibility of type A is Π^0_n, reducibility of type $A \Rightarrow A$ is of the form $\forall(\Pi^0_n \Rightarrow \Pi^0_n)$, thus Π^0_{n+1}.

Reducibility does not use, for a given term, any induction on types. Indeed, when one proves (R1)–(R3), one does it separately, type after type, for instance in 981 steps. To sum up, we have a schema of proof, for each term t, of the fact that t is sN, this only using induction on formulas of arithmetic, i.e., in **PA** (or **HA**).

On the other hand, for each type, one uses induction on the reducibility of this type, in relation to property (R3). We shall give an alternative version of this property, for which induction on reducibility is not needed.

Let E be a finite set of *simple* terms; immediate E-reduction is immediate reduction $t \rightsquigarrow u$, restricted to the case where $t \in E$; E-reduction $t \rightsquigarrow_E u$ is its reflexive/transitive closure. An E-normal form for t is any u such that $t \rightsquigarrow_E u$ and $u \notin E$. If t is E-sN, i.e., if all E-reduction sequences starting from t are bounded by an integer $|t|_E$, then t has finitely many E-normal forms.

(R′3): if E is a finite set of simple terms, if $t \in E$ is E-sN and all its E-normal forms are reducible, then t is reducible.

This condition generalises (R3) which is the particular case $E = \{t\}$; conversely, (R′3) is an iterated version of (R3).

We can replace (R3) with (R′3); for this we must show two lemmas on E-reductions:

Lemma 8.1. *If t and u are respectively E-sN and sN, then $(t)u$ is F-sN, with $F := \{(t')u'; t' \in E,\ u \rightsquigarrow u'\}$ and its F-normal forms are the $(t')u'$, where t' is an E-normal form of t and u' is the normal form of u.*

Lemma 8.2. *If t and u are respectively E-sN and sN, then $(\lambda x t)u$ is F-sN, with $F := \{(\lambda x t')u'; t' \in E,\ u \rightsquigarrow u'\}$ and its F-normal forms are the $t'[u'/x]$, where t' is an E-normal form of t and u' is the normal form of u.*

These two lemmas can be established without difficulty, by induction on very simple formulas. This enables one to obtain the following result:

Proposition 8. *The strong normalisation theorem is provable, for each term t, in the same finitely axiomatisable subsystem $\mathcal{T}$ of* **PA** *(or* **HA***).*

Proof. Essentially in any system in which we can perform the simple inductions of the two previous lemmas. □

Corollary 8.1. *The strong normalisation theorem is provable in* **PA** *(or* **HA***).*

Proof. The coding techniques used for the second incompleteness theorem enable one to prove an arithmetical formula which formalises the previous proposition and therefore says: « for each term t one can prove in $\mathcal{T}$ strong normalisation for t ». By the reflexion schema (Section 3.B.4), one deduces a uniform proof in **PA** (or **HA**). □

In general, if one takes a subsystem of **T** (resp. **F**) generated by finitely many recursions (resp. extractions), the previous technique can be applied *mutatis mutandis* to yield a global proof of strong normalisation in **PA** (resp. $\mathbf{PA_2}$). The more recursors we use, the more inductions we shall need; in the same way, the more extractions we have, the more comprehensions we need.

6.D.4 The Gandy method. The Gandy method consists in modifying immediate reduction in system **T** so as to make it « increasing ». One begins by defining, for any type, the successor and the sum pointwise, e.g., $t + u := \lambda x((t)x + (u)x)$. Reduction thus becomes

$$(\lambda xt)u \rightsquigarrow_g t[u/x] + SSSSSu. \tag{6.12}$$

One verifies that it is confluent. Moreover, it is strictly increasing, in the sense that the size of the term increases: here from « size of t » + « size of u » +4 to at least « size of t » + « size of u » +5. In particular if Gandy reduction is weakly normalising, it will also be strongly normalising, because of confluence.

This is sufficient in the absence of the recursor: weak normalisability comes from the possibility of recursively translating λxt into $\lambda xt + SSSSSx$ and from weak normalisation of the simply typed calculus. Observe that S and $+$ need not be the « actual » S and $+$, they might as well be variables; in particular the method extends – *modulo* Church–Rosser – to natural deduction. By the way, the Gandy method shows that the bounds on strong normalisation are of the same order as the bounds on weak normalisation.

But the original method was devised for system **T**; in case, one must also modify the reduction of the recursor so as to make it increasing. See also Section 7.B.2.

6.E Subtyping

6.E.1 Polymorphism. Polymorphism is the observation that the same λ-term can admit several types. Thus, in the simply typed calculus, Church integers admit the types $\boldsymbol{n} + \mathbf{2}$.

This is exploited in a typed programming language such as **ML**. The types are universally quantified simple types. When a λ-term is typable, one can give it a « principal type ». Polymorphism à la **ML** enables one to type functions such as the exponential by *unification* of the principal types. For instance, in order to type $(y)x$ where x, y are of type **nat**, one will solve the equation $(X + \mathbf{2}) \Rightarrow Z = Y + \mathbf{2}$, whose solution is $Z = X + \mathbf{2}$, $Y = X + \mathbf{1}$.

6.E.2 Subtyping. In system **F**, Church integers admit many more types than in **ML**, for instance **nat** + **2**. Note that, from the viewpoint of typable λ-terms, **nat** $\subset$ **nat** + **2**. There are therefore inclusions between types, this is *subtyping*.

It is legitimate to present second-order quantification as an intersection, the intersection of all $T + \mathbf{2}$. One can and that is what has been done by the « Torino school », enrich the language by means of « intersection types ».

An interpretation of system **F** in λ-calculus: one calls *type candidate* any set $\mathcal{T}$ of λ-terms enjoying:

(T1): any term of $\mathcal{T}$ is sN.

(T2): if $t \in \mathcal{T}$ and $t \rightsquigarrow t'$, then $t' \in \mathcal{T}$.

(T3): if t is simple and if all immediate reducts of t are in $\mathcal{T}$, then t is in $\mathcal{T}$.

This is a definition strictly inspired by reducibility candidates. If $\mathcal{T}$, $\mathcal{U}$ are type candidates, one defines

$$\mathcal{T} \Rightarrow \mathcal{U} := \{v \ ; \ \forall t \in T \ \ (v)t \in U\} \tag{6.13}$$

and proves that this is still a type candidate.

Type candidates are also closed under arbitrary intersections: this is obvious from the shape of the definition. Which enables one to define ΛXA as the intersection of all candidates $A[\mathcal{T}/X]$. Obviously, there is room for other kinds of intersections.

The reducibility proof can be seen as the fact that, if t is a closed term of type A, then t^- belongs to the candidate A. Conversely, if t is a closed λ-term and $t \in A$, is it the case that t is of the form t^-, with t of type A? This is wrong in general; but true in case[14] A is Π^1. The question of the typability of a λ-term in **F** is undecidable, a nice result of Wells, [100].

6.E.3 Subtyping and *spin*. In quantum mechanics, the *spin* of an electron represents a boolean, i.e., a system with two states noted $\pm 1/2$. We know that this system can be *measured* along an arbitrary axis; in case the electron acquires an *actual* spin $\pm 1/2$ along this axis. One will distinguish:

$\mathbf{Bool}_{\vec{z}}$**:** the *spins* which are definite along a given axis, say $\vec{z}$.

Bool: arbitrary *spins*.

There is an obvious inclusion $\mathbf{Bool}_{\vec{z}} \subset \mathbf{Bool}$. More in Section 17.4.2.

6.F Essence, existence and typing

6.F.1 Locative phenomena. There are indeed two readings of polymorphism, depending if one starts with essence or existence.

Essence. In this reading, the type is primitive, one constructs it, then one takes care of the objects. There is no real polymorphism, there is only one form (essence) for a given object. This is the viewpoint followed by the category-theoretic interpretations of logic. This is also the viewpoint I followed when making my

[14] I didn't check it, but this sounds reasonable.

category-theoretic interpretation of system **F** in coherent spaces (Section 8.3). For instance, the type $\forall X(X \Rightarrow X)$ is a coherent space with one point. Extraction enables one, from this point considered as a prototype, to build objects in all types $T \Rightarrow T$.

Existence. But one can instead contend that objects are anterior to their type, seen as an essence. This is the viewpoint of *subtyping*, this is also the viewpoint of *ludics*: an object may have several types, be representative of several essences. There, *locativity* becomes essential. Indeed one can define the product as an intersection, $A \wedge B := A \cap B$. Obviously, if $A = B$, one has little chance of getting a « real » cartesian product in this way. Unless one integrates the locative standpoint: when I write $A \wedge A$, I actually mean $A \wedge A'$, the conjunction of A and of a *copy* of A. We shall see this under the name of *mystery of incarnation* in Section 14.1.4.

The opposition locative/spiritual, which plays an important role in this book, has been widely used in novels and movies as a dramatic or comic figure. The most elaborate use of this opposition is perhaps to be found in Buster Keaton's *Our Hospitality*: depending on his location (inside or outside the house), the same character is treated according either to the noblest spiritual principles (southern hospitality) or to locative prejudices (taken as representative of a hated family).

6.F.2 Typing as essence. The difference between « pure » and typed objects is the distinction between things as they are and things as they should be. This is also the difference between handicraft and industry.

Imagine a radio receiver with a defective element; one can fix it by replacing the transistor SFK222E-28 with another transistor of the same type, of the same *specification*, SFK222E-28. Note that there is no need to understand the specification; and that the new transistor is by no means identical to the previous one. It may be of a different colour, of a different shape; but the machine has been conceived in function of the *idea* of this transistor, not at all around a peculiar individual.

For one can also use objects for what they are, not for their specification. In that case, the only possible replacement is that of an equal for an equal. For instance money is made to be exchangeable on the basis of its nominal value; an atypic use of money is to split a banknote in two parts, the two halves being unable to live separately: gangster technique. A note of 500 $ is an essence; a half-note is a unique, irreplaceable object, by no means 250 $.

6.F.3 Typing and computation. The activity consisting in manufacturing unique, irreplaceable – unless by themselves – objects, is called *handicraft* and does not stand mediocrity; in opposition to *industry* which contents itself with a not quite exciting contractual *minimum*, in the style of Mc Donald's. Industry is obviously (and unfortunately) right.

Witness computer science: one does not want programs written by little wizards, one wants structures, modules, that one can unscrew, modify, reassemble, etc. It is there that occurs the paradigm of *typed* functional programming. One considers a term of type, say, **nat** $\Rightarrow$ **bool** as a program yielding, in an implicit way, an answer yes/no to a query of the kind « does integer n satisfy P ? ». Here, the typing guarantees the termination of computations.

But one can do much better: one proves in a system in the style of **HA** the formula $\forall x\,(P \vee \neg P)$; then one translates the proof in second-order logic and takes the forgetful term in system **F**. In this way one gets a *realiser* which is a typed program, with a guarantee of termination and that it does what it is supposed to do. This is a nice idea, put forward by Krivine[15].

Note that industry brings as ever guarantees (termination) but at the price of a certain mediocrity (inefficient program). Thus, if one *proves* that France is connected using the familiar method of the « star domain », the underlying program yields a network centered on Paris. Which was by the way realised during the XIXth century by French railways: one applies the theorem and in order to go from Marseille to Tours one transits through Paris. One should therefore nuance this a bit: there is a tension between the guarantee coming from mathematical abstraction and the inefficiency of abstract methods.

Let us finally mention that there have been attempts at extending the paradigm of « proofs as programs » to classical logic, reduction to absurdity becoming a form of *control* instruction. It is abusive to put this type of approach on the same level as typed functional programming, since it is no longer a matter of modular input/output specifications.

6.F.4 Reducibility and essence. If one carefully looks at the proof of reducibility for system **F**, one discovers that the reducibility of type A closely mimics the formula A. So that the extraction on B – the only truly delicate point – is justified by a comprehension on something which is roughly B. Always this propensity for making circles, illustrated by the faulty normalisation proof given by Martin-Löf for its first system: the extraction on a rather dubious type was justified by a comprehension on more or less the same thing… but the system was nevertheless contradictory.

A systematic and inconsiderate application of normalisation techniques brings us back to the essentialist rut. This being said, is it as faulty as Tarskism? Yes and no:

Yes: if one tries to justify in this way systems through cut-elimination, it is not better than the justification by truth; without its healthy vulgarity.

[15]There was something of the like in logic programming, but it is not sure that the conceptors of PROLOG were ever aware of it.

No: this machine runs in circles at the level of comprehension, at the level of infinite objects and more generally at the level of contraction. On the other hand it is not vicious « outside of contraction ». One can see in it an honest, satisfying and original explanation of the *perfective* part of logic (Chapter 10).

We will eventually say the last word, through linear logic, on the perfect connectives $\otimes$, $\parr$, $\wedge$, $\oplus$. In the sense that we will have the feeling of having said everything, of having unscrewed everything. On the other hand, we will have more difficulties with exponentials and at the present moment, I am still using Sioux ruses to avoid the circularity, the essentialism that characterises them (Chapter 16).

Speaking of circularity, take for instance comprehension: this schema is represented by extraction, but the reducibility of extraction requires comprehension, roughly on the formula we started with. We arrive at a strange situation where we no longer know which is more primitive: does reducibility interpret term t, or is it that t would eventually be a way to enunciate its own reducibility? Again Nietzsche: « When you gaze long into an abyss, the abyss also gazes into you ».

One can say the same of the infernal pair essence/existence. Logic, surely born essentialist, began with manipulating universal rules. Long afterwards, at the end of a complex process, we eventually find an underlying structure for these rules; and thus, to require that a proof of $A \vee B$ be – with reservations of no interest in this discussion – a proof of A or a proof of B. One eventually rediscovers the existence under essence. But, while *studying* this existence, one reinstalls essence.

We started with the rules of logic; we arrived at the logic of rules and it turns out to be the logic we started with. This is slightly better than a plain circular promenade. Indeed, if we start with a bad logic, say « negation as failure » (Section 4.D.4), we can surely write rules, study their properties, even if it is not very exciting. But what we get, the logic of rules of negation as failure, bears no relation to the starting point.

Circularity is therefore not only a tarskian void, it is also a sign of harmony. But one cannot content oneself with that!

Chapter 7
The category-theoretic interpretation

7.1 The three layers

Instead of the usual explanation of logic with its infinity (transfinite, but *predicative*, they say: see Section 7.B.4) of *matrioshka*-turtles, one will modestly content oneself with three foundational layers, three undergrounds not at all (meta-)isomorphic. Layer -1 will be the level of *truth*, layer -2 the level of *functions*, layer -3 the level of *actions*.

7.1.1 The first underground. It is the level of truth, of provability, of formal consistency. For most people, foundations are wholly located in that layer.

Sense and denotation. Frege, the founder of modern logic, was surely a damned essentialist: witness his contempt for the geometrical ideas of Riemann – whose *Habilitationschrift* anticipated, in the middle of the XIXth century, the theory of general relativity[1].

His opposition between *sense* (implicit) and *denotation* (explicit) is typical of a not too hot approach to logic. For instance, the two expressions « the morning star » and « the evening star » have different *senses*, but the same *denotation*, Venus. In this line of thought, logic appears as a sort of « calculus of denotations »: a theorem (whose sense is anything except « true ») has the same denotation as « true »: the proof is a way to make this denotation explicit.

In the same way, one can say that the equality $t = u$ is interesting only because it is not an *identity*, that t and u are distinct *at the level of sense*.

This thought quickly finds its limitations which are those of the dichotomy subject/object. Everything takes place in a universe where the subject (which will become a formal system) and the object (a model, therefore a set) answer to each other without ever meeting. Completeness/soundness establishes a sort of *duality*, between proofs of A and models of $\neg A$:

Soundness: if one has both a proof of A and a model of $\neg A$, then… contradiction.

Completeness: proofs and models are *polar* in this duality.

---

[1] Not to speak of his political opinions that were mostly known by hearsay: his complete works are stalled in the beginning of the years 1920. Shortage of paper?

Pole and polars. Remember that, given a binary operation, noted $a, b \rightsquigarrow \langle a \mid b \rangle : A \times B \mapsto C$ and a subset $P \subset C$ (the « pole ») we can define the *polar* $X^p \subset B$ of a subset $X \subset A$ (resp. $Y^p \subset A$ of a subset $Y \subset B$) by

$$\begin{aligned} X^p &:= \{y \in B \,; \forall x \in X \ \langle x \mid y \rangle \in P\}, \\ Y^p &:= \{x \in A \,; \forall y \in Y \ \langle x \mid y \rangle \in P\}, \end{aligned} \tag{7.1}$$

and that

- The map « polar » is decreasing: $X \subset X' \Rightarrow X'^p \subset X^p$.
- The set $\mathrm{Pol}(A) \subset \wp(A)$ of *polar* sets, i.e., of the form Y^p, is closed under arbitrary intersections. In particular, A is polar and X^{pp} is the smallest polar set containing X.
- As a consequence, $X^{ppp} = X^p$.

Negation is the most important[2] connective of logic, classical up to this point. We shall interpret it on the basis of the paradigm:

$$\textit{negation} = \textit{polar}$$

Mathematicians use, abuse, of polarity; for instance, if $A = B$ is the space $\mathbb{R}^3$ and $\langle a \mid b \rangle$ is the scalar product, a current choice will be $P = \{0\}$, which leads to *orthogonality*. Polar sets become vector subspaces, the polar of a line is a plane, etc.

Classical logic is based on a duality corresponding to soundness: a proof and a model are *never* polar, i.e., $P = \emptyset$. This establishes a bleak duality: the only polar sets are $\emptyset$ and A (or B). In other words:

- Proofs cannot discriminate models.
- Symmetrically, models do not discriminate proofs.

Very concretely, a model of $\neg A$ cannot tell the difference between two proofs of A since the very existence of this model of $\neg A$ opposes the existence of a proof of A. In other terms, it is a matter of a definition of *provability*, not of proofs. In this perspective, a specific proof is only a bureaucratic *artifact*: computing a denotation, period. One must therefore seriously improve the duality proofs/models!

Layer -1 is conceptually very poor: truth, consistency. With a big effort, one arrives at *admissible rules*: « if A is provable, B is provable ». The \$1000 question: find the relation between admissible rules and logical implication… how bleak!

Although classical logic has nowadays (and I bear some responsibility for this!) a quite satisfactory category-theoretic reading (Sections 7.A.6, 12.A and 15.4),

[2]Not the most useful, which would rather be implication.

I have a propensity to believe that an interpretation confined to the « first underground », the layer « true/provable », is quite sufficient in that case. Indeed, classical logic rests upon a duality with an empty pole, which only recognises provable/consistent and succeeds in this way in justifying biased principles such as the excluded middle. It is therefore likely that the search for fine grain interpretation of classical proofs belongs to the realm of *methodological* mistakes... A non-dogmatic viewpoint, subject to contradictory discussion: I didn't say « technical baloney » or « triviality », since the works on classical proofs are anyway worthy of interest.

7.1.2 The second underground

The covenant. In the same order of thought, I think that it is a methodological mistake to seek semantics for intuitionistic or linear logics. It is however technically possible: Kripke or topological models in the intuitionistic case (Section 4.E); phase models in the linear case (Section 10.1). In the latter case, models even turned out to be *technically* useful, witness for instance certain results of Lafont [70]. This being said, technical usefulness is not a guarantee of sense: one should then take seriously the paraconsistent system used by Rosser in his symmetrisation of Gödel's theorem (Section 2.D.3). The question is not whether one has the *right* to use models outside classical logic, the answer being obviously « yes »; it is whether this kind of explanation is *appropriate*: the answer is clearly « no ».

Indeed, if we stay within the opposition true/provable, there is little, except consistency, to satisfy our hunger. But what is a consistent intuitionistic theory, which however admits a Kripke model? A nothing, a meaningless doohickey: for instance classical logic is a consistent extension of intuitionistic logic, so what? It is the place to introduce the idea of a *covenant* – which will eventually lead us to refine the duality sense/denotation.

The covenant of a formal system can be *plausibility*. It is a judiciary version of logic – « what I say is not false » – this is the one prevailing in front of a tribunal, every defendant being supposedly innocent; one should rather say *not-guilty*, since, among all those lifetime senators that escape jail to the benefit of doubt, there must surely be a couple of criminals...

Plausibility is the existence of a model, or, in an equivalent way, consistency: it is the *classical* covenant, but it is not the only possible one. Think for instance of a bank; if the bank says: « you have got $1000 », we don't only want it to be plausible, we also want to know that we can get these $1000. By the way, everybody knows people who are expert at promising without paying[3]: those are adepts of classical logic, since it is exactly what happens with the excluded middle:

[3]For instance, Bernard Madoff.

System: $A \vee \neg A$.

I: I don't believe in this.

System: If both are false, A is false, hence $\neg A$ is true.

I: Yes indeed!

System: But you told me that $\neg A$ is false.

I: I give up, you are too smart.

This discussion with an expert in sophisms leaves an unpleasant after-taste: indeed the contradictor gets mixed up, but the system does not argue earnestly.

A covenant better adapted to banking style realities is therefore the following: if one announces an existence, one must be able to find a witness. For instance, if one says that « there are weapons of mass destruction », one must be able to exhibit them, since one cannot be happy with the first underground, with the classical version: « he who says the contrary is part of the Axis of Evil ». The exigency of *testimony* must not be confused with a professed *explicit deduction*, of which we already exposed the oxymoronic character (Section 4.2.3). A bank is not supposed to keep money: it should make it circulate; otherwise it is styled differently: it is called a miser[4]. If one asks a bank for one's money, it should yield it, even if it takes some time; the failure to do this is known as *bankruptcy*, the financial form of inconsistency.

One therefore arrives at the following covenant: if I prove a disjunction $A \vee B$, I must be able to justify one of the two sides. This is why the only *methodologically sound* notion of intuitionistic consistency is that of a theory consistent in the usual sense, but also satisfying the properties of existence and disjunction[5].

Let us come back to the fregean paradigm – to divert it from its setting subject/object to a setting that would rather be subject/subject. A proof has a *sense* and a *denotation*; the denotation makes explicit the data linked to existence and disjunction. Logical operations should therefore be interpretable as operations on this implicit contents.

Category-theoretic reading. It is what is done by the functional interpretation of Chapter 5, of which we shall *restrict* the scope. Logic now belongs in a *category* whose « objects » are the formulas and whose *morphisms* are the proofs; the details will follow later. For the moment, we content ourselves with the observation that the pair morphism/object is clearly more interesting than the pair proof/model – subject/object – of the classical world. The rule of *Modus Ponens*, or rather the

[4]The analogue of circulation is deduction.

[5]See the definition of « saturated » in Section 4.E.

transitivity of implication, the *syllogism*, becomes the *composition* of morphisms:

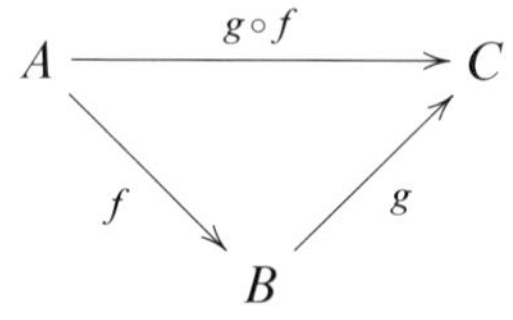

(7.2)

Let us compare this to the « first underground » reading. In the years around 1920, Łukasiewicz explained the transitivity of implication by the transitivity of inclusion: if $A \subset B \subset C$, then $A \subset C$. The height of derision: it is the transitivity of implication that explains the transitivity of inclusion, not the other way around!

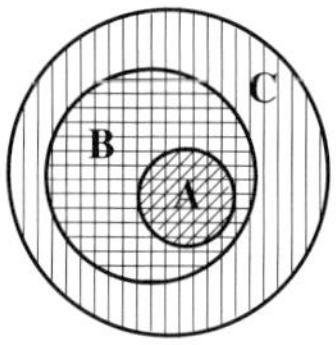

Figure 7.1. The syllogism according to Łukasiewicz.

Commutations. If classical logic, i.e., the interpretation by provability/consistency, were really satisfactory, we would have a general completeness theorem, not only for predicate calculus. Now, there is nothing of the like; the stumbling block being the incompleteness theorem, more precisely the fact that:

Provability does not commute with negation.

To make provability and negation commute is obviously a procedural, cognitive idea, since it opposes a strict dichotomy subject/object. It is even a good idea, provided one changes everything, from the cellar to the attic, only retaining a lax setting: we witnessed the ruination of epistemic, non-monotonic logics – not to speak of the procedural negation of PROLOG (Section 4.D.4) –, all based upon an uncouth commutation.

In general, the idea of making proofs and logical connectives commute is excellent and if one sticks to operations less « loaded » than negation, plausible. In this way, intuitionism realises – at the second underground – a commutation between *proof* and *disjunction*. To prove $A \vee B$ is to prove A or to prove B. W.r.t. Tarski, one has replaced truth with proof. The « or » of « or prove » is a *procedural* disjunction, operating on the proof itself. Starting with this idea, one can write deductive logical rules (the system **NJ**) and discover that these logical rules actually enjoy the disjunction property. In other terms, one has an equivalence between the *rules of logic* and the *logic of rules*.

This equivalence is not the result of a discretionary action: for instance one could not have declared (see *supra*) that « to prove $\neg A$ is not to prove A ». It results from a deep equilibrium expressed by the theorem of normalisation of system **NJ**.

7.1.3 The third underground

Dynamics. Let us come back for the last time to our detoured fregean paradigm. The second layer opposes, for a given proof, a sense to a category-theoretic denotation (a morphism). Let us be down to earth and remember that the first quality of a proof is to be *understandable*: which is achieved by its modular structuration in lemmas, intermediate results, significant concepts. Which cut-elimination destroys, by producing an explicit pulp, that a machine can maintain, but which is normally illegible. Thus, consider the camping manager who computes the bill in the following « cut-free » way: « 7 locations at \$5, 7 parkings at \$2, 7 TV supplements at \$3 gives $7 \times 5 + 7 \times 2 + 7 \times 3 = 35 + 14 + 21 = 70$, »... A limited amount of thought (the use of a lemma, distributivity) permits a quicker computation, $7 \times (5 + 2 + 3) = 7 \times 10 = 70$: this is the typical illustration that « cut-free » means « moron ».

The distinction sense/denotation eventually resolves into: with/without cuts. In other words, one must leave room for cut-elimination, for the *dynamics* of explicitation. Some have proposed 2-categories[6], to explain that a morphism can change. This viewpoint is well-adapted to algebraic topology – to speak of homotopy –, but in logic it has been so far barren. On the opposite side is the proposed « operational semantics » which is nothing but a paraphrase of the process of formal explicitation. Which is by the way my main objection to the expression « denotational semantics » to speak of the second underground: it suggests another panel, operational, which is by the way correct. The only problem is that « operational semantics » is a very connoted expression: it means *ad hoc*, boring, illegible and theoretically empty.

Take λ-calculus; some authors describe the rules of term formation as being *syntax*, reduction as being *semantics*. It is a correct intuition of the « third underground », the only thing one can save from the wreck of « operational semantics ».

Geometry of interaction. It will be necessary to find *mathematical* tools capable of speaking of this *procedural* layer. This will be *geometry of interaction* (GoI). Note that, in GoI, the identity axiom is an *extension cord*, implementing a delocation (to carry a current from one point to another), and that cut is the *plugging* of two complementary interfaces (Section 3.2.1). Cut-elimination thus becomes the

[6] Generalisation of categories in which there may be « 2-morphisms » between morphisms of the same $\mathbf{C}(A, B)$.

solving of an input/output equation on the Hilbert space:

$$\begin{aligned} f(x \oplus y) &= x' \oplus y', \\ g(y' \oplus z) &= y \oplus z'. \end{aligned} \tag{7.3}$$

The solution of this equation is written $h(x \oplus z) = x' \oplus z'$: it corresponds to syllogism, to *category-theoretic composition*. The component y corresponds to the *computation* in the actual sense of the term (Chapter 19).

Articulation. A legitimate question: we are digging deep into the underground, but how to be sure to lose nothing? Indeed, the category-theoretic viewpoint, the second underground, is not at ease with consistency: it interprets logic without taking care of this issue. Indeed a logical system may have a nice category-theoretic interpretation – this is the case for Prawitz's formulation of naïve set-theory – while being inconsistent: what interests us at the second underground is to distinguish between proofs, not between formulas. This being said, the viewpoint of consistency, although limited, is not stupid: one should not lose sight of it.

Everything works better at the third underground, which loses nothing of the two upper layers:

Consistency: geometry of interaction defines a duality that can be *metaphorically* read as a sort of *game*. In this setting, it is possible to say that one of the two players is the « winner ». There need not be a winner, but there are never two of them. In other words, in a duality « proofs of A/proofs of $\neg A$ », the two sides are not both systematic winners: this is consistency.

Composition: when building a category, the motor nerve is *associativity* of composition. Typically, Church–Rosser is a way to access associativity (Section 7.A.5). In GoI, associativity corresponds to a double syllogism:

$$\begin{aligned} f(x \oplus y) &= x' \oplus y', \\ g(y' \oplus z) &= y \oplus z', \\ h(z' \oplus w) &= z \oplus w', \end{aligned} \tag{7.4}$$

which can be decomposed, in two different ways, into two simple syllogisms. One must then prove that the « composition » $\bullet$ resulting from the solution of the syllogism equation satisfies

$$f \bullet (g \bullet h) = (f \bullet g) \bullet h. \tag{7.5}$$

This is the main property of GoI, a property so important that it even participates to the solution of the syllogism.

7.2 Closed cartesian categories

7.2.1 Categories. Remember that a category **C** consists of the following data:

Objects: a *class* of objects $|\mathbf{C}|$.

Morphisms: for all objects A, B of **C**, the *set* $\mathbf{C}(A, B)$ of the morphisms of *source* A and *target* B.

Composition: given $f \in \mathbf{C}(A, B)$, $g \in \mathbf{C}(B, C)$, the *composition* $g \circ f \in \mathbf{C}(A, C)$. One requires « $\circ$ » to be associative, with neutral elements, the identity morphisms $\iota_A \in \mathbf{C}(A, A)$.

The word « morphism », which refers to *form*, places categories under an *essentialist* patronage. By the way, this is not a crime, but one must notice it: the structure is anterior to the object.

One usually uses *diagrams* like (7.2), of which one always enunciates the *commutativity*[7]. Layer −2 rests upon commutative diagrams, which express the functional equations of Chapter 5.1; this is a bias, since, as in George Orwell's *Animal farm*, one side is more commutative than the other (Section 7.1.3).

A (covariant) *functor* of **C** to **D** is made of the following data: for any object $A \in |\mathbf{C}|$, of an object $\boldsymbol{F}(A) \in |\mathbf{D}|$ and for all morphisms $f \in \mathbf{C}(A, B)$ of a morphism $\boldsymbol{F}(f) \in \mathbf{D}(\boldsymbol{F}(A), \boldsymbol{F}(B))$ such that

$$\begin{aligned} \boldsymbol{F}(\iota_A) &= \iota_{\boldsymbol{F}(A)}, \\ \boldsymbol{F}(g \circ f) &= \boldsymbol{F}(g) \circ \boldsymbol{F}(f). \end{aligned}$$

There are also *contravariant* functors, reversing the sense of arrows, $\boldsymbol{F}(f) \in \mathbf{D}(\boldsymbol{F}(B), \boldsymbol{F}(A))$, $\boldsymbol{F}(g \circ f) = \boldsymbol{F}(f) \circ \boldsymbol{F}(g)$. There are also functors in several arguments, etc.

The functors of **C** to **D** are the objects of a category[8] whose morphisms are the *natural transformation* from **F** to **G**: it consists in giving, for all $A \in |\mathbf{C}|$, an object $\boldsymbol{T}(A) \in \mathbf{D}(\boldsymbol{F}(A), \boldsymbol{G}(A))$ such that the diagrams

$$\begin{array}{ccc} \boldsymbol{F}(A) & \xrightarrow{\boldsymbol{F}(f)} & \boldsymbol{F}(B) \\ {\scriptstyle \boldsymbol{T}(A)}\big\downarrow & & \big\downarrow{\scriptstyle \boldsymbol{T}(B)} \\ \boldsymbol{G}(A) & \xrightarrow[\boldsymbol{G}(f)]{} & \boldsymbol{G}(B) \end{array} \tag{7.6}$$

are commutative for all A, B and $f \in \mathbf{C}(A, B)$.

[7] What a practical joker summarised by the « theorem »: *All diagrams are commutative*.

[8] Provided we don't pay too much attention to hairsplittings concerning sets and classes.

7.2.2 Cartesian categories. Remember that the *cartesian product* is an instance of *projective limit*, a.k.a. plain limit. Given objects A, B, one seeks an object $A \times B$ and morphisms $\pi_l \in \mathbf{C}(A \times B, A)$, $\pi_r \in \mathbf{C}(A \times B, B)$ which are *universal*, i.e., the most general possible. Which is translated as follows: for any other choice C, f, g, there exist a *unique* $(f, g) \in \mathbf{C}(C, A \times B)$ rendering commutative the diagram:

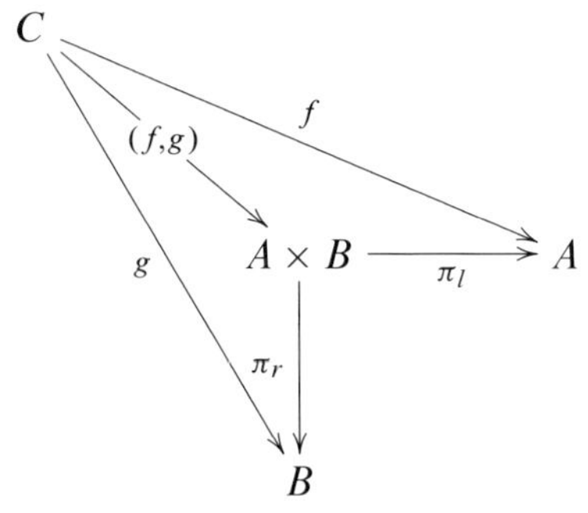

(7.7)

On this diagram, one could witness the product, the projections, the pair, indeed everything necessary to the interpretation of the rules of conjunction. Note that the eliminations (the projections) come before. This is because we are in the projective mode, in other terms, in negative *polarity*.

The only trouble with the definition is that, in categories, everything is up to isomorphism, so that $B \times A$ might as well claim the title of cartesian product of A and B. It is the delicate point fixed by universality, which enables one to untangle the skein of isomorphic solutions by *rigidifying* the isomorphisms between various solutions. Eventually, the solutions are isomorphic, but in a *univocal* way.

Unless we dedicate ourselves to categories, we should not panic in front of these diagrams. They express, in a very unnatural way, that a solution is natural, *canonical*. Concretely, truly natural constructions – *in the primal sense of the term* – are also natural in the sense of categories. In case of doubt, there remains the pedantic solution of diagrams: that's what they are good at.

There is a 0-ary analogue of the product, the projective solution to nothing. This means that one looks for an object T with the property that any object A is mapped in a unique way in T, i.e., $\mathbf{C}(A, \mathsf{T})$ has exactly one element. This is called a *terminal* object[9].

A *cartesian* category is a category with a terminal object in which a cartesian product has been given to us. This is therefore slightly more than requiring the mere existence of the cartesian product; in practice one hardly sees the difference. In particular, a cartesian category is equipped with the binary functor $\times$, with, for $f \in \mathbf{C}(A, A')$, $g \in \mathbf{C}(B, B')$, $f \times g := (f \circ \pi_l, g \circ \pi_r) \in \mathbf{C}(A \times B, A' \times B')$.

[9]The standard notation would rather be $\mathbf{1}$.

7.2.3 Closed cartesian categories. Cartesian categories are only an insignificant, but necessary, step in direction of the category-theoretic version of implication. Indeed, we want to express the following: if $A, B \in |\mathbf{C}|$, then $\mathbf{C}(A, B)$ is also an object of $\mathbf{C}$. Which means nothing *stricto sensu*. This being said, in a concrete category, this immediately makes an evident sense: can the morphisms of a certain kind of structure be equipped with the same structure? The answer is universally « no » and we must be grateful to D. Scott [92] for having provided the first non-trivial example[10] of a non-degenerate CCC... even if the thing is rather devalued after 40 years[11].

Let us consider the following universal problem: given A, B, find B^A and a morphism $\epsilon \in \mathbf{C}(B^A \times A, B)$. Universality (projective version) is that, given any other solution C, f, there exists a unique morphism $\lambda(f) \in \mathbf{C}(C, B^A)$ rendering commutative the diagram:

$$\begin{array}{ccc} C \times A & \xrightarrow{\lambda(f)\times\iota_A} & B^A \times A \\ & {\scriptstyle f}\searrow \quad \swarrow{\scriptstyle \epsilon} & \\ & B & \end{array} \tag{7.8}$$

This diagram contains everything necessary to the interpretation of the rules of implication. Note that we are still in the projective mode (negative polarity) since evaluation (the elimination) comes before the introduction.

This formulation is almost incomprehensible; personally, I never used it. Let us see what it means in « vulgar » terms, i.e., if objects are sets and morphisms are functions: ϵ is *evaluation*, i.e., a morphism taking as arguments the pair of an « application » y of A into B and of a point x of A and yielding the result $(y)x$. Universality expresses that any function $f[z, x]$ of two arguments, hence of $C \times A$ into B, factorises through evaluation as a function from C to B^A: one forms $\lambda x f$ and then $f = (\lambda x f)x$; this is nothing but the commutativity of the diagram. The only use of C is to replace « points » with morphisms. We require slightly more, i.e., that the solution is unique. To sum up, the diagram enunciates, in awfully pedantic terms, the equivalence between two notions:

- The set of morphisms $\mathbf{C}(A, B)$.
- The applicative object B^A.

In practice it is enough to establish a sufficiently natural correspondence. In case of doubt, we have to indulge in diagram-chasing, but only in case of absolute necessity...

[10] The trivial example being the category of sets.

[11] Which happened to much more famous discoveries, e.g., penicillin, of which nobody contests the historical import, but which is of no use today.

Definition 12 (CCC). A *closed cartesian category* (CCC) is a cartesian category in which the evaluation problem has a universal solution.

7.2.4 Intuitionistic isomorphisms. The canonical isomorphisms of a CCC are the following:

Product: commutativity, associativity, neutral element T.

Implication: $(C^B)^A \simeq C^{(A\times B)}$, $A^{\mathsf{T}} \simeq A$, $(A \times B)^C \simeq A^C \times B^C$.

These isomorphisms – morphisms invertible on the left and on the right – are obtained by twirling the machinery of universal problems; they are indeed invertible natural transformations, *equivalences* between functors.

Observe the nuance with the first underground: from the standpoint of provability, the previous isomorphisms obviously induce logical *equivalences*. But the converse is wrong; typically $A \Leftrightarrow A \wedge A$, while in any *honest* category, A is not isomorphic to its cartesian « square » $A \times A$.

7.3 Examples of CCC

7.3.1 Degenerate CCCs. The simplest CCC is obtained by considering the *cofinite* subsets of $\mathbb{N}$, ordered by inclusion. The terminal object is $\mathbb{N}$, the product is intersection, y^x is given by $y \cup \complement x$. If we didn't define morphisms, this is because it is a strange category, in which $\mathbf{C}(x, y)$ is non-empty exactly when $x \subset y$, in case there is exactly one morphism. We also speak of a *degenerate* category. Degenerate categories indeed bring us back to the first underground, the layer of truth values. They do not distinguish between morphisms (the proofs): it is not quite what we have in mind.

We will later see (Section 7.A.4) that classical logic is degenerate, in the sense that any category-theoretic interpretation is a preordered set, which does not distinguish between morphisms.

7.3.2 Sets. There is at least one non-degenerate CCC, the *category of sets* with, as morphisms, all functions from A to B. We clearly see that the set-theoretic product and its two projections will provide us with a cartesian product, any singleton $\{a\}$ playing the role of a terminal object. As to the evaluation problem, we can take for B^A the set (precisely noted in this way) of all functional graphs of A into B and for ϵ the function $y, x \rightsquigarrow (y)x$. This is so natural that there is nothing to check.

There is obviously something wrong here. We speak of *morphisms* while there is no *structure*: we therefore reach a plain *misunderstanding*, even if this one is not deadly like degeneracy. For instance, a simple consideration on the monstrous *cardinals* of X, X^X, X^{X^X}, … suggests that something is wrong.

7.3.3 Scott domains. This work [92] (and the contemporaneous work of Ershov) presents itself as a CCC of topological spaces. There is a simple historical reason for that: it closes a reflexion, mainly held in the years 1960 by Kleene, Kreisel, Gandy... around the *topological* interpretation of recursive *functionals*, i.e., computable functions of finite type. This being said, the result is dubious from the topological standpoint, since *no* interesting topological space is a Scott domain. Methodologically speaking, Scott domains should rather be presented as *continuous lattices*; which exposes the limitations of the concept: those of lattices, a second-zone mathematical notion.

I will present an explicit version based on sequents. I will go very fast: indeed, *coherent spaces* do the same, but in a simpler way: see Chapter 8, where all important details can be found. If one is absolutely fond of Scott domains, by nostalgia, dogmatism or love of technical complications, what follows is enough to reconstitute them.

A *Scott domain* $(X, \mathcal{F})$ is a set (usually denumerable) and an axiomatics $\mathcal{F}$ made of sequents $x_1, \dots, x_n \vdash x$ and $x_1, \dots, x_n \vdash$ $(x_i, x \in X)$ consistent w.r.t. logical consequence, here structural rules and cut: one cannot prove the sequent $\vdash$ from $\mathcal{F}$. A *coherent* subset of $(X, \mathcal{F})$ is a set $A \subset X$ such that $\mathcal{F} \cup \{\vdash x; x \in A\}$ is consistent. Thus, if $x, y \in X$, the axiom $x, y \vdash$ enunciates the incoherence of x, y and impedes a coherent subset from containing both x and y.

A coherent subset A is *saturated* when all y such that $\mathcal{F} \cup \{\vdash x; x \in A\}$ proves $\vdash y$ already belong to A. As a consequence, any coherent subset A generates a saturated subset $\bar{A}$. $A \sqsubset_{\mathcal{F}} X$ means that A is a saturated subset of $(X, \mathcal{F})$. In practice, saturation makes Scott domains unmanageable: saturated sets are most likely *infinite*.

If $(X, \mathcal{F})$ and $(Y, \mathcal{G})$ are domains, a morphism is a function φ such that:

- If $A \sqsubset_{\mathcal{F}} X$, then $\varphi(A) \sqsubset_{\mathcal{G}} Y$.
- If $A = \uparrow\bigcup_i A_i$ is a directed union, then $\varphi(A) = \uparrow\bigcup_i \varphi(A_i)$.

Remember that « directed » means that $\forall i, j \in I\ \exists k \in I\ A_i, A_j \subset A_k$ and also that the set I is non-empty.

It is immediate that one can take as cartesian product the disjoint union. Indeed a coherent subset of $\mathcal{F} + \mathcal{G}$ uniquely splits as a pair of coherent subsets: this is because there is no interaction between $\mathcal{F}$ and $\mathcal{G}$. The same is true of saturation: $C \sqsubset_{\mathcal{F}+\mathcal{G}} X + Y$ iff $C \cap X \sqsubset_{\mathcal{F}} X, C \cap Y \sqsubset_{\mathcal{G}} Y$.

If φ is a morphism from $(X, \mathcal{F})$ to $(Y, \mathcal{G})$, if $A \sqsubset_{\mathcal{F}} X$, if $y \in \varphi(A)$, let us write A as the directed union of its *finitely generated* saturated subsets A_i. Then $y \in \varphi(A_i)$ for a certain i. One concludes that there exists $a \subset A$, a finite, such that $y \in \varphi(\bar{a})$.

Let us consider the set $Z := X_{\text{fin}} \times Y$, where X_{fin} stands for finite coherent subsets of $(X, \mathcal{F})$. And let $\mathcal{H}$ be the set of sequents:

$$\begin{array}{rl} (a, y_1), \ldots, (a, y_n) \vdash & (y_1, \ldots, y_n \vdash \ \in \mathcal{G}) \\ (a, y_1), \ldots, (a, y_n) \vdash (a, y) & (y_1, \ldots, y_n \vdash y \ \in \mathcal{G}) \\ (a, y) \vdash (b, y) & (b \vdash a \ \in \mathcal{F}) \end{array} \tag{7.9}$$

In the last line, the notation $b \vdash a \in \mathcal{F}$ means that, for all $x \in a$, the sequent $b \vdash x$ is provable in $\mathcal{F}$; it is this last line which makes saturated subsets almost never finite, since, if $(a, y) \in A$ and $a \subset b$, then $(b, y) \in A$. One easily verifies that:

- If φ is a morphism of $(X, \mathcal{F})$ in $(Y, \mathcal{G})$, then the set of $(a, y) \in Z$ such that $y \in F(\bar{a})$ is saturated w.r.t. $\mathcal{H}$.
- Every saturated subset of Z comes from a (unique) morphism φ.

The previous correspondence enables one to internalise implication and thus obtain a CCC. If one is only concerned with obtaining a CCC and not with its fine grain properties, one can be happy with that.

7.4 Logic in a CCC

7.4.1 Interpretation. The category-theoretic interpretation of logic originates from syllogistics, i.e., from the transitivity of implication: $A \vdash B$ is read as $\mathbf{C}(A, B)$ and a proof of this sequent as a morphism of A into B. In order to pass from the restricted setting of sequents $A \vdash B$ to the general setting of sequents $\Gamma \vdash A$, one can use the cartesian product (and the terminal object for empty contexts).

Left rules: they correspond *grosso modo* to the solution of the problem.

Right rules: they correspond to the universality of this solution.

Reductions: they express the commutation of the universality diagram.

7.4.2 η-conversion. However, nothing corresponds to the *unicity* of the solution of the universal problem. This unicity can be expressed by means of the commutative diagram:

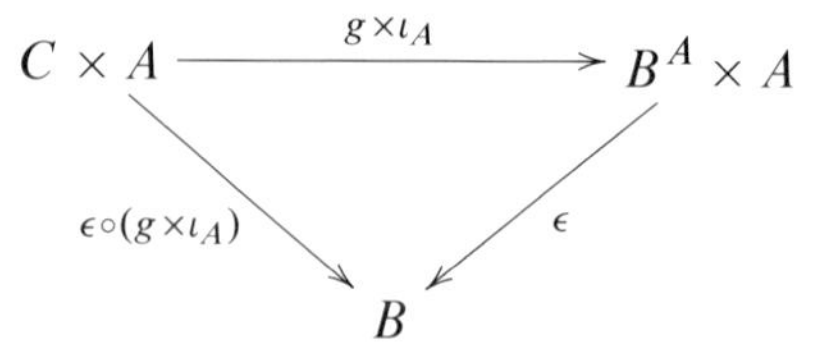

(7.10)

Comparing to (7.8), one gets by unicity $g = \lambda(\epsilon \circ (g \times \iota_A))$.

Translated in human terms, this yields the equation

$$\lambda x(g)x = g, \tag{7.11}$$

known under the code name « η »[12]. Observe that x is not free in g: this would mean nothing otherwise.

Similarly, the unicity in the case of a product:

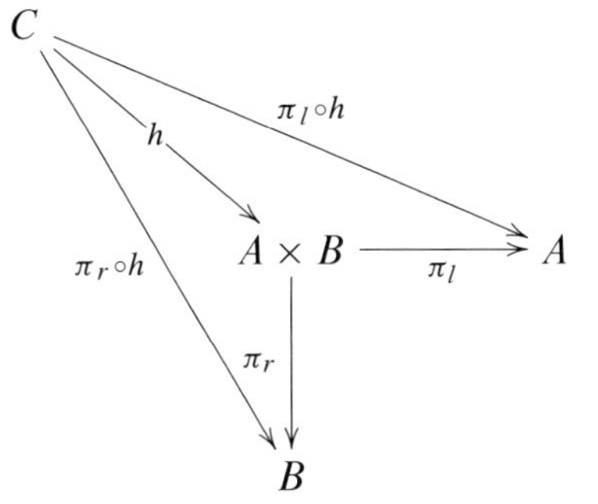

(7.12)

yields $(\pi_l \circ h, \pi_r \circ h) = h$ and in human language, the equation

$$(\pi_l a, \pi_r a) = a \tag{7.13}$$

which is called *surjective pairing*. These equations can be added to rewriting, but in which sense should we orient reduction? Tradition is rather for left to right (η-conversion) but this is not very glorious, especially for *surjective pairing*, the conjunctive case. If one revisits these equations in the logical spirit, one can pose the problem of *identity elimination*, in the style of *cut-elimination*. This is desperate, since identity is most likely the only axiom; but one can still try to *reduce* it, concretely by replacing an identity axiom on a complex formula with identity axioms on its subformulas. If one writes the corresponding reductions, they are exactly the above equations (7.11)–(7.13), oriented from right to left. This is known as « η-expansion ». This is a very technical and boring subject[13].

Let us come back for a while to the coding of data types. If we adopt, for integers, the type $\forall X((X \Rightarrow X) \Rightarrow (X \Rightarrow X))$, one finds a *duplicate*; indeed, besides the integer $\bar{1} := \Lambda X \lambda x^{X \Rightarrow X} \lambda y^X (x) y$ coexists $\Lambda X \lambda x^{X \Rightarrow X} x$. η-expansion reduces the duplicate to $\bar{1}$, but this is slightly irritating. Hence, when coding data types, one must put the X before, here $\forall X(X \Rightarrow ((X \Rightarrow X) \Rightarrow X))$, which avoids questions which are embarrassing but run more or less into emptiness.

[12] The smaller sister of « β », which is the nickname of the usual equation $(\lambda x t)u = t[u/x]$.

[13] Most liked by students, since it can be used to fill a not quite abundant PhD. a chapter on « η » brings its lot of foreseeable and tedious technical complications, 100% perspiration, 0% inspiration; in other words, it consumes paper.

The rule η is so robust that its « refutations » are dubious. For instance, one introduces *on purpose* mistakes in the coding of functions, see (7.20) in Section 7.B.1: this sort of joke reinforces the rule more than it criticises it. Fortunately, *quantum coherent spaces* will bring some fresh air in the academistic desert of the « η-rule » (Section 17.5.4).

7.A Classical logic

7.A.1 Direct sums. The direct sum is a typical example of an *inductive limit*, a.k.a. *colimit*. Given A, B one seeks $A + B$ and morphisms $\iota_l \in \mathbf{C}(A, A + B)$ and $\iota_g \in \mathbf{C}(B, A + B)$ in a universal way. This means that, for any other solution, C, f, g, there exists a unique morphism $f + g \in \mathbf{C}(A + B, C)$ rendering commutative the diagram:

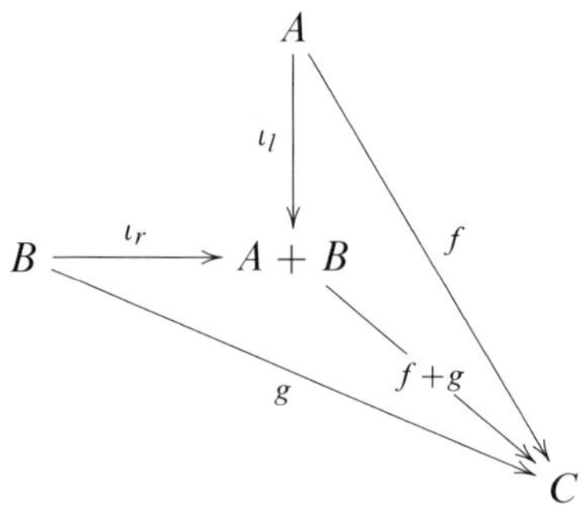

(7.14)

The 0-ary companion of sum is the *initial* object $\mathbf{0}$, which has the property that $\mathbf{C}(\mathbf{0}, A)$ has exactly one element for all A.

Sum and initial object obey the paradigm of *inductive* limits. It is a matter of *positive* polarity: first come the introductions, then the eliminations. The inversion of the sense of diagrams is the source of the fabricated aspect of the elimination rules of natural deduction.

7.A.2 Isomorphisms. In presence of disjunction, our list of canonical isomorphisms increases:

Sum: commutativity, associativity, neutral element $\mathbf{0}$.

Implication: $C^{A+B} \simeq C^A \times C^B$, $C^{\mathbf{0}} \simeq \mathsf{T}$.

Product: $A \times (B + C) \simeq A \times B + A \times C$, $A \times \mathbf{0} \simeq \mathbf{0}$.

7.A.3 Eta and disjunction. The version of « η » adapted to disjunction, i.e., with the notations of Section 5.C.2:

$$\delta(x^A \iota_l x_A)(y^B \iota_r y_B)a = a \tag{7.15}$$

expresses as usual the unicity part in the universal problem. In particular, this equation implies all commutative reduction (or rather the corresponding equalities), which are specific instances of this unicity.

7.A.4 Classical logic is degenerate. According to a brutal category-theoretic reading of classical logic, negation should appear as an involution (contravariant functor, whose square is the identity). Since $\mathbf{0} \times \mathbf{0} \simeq \mathbf{0}$, the involution yields $\mathsf{T} + \mathsf{T} \simeq \mathsf{T}$. Consider the commutative diagram:

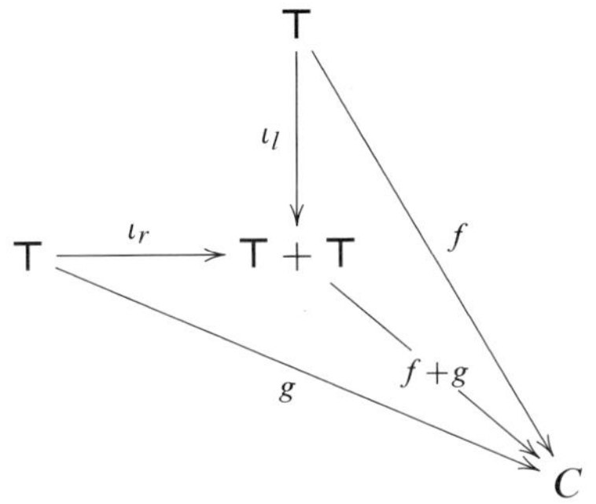

(7.16)

Since $\mathsf{T} + \mathsf{T}$ is terminal, $\iota_l = \iota_r$ and one concludes that $f = g$.

Using $\mathbf{C}(\mathsf{T}, B^A) \simeq \mathbf{C}(A, B)$, one concludes that the category is degenerate.

7.A.5 Digression: Loch Ness categories. A certain number of « solutions » to the degeneracy (inconsistency at layer -2) circulate. All those I have seen being faulty, I will not indulge in a teratology, especially since some people devote an incredible amount of energy in the production of new erroneous solutions. A few remarks:

- If there is a category-theoretic solution, one is liable to provide a *legible* category. And not to formulate the adjunction rules – say – of a professed « subtraction » – the typical connective of the category-theoretic *bricoleurs* – supposedly acting like implication, but on the left. Hence, one must provide a *concrete* category, or at least a translation into a system already having a non-degenerate category-theoretic interpretation. What the experts in « subtraction » carefully avoid doing... with good reasons.

- They prefer to fiddle with reduction rules. One thus witnesses classical logics where everything is confluent, and even strongly normalisable. One simply « forgot » the basic reductions that render commutative the diagrams. By the way, this is easy to understand: write the syntax of no matter which system *without* reduction rules and you will see that it is sN and confluent.

- Human perversity knowing no limits, one can also retain all rewritings, but as *leftmost reductions*: one can only normalise the leftmost *redex*. Church–Rosser works – for want of conflicts –, but what does this actually mean? One suspects a con in the style « paraconsistent logic » and one is right. In fact this « solution » satisfies the equation by sacrificing the *associativity* of composition.

Indeed, the deep sense, beyond technique, of Church–Rosser, is *compositionality*. Concretely this means that, in $(t)(u)(v)z$, I can as I please:

(i) Separately reduce $(t)(u)y$ into U and $(v)z$ into v', then reduce $U[v'/y]$.

(ii) Separately reduce $(u)(v)z$ into V and $(t)x$ into t', then reduce $t'[V/x]$.

This can be done without affecting the result, indeed it is the result of the normalisation of $(t)(u)(v)z$. One easily sees that a protocol of the kind « left first » does not operate in the same order of combinations (i) and (ii): a « fabricated » Church–Rosser produces nothing, but wind.

7.A.6 Polarised interpretation. The only known solution is based on *polarisation*, see [45]. This consists in:

- Selecting a *pole* P and replacing intuitionistic negation with $\neg A := A \Rightarrow P$.

- Using Gödel's translation while carefully distinguishing between *negative* formulas: simply negated, and *positive* formulas: doubly negated[14]. Classical negation will be interpreted as the exchange between $\neg A$ and $\neg\neg A$; it is involutive *a priori*.

Conjunction

$\wedge$	$\neg B$	$\neg\neg B$
$\neg A$	$\neg(A \vee B)$	$\neg\neg(\neg A \wedge B)$
$\neg\neg A$	$\neg\neg(A \wedge \neg B)$	$\neg\neg(A \wedge B)$

(7.17)

It is immediate that conjunction is commutative, associative with neutral element $\neg\neg\mathsf{T}$, all this in the sense of category-theoretic isomorphisms. There is a crucial use of $\neg(A \vee B) \simeq \neg A \wedge \neg B$.

[14]Note that this choice is not obvious, since it is possible to consider a doubly negated formula as simply negated.

Disjunction

$$\begin{array}{c|cc} \vee & \neg B & \neg\neg B \\ \hline \neg A & \neg(A \wedge B) & \neg(A \wedge \neg B) \\ \neg\neg A & \neg(\neg A \wedge B) & \neg\neg(A \vee B) \end{array} \tag{7.18}$$

This is also commutative, associative with neutral element $\neg\mathsf{T}$. To verify this, it is enough to remark that, since negation exchanges $\neg A$ and $\neg\neg A$, this tableau is the image of the previous one *modulo* De Morgan.

Finally observe that the intuitionistic distributivity of $\times$ over $+$ implies that $\wedge$ distributes over positive disjunctions and that $\vee$ distributes over negative conjunctions (still in the sense of category-theoretic isomorphisms). See also Section 12.A.

7.B Various interpretations

7.B.1 Pure λ-calculus. Scott domains have been introduced in order to interpret pure λ-calculus. We easily see that the requirement of self-application translates into the equation

$$D \simeq D^D \tag{7.19}$$

that we could surely not solve in set-theory, because of Cantor's paradox. We will try to solve the problem by means of an inductive limit (union) of domains; which seems problematic for questions of *variance*, since B^A is covariant in B and contravariant in A. Considering *embeddings*, we can succeed in making the functor covariant in both arguments. Since it moreover commutes with direct limits, we get the solution as the direct limit of a denumerable system. See the details in the setting of *coherent spaces* (Section 8.3.2).

The same construction might as well yield

$$D \simeq D^D \times A, \tag{7.20}$$

thus a « refutation » of « η ». What is encoded is not only a function, since there is a second component: the correspondence application $\rightarrowtail$ function is no longer injective. But is this counterexample *convincing*?

Strangely, Scott formulated his construction as a projective limit. You will argue that it is a matter of conventions on the sense of arrows. But this is forgetting that inductive limits, which are a variation on sums, usually correspond to union, an operation of finitary character, while projective limits correspond to products, usually infinite. But this is not the only weirdness: to see topology there is a bit abusive, we shall come back to that point.

7.B.2 Majorisability à la Howard. While dealing with interpretations, let us mention an interesting approach to system **T**. Starting with natural numbers, one builds (for instance set-theoretically) the functionals of finite type above $\mathbb{N}$. We then inductively define a binary relation $\preccurlyeq$:

Type nat: $m \preccurlyeq n$ iff $m \leq n$.

Type $A \Rightarrow B$**:** $f \preccurlyeq g$ iff for all $a \preccurlyeq b$ of type A, $f(a) \preccurlyeq g(b)$.

It is easily seen that $\preccurlyeq$ is transitive and antisymmetric. But it is not reflexive: typically, a function from $\mathbb{N}$ to $\mathbb{N}$ satisfies $f \preccurlyeq f$ when it is monotonic.

Definition 13 (Majorisability). A functional f is *hereditarily monotonic* (hC) iff it satisfies $f \preccurlyeq f$. It is *hereditarily majorisable* (hM) if there is an hC g such that $f \preccurlyeq g$.

Howard's result is that all functionals of system **T** are hM. The proof is easy. One recursively modifies the terms by replacing the recursor φ with φ_h:

$$\begin{aligned}(\varphi)Sx &= ((a)x)(\varphi)x,\\ (\varphi_h)Sx &= ((a)x)(\varphi_h)x + (\varphi_h)x,\end{aligned} \tag{7.21}$$

so as to make them increasing.

Note the similarity with Gandy (Section 6.D.4), or rather the similarity of Gandy with Howard. It is indeed plausible that Gandy took inspiration from Howard, which by the way takes nothing away from his idea.

Majorisability is part of these techniques that one should know better; here follows an application:

7.B.3 The continuity modulus. Let us come back to our domains. The basic domain for integers is $\mathbb{N}$, with as axioms $p, q \vdash$ (for $p \neq q$). $\mathbb{N}^{\mathbb{N}}$ is made of pairs $(\{p\}, q)$ or $(\emptyset, q)$, with an axiomatics not too hard to write; as to $\mathbb{N}^{\mathbb{N}^{\mathbb{N}}}$, it is made of pairs (A, r), with A a finite set of $(\{p_i\}, q_i)$ or reduced to $(\emptyset, q)$ and an axiomatics close to illegibility.

A functional Φ of type $(\mathbf{nat} \Rightarrow \mathbf{nat}) \Rightarrow \mathbf{nat}$, coming from, say, system **T**, will be interpreted by a set of pairs (A, r). Concretely, if $\Phi(f) = r$, then one can find a finite subgraph $A \subset f$ such that (A, r) is in the code of Φ[15]. Which means that one can associate to Φ a « continuity modulus », yielding a bound on the « piece of graph » of f actually used: typically the values $f(n)$ for $n \leq M$. The modulus is a recursive functional.

On the other hand, it is not *internalisable* in system **T**. Indeed, assuming that Φ takes values bounded by 2, I can surely majorise (in the sense hM) Φ by the

[15] With an extraordinary case: if f is the constant function $f(x) := q$, A can take the shape $\{(\emptyset, q)\}$.

constant functional 2, let us call it Ψ. If the modulus, as a functional of Φ and f, were definable in **T**, it would be hM and one would conclude that the modulus is majorised by an expression $\Theta(\Psi, f)$ independent of Φ. Which is clearly absurd: consider $\Phi_n(f) := 1 + (-1)^{f(n)}$.

7.B.4 Kreisel and predicativity. When Poincaré in *Science et Méthode* (1908) introduced *predicativity*, he had not the slightest idea of the dubious posterity of this idea. Poincaré, by contempt of logical paradoxes, summarily got rid of the question by requiring « predicative » definitions, i.e., not to define an object by means of a class or a set containing it. Poincaré wanted above all to react against the *abstract nonsense* of the Russell kind and recenter the debate on real mathematics, typically on geometry.

He was unfortunately taken literally by the most unbridled essentialists. His common sense – but not very deep – remark eventually became a slogan for those he was combatting: when a formalist claims to be a representative of Poincaré, it is like a creationist claiming the patronage of Darwin!

The normative activity of predicativists concentrated on second-order logic: indeed, the abstraction term $T = \{x; A\}$, when A contains second-order quantifiers, is « impredicative »: these quantifiers refer to a totality containing T. One therefore restricts comprehension to the case where A is arithmetical. Since this is not enough, one transfinitely iterates this « predicative » comprehension; but caution, the ordinal must in turn be predicative. One hardly knows why, but it is the ordinal Γ_0 which is the upper limit of predicativity: one dispenses with the details which are nothing but censorship inspired by divine revelation. By the way, one can wonder what is the status of the first non-predicative ordinal Γ_0; it is well-founded, but it is predicatively incorrect to mention it. This is reminiscent of old style libraries, with their « inferno » filled with forbidden books; in this world, there are true results that one cannot mention.

When I was young, I took this baloney seriously for a couple of years, while suspecting that it was running into sectarian activities. One day I spoke of this to Kreisel, who told me: « You know, in arithmetic, the definition of N, *the smallest integer such that* $A[n]$, is impredicative »; and I lost my ultimate doubts.

This remark of Kreisel is doubly interesting. First, in that it is typical of his defiance against ideologies. Then and above all, in that it displays this blind spot which is the subject of this book: one thinks (*modulo* a certain culture dating back to Cantor's paradox) that the notion of the set of integers is not well-defined and one bridles at it with predicative *bondage*. On the other hand, one thinks that the notion of integer is absolute. The remark of Kreisel casts a serious doubt: does this integer N really preexist its definition?

Part III

Linear logic

Chapter 8
Coherent spaces

8.1 *Grandeur* and misery of Scott domains

8.1.1 Recursive functionals. In the years around 1960, Kleene, Kreisel, Gandy and others studied recursive *functionals*, in other terms those computable functions that take as argument not an integer, but a function. This functional argument is given under the form of an *oracle*, i.e., of an arbitrary function, by no means computable[1]. The finiteness of computation suggests a notion of *continuity*: approximation by means of finite data. In fact, a functional « of type 2 », i.e., sending functions into integers, is continuous as a function of the product space $\mathbb{N}^{\mathbb{N}}$ into $\mathbb{N}$. Which is not topologically trivial: a set $X \subset \mathbb{N}^{\mathbb{N}}$ is open iff for all $f \in X$, there exists N such that, whenever g coincides with f on the arguments $0, \ldots, N-1$, then $g \in X$.

Beyond type 2, this approach no longer works.

8.1.2 *Grandeur.* Scott domains extend continuity to higher types at the cost of a somehow reasonable – especially in the recursive world – modification: the use of *partial* data.

Take for instance natural numbers: besides $0, 1, 2, \ldots$, there is an *indefinite* datum. Representing n by the singleton $\{n\}$, the indefinite datum will be the empty set. One sees the emergence of a partial order relation; in our presentation of Scott domains, the inclusion between saturated subsets. There are two possible presentations of the *domain* made of saturated subsets:

Partial order: the domain is a partially ordered set; morphisms are monotonic maps preserving directed suprema.

Topology: the domain is topologised by the fundamental open sets $\{A \,;\, a \subset A\}$, where a is a *finite* coherent set; morphisms are continuous functions.

Whatever approach we take, order, topology, or simply the explicit description I gave in Section 7.3.3, one must acknowledge that Scott managed to:

- Construct a CCC with objects of reasonable size. Indeed, domains are made of the (saturated) subsets of a denumerable set; the cardinal does not exceed that of the continuum, $2^{\aleph_0}$.
- Give an original modelisation of λ-calculus.

[1] When data come in the form of a « stream », one does not care whether or not they are produced by a machine. However, one wants to handle them *effectively*.

8.1.3 Misery. In a Scott domain, everything is under the sign of redundancy: if $(a, y) \in A$ and $a \subset b$, then $(b, y) \in A$; this primal redundancy « breeds », it accumulates nested redundancies: if $(\{(a_1, y_1), \ldots, (a_n, y_n)\}, z) \in A$ and $a_1 \subset b_1, \ldots, a_n \subset b_n$, then $(\{(b_1, y_1), \ldots, (b_n, y_n)\}, z) \in A$. One can therefore get nothing concrete from these structures, completely illegible at type 3 and beyond.

At a deeper level, one is entitled to ask whether it is quite topology. Yes, in the « legal » sense; but is this *good* topology?

One first observes that a Scott domain is *never* Hausdorff. Indeed, the only neighbourhood containing $\bar{\emptyset}$ is the full space. Scott domains enjoy the « poor man's separation » (property T_0), saying that, whenever $x \neq y$, there is an open set containing one but not the other. This property means little, since one can always ensure it by an appropriate quotient. Scott domains exclude all interesting topological spaces, in particular $\mathbb{R}$.

Grosso modo, everything works because the topological structure is very poor: it is indeed the structure of directed *suprema* styled « topology ». Everything is continuous, more or less; for instance a two-variable function separately continuous will be « Scott continuous ». Which is antagonistic to the whole topological tradition: if one distinguishes between simple, uniform, etc. forms of convergence, this is precisely because a separately continuous function is most likely *not* continuous. Except with Scott, but this is because we are not dealing with the « right » spaces.

However mathematicians make heavy use of directed suprema, typically when extending the Riemann integral to lower semi-continuous (l.s.c.) functions. But they don't call this « topology »: although one can contend that a l.s.c. function is a continuous function with values in $\mathbb{R}$ (where the open sets are now the $]r, +\infty[$, $a \in [-\infty, +\infty]$), they didn't find it worthwhile to introduce a topology « à la Scott » dedicated to l.s.c. functions. They preferred to keep the right definitions and say, in this very case, that the restriction to monotonic systems enables one to pass from a *simple* convergence to a *uniform* convergence (Dini's theorem). Instead of introducing a half-baked *ad hoc* topology, one stresses a phenomenon of *bonification*, from simple to uniform – or from weak to strong, see Section 19.4.3.

8.2 Coherent spaces

Coherent spaces are originally a simplification of Scott domains; the irredundant encoding due to *stability* enables a more limpid approach: we no longer content ourselves with the *possibility* of interpreting logic, one is able to *write down* this interpretation. One eventually discovers a logical layer finer than intuitionism, *linear logic* [37].

8.2.1 Coherent spaces. *Coherent spaces* appear as a very particular case of Scott domain, namely that of an axiomatics $\mathcal{F}$ reduced to sequents $x, y \vdash$, with $x \neq y$.

The complement of this set of pairs is *coherence*. Coherent subsets, renamed *cliques*, are always saturated: no more finiteness problems.

Definition 14 (Coherent spaces). A *coherent space* X consists of:

Web: an underlying set, its *web* $|X|$.

Coherence: a reflexive and symmetric relation, *coherence* $x \mathrel{\frown\!\!\!\!\smile}_X y$.

A *clique* $a \sqsubset X$, is a subset of $|X|$ made of pairwise coherent points.

8.2.2 A category-theoretic intuition. Remember that a partial order is a degenerate category. Instead of the dubious intuition of topology, one starts with the « partial order » presentation:

Coherent space: category $\mathbf{X}$.

Cliques: the objects of $\mathbf{X}$.

Inclusion between cliques: the unique morphism of $\mathbf{X}(a, b)$.

Monotonic function: functor from $\mathbf{X}$ to $\mathbf{Y}$.

Continuity w.r.t. directed suprema: preservation of direct limits.

Indeed, a functor from $\mathbf{X}$ to $\mathbf{Y}$ preserving *direct limits* is the same as a morphism in the sense of Scott. But let us come to direct limits:

8.2.3 Direct limits. The notion of direct limit is a particular case of *inductive* limit. A *direct system* in the category $\mathbf{C}$ consists of the following data (X_i, f_{ij}):

Objects: a family $(X_i)_{i \in I}$ indexed by a *directed* partially ordered set I.

Morphisms: for $i \preccurlyeq j \in I$, a morphism $f_{ij} \in \mathbf{C}(X_i, X_j)$.

We require that $f_{ii} = \iota_i$ and $f_{ik} = f_{jk} \circ f_{ij}$:

$$\begin{array}{ccc} X_i & \xrightarrow{\quad f_{ik} \quad} & X_k \\ & {\scriptstyle f_{ij}} \searrow \quad \nearrow {\scriptstyle f_{jk}} & \\ & X_j & \end{array} \tag{8.1}$$

The direct limit consists in completing the diagram by a « point at infinity », i.e., by an object X and morphisms $f_i \in \mathbf{C}(X_i, X)$, so as to render commutative the diagrams:

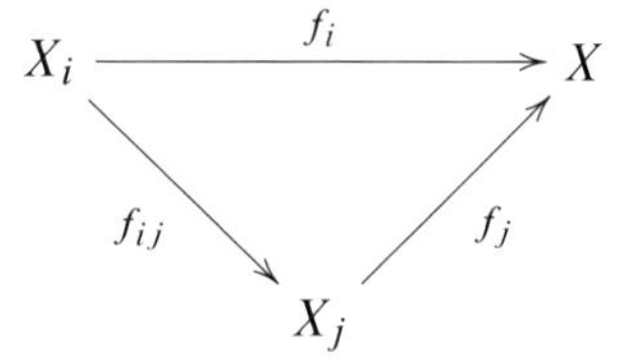

(8.2)

And this, in a universal way; in other terms, if (Y, g_i) is another candidate, there exists a unique $h \in \mathbf{C}(X, Y)$ rendering commutative the diagrams:

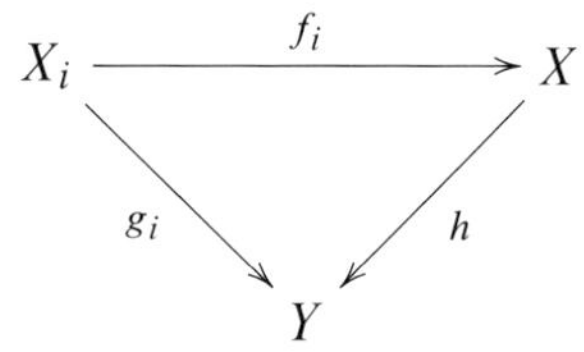

(8.3)

In a degenerated category, the direct limits are directed suprema.

8.2.4 Redundancy. If we stay there, we gain nothing; we only replace one nonsense with another. We even lose everything, since coherent spaces do not form a CCC – at least with the morphisms so far considered. Why? Because implication pushes us outside of the setting of coherent spaces.

The emblematic example, due to Plotkin [86], is the « parallel or », a function of two boolean arguments, which answers « true » when one of the two arguments is true, without even looking at the other. Anticipating the definition of a cartesian product, the coherent space **bool** × **bool** has four points, v, f, v', f', all coherent but (v, f) and (v', f'). One can define a monotonic map F from the cliques of **bool** × **bool** to the cliques of **bool**, as follows:

$$F(a) = \{v\} \quad (\text{if } v \in a \text{ or } v' \in a) \tag{8.4}$$

$$F(\{f, f'\}) = \{f\} \tag{8.5}$$

$$F(a) = \emptyset \quad (\text{in the other cases}) \tag{8.6}$$

The coding of F à la Scott is made of the pairs

$$(\{v\}, v),\ (\{v'\}, v),\ (\{v, v'\}, v),\ (\{v, f'\}, v),\ (\{v', f\}, v),\ (\{f, f'\}, f).$$

This coding is redundant: there are five data for the sole (8.4), two minimal choices $\{v\}, \{v'\}$, whose presence is anyway necessary and lax ones $\{v, v'\}, \{v, f'\}, \{v', f\}$. One could try to avoid redundancy by keeping only the minimal choices, but this does not work; indeed, let us consider:

$$G(\{v, v'\}) = \{v\}$$
$$G(\{f, f'\} = \{f\}$$
$$G(a) = \emptyset \qquad \text{(in the other cases)}$$

whose coding is made of $(\{v, v'\}, v), (\{f, f'\}, f)$. Observe that $G \subset F$ (pointwise); if one wants to reflect this inclusion on the encodings, one must retain the redundant code $(\{v, v'\}, v)$ of F.

8.2.5 Pull-backs. The playmate of direct limits is the *pull-back*, a form of projective limit.

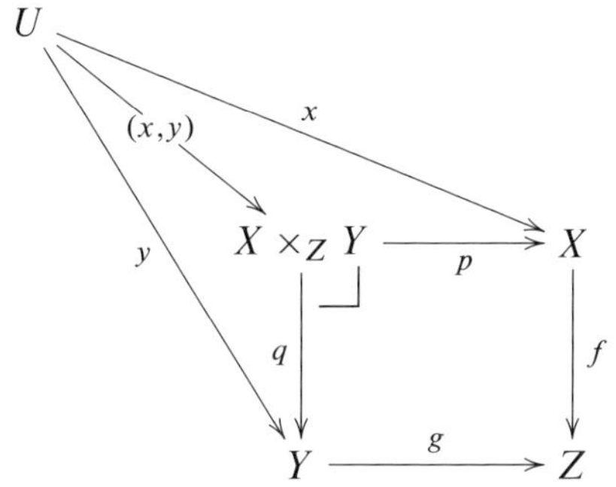

(8.7)

One seeks the universal completion of a diagram consisting of two morphisms $f \in \mathbf{C}(X, Z)$, $g \in \mathbf{C}(Y, Z)$. The solution of this problem is the *pull-back*; note the « corner » ⌟ which signals a pull-back. In a degenerate category, a pull-back is the infimum of two elements x, y such that there exists z with $x, y \preccurlyeq z$.

It is natural to ask that a functor commutes with direct limits and pull-backs. In this case one must consider *cartesian* natural transformations, i.e., such that:

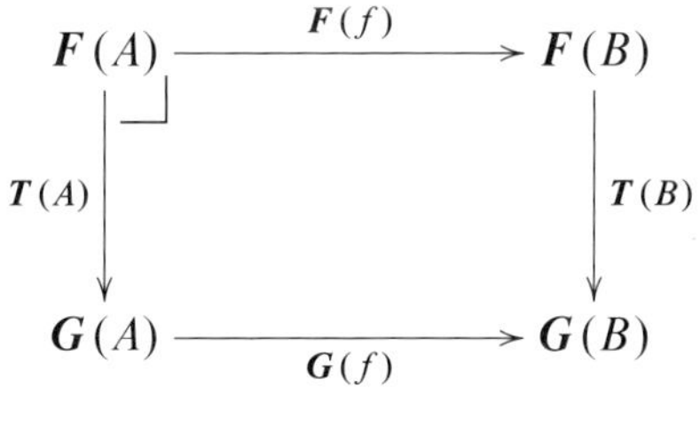

(8.8)

We eventually completed our definition:

Preservation of coherent intersections: preservation of pull-backs.

Berry order: cartesian natural transformations.

We shall soon detail all of these. The fundamental novelty is that pull-backs introduce an idea of « principal solution » in matters of coding. One must stress the fact that this sort of thing horrifies logicians: it is indeed in the name of a certain « esthetics of ugliness » that the preservation of pull-backs has been, not only neglected, but also decried. Since it plays against the arbitrariness of coding, it is a *structuring* property. It also knocks off balance certain dogmas, set-theoretic – or topological, since pull-backs do not make sense topologically. I also think that the habit – the abuse – of reductions in all senses, makes it possible that, as soon as he can reduce something to a finite setting, the logician can feel no longer involved: finite for finite, it makes no longer any difference as to logical complexity.

8.2.6 Stable functions and stable order. I now faithfully translate the category-theoretic ideas which precede; *stability* (in the setting of Scott domains) is due to Berry [12].

Definition 15 (Stability). If X and Y are coherent spaces, a *stable function* from X to Y satisfies:

Cliques: if $a \sqsubset X$, then $F(a) \sqsubset Y$.

Monotonicity: if $a \subset b \sqsubset X$, then $F(a) \subset F(b)$.

Continuity: $F(\uparrow\bigcup_i a_i) = \uparrow\bigcup_i F(a_i)$, provided the family (a_i) is directed.

Stability: if $a \cup b \sqsubset X$, then $F(a \cap b) = F(a) \cap F(b)$.

One defines the *stable* order between stable functions from X to Y:

Berry order: $F \sqsubset G$ iff for all $a \subset b \sqsubset X$, $F(a) = F(b) \cap G(a)$.

The function F of (8.4)–(8.6) is not stable, since $F(\{v\}) = F(\{v'\}) = \{v\}$, but $F(\emptyset) \neq \{v\}$, although $\{v, v'\}$ is a clique. Stability corresponds to a determinism of computation (not only of its result): when performing a computation, a well-defined part of the data is actually used. Which is not the case for the « parallel or », since there is an ambiguity as to the information actually needed: the answer is « true » when one of the two arguments is true, which yields two possible ways of acting when both are true.

The stable order appears as the necessary technical companion of stability, in view of the *adjunction* which defines the function space. Indeed, a stable function from $X \times Y$ to Z must appear as a stable function from X to Z^Y. In (8.4)–(8.6), compare the unary functions $F_\emptyset(\{v'\}) = \{v\}$, $F_\emptyset(a) = \emptyset$ $(a \neq \{v'\})$ and $F_{\{v\}}(a) = \{v\}$: since $F_\emptyset(\emptyset) = \emptyset \neq \{v\} = F_\emptyset(\{v'\}) \cap F_{\{v\}}(\emptyset)$, $F_\emptyset \sqsubset F_{\{v\}}$ fails. Indeed, these two stable maps are such that $F_\emptyset \subset F_{\{v\}}$, but the minimal data for $F_\emptyset$ are no longer minimal for $F_{\{v\}}$.

8.2.7 Coherent spaces as a CC. First of all, we have really defined a category. This is because stable functions are defined by preservation (e.g., if $a \cup b \sqsubset X$, then $F(a) \cup F(b) \subset F(a \cup b) \sqsubset Y$).

Next, we need a product and a terminal object. This last point is easy: it is enough to take the empty coherent space $\emptyset$. Indeed it has only one clique, $\emptyset$ and the function $F(a) := \emptyset$ associating to each clique of X the empty set is surely stable; moreover, it is the only choice.

The cartesian product is defined as follows[2]:

Web: $|X \mathbin{\&} Y| := |X| + |Y| = |X| \times \{1\} \ \cup \ |Y| \times \{2\}$.

Coherence:

$$\begin{aligned}
(x,1) \frown_{X\&Y} (x',1) &\iff x \frown_X x' \\
(y,2) \frown_{X\&Y} (y',2) &\iff y \frown_Y y' \\
(x,1) \frown_{X\&Y} (y,2) &
\end{aligned}$$

The following property explains everything:

Proposition 9. *The cliques $c \sqsubset X \mathbin{\&} Y$ are exactly the disjoint unions $a + b$ of a clique $a \sqsubset X$ and a clique $b \sqsubset Y$, this decomposition being unique.*

Proof. Obvious: there is no incoherence between X and Y. □

One can define the two projections by

$$\begin{aligned}
\pi_l(a+b) &= a, \\
\pi_r(a+b) &= b.
\end{aligned} \tag{8.9}$$

It is obvious that projections are stable (and even linear, see Chapter 9).

Let us proceed with our studious checking: if F, G are stable functions from Z into X, Y, then one can define (F, G) from Z into $X \mathbin{\&} Y$ by

$$(F,G)(c) := F(c) + G(c) \quad (c \sqsubset Z). \tag{8.10}$$

It is immediate that $\pi_l \circ (F,G) = F, \pi_r \circ (F,G) = G$. Unicity is so obvious that one does not even dare to justify it.

8.2.8 Coherent spaces as a CCC

Proposition 10. *Let F, G be stable functions from X into Y, let $A \sqsubset X$ and $y \in F(A)$. Then:*

(i) *There exists $a \subset A$, a finite, such that $y \in F(A)$.*

[2] Anticipating linear logic, I introduce the notation & instead of × that should be more natural.

(ii) *If a is chosen **minimal**, it is **minimum**, i.e., unique.*

(iii) *If $F \sqsubset G$, then $y \in G(a)$ and a remains the minimum choice.*

Proof. Write A as the directed union of its finite subsets. Then

$$F(A) = \uparrow\bigcup\{F(a)\,;\, a \subset A,\ a \text{ finite}\}.$$

Which proves (i). Another choice $b \subset A$ would yield by stability, since $a \cup b \sqsubset A$, $y \in F(a) \cap F(b) = F(a \cap b)$: if a is minimal, then $a = a \cap b$, hence $a \subset b$. Which proves (ii). Finally, if $b \subset a$ and $y \in G(b)$, then $y \in G(b) \cap F(a) = F(b)$ hence $a = b$. Which proves (iii). □

Definition 16 (Skeleton). If F is a stable map from X into Y, one defines its *skeleton*[3] $\mathrm{Sk}(F)$:

$$\mathrm{Sk}(F) := \{(a, y)\,;\, y \in F(a) \wedge \forall b \subsetneq a \;\; y \notin F(b)\}. \tag{8.11}$$

This is not a matter of graphs. For instance, take the most trivial stable function, the identity function ι_X from X into X: its graph is made of the pairs (a, a), where $a \sqsubset X$; while its skeleton corresponds to the minimal solutions to $y \in F(a) = a$: this yields $\mathrm{Sk}(\iota_X) = \{(\{x\}, x); x \in |X|\}$.

One now defines the coherent space $X \Rightarrow Y$. Let us introduce the notation $x \frown x'$ for *strict* coherence, i.e., for $x \coh x' \wedge x \neq x'$.

Web: $|X \Rightarrow Y| := X_{\mathrm{fin}} \times |Y|$, X_{fin} being the set of *finite* cliques of X.

Coherence:

$$\begin{aligned}(a, y) \coh_{X \Rightarrow Y} (a', y') \iff\ & (a \cup a' \sqsubset X \Rightarrow y \coh_Y y') \\ & \wedge\ (a \cup a' \sqsubset X \wedge a \neq a' \Rightarrow y \frown_Y y')\end{aligned}$$

Theorem 24 (Representation). Sk *defines a bijection between the stable functions from X into Y and the cliques of $X \Rightarrow Y$. The reciprocal bijection associates to a clique $C \sqsubset X \Rightarrow Y$ the stable function $(C)\cdot$ defined by*

$$(C)A := \{y\,;\, \exists a \subset A \;\; (a, y) \in C\}. \tag{8.12}$$

Moreover, the bijection exchanges the Berry order and inclusion.

Proof. We can easily see that $\mathrm{Sk}(F)$ is a clique. If $(a, y), (a', y') \in \mathrm{Sk}(F)$ and if $a \cup a' \sqsubset X$, then $y, y' \in F(a \cup a')$, hence $y \coh_Y y'$. Moreover, if $a \neq a'$, we cannot

[3]Which was for a long time called « trace », a hateful terminology, moreover untenable in the presence of a *genuine* use of linear algebra.

have $y = y'$, since there would thus be two minimal solutions $y \in F(a), y \in F(a')$ with $A = a \cup a'$.

Conversely, observe that $(C)\cdot$ is stable: $(C)A$ is a clique, since if $a, a' \subset A$, $a \cup a' \sqsubset X$, hence if $(a, y) \in C$, $(a', y') \in C$, we must have $y \circ_Y y'$. The definition is obviously monotonic, and since it only makes use of an existential quantification, it commutes to directed unions. Finally, it is stable: if $A \cup B$ is a clique and $y \in (C)A \cap (C)B$, this is because $(a, y), (b, y) \in C$. But $a \cup b \subset A \cup B$ is a clique and coherence implies that $a = b$: we therefore deduce that $y \in (C)(A \cap B)$, the non-trivial part of stability.

Finally, we verify without the slightest problem that $(\mathrm{Sk}(F))A = F(A)$ and $\mathrm{Sk}((C)\cdot) = C$.

The part concerning the order is nothing but point (iii) of Proposition 10. □

Proposition 11. *Coherent spaces do form a CCC.*

Proof. The function ϵ from $(X \Rightarrow Y)\ \&\ X$ into Y, defined by $\epsilon(C + A) := (C)A$, is stable: we did check it with C constant while proving the theorem. If F is a stable map from $Z \times X$ into Y, we show that the map $D \rightsquigarrow \mathrm{Sk}(F(D + \cdot))$ is stable. The commutativity of (7.8) means that $(\mathrm{Sk}(F(D + \cdot)))A = F(D + A)$. The unicity of the solution means that, if Φ is a stable function from Z into $X \Rightarrow Y$, then $\mathrm{Sk}((\Phi(D)\cdot) = \Phi(D)$ (this is « η »). □

Personally, I find this sort of result illegible. One must write it, but not read it, under the penalty of becoming a bureaucrat. The real result is the theorem, which does establish the right correspondence. Since this correspondence is natural in the natural sense of the term, it is also natural in the category-theoretic sense.

8.3 Interpretation of system F

Remember that this is an *essentialist* interpretation, types first.

8.3.1 Embeddings. Our problem is to work with *variable* types in the style of $X \Rightarrow X$; one would like to make functors out of them, but there is a problem of *variance*. Indeed, $X \Rightarrow Y$ is covariant in Y, contravariant in X: from stable functions f from X' into X and g from Y into Y', one can pass from $X \Rightarrow Y$ to $X' \Rightarrow Y'$ – *modulo* the coding of morphisms – by composition:

$$C \rightsquigarrow \mathrm{Sk}(g \circ (C) \cdot \circ f). \tag{8.13}$$

Everything works « in the wrong direction » for f. We will try to modify the variance by replacing f with its inverse, which obviously requires certain restrictions on f. The idea is that, if $|X| \subset |X'|$, then $f(a) := a \cap |X|$ is invertible (on one side) with inverse $\pi(a) := a$. Our attention is thus attracted by the *inclusions*

$|X| \subset |X'|$. A very important thing is that $\bigcirc_X$ is the restriction of $\bigcirc_{X'}$ to $|X|$: it is only in this way that π can be the inverse of something; it is what we shall simply note $X \subset X'$. In order to make the category-theoretic machinery work, inclusions are not enough, we need them *up to isomorphism*, which yields:

Definition 17 (Embeddings). An *embedding* of X into Y is an *injective* function from $|X|$ into $|Y|$ such that $x \frown x'$ iff $f(x) \frown f(x')$.

If one takes embeddings as morphisms, & and $\Rightarrow$ become *covariant* functors. Rather than a verification without interest, let us remark that in – say – the case of implication:

- If $|X| \subset |X'|$ and $|Y| \subset |Y'|$, then we have $|X \Rightarrow Y| \subset |X' \Rightarrow Y'|$ and $\bigcirc_{X \Rightarrow Y} = \bigcirc_{X' \Rightarrow Y'} \upharpoonright |X \Rightarrow Y|$. In other words, $\Rightarrow$ preserves inclusions.
- For this reason, it preserves embeddings.

8.3.2 The « Scott model ». Let us get at direct limits. A direct system (X_i, f_{ij}), where the f_{ij} are inclusions admits the direct limit (X, f_i), where $|X| = \uparrow\bigcup |X_i|$, $\bigcirc_X = \uparrow\bigcup \bigcirc_{X_i}$ and f_i is the inclusion of $|X_i|$ into $|X|$. Note that this limit only exists because the system is *directed*; if $x \in |X_i|$, $y \in |X_j|$, then $x, y \in |X_k|$ for an appropriate k, which enables one to define $x \bigcirc_X y :\Leftrightarrow x \bigcirc_{X_k} y$. By the way, note that functors like &, $\Rightarrow$ will preserve directed unions. This is obvious on the basis of their *existential*, expansive, expression.

As a consequence of the existence of direct limits of inclusions, direct limits of embeddings do exist; I give their definition without justification:

$|X|$: the set $\{(x, i); x \in |X_i|\}$, quotiented by the relation $(x, i) \sim (y, j) \Leftrightarrow \exists k(i, j \preccurlyeq k \wedge f_{ik}(x) = f_{jk}(y))$. It is the directedness of I which makes $\sim$ an equivalence.

f_i: $f_i(x) := (x, i)$ (or rather its equivalence class).

$\bigcirc_X$: $(x, i) \bigcirc_X (y, j) \Leftrightarrow \exists k(i, j \preccurlyeq k \wedge f_{ik}(x) \bigcirc_{X_k} f_{jk}(y))$

From what precedes, our functors do preserve direct limits.

Let us now take a coherent space X with $|X| = \{x\}$. One can embed $X_0 := X$ into $X_1 := X \Rightarrow X$, by sending the point x to the point $(\emptyset, x)$. If one has constructed $X_0, \ldots, X_{n+1}$, together with embeddings π_i from X_i into X_{i+1}, one can define $X_{n+2} := X_{n+1} \Rightarrow X_{n+1}$ and an embedding π_{n+1} of X_{n+1} into X_{n+2}, $\pi_{n+1} := \pi_n \Rightarrow \pi_n$. If, for $m \leq n$ one defines $\pi_{mn} := \pi_{n-1} \circ \cdots \circ \pi_m$, then it is clear that (X_m, f_{mn}) is a direct system, with limit (X, f_m). Since the functor $\Rightarrow$ commutes with direct limits, we get:

$$\begin{aligned}(X \Rightarrow X, f_m \Rightarrow f_m) &\simeq \varinjlim(X_m \Rightarrow X_m, f_{mn} \Rightarrow f_{mn}) \\ &= \varinjlim(X_{m+1}, f_{m+1n+1}) = (X, f_{m+1}).\end{aligned}$$

Theorem 25 (Scott's theorem). *The equation $X \simeq X \Rightarrow X$ admits a non-empty solution.*

Of course Scott proved the result for his *domains*, not for coherent spaces. But this makes no difference, the only thing he uses is the *expansive* character of implication, i.e., commutation to direct limits.

8.3.3 Pull-backs. The main problem with the interpretation of system **F** is the definition of extraction $\{A\}X$. Indeed, an object of type $\forall XA$ must represent a function capable of associating to each coherent space X a clique $A_X \sqsubset A[X]$. Which poses obvious problems of circularity: the idea is to circumvent it by means of direct limits. Suppose that one can approximate coherent spaces by *finite* coherent spaces. One could then define – at least try to define – the extraction $\{A\}X$ on arbitrary spaces from the extraction on finite ones.

This is enough to explain the relinquishment of Scott domains. A Scott domain cannot be approximated by finite ones: since the embeddings are faithful, a subdomain $X_0 \subset X$ containing x must also contain all y such that $x \vdash y$; this makes X_0 infinite. This constraint is at the basis of coherent spaces: no saturation!

What is a pull-back of inclusions? If $|X|, |Y| \subset |Z|$, we get the pull-back by restricting Z to $|X| \cap |Y|$. It is immediate that the functors &, $\Rightarrow$ preserve the pull-backs of inclusions. By the way, the preservation of direct limits and pull-backs corresponds to existential definitions (direct limits) with *unicity* of witnesses, see for instance the definition of $(C)A$ in (8.12), where the witness a is indeed unique.

The pull-back of embeddings π from X into Z and π' from Y into Z is a variation on the theme of inclusion: let $T = \{(x, y); \pi(x) = \pi'(y)\}$, with $(x, y) \circ_T (x', y') \Leftrightarrow x \circ_X x' \ (\Leftrightarrow y \circ_Y y')$ and the embeddings $(x, y) \rightsquigarrow x$, $(x, y) \rightsquigarrow y$. One concludes that the functors &, $\Rightarrow$ preserve pull-backs.

8.3.4 Variable spaces and cliques. This leads to the following definition:

Definition 18 (Variable coherent space). A (one parameter) *variable coherent space* is a functor Φ from the category of coherent spaces (with embeddings) in itself preserving direct limits and pull-backs.

Suppose, for legibility, (it really changes nothing) that Φ also preserves inclusions. One calls *minimal datum* (w.r.t. Φ) a pair (X_0, x) consisting of a finite coherent space X_0 and of $x \in |\Phi(X_0)|$ such that $x \notin |\Phi(Y)|$ when $Y \subsetneq X_0$. Obviously:

Proposition 12. *If $x \in |\Phi(X)|$, there is a unique minimal datum (X_0, x) such that $X_0 \subset X$ and $x \in \Phi(X_0)$.*

Minimal data from a proper class, but they are indeed very scarce *if one ignores duplicates*. (X, x) and (Y, y) are said to be *equivalent* when there is an isomorphism π of X onto Y such that $\Phi(\pi)(x) = y$.

One will equip the equivalence classes of minimal data of a coherence relation: two classes c, d are (strictly) incoherent when there are representatives $(X, x) \in c$, $(Y, y) \in d$ such that X, Y are subspaces of the same space Z and such that x and y are strictly incoherent in $\Phi(Z)$. This definition makes appear the possibility – by far the most frequent – of *schizophrenic*, i.e., self-incoherent, classes.

Let us indulge in a small computation, with $\Phi(X) := (X \ \& \ X) \Rightarrow X$. The minimal data are of the form $(X, (a + b, x))$, with $a \cup b \cup \{x\} = |X|$. There are only three self-coherent classes, indeed:

- $(\{x\}, (\emptyset + \emptyset, x))$ is schizophrenic: it is equivalent to $(\{y\}, (\emptyset + \emptyset, y))$; if $|Z| = \{x, y\}$, with x, y incoherent, then $(\emptyset + \emptyset, x)$ and $(\emptyset + \emptyset, y)$ are incoherent in $\Phi(Z)$.

- $(X, (a + b, x))$ is schizophrenic when, say, $y \in a$ with $y \neq x$. Let us form Z by duplicating the point y: one adds y' and we only need to make precise the relation between the « twins »: they are coherent. If X' is the subspace of Z obtained by replacing y by its twin y', one gets another element of the class, $(X', (a' + b', x))$, by replacing in a (and, in case, in b) y with y'. Since $a \cup a', b \cup b'$ are cliques and $a \neq a'$, it follows that $(a + b, x)$ and $(a' + b', x)$ are incoherent in $\Phi(Z)$.

We eventually found three self-coherent classes, pairwise incoherent, those of $(\{x\}, (\{x\} + \emptyset, x))$, $(\{x\}, (\emptyset + \{x\}, x))$ and $(\{x\}, (\{x\} + \{x\}, x))$.

One defines, by restricting to self-coherent classes, a coherent space $\mathrm{Sk}(\Phi)$.

8.3.5 Variable cliques. A *variable clique* of Φ, is the datum, for any coherent space X, of a clique $A_X \sqsubset \Phi(X)$. Assuming, for legibility, that Φ preserves inclusions, we require

$$X \subset Y \ \Rightarrow \ A_X = A_Y \cap |X| \tag{8.14}$$

with the following consequence: A can be represented by its minimal data, that is, by $\mathrm{Sk}(A) := \{(X, x) ; x \in A_X \text{ and } (X, x) \text{ minimal }\}$. Indeed, we have $A_Y = \{x ; \exists X \subset Y \ (X, x) \in \mathrm{Sk}(A)\}$. Which can be rigourously written:

Definition 19 (Variable cliques). A *variable clique* of Φ, it is the datum, for any coherent space X, of a clique $A_X \sqsubset \Phi(X)$ such that, if π is an embedding of X into Y, then

$$A_X = \Phi(\pi)^{-1}(A_Y). \tag{8.15}$$

With such a definition one sees that:

- The skeleton $\mathrm{Sk}(A)$ is a union of classes of minimal data.
- These classes are pairwise coherent, in particular self-coherent.

We thus get a bijection between cliques of $\mathrm{Sk}(\Phi)$ and *variable cliques* of Φ.

In the case of $\Phi(X) := (X \,\&\, X) \Rightarrow X$, there are four variables cliques:

$$\begin{aligned}
\emptyset\colon &\quad A_X = \emptyset, \\
\{(\{x\}, (\{x\}, \emptyset, x))\}\colon &\quad (A_X)(a+b) = a, \\
\{(\{x\}, (\emptyset, \{x\}, x))\}\colon &\quad (A_X)(a+b) = b, \\
\{(\{x\}, (\{x\}, \{x\}, x))\}\colon &\quad (A_X)(a+b) = a \cap b.
\end{aligned}$$

8.3.6 Finalisation. The idea can be carried away without a problem up to the end [36]. Observe that we do not ask to *analyse* the functors corresponding to variable types, which can be very complicated. These functors are *given* to us, which is enough to define the coding.

8.A Asymmetric interpretations

8.A.1 Generalities. The three-valued interpretations of **LK** (Section 3.3.3), have unexpected scions in the functional universe. I shall give an interpretative scheme, then apply it to a particular case.

Suppose that one disposes, in the setting of a concrete category-theoretic interpretation, of two variants: a weak one and a strong one, noted wA, sA of each object A. And that

$$\begin{aligned}
sA &\subset wA, \\
sA \Rightarrow wB &= w(A \Rightarrow B), \\
s(A \Rightarrow B) &= wA \Rightarrow sB, \\
wA \,\&\, wB &= w(A \,\&\, B), \\
s(A \,\&\, B) &= sA \,\&\, sB.
\end{aligned} \tag{8.16}$$

Then it is possible to prove that all normal terms satisfy the weak version: this is nothing but the functional rereading of « cut-free implies not false ». This is obviously very interesting in the case of systems not enjoying normalisation, as we shall see it. Note that I wrote the equations that seemed the most natural to me, but that we only need the inclusions left $\subset$ right. Let us give an example:

8.A.2 Quantitative interpretation. My first incursion into category-theoretic interpretations [40] would not deserve inclusion here if it were not for this example. A *quantitative domain* is a set X, at most denumerable. The objects of the domain are expressions $\sum_{x \in X} n_x \cdot \vec{x}$, where the coefficients n_x are either natural numbers, or the value ∞. Morphisms from X to Y are given by *analytic*[4] functions:

$$\Phi\Big(\sum_{x \in X} n_x \cdot \vec{x}\Big) = \sum_{y \in Y} \Big(\sum_{a \in X_{\text{fin}}} N_{(a,y)} \cdot \prod_{x \in a} n_x\Big) \cdot \vec{y} \tag{8.17}$$

Here X_{fin} denotes the finite multisets of X, which makes $\prod_{x \in a} n_x$ a monomial: if $a = 2x + 3y + z$, $\prod_{x \in a} n_x = n_x^2 n_y^3 n_z$. One sees that analytic functions from X into Y can be represented by means of the domain $X_{\text{fin}} \times Y$.

One naturally gets interested in the *finiteness* of coefficients. Apropos of finiteness, there are two notions:

Weak finiteness: all coefficients are finite.

Strong finiteness: moreover, they are almost all null.

We can easily check that the pair weak/strong actually satisfies the conditions of the previous section. For instance: if Φ is strongly finite and $\sum_{x \in X} n_x \cdot \vec{x}$ is weakly finite, then $\Phi(\sum_{x \in X} n_x \cdot \vec{x})$ is strongly finite. Also, if $\Phi(\sum_{x \in X} n_x \cdot \vec{x})$ is weakly finite for all strongly finite $\sum_{x \in X} n_x \cdot \vec{x}$, then Φ is weakly finite.

Conclusion: in a quantitative interpretation of λ-calculus, normal terms (and therefore normalisable terms) have a weakly finite interpretation.

8.A.3 Hexagons. Hexagons yield an asymmetric interpretation of logic when one retains the usual morphisms. The functors such as $X \Rightarrow X$ don't know what to do, so they become schizophrenic: they split their covariant part from their contravariant part [8].

Let F, G be two functors, contravariant in X, covariant in Y. A *hexagonal* transformation from F to G is a family A_X of morphisms rendering commutative the following diagram, for all X, Y and $f \in \mathbf{C}(X, Y)$:

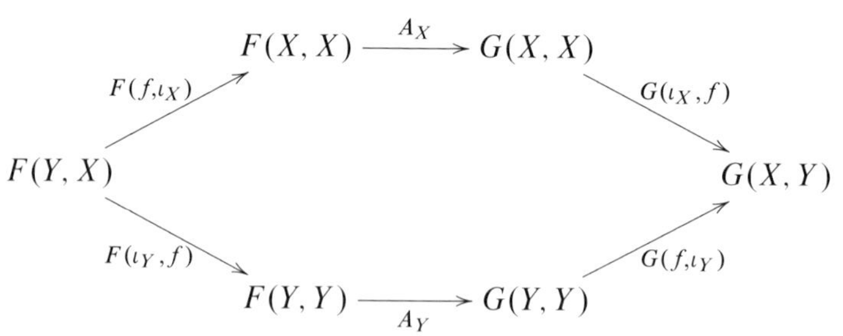

(8.18)

[4]This work is the genuine source of linear logic, since the linear case is spectacular: the monomials are of degree 1.

If $F(X,Y) = G(X,Y) = X \Rightarrow Y$, if A_X is the morphism $\bar{n}_X(h) := h^n$, the diagram says that, for $g \in \mathbf{C}(Y,X)$ and $f \in \mathbf{C}(X,Y)$:

$$f \circ \bar{n}_X(g \circ f) = \bar{n}_Y(f \circ g) \circ f \tag{8.19}$$

Both sides can indeed be written $f \circ g \circ f \circ \cdots \circ g \circ f$.

This interpretation is only valid for *normal* terms: indeed it stumbles on the impossibility to *compose* hexagons, in conformity with the limitations of asymmetrical interpretations: in the following diagram, the « central diamond » is ill-disposed, nothing can be done.

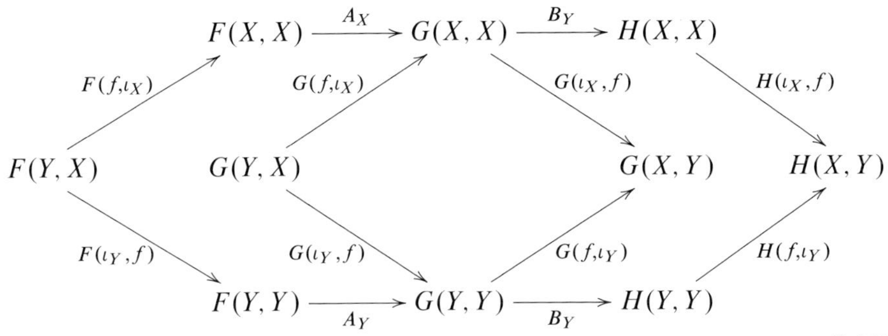

(8.20)

One should not think that a gimmick can reglue these two diagrams: thus, if the « central diamond » were a pull-back, an appropriate morphism from $F(Y,X)$ to $G(Y,X)$ would prompt the desired gluing; but this hypothesis is not very robust. We have just reached the heart of darkness of cut-elimination! In spite of its obvious interest, hexagonality never bore that much fruit; presumably for want of the right methodological setting.

Chapter 9
Linear logic

9.1 Linearity in coherent spaces

9.1.1 Definition and examples. If we consider the elimination rules as functions from the main premise to the conclusion: $\pi_l : X \ \& \ Y \mapsto X$, $\pi_r : X \ \& \ Y \mapsto Y$, $(\cdot)a : (X \Rightarrow Y) \mapsto Y$, it turns out that they enjoy an additional property, *linearity*:

Definition 20 (Linearity). A stable function F is *linear* when it preserves arbitrary coherent unions.

For instance, $(C \cup D)a = (C)a \cup (D)a$, which can be seen in equation (8.12).

Proposition 13. *F stable from X to Y is linear iff $F(\emptyset) = \emptyset$ and $F(a \cup b) = F(a) \cup F(b)$ for all cliques $a, b \in X$ such that $a \cup b \sqsubset X$.*

Proof. Any union is a *directed* union of finite unions. Finite unions can be handled by means of the 0-ary ($\emptyset$), unary (trivial) and binary ($a \cup b$) cases. □

There is another important example of a linear function, the identity function: indeed linearity is a preservation and identity is the only function preserving *everything*. More generally, *embeddings* are linear.

Linearity can alternatively be presented as the preservation of:

- Coherent differences: if $a \cup b \subset X$, $F(a \setminus b) = F(a) \setminus F(b)$. Indeed, if $c := a \setminus b$, then $a = c \cup (a \cap b)$ hence $F(c) \cap F(a \cap b) = F(c) \cap F(a) \cap F(b) = \emptyset$, but also $F(a) = F(c) \cup F(a \cap b) = F(c) \cup (F(a) \cap F(b))$.
- Disjoint unions: if $a = \sum_i a_i$, then $F(a) = \sum_i F(a_i)$.

These alternative preservations encompass stability, indeed:

- $a \cap b = a \setminus (a \setminus b)$.
- $a = a \cap b + a \setminus b$.

The preservation of sums and intersections extends to that of differences; which can be algebrised; this is why the interpretation can be reformulated in terms of vector spaces, e.g., Banach spaces (Section 15.A).

9.1.2 The category COH

Definition 21 (COH). One defines the category **COH** by:

Objects: coherent spaces.

Morphisms: $\mathbf{COH}(X, Y)$ consists of the *linear* functions from X to Y.

This category satisfies almost everything expected from a category. It is not a CCC, but it is a closed *monoidal* category; moreover, one can *reconstitute* a structure of CCC in it.

9.1.3 Linear implication

Proposition 14. *F from X to Y is linear iff its skeleton is made of pairs $(\{x\}, y)$.*

Proof. If F is linear, let us write $a \sqsubset X$ as the disjoint union of the singletons $\{x\}$, for $x \in a$. If $y \in F(a)$, then $y \in F(\{x\})$ for a certain x, which is a minimal choice, since $F(\emptyset) = \emptyset$. Conversely, if $C = \mathrm{Sk}(F)$ is made of pairs $(\{x\}, y)$, then $(C)\bigcup_i a_i = \{y\,;\, \exists x \in \bigcup_i a_i\ (\{x\}, y) \in C\} = \bigcup_i \{y\,;\, \exists x \in a_i\ (\{x\}, y) \in C\} = \bigcup_i F(a_i)$ □

If we restrict ourselves to linear functions, the expression $(\{x\}, y)$ is too pedantic: (x, y) is preferable. Linear implication is defined (*modulo* the embedding $(x, y) \rightsquigarrow (\{x\}, y)$), as a *subspace* of intuitionistic implication.

Definition 22 (Linear implication). If X, Y are coherent spaces, we define $X \multimap Y$ by:

Web: $|X \multimap Y| = |X| \times |Y|$.

Coherence:

$$\begin{aligned}(x, y) \frown_{X \multimap Y} (x', y') \iff & (x \frown_X x' \Rightarrow y \frown_Y y') \\ & \wedge (x \smallfrown_X x' \Rightarrow y \smallfrown_Y y').\end{aligned} \tag{9.1}$$

The definition of coherence can also be written in a single line:

$$(x, y) \smallfrown_{X \multimap Y} (x', y') \iff (x \frown_X x' \Rightarrow y \smallfrown_Y y'). \tag{9.2}$$

This is only to save space, since the version (9.1) is more instructive.

9.1.4 Linear negation. It is time to introduce the abbreviations:

Incoherence: $x \asymp_X x' :\Leftrightarrow x \not\succ_X x'$.

Strict incoherence: $x \smile_X x' :\Leftrightarrow x \not\supset_X x'$.

With respect to X, an *anticlique* is a subset made of pairwise *incoherent* points. Indeed, the anticliques of X are the cliques of its *linear negation* $\sim X$. Originally, I used the notation $X^{\perp}$, which I am compelled to relinquish, at least in the setting of this book: it conflicts with the *real* orthogonality, that of euclidian or hilbertian spaces.

Definition 23 (Linear negation). If X is a coherent space, its *linear negation* is the space $\sim X$ with the same web, where coherence has been replaced with incoherence.

$$x \supset_{\sim X} x' : \iff x \asymp_X x'$$

Theorem 26 (Linear negation). *Linear negation extends into a contravariant involution of the category* **COH** *of coherent spaces.*

Proof. Since $(x, y) \supset_{X \multimap Y} (x', y') \Leftrightarrow (y, x) \supset_{\sim Y \multimap \sim X} (y', x')$, one may define, given $F \sqsubset X \multimap Y$, its *adjoint* $\sim F \sqsubset \sim Y \multimap \sim X$ by $\sim F := \{(y, x); (x, y) \in F\}$. Alternative approach: $\sim X$ is isomorphic to $X \multimap \curlywedge$, where $\curlywedge$ is a one-point space: the adjoint is isomorphic to the contravariant functor $\cdot \multimap \curlywedge$. $\square$

The definition of coherence in $X \Rightarrow Y$ splits into two parts.

(i) The function sends cliques to cliques.

(ii) The function is stable.

Linearly, (i) takes the form of an implication $x \supset x' \Rightarrow y \supset y'$, while (ii) takes the form of an implication $x \frown x' \Rightarrow y \frown y'$. Negation exchanges (i) and (ii): (i) rewrites as $y \frown y' \Rightarrow x \frown x'$, while (ii) rewrites as $y \supset y' \Rightarrow x \supset x'$. One sees that stability is nothing but the possibility of defining an *adjoint* function, by associating to $y \in |Y|$ the minimal datum $\{x\}$ such that $y \in F(\{x\})$.

This is the main interest of linear maps: indeed, if F is plain stable, we cannot build any reasonable adjoint map. This is easy to understand: the combination $X_{\text{fin}} \times |Y|$ is lopsided in favour of X. By the way, we had the opportunity to explain (Section 4.1.3) that the absence of an involutive intuitionistic negation is not « bad will », but a consequence of left structural rules, of which we shall precisely see that they *contradict* linearity.

9.1.5 Coherence and duality. If $a \sqsubset X, \sim X$, then $\sharp(a) \leq 1$: two points cannot be both coherent and incoherent. Hence a clique and an anticlique intersect in at most one point. Which enables one to enunciate adjunction under the splendid form:

Theorem 27 (Adjunction). *The adjoint $\sim F$ is uniquely defined by the condition*

$$\sharp(F(a) \cap b) = \sharp(a \cap \sim F(b)) \quad (a \sqsubset X, b \sqsubset \sim Y). \tag{9.3}$$

Proof. Both sides of (9.3) are equal to $\sharp(\mathrm{Sk}(F) \cap (a \times b))$. □

This suggests a partial « desessentialisation » of coherent spaces: one calls into question the *absoluteness* of the coherence relation.

This should not be taken as a sort of *existentialist obsession*. The question of understanding coherence outside the set-theoretic setting (i.e., graphs) is crucial from the *technical* standpoint. I take this opportunity to affirm that one can only progress by means of a *dialectics*, a dialogue, between philosophy and technique. Indeed, technique can run in circles, typically by solving complicated questions that have not the slightest interest. Reciprocally, philosophy easily turns into sophistry in the absence of technical input. Thus, linearity was never adumbrated by the philosophical approach[1], while it is frankly more exciting than the ninja turtles of the nested metas. A technical breakthrough can change our perception: for instance one now understands that linguistic sophisms about coding are just baloney. An example taken from the previous chapter: the idea that it is enough to translate a system into another: from this viewpoint, coherent spaces do no better – since there are *translations* in both directions – than Scott domains. But one would never have found linearity by confining oneself to Scott domains.

Let us come back to technique. The function $\langle a \mid b \rangle := \sharp(a \cap b)$ defines a *duality*, in the sense of Section 7.1.1. Two subsets $a, b \subset |X|$ are *polar* when

$$a \perp b :\iff \sharp(a \cap b) \leq 1. \tag{9.4}$$

Definition 24 (Coherent spaces, alternative). A *coherent space* of web $|X|$ is a subset of $\wp(|X|)$ equal to its bipolar.

This definition is strictly equivalent to the « official » one, that one will prefer in practice. Indeed, if X is a coherent space (in the sense of Definition 24):

- If $a \in X$ and $b \subset a$, then $b \in X$.
- If $a \notin X$, then $\exists x, y \in a$ such that $\{x, y\} \in \sim X$.
- Conversely, if $\{x, y\} \notin \sim X$, then $\{x, y\} \in X$.

[1]Except perhaps Wittgenstein, an inspired guy, in the work of which everything can be found, but – like Nostradamus – only *afterwards*.

One concludes, by introducing $x \frown_X y :\Leftrightarrow \{x, y\} \in X$, that $a \in X$ (in the sense of Definition 24) iff $a \sqsubset X$ (in the sense of Definition 14).

The definition I just gave has the immense advantage of being generalisable to vector spaces (Chapter 17).

9.2 Perfect linear connectives

9.2.1 Multiplicatives. Using the De Morgan laws, we define a conjunction (*times*, or *tensor*) $X \otimes Y := \sim(X \multimap \sim Y)$ and a disjunction $X \parr Y := \sim X \multimap Y$ (*par*, or *cotensor*). These connectives (as well as implication) are called *multiplicative*, since they are based upon the cartesian product of the webs.

Definition 25 (Multiplicatives). If X, Y are coherent spaces, we define the coherent spaces $X \otimes Y$, $X \parr Y$:

$$|X \otimes Y| = |X \parr Y| := |X| \times |Y|,$$
$$(x, y) \frown_{X \otimes Y} (x', y') :\Leftrightarrow x \frown_X x' \wedge y \frown_Y y',$$
$$(x, y) \smile_{X \parr Y} (x', y') :\Leftrightarrow x \smile_X x' \vee y \smile_Y y'.$$

The two definitions are related *modulo* De Morgan:[2]

$$\sim(X \otimes Y) = \sim X \parr \sim Y,$$
$$\sim(X \parr Y) = \sim X \otimes \sim Y,$$
$$X \multimap Y = \sim X \parr Y = \sim(X \otimes \sim Y).$$

which exchanges conjunction and disjunction, $\frown$ and $\smile$.

One verifies certain canonical isomorphisms:

Commutativity: $X \otimes Y \simeq Y \otimes X$, $X \parr Y \simeq Y \parr X$, to which one can relate $X \multimap Y \simeq \sim Y \multimap \sim X$.

Associativity: $X \otimes (Y \otimes Z) \simeq (X \otimes Y) \otimes Z$, $X \parr (Y \parr Z) \simeq (X \parr Y) \parr Z$, to which one can relate $X \multimap (Y \multimap Z) \simeq (X \otimes Y) \multimap Z$, $X \multimap (Y \parr Z) \simeq (X \multimap Y) \parr Z$.

Neutral elements: the one-point space, denoted by, depending on the context, $\mathbf{1}$ or $\bot$, is neutral, i.e., $X \otimes \mathbf{1} \simeq X$, $X \parr \bot \simeq X$, to which one relates $\mathbf{1} \multimap X \simeq X$ and $X \multimap \bot \simeq \sim X$.

In relation to our considerations about the second underground (or layer), note that two isomorphic coherent spaces ($\mathbf{1}$ and $\bot$) have a very different status at layer -1. At the categorical layer, the distinction between them is only a preciosity. The same problem of *unfaithfulness* will be found again with the additive neutrals; more dramatically, since the identification between $\mathbf{0}$ and $\top$ would produce a logical inconsistency.

[2]These are equalities, not plain isomorphisms.

9.2.2 Additives. There are also *additive* connectives, based upon the direct sum of webs $|X| + |Y| := (|X| \times \{1\}) \cup (|Y| \times \{2\})$: another conjunction (*with*, or *cartesian product*) $X \,\&\, Y$ whom we already met and a second disjunction (*plus*, or *direct sum*) $X \oplus Y$.

Definition 26 (Additives). If X, Y are coherent spaces, one defines the coherent spaces $X \,\&\, Y$, $X \oplus Y$:

$$\begin{aligned}
|X \,\&\, Y| = |X \oplus Y| :&= |X| + |Y|,\\
(x,1) \coh_{X\&Y} (x',1) \Leftrightarrow (x,1) \coh_{X\oplus Y} (x',1) :&\Leftrightarrow x \coh_X x',\\
(y,2) \coh_{X\&Y} (y',2) \Leftrightarrow (y,2) \coh_{X\oplus Y} (y',2) :&\Leftrightarrow y \coh_Y y',\\
(x,1) \frown_{X\&Y} (y,2)&,\\
(x,1) \smile_{X\oplus Y} (y,2)&.
\end{aligned}$$

In other words, the two additives reglue two coherent spaces: in the « with », X and Y are coherent, while in the « plus », they are incoherent: these are the only two possibilities for two *a priori* unrelated spaces. These two choices are swapped by De Morgan:

$$\begin{aligned}
\sim(X \,\&\, Y) &= \sim X \oplus \sim Y,\\
\sim(X \oplus Y) &= \sim X \,\&\, \sim Y.
\end{aligned}$$

Proposition 9 finds an analogue for the *plus*:

Proposition 15. *The cliques $c \sqsubset X \oplus Y$ are the $a + \emptyset$, with $a \sqsubset X$ and the $\emptyset + b$, with $b \sqsubset Y$.*

This is immediate; but this expression is not quite unique, since $\emptyset$ can be written, as we please, $a + \emptyset$ or $\emptyset + b$. This means, among other things, that $\oplus$ is not an adequate choice for intuitionistic disjunction: there are two canonical inclusions of X and Y in $X \oplus Y$ (indeed the adjoints $\iota_l = \sim\pi_l$, $\iota_r = \sim\pi_r$ of the two projections); on the other hand, one cannot define the *conditional*, since, if F, G are stable functions from X, Y to Z, the definition

$$\begin{aligned}
H(a + \emptyset) &:= F(a),\\
H(\emptyset + b) &:= G(b)
\end{aligned} \tag{9.5}$$

turns out to be inconsistent in the case $a = b = \emptyset$. This drawback of « plus » is one of the origins of linear logic: indeed, how to reglue F and G, except by assuming that they take the same value on $\emptyset$ – which can reasonably be only $\emptyset$. Mathematical culture does the rest and from a tooth reconstitutes the full dinosaur: the condition $F(\emptyset) = \emptyset$ immediately suggests the generalisation $F(a \cup b) = F(a) \cup F(b)$.

Note the canonical isomorphisms:

Commutativity: $X \mathbin{\&} Y \simeq Y \mathbin{\&} X$, $X \oplus Y \simeq Y \oplus X$.

Associativity: $X \mathbin{\&} (Y \mathbin{\&} Z) \simeq (X \mathbin{\&} Y) \mathbin{\&} Z$, $X \oplus (Y \oplus Z) \simeq (X \oplus Y) \oplus Z$.

Neutral elements: the empty space, denoted by, according to the context, $\top$ or $\mathbf{0}$, is neutral, i.e., $X \mathbin{\&} \top \simeq X$, $X \oplus \mathbf{0} \simeq X$.

Distributivity: $X \otimes (Y \oplus Z) \simeq (X \otimes Y) \oplus (X \otimes Z)$, $X \parr (Y \mathbin{\&} Z) \simeq (X \parr Y) \mathbin{\&} (X \parr Z)$, $X \multimap (Y \mathbin{\&} Z) \simeq (X \multimap Y) \mathbin{\&} (X \multimap Z)$ and $(X \oplus Y) \multimap Z \simeq (X \multimap Z) \mathbin{\&} (Y \multimap Z)$.

Absorption: $X \otimes \mathbf{0} \simeq \mathbf{0}$, $X \parr \top \simeq \top$, to which one can relate $X \multimap \top \simeq \top$ and $\mathbf{0} \multimap X \simeq \top$.

9.2.3 Digression: notations. Here, an explanation as to the notations: the disclosure of perfective (I will explain the terminology later) connectives, totally unknown so far, posed a notational problem: whereas a second symbol imposed itself for negation and implication, it was obviously impossible to keep the usual notations for the two conjunctions (and disjunctions). Indeed, while « with » corresponds (if one does not look too closely) to intuitionistic conjunction, the disjunctive connectives are novel, even if the « plus » is reminiscent of intuitionistic disjunction. The notation has been built around a mnemonics enabling one to remember the isomorphisms, essentially the distributivities, the neutrals and the absorbers. There are two graphical groups:

« Algebraic » style: $\otimes$, $\oplus$, $\mathbf{1}$, $\mathbf{0}$. Inside this group, the algebraic isomorphisms suggested by the notation hold.

« Logical » style: $\parr$, &, $\bot$, $\top$. This group is less legible than the other, but if one remembers that $\parr$, $\bot$ are multiplicative, &, $\top$ are additive, one easily recovers the isomorphisms.

When, later, following the works of Andreoli [5], the notion of *polarity* became central, it was discovered that the graphical style exactly matched polarity. What could look fishy was indeed very natural: isomorphisms are *commutations* of logical operations and only operations with the same polarity do commute (Section 10.A.4).

These notations have been vigourously attacked. There are two types of contest, not necessarily independent:

The fatherhood lawsuit: certain adepts of « relevant » logics, wanted to insist on distinctions prefiguring $\otimes$/ &, although they completely fumbled. Behind their choice of symbols, one could hear « he stole it from us ». This is the story of the old maid shouting « stop rapist ! »: between us, who ever thought of stealing whatsoever from *relevant* logic? This system (Section 10.2.1), based upon the rejection of weakening, but without *Hauptsatz*, is one of those bureaucratic mistakes in the style of **S5**, of which the best thing one can say is that there exist even worse.

There have been more legitimate – thus less aggressive – contests coming from the tradition of monoidal categories, for instance I for the multiplicative neutral instead of **1**.

The formalist *credo*: *Notations mean strictly nothing; no matter which one we choose...* One thus saw « $\oplus$ » in the role of the « par », which, with « $\times$ » for « with », yields $X \oplus (Y \times Z) \simeq (X \oplus Y) \times (X \oplus Z)$, which one must reread three times, not to speak of using it. This is plain ideology, the *esthetics of nonsense*: indeed, if notations actually mean nothing, why refuse *manageable* ones? It is clear that the authors of those *aggressively* illegible notations are belated formalists. For them, the universe is governed by the arbitrariness of linguistic definitions and the fact that $\otimes$ distributes over $\oplus$ is nothing but a discretionary remark, presumably the fruit of definitions which are discretionary as well: next week, we shall change them and this will be the turn of $\oplus$ to distribute over $\otimes$!

The obscurity of notations, seen as a virtue, is widely exploited by all sorts of con men. For instance, the childish ideas ruling paraconsistent logics (Section 2.3.2) are defended, concealed by an impenetrable formal hedge: the definitions are stated in a (voluntarily) incomprehensible way, with a debauchery of symbols... Everything is done in order to keep nosy people away.

9.3 Imperfect connectives

9.3.1 Stability strikes back. We could try to give a categorical interpretation of intuitionistic logic in the linear world, thus reading a proof of $A_1, \ldots, A_n \vdash B$ as a *multilinear* function. Since we shall soon do it in earnest, let us forget the details and observe that everything would work well, if not for the *structural* rules of weakening and contraction. In what follows, A and B are supposedly interpreted by coherent spaces X and Y:

Weakening: reduced to its simplest expression, weakening corresponds to « material implication »: if I have B, then I still have B under hypothesis A. If a proof of B has been interpreted by a clique $b \sqsubset Y$, then « B under hypothesis A » will be the constant function $F(a) = b$. Such a function is stable, but not linear: indeed $F(\emptyset) \neq \emptyset$.

Contraction: reduced to its simplest expression, contraction corresponds to the reuse of hypotheses: if I got B under the hypotheses A and A, then I can get B under hypothesis A. In other terms if $f(x, y)$ is a bilinear function from

X, X into Y, then $f(x, x)$ should be linear… Baloney! Everybody knows that it is *quadratic*.

At the level of the skeleton, constant functions induce elements of the form $(\emptyset, y)$ and quadratic functions induce elements of the form $(\{x, x'\}, y)$; more generally, the unbridled use of structural rules produces elements of the form (a, y), where a is a finite clique of X. A stable function is a sort of polynomial of unknown degree, this is why one easily reaches *analytic* functions (Sections 8.A.2 and 15.A).

9.3.2 *Pons Asinorum.* The *Pons Asinorum*, the « bridge of asses », is a rhetorical figure of medieval pedagogy: the student (the ass) is brought to the middle of the bridge of knowledge with the help of a simple, but striking, example.

Linear logic, whose main value rests in its *perfect fragment* – whose category-theoretical structure we just described –, would be no more than another *paralogic* – not as hateful as paraconsistent, epistemic, non-monotonic or fuzzy logics, but a paralogic anyway – if it were reduced to its perfective, perfect, part. The absence of relation to usual logic, classical or intuitionistic, fatally leads to sectarianism and marginalisation: witness the fate of the aforementioned paralogics.

The climacteric remark is that usual (i.e., intuitionistic) implication is a *particular case* of linear implication.

Definition 27 (Of course!). If X is a coherent space, we define $!X$ as follows:

$$\begin{aligned} |!X| &= X_{\text{fin}}, \\ a \coh_{!X} a' &\Leftrightarrow a \cup a' \sqsubset X. \end{aligned} \tag{9.6}$$

Theorem 28 (Pons Asinorum).

$$X \Rightarrow Y = !X \multimap Y.$$

Proof. Obvious, a plain equality indeed! □

Symmetrically:

Definition 28 (Why not?). If X is a coherent space, we define $?X$ as follows:

$$\begin{aligned} |?X| &= (\sim X)_{\text{fin}}, \\ a \frown_{?X} a' &\Leftrightarrow a \cup a' \not\sqsubset \sim X, \end{aligned} \tag{9.7}$$

which is not a legible definition; I only fabricated a dual:

$$\begin{aligned} \sim !X &= ?\sim X, \\ \sim ?X &= !\sim X. \end{aligned}$$

The isomorphisms (9.8) explain why « ! » and « ? » are styled *exponentials*:

$$\begin{aligned} !(X \mathbin{\&} Y) &\simeq\ !X \otimes !Y, \\ ?(X \oplus Y) &\simeq\ ?X \parr ?Y. \end{aligned} \tag{9.8}$$

Since a (finite) clique of $X \mathbin{\&} Y$ decomposes as $a+b$ where a, b are (finite) cliques of X and Y. To these isomorphisms we can relate the 0-ary case:

$$\begin{aligned} !\top &\simeq \mathbf{1}, \\ ?\mathbf{0} &\simeq \perp. \end{aligned} \tag{9.9}$$

Indeed, I know no other canonical isomorphism in **COH** (except mistakes of logic like $\top \simeq \mathbf{0}$). Thus, the *adjunction* & / $\Rightarrow$ is a consequence of our list of isomorphisms:

$$\begin{aligned} X \Rightarrow (Y \Rightarrow Z)) &=\ !X \multimap (!Y \multimap Z) \\ &\simeq\ !X \otimes !Y \multimap Z \\ &\simeq\ !(X \mathbin{\&} Y) \multimap Z \\ &\simeq (X \mathbin{\&} Y) \Rightarrow Z. \end{aligned}$$

9.4 The logical system

9.4.1 Generalities. Contrary to classical logic, linear logic admits a non-degenerate category-theoretic interpretation. But, due to the left/right symmetry expressed by linear negation, it cannot be written in « natural deduction » style. We are therefore led to express linear logic in the setting of sequent calculus, which appears, at least at first sight, as a regression.

Constructive linear negation – i.e., the symmetry left/right recovered – enables one to understand differently intuitionistic logic. Before linear logic, one thought that the restriction *one formula on the right* was the cause of phenomena of the style « disjunction property ». There is now a much better explanation: *the absence of structural rules*, especially contraction. A prohibition ensured by the intuitionistic maintenance: one must be two to contract. This is why linear logic, with its calculus « everything on the right », will still enjoy the existence and disjunction ($\oplus$) properties. We also better understand the reduction at absurdity, *contraposition*: it is wrong in the intuitionistic regime, because of the left contractions/weakenings which produce stable functions which are *non-linear*, hence with no adjoint. In other words, what is « reprehensible » in the reduction at absurdity is not the fact of assuming $\neg B$ to get $\neg A$, it is assuming it *twice* or more.

Linear logic is truly issued from the category-theoretic interpretation in coherent spaces. This interpretation, wholly in the second underground, yields no logical indication in the usual sense; for instance, it does not distinguish between the empty space and its negation, while, logically speaking, their identification causes an inconsistency[3]. In other words, the sequent calculus which follows is only *approximately* founded upon coherent spaces.

9.4.2 The language of LL. Since there are twice more connectives than usual, we will choose a right version. Concretely: formulas are built from *literals* $p, q, r, \sim p$, $\sim q$, $\sim r$, ..., i.e., of atomic formulas and their negations and the constants $\mathbf{1}$, $\bot$, $\top$, $\mathbf{0}$, by means of the connectives « ! » and « ? » (unary) and $\otimes$, $\parr$, $\oplus$, & (binary) and the quantifiers $\forall x A$ and $\exists x A$. We can also consider, *mutatis mutandis*, second-order quantifications. We shall not insist too much on the aspect « quantifiers », which is the less innovative aspect of linear logic[4].

Linear negation is *defined* by De Morgan style equations:

$$\begin{aligned}
\sim\mathbf{1} &:= \bot, & \sim\bot &:= \mathbf{1},\\
\sim\mathbf{0} &:= \top, & \sim\top &:= \mathbf{0},\\
\sim(p) &:= \sim p, & \sim(\sim p) &:= p,\\
\sim(A \otimes B) &:= \sim A \parr \sim B, & \sim(A \parr B) &:= \sim A \otimes \sim B,\\
\sim(A \oplus B) &:= \sim A \,\&\, \sim B, & \sim(A \,\&\, B) &:= \sim A \oplus \sim B,\\
\sim(!A) &:= ?\sim A, & \sim(?A) &:= !\sim A,\\
\sim(\exists x\, A) &:= \forall x \sim A, & \sim(\forall x\, A) &:= \exists x \sim A.
\end{aligned}$$

As to linear implication, it is also defined:

$$A \multimap B \;:=\; \sim A \parr B.$$

The *sequents* are of the form $\vdash \Delta$; bilateral sequents $\Gamma \vdash \Delta$ can be translated as $\vdash \sim\Gamma, \Delta$.

9.4.3 The calculus LL

Identity/Negation

$$\frac{}{\vdash \sim A, A}\ (\textit{identity}) \qquad\qquad \frac{\vdash \Gamma, A \qquad \vdash \sim A, \Delta}{\vdash \Gamma, \Delta}\ (\textit{cut})$$

Structure

$$\frac{\vdash \Gamma}{\vdash \Gamma'}\ (\textit{exchange})$$

[3]Inconsistency at the first underground, but not at the second.
[4]Except for the striking fact that $\exists$ is now the linear negation of $\forall$.

Logic

$$\frac{}{\vdash \mathbf{1}}\ (\textit{one}) \qquad\qquad \frac{\vdash \Gamma}{\vdash \Gamma, \bot}\ (\textit{false})$$

$$\frac{\vdash \Gamma, A \qquad \vdash B, \Delta}{\vdash \Gamma, A \otimes B, \Delta}\ (\textit{times}) \qquad\qquad \frac{\vdash \Gamma, A, B}{\vdash \Gamma, A \parr B}\ (\textit{par})$$

(*no rule for zero*)

$$\frac{}{\vdash \Gamma, \top}\ (\textit{true})$$

$$\frac{\vdash \Gamma, A}{\vdash \Gamma, A \oplus B}\ (\textit{left plus}) \qquad \frac{\vdash \Gamma, B}{\vdash \Gamma, A \oplus B}\ (\textit{right plus}) \qquad\qquad \frac{\vdash \Gamma, A \qquad \vdash \Gamma, B}{\vdash \Gamma, A \,\&\, B}\ (\textit{with})$$

$$\frac{\vdash ?\Gamma, A}{\vdash ?\Gamma, !A}\ (\textit{of course}) \qquad\qquad \frac{\vdash \Gamma}{\vdash \Gamma, ?A}\ (\textit{weakening})$$

$$\frac{\vdash \Gamma, A}{\vdash \Gamma, ?A}\ (\textit{dereliction}) \qquad\qquad \frac{\vdash \Gamma, ?A, ?A}{\vdash \Gamma, ?A}\ (\textit{contraction})$$

$$\frac{\vdash \Gamma, A[t/x]}{\vdash \Gamma, \exists x\, A}\ (\textit{there exists}) \qquad\qquad \frac{\vdash \Gamma, A}{\vdash \Gamma, \forall x\, A}\ (\textit{for all})$$

9.4.4 Interpretation of proofs. Officially, one must associate to a proof π of the sequent $\vdash \Delta$ a clique in the « par » of the elements of Δ, or rather, of the associated coherent spaces. This way is very tedious, especially the first time. Let us take another way, a method not that rigourous, but absolutely transparent[5]. We will write $\vdash \Delta$ under the form $\Gamma \vdash A$, by distinguishing a formula of Δ and « migrate » the other formulas to the left. For instance, $\vdash A, B, C$ will become $\sim B, \sim C \vdash A$. Instead of constructing a clique in the « par » of Δ, we will build a multilinear function from Γ into A, for instance from $\sim B, \sim C$ into A. In some cases[6], we find nobody to keep on the right: we put there the constant $\bot$, which corresponds to a 0-ary « par ». If you don't find this rigourous, you can anyway go back to the « official » definitions, which by the way, offer no *technical* difficulty.

[5] *Dixit* Kreisel, speaking of certain colleagues: « not to confuse rigour with *rigor mortis* ».

[6] For instance, in Section 9.4.6, when everybody is underlined.

Identity/Negation:

$$\frac{}{A \vdash A}\ (identity) \qquad \frac{\Gamma \vdash A \quad \Delta, A \vdash B}{\Gamma, \Delta \vdash B}\ (cut)$$

The identity axiom corresponds to the identity function from A into A. Cut is the composition of a multilinear function f from Γ into A and a multilinear function g of Δ, A into B and the result is a multilinear function from Γ, Δ into B.

Exchange: nothing important happens, for instance a function from A, B into C becomes a function from B, A into C, $g(y, x) := f(x, y)$. Between us, this does not mean much here, since we are working *modulo* left/right moves, which are surely forms of exchange[7].

Multiplicatives:

$$\frac{\Gamma \vdash A \quad \Delta \vdash B}{\Gamma, \Delta \vdash A \otimes B}\ (times) \qquad \frac{\Gamma, A, B \vdash C}{\Gamma, A \otimes B \vdash C}\ (par)$$

If f (resp. g) is a multilinear function from Γ (resp. Δ) into A (resp. B), then one can define a multilinear function $f \otimes g$ from Γ, Δ into $A \otimes B$. What is a composition with the bilinear function from A, B into $A \otimes B$: $a \otimes b := \{(x, y); x \in a,\ y \in b\}$. The rule of « par » expresses the universality of the tensor w.r.t. bilinear functions: if f is bilinear from A, B into C, then we can find a (unique) linear g from $A \otimes B$ into C, satisfying $g(a \otimes b) = f(a, b)$. If one were to follow the « official » interpretation in terms of cliques, this steps would translate as a rebracketing.

$$\frac{}{\vdash \mathbf{1}}\ (one) \qquad \frac{\Gamma \vdash A}{\Gamma, \mathbf{1} \vdash A}\ (false)$$

These rules, 0-ary versions of the previous, express the neutrality of $\mathbf{1}$, a one-point coherent space 1. I called « false » the constant $\perp$ for mnemonic reasons. But this constant is not contradictory in the absence of weakening. In the same way, the empty sequent « $\vdash$ » is not quite absurd.

[7] The rule of exchange does not go without saying, this the problematic of non-commutativity. But it is delicate to discuss it in a functional setting.

Additives:

$$\frac{\Gamma \vdash A \quad \Gamma \vdash B}{\Gamma \vdash A \,\&\, B}\ (with) \qquad \frac{\Gamma \vdash A}{\Gamma \vdash A \oplus B}\ (left\ plus) \qquad \frac{\Gamma \vdash B}{\Gamma \vdash A \oplus B}\ (right\ plus)$$

The rule of the « with » takes two multilincar functions of the « same » arguments Γ to build a new one with values in the cartesian product. The rules of the « plus » are nothing but compositions with the canonical embeddings of A and B into $A \oplus B$.

One could have polarised differently:

$$\frac{\Gamma, A \vdash C \quad \Gamma, B \vdash C}{\Gamma, A \oplus B \vdash C}\ (with) \qquad \frac{\Gamma, A \vdash C}{\Gamma, A \,\&\, B \vdash C}\ (left\ plus) \qquad \frac{\Gamma, B \vdash C}{\Gamma, A \,\&\, B \vdash C}\ (right\ plus)$$

making use of the conditional – which works well in the linear setting – and the two projections, which are the adjoint of the embeddings.

$$\frac{}{\Gamma \vdash \top}\ (true)$$

is interpreted by a multilinear map constantly equal to $\emptyset$, the only clique of $\top$. Note that **0** (which is category-theoretically isomorphic to $\top$) plays the role of the absurdity; this is why it has no rule. This is by the way, a procedural formulation: the absurdity, the logical vacuum, has an empty set of rules.

Quantifiers: I give up, this is never very exciting and, in the linear setting, not that surprising. It is superficially a variation on the additives, unless one moves to second order, in case it is a variation on system **F**.

One should proceed with exponentials, but one stumbles on a difficulty, at least if one wants to keep a functional intuition. A linear function from $!A$ into B is indeed a stable function from A into B. In order to describe what happens in the exponential case, it would be nice to have a *mixed* calculus, linear/stable.

9.4.5 Mixed calculus. Sequents are still of the form $\vdash \Delta$, but amongst the formulas of Δ, some are underlined, in order to signal a « classical » maintenance. The writing A suggests that A is not underlined. On the other hand, Δ may contain both underlined and non-underlined formulas; on the other hand, $\underline{\Delta}$ means that *all* formulas of Δ are underlined.

Identity/Negation

$$\frac{}{\vdash {\sim}A, A}\ (identity)$$

$$\frac{\vdash \Gamma, A \quad \vdash {\sim}A, \Delta}{\vdash \Gamma, \Delta}\ (cut)$$

$$\frac{\vdash \underline{\Gamma}, A \quad \vdash \underline{{\sim}A}, \Delta}{\vdash \underline{\Gamma}, \Delta}\ (\underline{cut})$$

Structure

$$\frac{\vdash \Gamma}{\vdash \Gamma'}\ (exchange)$$

$$\frac{\vdash \Gamma, A}{\vdash \Gamma, \underline{A}}\ (dereliction)$$

$$\frac{\vdash \Gamma}{\vdash \Gamma, \underline{A}}\ (weakening)$$

$$\frac{\vdash \Gamma, \underline{A}, \underline{A}}{\vdash \Gamma, \underline{A}}\ (contraction)$$

Logic

$$\frac{}{\vdash \mathbf{1}}\ (one)$$

$$\frac{\vdash \Gamma}{\vdash \Gamma, \bot}\ (false)$$

$$\frac{\vdash \Gamma, A \quad \vdash B, \Delta}{\vdash \Gamma, A \otimes B, \Delta}\ (times)$$

$$\frac{\vdash \Gamma, A, B}{\vdash \Gamma, A \parr B}\ (par)$$

(*no rule for zero*)

$$\frac{}{\vdash \Gamma, \top}\ (true)$$

$$\frac{\vdash \Gamma, A}{\vdash \Gamma, A \oplus B}\ (left\ plus)$$

$$\frac{\vdash \Gamma, B}{\vdash \Gamma, A \oplus B}\ (right\ plus)$$

$$\frac{\vdash \Gamma, A \quad \vdash \Gamma, B}{\vdash \Gamma, A \,\&\, B}\ (with)$$

$$\frac{\vdash \underline{\Gamma}, A}{\vdash \underline{\Gamma}, !A}\ (of\ course)$$

$$\frac{\vdash \Gamma, \underline{A}}{\vdash \Gamma, ?A}\ (why\ not)$$

$$\frac{\vdash \Gamma, A[t/x]}{\vdash \Gamma, \exists x\, A}\ (there\ exists)$$

$$\frac{\vdash \Gamma, A}{\vdash \Gamma, \forall x\, A}\ (for\ all)$$

9.4.6 Interpretation, concluded. Our functional paradigm is changed: the sequent $\Gamma \vdash A$ is interpreted by a stable function with arguments chosen in Γ and moreover linear in those which are not underlined. Most rules are interpreted as before, the only novelty being a stable dependency (upon the underlined part of the context).

Cut:

$$\frac{\underline{\Gamma} \vdash A \quad \underline{A}, \Delta \vdash B}{\underline{\Gamma}, \Delta \vdash B} \ (\underline{cut})$$

This rule is still interpreted by a composition $g \circ f$; but g is not linear w.r.t. A and f is not linear at all. The composition is linear only w.r.t. those elements of Δ which are not underlined. One sees that it is necessary that Γ be fully underlined, since possible linear dependencies would be « killed » by the taking of f as *argument* of a non-linear function.

Structure:

$$\frac{\Gamma, A \vdash B}{\Gamma, \underline{A} \vdash B} \ (dereliction)$$

Dereliction consists in forgetting linearity.

$$\frac{\Gamma \vdash B}{\Gamma, \underline{A} \vdash B} \ (weakening) \qquad \frac{\Gamma, \underline{A}, \underline{A} \vdash B}{\Gamma, \underline{A} \vdash B} \ (contraction)$$

Weakening corresponds to constant function, contraction to the identification of two arguments.

Exponentials:

$$\frac{\underline{\Gamma} \vdash A}{\underline{\Gamma} \vdash !A} \ (of\ course) \qquad \frac{\Gamma, \underline{A} \vdash B}{\Gamma, !A \vdash B} \ (why\ not)$$

In the first case, a stable function f with values in A becomes a stable function with values in $!A$: $!f(\vec{\gamma}) := \wp_{\mathrm{fin}}(f(\vec{\gamma}))$. The second case is nothing but the coding of a stable function by a function linear on $!A$: this is our *Pons Asinorum*.

9.4.7 Cut-elimination. The interpretation we just gave is compatible with cut-elimination. This is almost obvious, since proofs have been interpreted by functions (linear or plain stable) and that cut is composition, i.e., the *activation* of functions.

The linear sequent calculus, in all its versions (first, second-order), enjoys cut-elimination. This is not earth-shaking, since it has been built *around* an invariant of normalisation, coherent spaces. We admit it, since we shall soon prove a *Hauptsatz* (Section 10.1.6); not to speak of *proof-nets* of Chapter 11 will enable one to pose the problem in a setting friendlier than sequent calculus.

In case we want to prove normalisation for the linear analogue of system **F**, we need to adapt *reducibility* (Section 6.2). We introduce a *duality*: a contextual proof π of A, i.e., of some $\vdash \Gamma, A$ and a contextual proof of π' of $\sim A$, i.e., of some $\vdash \sim A, \Delta$, yielding by cut a proof $\langle \pi \mid \pi' \rangle$ of the context Γ, Δ. If we take as *pole* the set of strongly normalisable proofs, we get a notion of *reducibility candidate* of type A: a set $\mathcal{A}$ of contextual proofs of A which is *polar* (Section 7.1.1) and such that the identity axiom $\vdash A, \sim A$ belongs to $\mathcal{A} \cap \mathcal{A}^p$. Of course, to do this rigourously, we must replace sequents with a structure enjoying Church–Rosser; which the proof-nets of Chapter 11 will provide.

9.A Monoidal categories

9.A.1 The tensor product according to Bourbaki. There is an analogue of cartesian categories adapted to linear logic, namely *symmetrical monoidal categories*. This would normally be the place for the appropriate definitions. But it is a lot of work for a rather marginal output, at least in the case that interests us: personally, I finalised coherent spaces and linear logic without having ever heard of monoidal categories; on the other hand I knew the definition of the tensor product due to Bourbaki, as the universal solution of a factorisation problem for bilinear functions:

We seek a space $A \otimes B$ and *bilinear* function $\otimes: A, B \mapsto A \otimes B$, with the following *universal property*: if $G: A, B \mapsto C$ is another solution (bilinear function), then there is a unique *linear* $H: A \otimes B \mapsto C$ such that $G(a, b) = H(a \otimes b)$ for all $a \in A$, $b \in B$. We see by the way that the tensor product is of « inductive », positive, style.

My advice is to work with multilinear functions in an appropriate setting, without trying to delve into the category-theoretic framework. This may become necessary in certain cases, but one should not forget that ideas are not to be found in diagrams.

9.A.2 Symmetric monoidal categories. Let us say a word, since I mentioned them, about *symmetrical*[8] *monoidal categories*: the tensor product is bluntly given *as primitive*, under the form of a covariant binary functor $\otimes$ of $\mathbf{C} \times \mathbf{C}$ into itself, as well as a « neutral element » $I \in \mathbf{C}$. One must state commutativity, associativity and neutrality. It is a matter of invertible natural transformations, which are also *given*[9]:

$$\begin{aligned} \operatorname{ass}(A, B, C) &:= A \otimes (B \otimes C) \mapsto (A \otimes B) \otimes C, \\ \operatorname{com}(A, B) &:= A \otimes B \mapsto B \otimes A, \\ \operatorname{neu}(A) &:= A \otimes I \mapsto A. \end{aligned}$$

[8] Symmetrical for « commutative ».

[9] Santa Claus is generous with category theory.

They are subject to the *Mac Lane–Kelly* equations [73], the most noted of which being:

$$\begin{array}{ccccc}
A \otimes (B \otimes (C \otimes D)) & \xrightarrow{\text{ass}} & (A \otimes B) \otimes (C \otimes D) & \xrightarrow[\text{ass}]{} & ((A \otimes B) \otimes C) \otimes D \\
\Big\downarrow{\scriptstyle \iota\otimes\text{ass}} & & & & \Big\uparrow{\scriptstyle \text{ass}\otimes\iota} \\
A \otimes ((B \otimes C) \otimes D) & & \xrightarrow{\qquad\qquad\text{ass}\qquad\qquad} & & (A \otimes (B \otimes C)) \otimes D
\end{array} \tag{9.10}$$

This equation expresses the associativity of… associativity. One similarly states the commutativity of commutativity:

$$\begin{array}{ccc}
A \otimes B & \xrightarrow{\qquad\iota\qquad} & A \otimes B \\
 & {\scriptstyle\text{com}}\searrow \quad \nearrow{\scriptstyle\text{com}} & \\
 & B \otimes A &
\end{array} \tag{9.11}$$

Etc. One can define *closed symmetric monoidal categories* by means of an adjunction $\otimes/\multimap$ strictly analogous to the adjunction $\& \, /\Rightarrow$ of CCC, see equation (7.8) in Section 7.2.3.

Between us, if one has the use of it, OK. But it should not be a way to complicate simple things, nor to hide weaknesses. Again, I never used that, I keep this sort of thing on a shelf, like a fire extinguisher. Knowing that it exists is however, if not useful, at least soothing.

9.A.3 $*$-autonomous categories. Barr defined, long before my discovery of linear logic, « $*$-autonomous » categories [10], which resemble, in its main lines, the perfective part of linear logic. In its main lines only, since the tensor will have a tendency to be the same as the cotensor, etc. This does not matter, it is an approach going in the same direction and, by the way, coherent spaces form a $*$-autonomous category, see [11].

9.A.4 Non-commutativity. Non-symmetrical (i.e., non-commutative) monoidal categories are very interesting. They are obtained, not by relinquishing the isomorphism between $A \otimes B$ and $B \otimes A$, but by relinquishing the corresponding diagram, the *commutativity of commutativity*. Then phenomena of *braiding* do occur.

There is also a non-commutative logic, dating back to an old paper of Lambek [72]. To tell the truth, this was hardly more than a calculus without involution, reduced to the tensor and two implications, one on the left, the other on the right. This system, rudimentary, but nice, had a certain success in linguistics. I extended it into a system of « cyclic » logic, with an involutive negation, where exchange is restricted to *circular* permutations (Section 11.1.3). More recently, Ruet and Abrusci extended cyclic logic into a system in which a commutative and a non-commutative tensor coexist harmoniously (Section 18.B).

In spite of a promising start, the « non-commutative » never really took off. It has problems dwelling at layer -2, since the *braided* – i.e., non-commutative – monoidal categories are not of much help here. Indeed, this logic refuses commutativity, while a braided monoidal category is commutative, but *in a non-canonical way*. In a general way, the categorical explanations of non-commutative logic have a fabricated, artificial, flavour. Could it indeed be the case that this animal belongs in the third underground?

Chapter 10
Perfection vs. imperfection

10.1 Phase semantics

10.1.1 Generalities. We shall give a *semantics* of linear logic, i.e., an interpretation at layer -1, thus in terms of *provability*[1]. Granted the radical novelty of linear logic, it was important (1986) to establish safeguards with respect to dubious category-theoretic isomorphisms of the style $\mathbf{0} \simeq \top$. Phase semantics was able to seat linear logic on more conformist, more easily accessible grounds. Finally, is it deep or is it *ad hoc* like most of algebraic semantics? We shall eventually discuss the epistemological value of this controversial tool.

10.1.2 Phase models. If $\mathcal{M}$ is a commutative monoid, noted multiplicatively and with neutral element 1, we can define, if $X, Y \subset \mathcal{M}$:

$$X \multimap Y := \{m; \forall x \in X\ mx \in Y\}. \tag{10.1}$$

A formula reminiscent of the quotient of ideals in algebra.

A *phase space* is the pair $(\mathcal{M}, \perp\!\!\!\perp)$ of a commutative monoid and a *pole* $\perp\!\!\!\perp \subset \mathcal{M}$. We define the *negation* $\sim X$ of a subset $\mathcal{M}$ as $X \multimap \perp\!\!\!\perp$:

$$\sim X := \{y; \forall x \in X\ yx \in \perp\!\!\!\perp\}. \tag{10.2}$$

A *fact* is a set which is polar (w.r.t. the *duality* we just defined, see Section 7.1.1), i.e., such that $X = \sim\sim X$, in other terms such that X is of the form $\sim Y$. If we choose an empty pole, there are only two facts, $\emptyset$ (false) and $\mathcal{M}$ (true) and everything reduces to a bleak classical truth computation.

If Y is a fact, it is immediate that $X \multimap Y$ is a fact, since

$$\begin{aligned} X \multimap Y &= \{m; \forall x \in X\ mx \in Y\} \\ &= \{m; \forall x \in X \forall y \in \sim Y\ mxy \in \perp\!\!\!\perp\} = \sim(X \cdot \sim Y). \end{aligned}$$

10.1.3 Interpretation of logic. Let us restrict to the propositional case. We already know how to interpret the constant $\perp\!\!\!\perp$, linear implication and negation. If

[1] The name *phase semantics* is an accident: it refers to a diffuse feeling of kinship with quantum physics. The mistake was to believe that one could establish a link logic/quantum at layer -1: one needs at least layer -2.

$X, Y \subset \mathcal{M}$, we define $X \cdot Y := \{xy \ ; \ x \in X \wedge y \in Y\}$.

$$\begin{aligned} X \otimes Y &:= \sim\sim(X \cdot Y) \\ X \parr Y &:= \sim(\sim X \cdot \sim Y) \\ \mathbf{1} &:= \sim\sim\{1\} \\ X \oplus Y &:= \sim\sim(X \cup Y) \\ X \,\&\, Y &:= X \cap Y \\ \mathbf{0} &:= \sim\sim\emptyset \\ \top &:= \mathcal{M} \\ !X &:= \sim\sim(X \cap \mathcal{I}) \\ ?X &:= \sim(\sim X \cap \mathcal{I}) \end{aligned} \tag{10.3}$$

Where $\mathcal{I}$ is the set of *idempotents* of $\mathcal{M}$ which belong to $\mathbf{1}$.

10.1.4 Soundness. Suppose that the formulas $A, B, C, \ldots$ have been interpreted in a phase model $(\mathcal{M}, \maltese)$ by facts $[A], [B], [C], \ldots$.

- A is *accepted* by the model iff $1 \in [A]$. Equivalently, $[\sim A] \subset \maltese$.
- For a sequent $\vdash A, B$, acceptance can be written as we please $1 \in [A] \parr [B]$, $[\sim A] \subset [B]$, $[\sim B] \subset [A]$, $[\sim A] \cdot [\sim B] \subset \maltese$.
- One can also replace negations with *prenegations*, i.e., $A', B', C', \ldots$ such that $[A] = \sim A'$, $[B] = \sim B'$, $[C] = \sim C', \ldots$. Thus, the acceptance of $\vdash A, B, C$ can be expressed by the inclusion $A' \cdot B' \subset [C]$.

There is not the slightest difficulty in checking the validity of all logical rules. We will rather focus on some structuring results:

(i) Negation exchanges definitions by De Morgan. This is more or less immediate, for instance, $\sim\{1\} = \{m; 1m \in \maltese\} = \maltese$, hence $\mathbf{1} = \sim\maltese$.

(ii) If $X, Y \subset \mathcal{M}$, if $z \in \sim(X \cdot Y)$ and $y' \in Y$, then $zy' \in \sim X$; if $x \in \sim\sim X$, then $zy'x \in \maltese$. y' being arbitrary, $zx \in \sim Y$. If $y \in \sim\sim Y$, then $zxy \in \maltese$, hence, $xyz \in \maltese$. This proves that $\sim\sim X \cdot \sim\sim Y \subset \sim\sim(X \cdot Y)$. Since this argument is more or less illegible, I rewrite the same thing in natural deduction. Note that the discharges of hypotheses are unambiguous.

$$\dfrac{\dfrac{\dfrac{\dfrac{\dfrac{\dfrac{\dfrac{[x' \in X] \quad [y' \in Y]}{x'y' \in X \cdot Y} \qquad [z \in {\sim}(X \cdot Y)]}{x'y'z \in \bot}}{y'z \in {\sim}X} \qquad x \in {\sim\sim}X}{xy'z \in \bot}}{xz \in {\sim}Y} \qquad y \in {\sim\sim}Y}{xyz \in \bot}}{xy \in {\sim\sim}(X \cdot Y)}$$

(iii) Under the same hypotheses, ${\sim\sim}X \cup {\sim\sim}Y \subset {\sim\sim}(X \cup Y)$.

With the following remarkable consequences:

(i) The associativity, the commutativity of $\otimes$, $\oplus$ together with the appropriate neutral elements. For instance, for the tensor

$$\begin{aligned}(X \otimes Y) \otimes Z &:= {\sim\sim}({\sim\sim}(X \cdot Y) \cdot Z) = {\sim\sim}((X \cdot Y) \cdot Z) \\ &= {\sim\sim}(X \cdot (Y \cdot Z)) = {\sim\sim}(X \cdot {\sim\sim}(Y \cdot Z)) \\ &=: X \otimes (Y \otimes Z).\end{aligned}$$

(ii) In the same way, $\otimes$ distributes over $\oplus$ and $\mathbf{0}$ is absorbing for $\otimes$.

(iii) Negation yields the dual associativities and distributivities, e.g., $⅋$ / &.

Positive polarity is indeed hiding behind this simplification of double negations. A surprising interpretation of this, *focalisation*, is given in Section 10.A.

10.1.5 Completeness. We establish completeness by means of the *tautological* model. $\mathcal{M}$ is the monoid of *mixed* contexts (Section 9.4.5): this means multisets of formulas, not taking care of multiplicities in case of underlining. The neutral element is the empty context, the product is defined in the obvious way:

$$(2A + B + \underline{C})(4B + \underline{C} + D) = 2A + 5B + \underline{C} + D.$$

The pole is the set of *provable* contexts. The language is interpreted as follows: $[p] := {\sim}\{p\}$, i.e., the set of contexts Γ such that $\vdash \Gamma, p$ is provable.

Let us introduce the notation $x \perp y$ for $xy \in \bot$; in the same way, $X \perp Y$ when $x \perp y$ $(\forall x \in X, \forall y \in Y)$, $x \perp Y$ when $x \perp y$ $(\forall y \in Y)$.

Theorem 29 (Linear completeness). *In the tautological model, the interpretation* $[A]$ *of* A *is equal to* $\sim\{A\}$.

Proof. See next section. The inclusion $[A] \subset \sim\{A\}$ is sufficient: $\sim\{A\} \perp\!\!\!\perp \sim\{\sim A\}$ (use the cut rule), hence $[A] \subset \sim\{A\} \subset \sim\sim\{\sim A\} \subset \sim[\sim A] = [A]$. □

Corollary 29.1. *In particular, if* A *is accepted by this model,* A *is provable.*

Proof. A is accepted iff $\emptyset \perp\!\!\!\perp A$, i.e., if A is provable. □

10.1.6 The *Hauptsatz*. The previous method yields actually (a remark of Okada) an elegant proof of (non-effective) cut-elimination. We modify the tautological model by replacing *provable* with *cut-free provable*. Atoms are interpreted by the $\sim\{p\}$, as before. We thus get an analogue of Theorem 29, the completeness of the cut-free system:

Theorem 30 (Cut-free linear completeness). *In the cut-free tautological model, the interpretation of any formula* A *is included in the set* $\sim\{A\}$.

Proof. This is indeed the non-trivial part of Theorem 29, the one which does not use cut. We first observe that $[A] \subset \sim\{A\}$ iff $A \in [\sim A]$. We then proceed by induction, assuming the result to hold for subformulas: here, when I say « provable », I imply *cut-free*.

$\sim p$: $\sim p \in [p]$ because of the identity axiom.

$\boldsymbol{\perp\!\!\!\perp}$: if $\Gamma \in [\boldsymbol{\perp\!\!\!\perp}]$, i.e., if $\vdash \Gamma$ is provable, then $\vdash \Gamma, \boldsymbol{\perp\!\!\!\perp}$ is provable too.

1: the axiom for **1** says that $\mathbf{1} \in \boldsymbol{\perp\!\!\!\perp} = [\sim\mathbf{1}]$.

$\parr$: $A \in [\sim A]$ and $B \in [\sim B]$ by hypothesis; if $\Gamma \in [A \parr B]$, $\Gamma \perp\!\!\!\perp A \cdot B$, hence $\vdash \Gamma, A, B$ is provable. By the rule of « Par », $\vdash \Gamma, A \parr B$ is provable too.

$\otimes$: $[A] \subset \sim\{A\}$, $[B] \subset \sim\{B\}$, yields, by a « Tensor » rule, $[A] \cdot [B] \subset \sim\{A \otimes B\}$, hence $[A \otimes B] \subset \sim\{A \otimes B\}$.

&: from $[A] \subset \sim\{A\}$, $[B] \subset \sim\{B\}$ and the rule of « With », we easily deduce $[A \,\&\, B] := [A] \cap [B] \subset \sim\{A \,\&\, B\}$.

$\oplus$: $[A] \subset \sim\{A\}$, $[B] \subset \sim\{B\}$, yields, using the rules of « Plus », $[A] \cup [B] \subset \sim\{A \oplus B\}$, hence $[A \oplus B] \subset \sim\{A \oplus B\}$.

$!A$: observe that the set $\mathcal{I}$ of the definition consists exactly of the underlined contexts $\underline{\Delta}$. Here occurs a hidden use of contraction: we do not count the repetitions of underlined formulas; and weakening: if $\Gamma \in \boldsymbol{\perp\!\!\!\perp}$, then $\Gamma, \underline{\Delta} \in \boldsymbol{\perp\!\!\!\perp}$ (in other terms, $\underline{\Delta} \in \boldsymbol{\perp\!\!\!\perp} \multimap \boldsymbol{\perp\!\!\!\perp} = [\mathbf{1}]$). If $\Gamma \in [A] \cap \mathcal{I}$, then $\Gamma \in \sim\{A\}$ and is underlined. By the rule « Of course », $\Gamma \in \sim\{!A\}$, which shows that $[A] \cap \mathcal{I} \subset \sim\{!A\}$ and eventually $[!A] \subset \sim\{!A\}$.

$?A$**:** if $[A] \subset \sim\{A\}$, then by dereliction, $[A] \subset \sim\{\underline{A}\}$, i.e., $\underline{A} \in [\sim A]$ and since $\underline{A} \in \mathcal{I}$, we conclude that $\underline{A} \in [!\sim A]$. Hence $[?A] \subset \sim\{\underline{A}\}$. By the rule of « Why not », we conclude that $[?A] \subset \sim\{?A\}$. □

Observe that each case uses exactly one rule, « its rule »; except exponentials which are indeed the succession of two operations (Section 10.A.6).

Corollary 30.1. *Linear logic enjoys the* ***Hauptsatz****.*

Proof. If A is provable, it is validated by any model, including the cut-free tautological model; $\emptyset \perp A$ in this model hence A is cut-free provable. □

10.1.7 Discussion. It is time to discuss algebraic semantics and to see what is *ad hoc* in these matters. The common point is that completeness is obtained by means of a structure based upon logical consequence. Which opens the door to all possible abuses: say that we interpret logic $\mathcal{L}$ by means of « L-algebras ». The structure of provability will yield, as if by chance, an L-algebra: not very glorious!

Phase semantics seems to be no exception: let us consider the tautological model corresponding to an « extension » $\mathcal{L}$ of linear logic by means of a new axiom schema. The interpretation of a formula A remains equal to $\sim\{A\}$. One gets a completeness theorem w.r.t an *ad hoc* modification of the notion of model. Summing up, completeness obeys *perinde ac cadaver*!

One begins to see the tail of the wolf when proceeding with *soundness*: with algebraic semantics, it may be the case that one is *unable* to construct an L-algebra other than the one based upon provability in $\mathcal{L}$!

The same phenomenon occurs with phases: if an algebraic *diktat* is enough for completeness, it may turn out to be a misfit w.r.t. *soundness*: in other words, outside the tautological model, one will have extreme difficulty in checking the extra axioms, even with a fabricated notion. One will be forced to require that the facts (not only the points of the space $\mathcal{M}$) enjoy the extra property, i.e., one will conclude with a walloping.

We shall now study several modifications of logic: in certain cases we will naturally get soundness; in other cases, there will be no honest way to get it[2]:

Weakening: the weakening rule « if $\vdash \Gamma$ is provable, then $\vdash \Gamma, \Delta$ is provable too » suggests the condition: $x \in \perp \Rightarrow xy \in \perp$, i.e., $\sim\perp = \mathcal{M}$, in other terms $\mathbf{1} = \top$, which is a way of expressing weakening.

Contraction: we ask in the same way that $x^2y \in \perp \Rightarrow xy \in \perp$. This indeed implies $x^2 \in Y \Rightarrow x \in Y$ for any *fact* X. If $x \in X$, then $x^2 \in X \cdot X \subset X \otimes X$. By taking $Y := X \otimes X$, we conclude that $X \subset X \otimes X$, which is a way of expressing contraction.

[2]On these matters, some progress has been made by Terui [98].

Reverse contraction: a dumb rule, enabling weakening of formulas already present: $xy \in \perp\!\!\!\perp \Rightarrow x^2y \in \perp\!\!\!\perp$. Each fact then satisfies $x \in X \Rightarrow x^2 \in X$, but this is not enough to ensure $X \otimes X \subset X$. This falls short, since the points x^2, with $x \in X$ are in X; but we need the xy, with $x, y \in X$.

Non-commutative: what about a non-commutative monoid? We would thus get two negations, a left one and a right one, etc. Obviously this begins to generate vertigo when one observes that there will be left facts and right facts: the sets which are left or right negations. Without that, even staying within left facts, it does not work, since we do not get in this way an *associative* tensor. Indeed, the associativity of tensor requires the simplifications of certain double negations, but the argument of Section 10.1.4 no longer holds. We discover that a non-commutative phase space induces a non-associative space of facts.

Cyclic: if the pole is *cyclic*, i.e., if $xy \in \perp\!\!\!\perp \Rightarrow yx \in \perp\!\!\!\perp$, everything works well, the two negations coincide and the tensor becomes associative. This corresponds to *cyclic* linear logic (Section 11.1.3).

We conclude that the « logics » based on reverse contraction or on blunt non-commutativity are unsound. Phase semantics is therefore not that dumb! Anyway, less tolerant than Kripke models which accept without cringing the most fabricated modalities.

10.2 A perfect world?

10.2.1 Implication and causality. The implication between « All men are mortal » and « Socrates is mortal » is everything but a causality; let us go into details.

Perfection. To speak of *causality* is in particular to speak of *effects*. An effect takes place in time: it has a *duration*, at least a beginning. It is not like a *truth* which exists, without any cause. If one seeks a « Socrates effect », one will rather choose: « Socrates is dead », or better, « Socrates died ». In both cases, one would use a *perfect* past. By no ways, the imperfect or progressive « Socrates was dying » which is only suited for a narrative situated at the time of Socrates.

The distinction perfect/imperfect (or perfective/imperfective) is present in various languages, e.g., Russian where perfect is (regularly) formed upon imperfect with the help of the prefix « по ».

The perfect world is a world of *actions*, that one *performs*, in the sense that one concludes, one *finishes* them; this finiteness may also be understood set-theoretically. A world rather explicit and concrete, in contrast to the imperfect world, rather implicit and abstract.

« All men are mortal » is not of this form and, by the way, the implication is not causal[3]. A cause for the death of Socrates would rather be « Socrates drank hemlock », a *preterite*. The imperfect forms « Socrates was drinking hemlock », or « Socrates used to drink hemlock » remains inadequate.

Causality and linearity. Although causality is a very complex thing, one can make a few remarks:

(i) Generally, the cause is part of the effect. This is why, since time immemorial, « material implication » was criticised. Technically speaking, this amounts to a refusal of weakening; but « old style logic » was not able to go beyond that point, which is far from sufficient. As an example, let us take « relevant logics ». They were fabricated in the years 1960 around a refusal of weakening, while trying to stay as close as possible to classical logic. Which suggests a joke: « Paradise is not that different from Earth, there are streets, brothels, banks, petrol wells, Saddam has just been removed »; without minimising the nefariousness of the guy, one cannot rebuild the world around such a minor point. In the same way, the rejection of weakening should have been accompanied with a complete recentring of logic along new axes; an attempt was made instead to recover on one hand what had been relinquished by the other. Finally, « relevant » logicians accumulated logical *diktats*; and their systems never quite recovered from the absence of *Hauptsatz*.

(ii) In general, the cause does not survive to the consequence. Take the hemlock: Socrates took it and there was no more of it. The same in the typical causality, the *purchase*: the money which is the cause of it changes owner, and cannot be reused by the same person. This is why *contraction* is usually wrong in the causal universe.

Measure in quantum mechanics provides us with a nice example: when I measure the *spin* of an electron, it changes state (it orients along the axis $\vec{z}$ just to please me). It is by the way the quantum world which seems the most purely causal in the sense of linear logic.

But there are causalities where the cause very well survives to its effect, typically radio reception: one cannot say that the fact of opening one's receiver changes whatsoever to the emission. Those are actions without reaction, see *infra*.

Action vs. reaction. We are in the process of interpreting causality by linear implication « $\multimap$ ». *Quid* of negation? It can be seen as an *observation*, as a question that

[3]One could say: « Socrates is dead, hence he is mortal »; on the other hand « Socrates is dead, hence he drank hemlock » is incorrect.

one asks – what, in the quantum world, is close to a « destruction ». The adjunction which enables one to pass from $A \multimap B$ to $\sim B \multimap \sim A$ is that a question posed to B is transmitted to A. If one thinks that observation means destruction, one sees that it is a matter of *reaction* which, by destroying the premise, impedes its reuse.

Late considerations. Let us consider typical action/reaction pairs, of the style write/read, send/receive (much different from the tarskian pair speak-true/lie). There is no strict equality between the partners: one is active, the other passive.

From the logical viewpoint, formulas will be split into active (positive) and passive (negative) ones. During the interaction – cut-elimination – the hand can change. *Polarity* therefore only deals with the *initial* situation: think of an epistolary exchange.

The third underground. Due to the procedural, dynamical, character of actions, we are indeed at layer -3. Layer -2 remains practicable (monoidal categories), as well as the phases of layer -1. But I however believe that, in the same way classical logic essentially belongs in layer -1 and intuitionistic logic belongs in layer -2, it is at the third underground where linear logic truly belongs. And I think that the futurist systems like **LLL**, **ELL** (Chapter 16) make perhaps no natural sense in the higher layers.

10.2.2 Resources

The two conjunctions. Phases suggest an approximate analogy with money, time, space, food, indeed everything that can be styled a *resource*. The main quality of a resource is that it cannot be duplicated, in other words that the rule of contraction does not apply. To such an extent that one speaks of the *miracle* of loaves and fishes and that the law severely *represses* the duplication of money. The notion of resource is less clearly antagonistic to weakening. Everything depends on whether we read weakening as destruction or as non-use: physics opposes the blunt destruction of matter and destruction of money is a criminal offense, although possible; but one can decide not to use all the resources.

In an interpretation in the style « money », $\vdash B$ means that I have got B *free of charge* (for the price of the neutral element, 0 \$). $A \vdash B$ means that one can get B by paying A. The tensor product $A \otimes B$ is the conjunction of both, to pay A and to pay B, or to get A and to get B.

Which explains this slightly worn out example: if A stands for 1 \$, B and C stand for two brands of cigarettes, then $A \vdash B$ (resp. $A \vdash C$) means that, for one \$, one can get a pack of *Nazionali* (resp. a pack of *MS*). Linear logic inference yields $A \otimes A \vdash B \otimes C$, but not $A \vdash B \otimes C$: in other terms, to get the two packs, one must pay $A \otimes A$, i.e., 2 \$. One could play in the same way with chemical equations and remember that the non-respect of *proportions* in chemical equations (once more, the contraction rule) often causes deadly accidents, explosions, etc.

One can, in the same conditions, conclude $A \vdash B \,\&\, C$. The cartesian product $A \,\&\, B$ does not mean « both », but only « the one you please ». Think of a vending machine: one inserts a coin and one will get, as one pleases, either the *Nazionali* or the *MS*, but not both. One sees that the conjunction & has a disjunctive aspect: one or the other.

The connective $\oplus$ provides us with another idea, purely disjunctive. $A \oplus B$ means one or the other, but we cannot choose. This is the case of the result of a lottery, etc.

Is this serious? Yes, it enables us to represent in a faithful way, by means of logical consequence, a lot of concrete computation processes. Thus, *Petri nets*; we wire cells of the form:

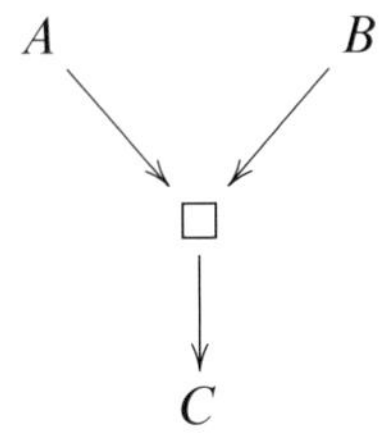

(10.4)

If tokens are put in compartments A and B, this fires a *transition*: the two tokens disappear and a new one appears in C. Linear logic represents this as an axiom $A \otimes B \multimap C$ (or $A, B \vdash C$). This representation is faithful, contrary to intuitionistic consequence, which would rightly conclude the presence of a token in C, but would not « free » compartments A and B.

Let us say at once that – in spite of a certain infatuation when Asperti [6] discovered this link in the early days of linear logic – this logical expression, although faithful, was not good for much.

Complexity of fragments. We gathered that linear logic is able to represent rather subtle things. This was applied to various types of abstract machines, let us especially quote [65], [77].

MLL: Kanovich has shown that the multiplicative fragment of linear logic is NP-complete. To be put side-by-side with the complexity of classical propositional calculus, which is coNP-complete.

MALL: with additives, the propositional calculus becomes PSPACE-complete. To be put side-by-side with the complexity of intuitionistic propositional calculus, which is PSPACE-complete too.

MAELL: the full propositional calculus is undecidable.

This last result is obtained by encoding *Minski machines*, a particular kind of abstract machine. The proof may be seen as a rigourous version of the observations concerning money, cigarettes, etc.

Limits of the explanation: the $\parr$. If one understands negation as an exchange between production and demand, one can try to *transpose* the Petri cell (10.4) into:

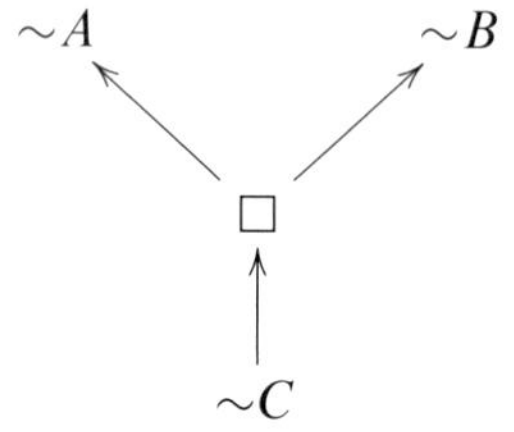

(10.5)

If we started with a coffee machine, which, from A (50 cts.) and B (20 cts.) produces a coffee C, this means that a demand for coffee translates into a demand for 50 cts. and for 20 cts. $A \otimes B \multimap C$ transposes as $\sim C \multimap \sim A \parr \sim B$, hence some people were tempted to write an equation in the style « $\otimes = \parr$ », which is – at least logically – an *atrocity* (Section 11.2.7).

If one looks closely, one finds a confusion between, on one hand « a demand for A and for B » and « a demand for A and a demand for B ». This is not the same thing, since « and » does not commute with « demand ». When one asks for two coins, one really needs both of them: getting only one would be of no use. A more correct version would be, for instance: « if I got A, then I demand B » and by no way a Petri net with two outputs.

One sees that, with « Par », one reaches the limits of the usual language. This is rather banal: thus, I had problems when writing down the part concerning sense and denotation, Section 7.1.1: the denotation, which is *explicit*, is the *implicit* part of sense. Hence the same word can be upturned like a glove depending on the sense in which one takes it, production or consumption.

Finally, a « Par » is like communicating vessels (Figure 10.1).

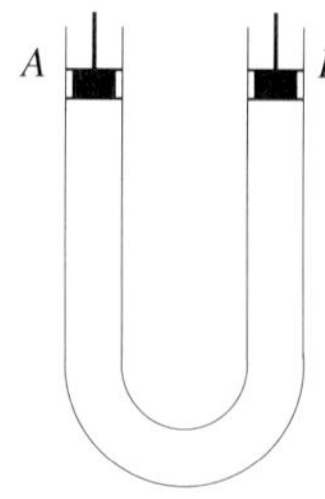

Figure 10.1. « Par » as communicating vessels.

There is a conjunction between A and B, in the sense that an action A is *concomitant* to an action B. But they don't exist independently. Thus, the extension cord (Figure 10.2) establishes a concomitance between A (right, to give 220 volts)

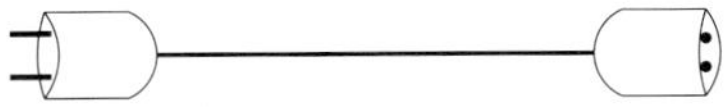

Figure 10.2. The identity as an extension cord.

and $\sim A$ (left, to receive, to consume, 220 volts), without the presence, at rest, of the slightest electric current.

Mode and time. In fact, to come back to a linguistic reading, we could see the multiplicative world as a conjunctive *mode*, since in its operationality, the two partners are present. This is contrary to the disjunctive mode which is that of additives, where only one partner will eventually act. Inside each mode, we will distinguish two *times*, conjunctive for $\otimes$, &, disjunctive for ⅋, $\oplus$. We see that $\otimes$ is plainly conjunctive, $\oplus$ plainly disjunctive. As to ⅋ and &, they are in-between, a disjunctive time inside a conjunctive mode or a conjunctive time inside a disjunctive mode.

Not to go too far. There are limits not to step over when making informal explanations:

- We cannot remain at the metaphoric level. The metaphor (which « carries next to »), is a *Pons Asinorum* which leads to something else. The greatest ridiculousness of the story of cuckolds (Section 2.3.3), is that the theory is about this riddle, together with a couple of the same character: the metaphor eventually appears as a metaphor of itself.

- We do not speak of the building of a coherent metaphorical system. Each metaphor is adapted to a precise point; to seek a *system* of approximations taken from current life is to believe that one can dispense with rigour.

- Never forget that this approach wholly rests upon the *Hauptsatz*. By puttering around with an *ad hoc* system, it is relatively easy to monkey with the phenomena I explained in terms of purchasing goods… as long as one remains cut-free. But one should « make operate » the interpretation, i.e., close it under logical consequence. If the system does not enjoy cut-elimination, the cut-free part can perform wonders: it will just be a bad joke.

By the way, let us come back to the *procedural* aspect of logic. It is cut-elimination which *activates* the descriptions I have given. The simplest is to take the additive conjunction: A & B means *the one that we choose*. How do we choose it? By means of an implication $(A \ \& \ B) \vdash C$, in other terms $\vdash \sim A \oplus \sim B, C$. The choice is performed during normalisation, according to the rule (left plus or right plus) introducing $\sim A \oplus \sim B$. The conjunctive character of $A \ \& \ B$ is that there are *two irons in the fire*, one for the left, one for the right; its disjunctive side is that only one of them will be used.

The viewpoint of the negation $\sim A \oplus \sim B$ is that of a situation where we do not know what is « in the box ». In order to « open the box » without risks, we need $\vdash A \ \& \ B, C$, with its « two irons in the fire », one to take care of the left case, the other to take care of the right case.

10.3 The world is imperfect

10.3.1 Imperfection. We come back to the traditional world, the *imperfect* world. Imperfection corresponds to actions that last, that repeat themselves. For us, it will be synonymous of « eternal », « intangible ». In other words, of *truth* and, eventually, of infinity, of essence.

10.3.2 Of course! We can see $!A$ as a passage to the limit, $A \otimes A \otimes \cdots \otimes A$ infinitely many times or, better, *ad libitum*. We can think of an absence of reaction, or at least, of a negligible one: thus, the level of the waters does not lower when I drink a glass.

In this perspective, the fundamental principle could be $!A \vdash A \otimes !A$, i.e., if A *ad libitum*, then A and A is still available *ad libitum*. Which is only an approximation: indeed the principles are organised into *dereliction*, *weakening*, *contraction* and *promotion*:

$$\begin{aligned} &!A \vdash A, \\ &!A \vdash \mathbf{1}, \\ &!A \vdash !A \otimes !A, \\ &!\Gamma \vdash !A \qquad \text{if } !\Gamma \vdash A. \end{aligned} \tag{10.6}$$

The exponential is a sort of copying machine, which can yield one copy (dereliction), be erased (weakening), or duplicate itself (contraction). Thinking of computation, $!A$ corresponds to a static part, that will not change, typically something in ROM memory.

10.3.3 Modalities

The modal logic S4. The modal logic **S4** is obtained from classical logic by adding two dual modalities; which is written in right formulation:

$$\frac{\vdash\ \Diamond\,\Gamma, A}{\vdash\ \Diamond\,\Gamma, \Box A}\ (necessity) \qquad \frac{\vdash \Gamma, A}{\vdash \Gamma, \Diamond A}\ (possibility)$$

This system is not too bad, since the *Hauptsatz* holds.

Exponentials are therefore a particular case of *modalities*. This sort of connective comes back to Aristotle, but owes much to the Middle Age Thomist logicians. Modality represents the *essentialist* part of logic... as if logic were not already essentialist! There is indeed a contradiction in speaking of necessity in a world – classical logic – already governed by truth. In other words, although **S4** is not ridiculous, modality seems a superfluous addition: it brings nothing new, since structural rules are for free.

The faithful translation of **NJ** in **S4** given by $A \Rightarrow B := \Box(A \Rightarrow B)$, $\forall x A := \Box\forall x A$ is not that interesting: why change a good system belonging in layer -2 (intuitionistic logic) for a system only belonging in layer -1 that inherited the drawbacks of **LK**?

Linear modalities. The linear approach is more satisfactory: modality is the passage from the finite – concrete existence – to the infinite – abstract essence. This is the task of *dereliction* which, from an essence, produces an object, a being, a mortal indeed.

Exponentials root in a stable world, the world of *perenniality*. This being said, those who founded this stable world, the Thomist philosophers, should we take them seriously? These perenniality principles, aren't they too simplistic, isn't there more than plain perenniality, a second degree perenniality, the *perenniality* of *perenniality*? It is indeed strange that the only alternative to the perfect, dynamic, lively, world should be deep freezing at[4] $-273°C$.

In other words, either one stays in the finite, or one opens the Pandora's box of *actual infinity*, with its non-computable monsters, its towers of towers of exponentials[5]. There should exist an intermediate zone, surely infinite, but not *infinitely infinite*. This is the theme of « light exponentials » (Chapter 16).

10.3.4 *Il Menù del Cavaliere*. To be finished with linear logic metaphors, the following gastronomic menu, whose idea is due to Lafont, is rather amusing.

Menù del Cavaliere

Prezzo: 27 euros.

Antipasto: Prosciutto e Melone /e Fichi (depending upon the market)

[4] $-459°F$.

[5] A tower of exponentials whose height is a tower of exponentials.

Primo: Spaghetti/Gnocchi.

Bevande: Acqua del Tevere "SPQR", a piacere.

Which decomposes as a linear implication $Pz \multimap \ldots$ between the price and the tensor of three things: the *antipasto* which is a $\oplus$, since the choice is made by the restaurant, the *primo*, which is an &, since one chooses one's dish, the drink which is a « ! », since it is « as you please ». Which yields

$$Pz \multimap ((P \otimes M) \oplus (P \otimes F)) \otimes (S \,\&\, G) \otimes\, !A.$$

10.A Focalisation

10.A.1 Linear logic programming. This is an idea of Andreoli, see [5]. There is an essential difference between proof-search in the intuitionistic and linear regimes. Compare:

$$\frac{\Gamma, A \wedge B, A \vdash C}{\Gamma, A \wedge B \vdash C} \qquad\qquad \frac{\Gamma, A \vdash C}{\Gamma, A \,\&\, B \vdash C}$$

In the intuitionistic case, the left part Γ always increases when one moves upwards, since one must perform a contraction to keep a « safety copy » of the formula: this copy is no problem since there is no obligation to use it (this is the meaning of weakening). On the other hand, in the linear case, what happens is a *plain* replacement.

If the idea is to *display* the current state of a proof-search, by means of coloured points, one sees that the intuitionistic mode will gradually fill the screen, up to the moment where it will be totally white; while the linear mode will know how to erase, i.e., replace.

The proof-search paradigm, a dynamics rather distinct from normalisation, follows the same type of operationality[6]. Thus, one can, *mutatis mutandis*, program various dynamical effects using proof-search in linear logic.

It is known that logic programming (in the sense of proof-search) is a rather inefficient algorithmics. The main reason is that one can make *mistakes*, due to the non-unicity of the last rule.

10.A.2 Negative connectives. Let us come back to sequent calculus, in the mixed version of Section 9.4.5. The connectives $\bot$, $⅋$, $\top$, &, ?, $\forall$ are *invertible*; this means that:

[6] In proof-search, the *Hauptsatz* ensures the *coherence* of the answers, for instance the answer to two equivalent formulas will be the same.

- These connectives have a unique rule: every time one finds in a conclusion a *negative* formula, i.e., beginning with one of these connectives, one can write a univocal last rule.

- Nothing is lost, since one can prove the equivalence between the conclusion and the premise(s). This can be achieved by a cut with

$$\begin{aligned} \bot : &\quad \vdash \mathbf{1}, \\ A \parr B : &\quad \vdash {\sim}A \otimes {\sim}B, A, B, \\ \top : &\quad \text{no premise}, \\ A \,\&\, B : &\quad \vdash {\sim}A \oplus {\sim}B, A \text{ and } \vdash {\sim}A \oplus {\sim}B, B, \\ ?A : &\quad \vdash {\sim}?A, \underline{A}, \\ \forall x A : &\quad \vdash {\sim}\forall x A, A. \end{aligned}$$

In the very details, this amounts to working *modulo* « η » (Section 7.4.2), since one does allow the identity axiom on non-atomic formulas. Note that the inversion of rules, operated by means of cuts, can indeed be done by a simple induction on proofs. Moreover, inversion is a bijection at the category-theoretic layer (which explains η).

From the viewpoint of proof-*search*, we now get a simple idea: « invert the negative ». In other terms, if I must prove a sequent $\vdash \Gamma$, I can, w.l.o.g., recursively invert all negative connectives, up to the obtainment of sequents made of formulas which are either atomic, or positive, nay underlined.

Invertibility is the best proof of the following:

Theorem 31 (Commutativity). *Negative connectives « commute ». In other words, an expression NN' can isomorphically be written $N'N$, which, depending upon cases, will be called commutativity, associativity, neutrality, distributivity, absorption.*

Proof. Let us give one example, the « commutation » $\parr$ / &. By inversion, one sees that $\vdash \Gamma, A \parr (B \,\&\, C)$ inverts into $\vdash \Gamma, A, B$ and $\vdash \Gamma, A, C$ and $\vdash \Gamma$, $(A \parr B) \,\&\, (A \parr C)$ as well. Which shows the equivalence (isomorphism: the same proofs) between $A \parr (B \,\&\, C)$ and $(A \parr B) \,\&\, (A \parr C)$. □

10.A.3 Focalisation. If negative connectives « commute », so must their positive duals, but what could be the *procedural* contents of this?

Observe that positive rules are in no way invertible:

- There is not always an available rule to introduce a connective: typically, in $\vdash \Gamma, !A$, the $!A$ can be introduced only if Γ is underlined.

- In particular there are cases of failure, e.g., $\vdash \mathbf{1}, \mathbf{1}$.

- More generally, several rules are available, e.g., the two rules of a « Plus », or different choices of *focus*, i.e., of a formula to decompose by means of a positive rule. Each choice restricts the possibilities, it is *irreversible*.
- This is why any limitation of possible choices without irreversible consequence is very valuable.

Positive formulas have a peculiarity: each premise contains exactly one subformula of the focus. Which can even be extended, by adopting the set-theoretic variant (Section 3.1.2) for underlined formulas: there remains only dereliction.

Definition 29 (Focus). The *focus* of a positive rule is the formula created by the rule. The *focalisation* constraint is as follows: above a positive rule, each premise corresponding to a *positive* subformula of the focus in turn focuses on this very sub-focus; this recursively, up to the exhaustion of the hereditary positive subformulas of the focus.

The following example of a proof of $\vdash A \otimes B, C \otimes (D \oplus E)$ is a rather inefficient « obstacle race »; this proof hesitates between the left and right foci. It thus multiplies the possible causes of failure, without compensation. This is a typical case of non-focalised proof.

$$\dfrac{\begin{array}{c}\vdots \\ \vdash C\end{array} \qquad \dfrac{\begin{array}{c}\vdots \\ \vdash A\end{array} \qquad \dfrac{\begin{array}{c}\vdots \\ \vdash B, E\end{array}}{\vdash B, D \oplus E}}{\vdash A \otimes B, D \oplus E}}{\vdash A \otimes B, C \otimes (D \oplus E)} \tag{10.7}$$

Focalisation is precisely the prohibition of the « obstacle race » in the style of (10.7). In other terms, if a sequent $\vdash \Gamma$ contains no negative formula, one can, without compromising whatsoever, take as focus one of the formulas of Γ and recursively its subformulas, up to a change of polarity or atoms.

Focalisation is established by induction on the size of a cut-free proof. If the last rule is, say, a $\otimes$, with premises $\vdash \Delta, A$ and $\vdash \Pi, B$, we get the cases:

(i) A is positive and focalisation works on a formula C of the context Δ; it is enough to permute the « Tensor » in a topmost way.

(ii) A is negative, or is an atom; or A is positive and focalisation works on A: one looks at the other premise.

  (a) B is positive and focalisation works on a formula D of the context Π: the same as (i).

(b) B is negative or is an atom; or B is positive and focalisation works on B: one has indeed achieved focalisation on $A \otimes B$.

We can understand focalisation from the inclusions:

$$\begin{aligned} \sim\sim X \cdot \sim\sim Y &\subset \sim\sim(X \cdot Y), \\ \sim\sim X \cup \sim\sim Y &\subset \sim\sim(X \cup Y). \end{aligned} \tag{10.8}$$

Indeed, going back to the tautological model, the interpretation of $[A]$ is the set of all contexts enabling us to prove A. Hence $[A] \cdot [B]$ is the set of all contexts enabling us to prove $A \otimes B$ with a « Tensor » rule as last rule; similarly, $[A] \cup [B]$ is the set of contexts enabling us to prove $A \oplus B$ with a « Plus » rule as last rule. But then, the double negation $\sim\sim$ corresponds to the case of a proof whose last rule may focus on the context. In other words, when I write $\sim\sim([A] \cdot [B]) \cdot [C]$, I mean a proof of $(A \otimes B) \otimes C$ ending with a *tensor* rule, but whose premise containing $A \otimes B$ need not be proved with a *tensor* as last rule. The inclusions (10.8) do express that we can avoid the obstacle race focus/context: this is focalisation in the sense of Andreoli.

10.A.4 Polarity. Negatives commute, positives commute as well by duality. But what about positive and negative? This is quite simple, there is a *semi-commutation*: one can always replace a group PN with a group NP. This is easy to understand from the *procedural* standpoint: negatives are passive while positives are active. Everybody knows that, in real life, one can always postpone decisions, actions: this is known as procrastination. In mathematics, the most familiar example is the replacement of $\exists\forall$ with $\forall\exists$. The tarskian explanation, never short of truisms, says that, in $\forall x \exists y$, y depends upon x, which is not the case of $\exists y \forall x$. Thank you, Monsieur de la Palice[7]! I prefer the explanation in terms of procrastination, since it works in a more general setting:

$$\begin{aligned} \otimes/ \parr : &\quad A \otimes (B \parr C) \vdash (A \otimes B) \parr C, \\ \otimes/ \& : &\quad A \otimes (B \& C) \vdash (A \otimes B) \& (A \otimes C), \\ \oplus/ \parr : &\quad (A \parr B) \oplus (A \parr C) \vdash A \parr (B \oplus C), \\ \oplus/ \& : &\quad A \oplus (B \& C) \vdash (A \oplus B) \& (A \oplus C), \\ \oplus/ \& : &\quad (A \& B) \oplus (A \& C) \vdash A \& (B \oplus C). \end{aligned}$$

The general remark $PN \vdash NP$ does not dispense with verifications; but, those being done, it yields an extraordinary mnemonic tool to remember the semi–commutations.

[7]French logician (1470-1525), the unavowed precursor of Tarski: *Un quart d'heure avant sa mort, il était encore en vie*. By the way, the expression « vérité de la Palice » was in use long before the definition of truth by Tarski.

We have heard about temporal logic(s), this bleak police of time; cops who overlook *logical time*, the time of consequence, of causality. Time occurs when we cannot permute two rules, since one must be performed before the other, for fear of a procedural catastrophe. This is therefore the alternation positive/negative, answer/question, explicit/implicit.

10.A.5 Synthetic connectives. Logicians sometimes are concerned, with little imagination, with the idea of a generalised connective, say, ternary. The answer usually looks as follows: an abbreviation for a formula $\Phi[A, B, C]$. They didn't stray themselves![8]

Indeed, to speak of a connective supposes the existence of a set of introduction rules for the connective (and its negation). These rules are fatal abbreviations for pieces of proof in the « usual » system. To assume the existence of portions of proof enabling one to directly pass from sequents involving A, B, C to a sequent involving $\Phi[A, B, C]$, is a *focalisation* hypothesis. This hypothesis is reasonable only if Φ is of a single piece, without change of polarity.

Take, for instance, $\Phi[A, B, C] := A \oplus (B \otimes C)$. This connective (and its dual $\Psi[A, B, C] := A \,\&\, (B \parr C)$) admits rules, i.e.:

$$\frac{\vdash \Gamma, A}{\vdash \Gamma, \Phi[A, B, C]} \qquad \frac{\vdash \Gamma, B \quad \vdash \Delta, C}{\vdash \Gamma, \Delta, \Phi[A, B, C]} \qquad\qquad \frac{\vdash \Gamma, A \quad \vdash \Gamma, B, C}{\vdash \Gamma, \Psi[A, B, C]}$$

This is a genuine synthetic connective, which brings nothing new, but which takes its place among the others.

Let us now take $\Phi[A, B, C] := A \parr (B \otimes C)$. This « connective » (and its dual $\Psi[A, B, C] := A \otimes (B \parr C)$) admits rules too, i.e.:

$$\frac{\vdash \Gamma, A, B \quad \vdash \Delta, C}{\vdash \Gamma, \Delta, \Phi[A, B, C]} \qquad \frac{\vdash \Gamma, B \quad \vdash \Delta, A, C}{\vdash \Gamma, \Delta, \Phi[A, B, C]} \qquad\qquad \frac{\vdash \Gamma, A \quad \vdash \Delta, B, C}{\vdash \Gamma, \Delta, \Psi[A, B, C]}$$

Each rule is a combination $\parr / \otimes$. But there is not enough of them. Typically, they are unable to prove $\vdash \Psi[{\sim}A, {\sim}C, {\sim}B], \Phi[A, B, C]$. This sequent is indeed

[8] Not to speak of quantifiers: for instance, according to Mostowski, a quantifier is a function from $\wp(X^{n+1})$ into $\wp(X^n)$. Earth shaking!

provable in the « ordinary » system:

$$\dfrac{\dfrac{}{\vdash \sim A, A} \qquad \dfrac{\dfrac{\dfrac{}{\vdash \sim B, B} \qquad \dfrac{}{\vdash \sim C, C}}{\vdash \sim C, \sim B, B \otimes C}}{\vdash \sim C \parr \sim B, B \otimes C}}{\dfrac{\vdash \sim A \otimes (\sim C \parr \sim B), A, B \otimes C}{\vdash \sim A \otimes (\sim C \parr \sim B), A \parr (B \otimes C)}}$$

by means of a proof admitting no permutation of rules. In particular, the rules do not follow an order compatible with the synthetic rules we just wrote.

There is no classical propositional counterexample, since classical connectives can, as we please, be written positively or negatively. All that remains classically is that one can define a synthetic quantifier only in case it is purely universal (negative) or existential (positive).

10.A.6 Exponentials and polarity. The exponential $?A$ is performed in two steps, first $\underline{A}$ (positive), then $?A$ (negative) and similarly, although it does not show on the formalism, $!A$ gathers two operations, a negative one followed by a positive one. This double change of polarity causes exponentials to commute with nothing, with one exception: $!(A \,\&\, B) = !A \otimes !B$ and its dual: here one changes conjunction.

The same holds in modal logic; the classical substrate yields $\otimes = \& = \wedge$, hence $\Box(A \wedge B) \Leftrightarrow \Box A \wedge \Box B$. The attempts in the style of **S5** (Section 4.E) to force extra commutations such as $\Box/\forall$ are complete failures. Logic does not accept that[9]: one loses the *Hauptsatz*.

[9]But Kripke models do: they are very compliant.

Chapter 11
Proof-nets

11.1 ILL

In this first section, I shall *regress* and forget for a while our major breakthrough: the regained symmetry expressed by linear negation. I shall introduce, for purely didactic reasons, an *intuitionistic linear logic*, of which we shall see that it brings strictly nothing to the real linear logic: it is only its *teleological* version. This will enable us to approach *nets* through natural deduction. Of course, nets do not need this detour.

11.1.1 A regression. Can one speak of an « intuitionistic linear logic »? Linear + intuitionistic, what a heavy covenant!

ILL is defined as the system without $\sim$, $\perp$, ⅋, ? (but with $\multimap$): exactly what I used for the multilinear interpretation (Sections 9.4.4 and 9.4.6). I will not expatiate upon this parasitic system, which is only a *Pons Asinorum*. Note that:

- The abuse of adjectives – here, « intuitionistic linear » – often betrays a drop in quality. It also corresponds to a widespread taste for marginal structures. Thus, I propose to remove « linear » from « non-commutative linear logic »: roughly speaking, linearity calls into question the idempotency of conjunction (i.e., contraction and weakening) and algebra tells us that a non-commutative monoid is hardly idempotent.

- One can legitimately retain the sole functional aspect of linear logic, thus leaving no role for an involutive negation. I contend that, in that case, there is no motive for a bastard system: it is enough to *teleogise* the old system, the « real » LL; a conservation result will suffice.

11.1.2 A conservation result. In order to describe (without wasting paper) the « intuitionistic » version of LL, it suffices to rewrite the mixed version in left/right style, with exactly one formula on the right, this formula being non-underlined; forgetting the connectives $\sim$, $\perp$, ⅋, ?, which would come off badly anyway. This is therefore a *teleological* version of linear logic, since one distinguishes a *goal*, the formula on the right. And this system is not without practical virtues (Sections 9.4.4 and 9.4.6).

The only real problem is to know whether the teleological version is faithful. In other words, if I prove a formula of the fragment without $\sim$, $\perp$, ⅋, ?, but with $\multimap$, in linear logic, can I prove it in the teleological version?

The answer is yes, up to a small restriction, a remark of Schellinx: one should exclude the constant **0**. Between us, this is not that dramatic, since it essentially occurs in intuitionistic negation, a rather calamitous connective, if any. The only delicate case in teleologisation is that of the left rule for $\multimap$: if $\Gamma, \Delta, A \multimap B \vdash C$ (i.e., $\vdash \sim\Gamma, \sim\Delta, A \otimes \sim B, C$) follows from $\vdash \sim\Gamma, A$ and $\vdash \sim\Delta, \sim B, C$, i.e., $\Gamma \vdash A$ and $\Delta, B \vdash C$, one can teleologise by means of the rule:

$$\frac{\Gamma \vdash A \quad \Delta, B \vdash C}{\Gamma, \Delta, A \multimap B \vdash C}$$

But this neglects an alternative possibility: $\vdash \sim\Gamma, \sim\Delta, A \otimes \sim B, C$ follows from $\vdash \sim\Gamma, A, C$ and $\vdash \sim\Delta, \sim B$. It is easy to see that, if Δ, B do not contain **0**, there is no chance to ever reach $\vdash \sim\Delta, \sim B$ (take a classical semantics in which the atoms are true). A concrete counterexample is given by $(p \multimap \mathbf{0}) \multimap \mathbf{0} \vdash p \otimes \top$, which is provable, but not teleologically.

11.1.3 Cyclic logic vs. Lambek calculus. We might give attention to the *syntactical calculus* of Lambek [72]. And let us settle the irritating question of its relation to my own *cyclic*[1] *logic* and therefore, of its conservative extension, the *non-commutative logic* [4].

Lambek's calculus is scarcely a logic, since there are only three binary connectives, $\otimes, \multimap, \circ\!-$. The sequents are of the form $\Gamma \vdash A$, where Γ is totally ordered. The rules are the following:

Identity/Negation

$$\frac{}{A \vdash A}\ (identity) \qquad \frac{\Gamma \vdash A \quad \Delta, A, \Pi \vdash B}{\Delta, \Gamma, \Pi \vdash B}\ (cut)$$

Logic

$$\frac{\Gamma \vdash A \quad \Delta \vdash B}{\Gamma, \Delta \vdash A \otimes B} \qquad \frac{\Delta, A, B, \Pi \vdash C}{\Delta, A \otimes B, \Pi \vdash C}$$

$$\frac{A, \Gamma \vdash B}{\Gamma \vdash A \multimap B} \qquad \frac{\Gamma \vdash A \quad \Delta, B, \Pi \vdash C}{\Delta, \Gamma, A \multimap B, \Pi \vdash C}$$

$$\frac{\Gamma, A \vdash B}{\Gamma \vdash B \circ\!- A} \qquad \frac{\Gamma \vdash A \quad \Delta, B, \Pi \vdash C}{\Delta, B \circ\!- A, \Gamma, \Pi \vdash C}$$

[1]The name is due to Yetter [102].

We must understand $\Gamma \vdash A$ as the implication between the « Tensor » of the Γ, in the order of enumeration and A: $\bigotimes \Gamma \multimap A$. The rules are rather natural but $(\vdash \multimap)$ and $(\vdash \circ\!-)$, which are the exact opposite of what one would like to write. But nobody is perfect.

Let us come down to *cyclic logic*. The language is that of linear logic: one must only be careful as to negation, and reverse the order in the multiplicative case:

$$\begin{aligned} \sim(A \otimes B) &:= \sim B \parr \sim A, \\ \sim(A \parr B) &:= \sim B \otimes \sim A. \end{aligned}$$

The sequent calculus is a variant of mixed calculus. Indeed, I took pains to write it directly (page 191) in a way compatible with the cyclic case. The only novelty is the rule of exchange, which is restricted to *circular* permutations of the non-underlined formulas – the other retaining freedom of movement.

$$\frac{\vdash \Gamma, \Delta}{\vdash \Delta, \Gamma} \qquad\qquad \frac{\vdash \Gamma, \underline{A}, \Delta}{\vdash \Gamma, \Delta, \underline{A}}$$

Lambek's calculus *embeds* in cyclic logic by means of the translation:

$$\begin{aligned} A \multimap B &:= \sim A \parr B = \sim(\sim B \otimes A), \\ B \circ\!- A &:= B \parr \sim A = \sim(A \otimes \sim B), \\ A_1, \ldots, A_n \vdash B &:= \vdash \sim A_n, \ldots, \sim A_1, B. \end{aligned}$$

The cyclic exchange rule enables *teleologisation*, i.e., to put at the rightmost location – standing for the right part of the sequent – no matter what formula one wants. But, this being done, no degree of freedom remains; the structure becomes thus rigid, just as in Lambek's calculus.

I really said « embeds », since it is a faithful translation: if $\Delta, B \circ\!- A, \Pi \vdash C$ is provable under the form $\vdash \sim\Pi, A \otimes \sim B, \sim\Delta, C$ by a « Tensor » rule, the cutting off of the context (*modulo* cyclic permutations) between A and $\sim B$ must attribute C to $\sim B$: indeed, if Γ is made of formulas of Lambek's calculus, $\vdash \sim\Gamma$ is not provable in cyclic logic – not even in classical logic. Which cuts off $\sim\Pi$ into $\sim\Pi', \sim\Gamma$ and one eventually gets the rule:

$$\frac{\Gamma \vdash A \quad \Delta, B, \Pi' \vdash C}{\Delta, B \circ\!- A, \Gamma, \Pi' \vdash C}$$

In particular there is no « intuitionistic » specificity of the Lambek calculus that cyclic logic would not respect: the syntactic calculus is only a teleologised fragment of cyclic logic, not a distinct system.

The phase models of cyclic logic are given by pairs $(\mathcal{M}, \perp\!\!\!\!\!\perp)$ with $\mathcal{M}$ non-commutative and $\perp\!\!\!\!\!\perp$ *cyclic*, i.e., such that $xy \in \perp\!\!\!\!\!\perp \Rightarrow yx \in \perp\!\!\!\!\!\perp$.

11.1.4 Digression: non-associative logic. Under linguistics pretexts, we have witnessed the springing up of professed *non-associative* logics. I am in no position to judge the linguistic motivation; on the other hand, I want to be clear as to « non-associativity »: it cannot work. Indeed, logic is not a formalist world where everything is decided by a play on language. The adjective « non-associative » applies to logic in the procedural sense. But what is associativity from the procedural viewpoint? It is the existence of a category-theoretic – in other terms, compositional – interpretation. To relinquish associativity is to relinquish Church–Rosser, which, by the way, is not a dogma: it is an avenue, a vehicle for reasoning about normalisation. I know nothing that can replace it – and I doubt that the « non-associativists » can find something. Concretely, this means that the « non-associative » systems do not enjoy cut-elimination. By the way, imagine the catastrophe: instead of a comma, one has brackets. How can one speak of the context of A in $(B(CA)(DE))$? This is hopeless. Of course, one can still say that it is something like $(B(C\cdot)(DE))$ but try to build a whatsoever sensible definition out of this… . In particular, no help can be expected from non-associative monoids.

By the way, observe that the weak point of the *non-commutative*, e.g., of the cyclic is not the absence of phases, of categories, etc. It is the absence of a convincing *procedural* explanation of non-commutation. It cannot be that the order of cuts influences the result: this is non-association, contradicts Church–Rosser and kills layer -2. It is something else, but what? It could be a constraint of a temporal nature, « one does this before doing that », or of a spatial nature « A obstructs B ». One must try, something original must be found… but also simple enough so that one can reconstruct logic from it. Anyway, I don't think that the solution can be found at the category-theoretic layer – although I am fully confident in the ability of category-theorists to explain it *afterwards*.

11.1.5 A natural deduction. **ILL** can be formulated in natural deduction style; this is indeed its only interest. Let us quickly write a few rules:

$$\frac{\begin{array}{c}[A]\\ \vdots\\ B\end{array}}{A \multimap B}\ (\multimap I) \qquad\qquad \frac{\begin{array}{cc}\vdots & \vdots\\ A & A \multimap B\end{array}}{B}\ (\multimap E)$$

Obviously we must take linearity into account. If we stick to the sole multiplicatives, this is quite simple: there is one and only one discharged hypothesis.

Complications start with additives; thus, in

$$\frac{\begin{array}{cc}\vdots & \vdots\\ A & B\end{array}}{A \,\&\, B}\ (\& I)$$

we must require that both premises have the same active hypotheses which are counted as a single hypothesis. This is why

$$\dfrac{\dfrac{[A] \quad [A]}{A \,\&\, A}}{A \multimap A \,\&\, A}$$

can be accepted. We see that this is rather complicated. Morality: stick to multiplicatives.

The tensor accepts the following rules:

$$\dfrac{\begin{array}{c}\vdots\\ A\end{array} \qquad \begin{array}{c}\vdots\\ B\end{array}}{A \otimes B}\,(\otimes I) \qquad\qquad \dfrac{\begin{array}{c}\vdots\\ A \otimes B\end{array} \qquad \begin{array}{c}[A] \quad [B]\\ \vdots\\ C\end{array}}{C}\,(\otimes E)$$

including an awfully fabricated elimination, which is reminiscent of that of intuitionistic disjunction and which therefore requires commutative reductions. The partisans of « intuitionistic linear logic » never linger on this *detail*.

11.2 Multiplicative nets

11.2.1 Critique of rules. Putting side by side the four natural deduction rules we just found:

$$\dfrac{\begin{array}{c}[A]\\ \vdots\\ B\end{array}}{A \multimap B}\,(2) \qquad \dfrac{\begin{array}{c}\vdots\\ A\end{array} \qquad \begin{array}{c}\vdots\\ A \multimap B\end{array}}{B}\,(1) \qquad \dfrac{\begin{array}{c}\vdots\\ A\end{array} \qquad \begin{array}{c}\vdots\\ B\end{array}}{A \otimes B}\,() \qquad \dfrac{\begin{array}{c}\vdots\\ A \otimes B\end{array} \qquad \begin{array}{c}[A] \quad [B]\\ \vdots\\ C\end{array}}{C}\,(1,2,3)$$

These rules have several drawbacks:

(1): some premises are indeed hidden conclusions. That's not dramatic, but this complicates a lot of matters, think of the *main hypothesis* (Section 4.4.1). This concerns both eliminations.

(2): in the introduction of $\multimap$ the hypothesis A is « held at bay ». To the point that a gimmick is needed to physically link it to the rule, e.g., De Bruijn indices. This *non-locality* also concerns the elimination of $\otimes$.

(3): in the elimination of $\otimes$, one finds a formula C which does not belong here. It is a context, a rather arbitrary one and commutative reductions are introduced to maintain this arbitrariness, but at such a heavy price! It is, among the three, the only crippling drawback.

To sum up, only the introduction of ⊗ is perfect; on the other hand, its elimination cumulates the three drawbacks.

11.2.2 Putting things right side out. The fundamental idea is to « put everything right side out ». Indeed, to give up active hypotheses, keeping only conclusions.

The two following *links*, respectively called *Axiom* and *Cut*, enable us to replace a hypothesis with a conclusion and *vice versa*:

Those are indeed rules: the axiom has no premise and two conclusions; the cut has two premises and no conclusion. The same is true of the links *Tensor* and *Par*, which are indeed binary rules in the usual sense:

These four links enable us to put everything right side out:

Hypothesis: the hypothesis A becomes an axiom link:

⊸ I: the hypothesis A is now a conclusion, which enables us to write:

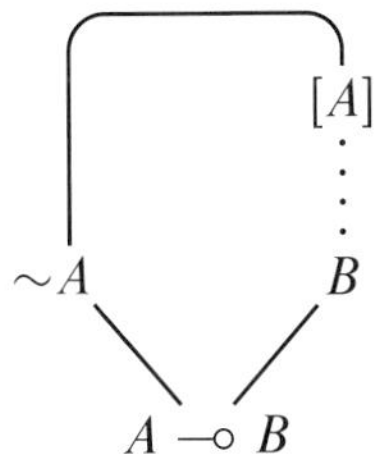

The fact that only one A is discharged (this is linearity) is crucial.

$\multimap$ **E:** we use cut to topsy-turvy the sense of deduction:

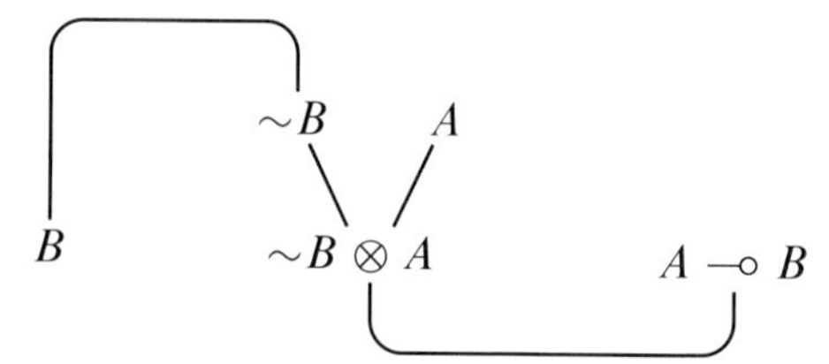

A subtler translation could dispense with cut in case the deduction is normal. But we are not presently gilding the lily.

$\otimes$ **I:** nothing to object, hence nothing to change:

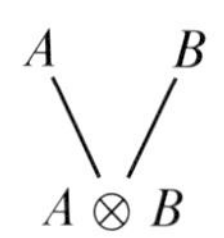

$\otimes$ **E:** this is the most complex case:

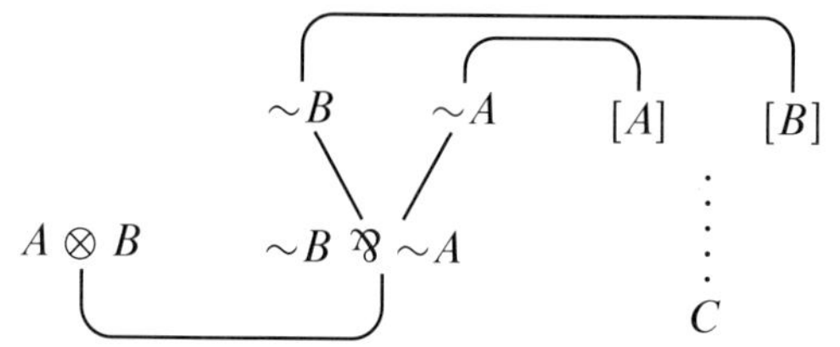

Which by the way shows, once everything has been put as a hypothesis, that

$$\multimap\text{-intro} = \otimes\text{-elim} = \parr\text{-link};$$
$$\multimap\text{-elim} = \otimes\text{-intro} = \otimes\text{-link}.$$

which is indeed only a half-surprise.

Coming back to the three drawbacks of natural deduction:

(1): disappears, since formulas increase from hypothesis to conclusion.

(2): discharged hypotheses become conclusions: no more holding at bay.

(3): the most spectacular improvement: the translation makes disappear any connexion between $A \otimes B$ and C.

It is time to relinquish our *Pons Asinorum* and directly translate the real thing, *sequent calculus*.

11.2.3 Translation of sequent calculus. This is more or less immediate: to each proof of a sequent $\vdash \Delta$ in the multiplicative fragment, one associates a *net*, i.e., a graph using links and whose conclusions are Δ. In fact, the sequent calculus rules enable one to build:

Identity: a net with two conclusions out of nothing (axiom link).

Cut, $\otimes$: reglue two distinct nets with conclusions Γ, A and $\Delta, \sim A$ (resp. and Δ, B) to obtain a single net of conclusions Γ, Δ (resp. $\Gamma, \Delta, A \otimes B$), by means of a cut-link (resp. a $\otimes$-link).

⅋: attach two conclusions A, B of a net $A ⅋ B$ by means of a ⅋-link.

Note the absence of exchange in this list. In terms of graphs, exchange is a blank operation. Which is no longer the case if one displays nets on a plane, since exchange will induce *crossings* of links. Compare:

The proof on the left comes from cyclic logic, i.e., does not quite use exchange, while the proof on the right (which says that $B \otimes A \multimap A \otimes B$) is valid only commutatively. One sees that cyclic logic induces planar nets. As to the crossing observed on the right, it would be tempting to introduce a *braiding*, but this never gave much: in particular, we can hardly dig out the possible *procedural* sense of braiding.

11.2.4 Nets and structures. If one calls any graph obtained by translating a sequent calculus proof a *proof-net*, it would be interesting to determine what sort of animal we just encountered.

We introduce below the notion of a *proof-structure*, a very lax approximation to the notion of net. We basically mean something looking like a proof without hypotheses, using links as rules. Remember (Section 5.1.5) that the notion of *occurrence* has no theoretical sense and that the formulas written are supposed to be *distinct* (but perhaps isomorphic): the fact that they have distinct locations is enough to make them different.

Definition 30 (Proof-structures). A *proof-structure* is a non-empty finite set of formulas and links such that:

- Every formula is a conclusion of exactly one link.

- Every formula is a premise of at most one link.

One calls a formula which is not the premise of a link a *conclusion*.

Contrary to intuition, there are structures without conclusions, the most typical example being the *vicious circle*:

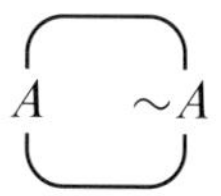

(11.1)

Remembering that identity is an extension cord, that cut is a plugging, we have just shown the incestuous plugging of an extension cord into itself; a proof-structure of utmost importance… and which is not a net, since the empty sequent is not provable.

11.2.5 Normalisation for structures. Structures provide us with a graphic setting where we can try to normalise. Every time we meet a cut we can reduce it either[2] as

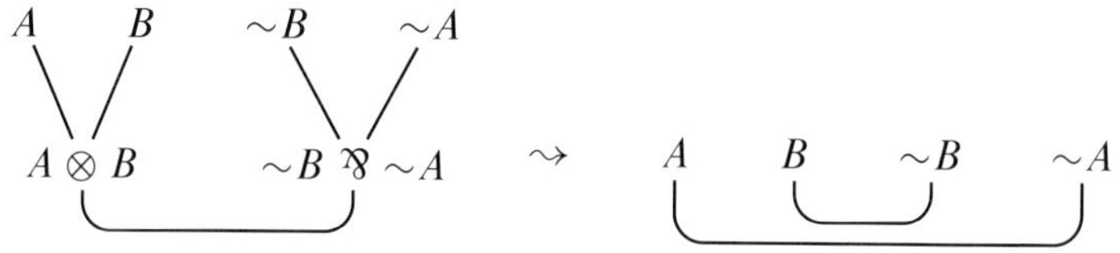

(11.2)

or as

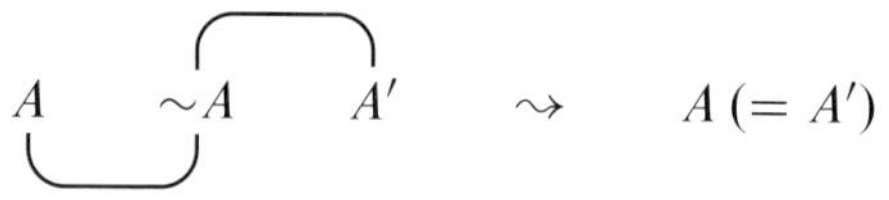

(11.3)

Note that it is a matter of identification of two « occurrences » of A, if one hangs onto this old-fashioned – albeit comfortable like old slippers – vision. Something more modern: cut is a plugging and the identity is an extension cord. The rule expresses the procedurality of the cord: it puts into contact, it *delocates* A in A' (or A' in A).

[2] I chose the non-commutative version of negation, to avoid unesthetic crossings.

Let us see in one example how these rules reflect the reduction of an implicative cut:

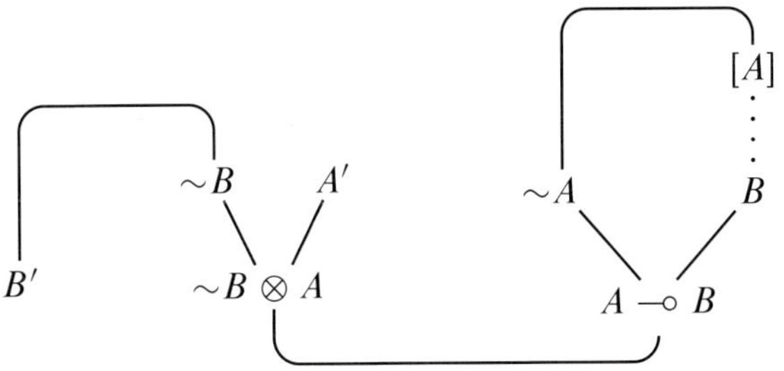

reduces, using (11.2), into

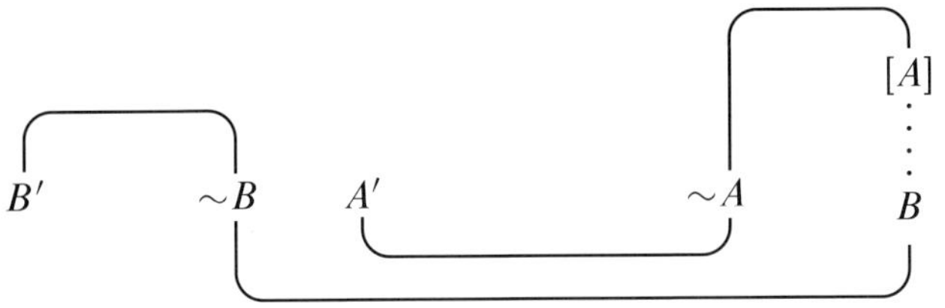

which becomes in turn, using (11.3),

$$\begin{array}{c} A' \ (= A) \\ \vdots \\ B' \ (= B) \end{array}$$

One sees that normalisation in fact decomposes into a simplification (11.2) followed by a regluing (11.3).

11.2.6 A normalisation « theorem ». First observe that the normalisation of proof-structures enjoys Church–Rosser. In fact, the proof reduces to checking the absence of *conflicts*, i.e., of substructures with two possible rewritings. There is an ambiguous situation where one can apply (11.3) in two possible ways:

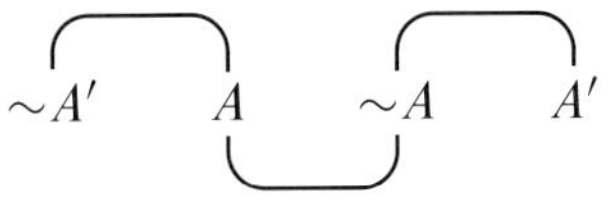

which yields as we please

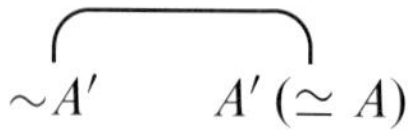

or

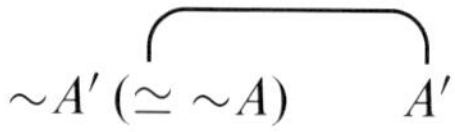

i.e., two identical results. Nothing surprising here, since reduction (11.3) corresponds to the procedurality of an extension cord. In other words, two cords together are like a single one; the only difference between left and right is that we did not « stretch out » the same cord: the output is anyway the same.

Theorem 32 (Faulty!). *Proof-structures enjoy strong normalisation.*

Proof. In steps (11.2) and (11.3), the number of links strictly decreases:

(11.2): three links are replaced with two cuts: deficit, 1 link.

(11.3): one axiom and one cut disappear: deficit 2 links. □

This is a nice proof, betrothed to a brilliant future, but unfortunately wrong. In fact, reduction (11.2) diminishes the size of the net only when $A' \neq A$. So that the structure without conclusions (11.1) reduces in itself. However, if such *vicious circles* are considered to be normal, then strong normalisation really holds.

I said « vicious circle ». It is an anomaly quite different from the non-termination coming from fixed points, from the antinomies in the style of Russell. This structure is autistic, since it does not communicate with the outside, to which it gives nothing – but takes nothing either.

11.2.7 Digression: compact linear « logic ».

The vicious circle is the typical structure encountered in a professed « compact linear logic », based upon a faulty reading of retrocausality in Petri nets (Section 10.2.2). This is the blunt identification $\otimes = \parr$. Which is expressed by the following proof-structure

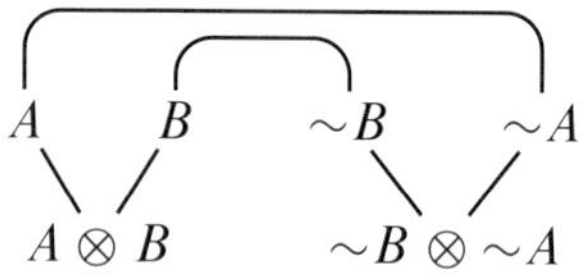

(11.4)

that we can toy with « cut », making $A = \sim B$, with $A \parr \sim A$

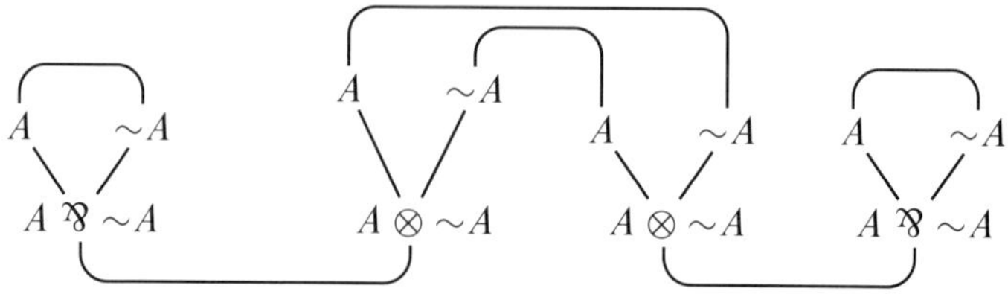

which normalises into the vicious circle. Compact logic proves the empty sequent « $\vdash$ » and avoids inconsistency *in extremis*. We must leave it in the department of atrocities, in the company of its colleagues, paraconsistent, epistemic, non-monotonic, fuzzy, quantum, relevant… « logics ».

Here we witness a frank divorce between the logical viewpoint and the category-theoretic viewpoint, for which $\otimes = \parr$ is not absurd. Thus, in algebra, the tensor is often equal to the cotensor, for instance in finite-dimensional vector spaces: $\mathcal{L}(E) \simeq E' \otimes E$, a canonical isomorphism devoid of any logical significance. This isomorphism enables us to define the *trace* $\mathrm{tr}(\sum_i x'_i \otimes x_i) := \sum_i x'_i(x_i)$: this linear form on $\mathcal{L}(E)$ is nothing but the transposition of the identity. By the way, an interesting remark of Freyd: $\mathrm{tr}(\iota_E) = \dim E$ is a scalar[3] depending on E. As if, in system **F**, we had found a term of type **nat** depending upon a type variable X and whose value n_X would vary with X. This remark illustrates the gap separating logic and categories, by the way both quite legitimate activities; one should not try to crush one upon another.

11.3 The correctness criterion

11.3.1 Sequentialisation. Proof-nets pose a novel problem in logic, that of *sequentialisation*. It is a complicated way of saying « characterising nets ». Indeed, we have an inductive definition of nets, which follows the steps of sequent calculus. But the writing with several conclusions does not make precise which one is the *last rule*: there might be none, as in the vicious circle (11.1). Among all proof-structures, one must thus distinguish those which are nets, i.e., those coming from sequent calculus. These structures are *sequentialisable* in a double sense: they come from sequents, but also one can build them *sequentially*, by a succession of steps corresponding to the rules of the calculus.

Additionally, we should determine whether the oblivion of sequentiality (the order of rules) was not pushed too far. It could be the case that two proofs in sequent calculus with the same « underlying » net, reduce, using cut-elimination, into proofs with distinct underlying nets. This is indeed not the case; translating the cut-elimination steps inside nets:

- Commutations do not affect the underlying net.
- The key case is translated into (11.2).
- The elimination of a cut with an identity translates as (11.3).

and Church–Rosser does the rest.

Sequentialisation is a most stimulating methodological challenge. Proof-structures are completely symmetrical in $\otimes$, $\parr$, two binary links. Since conjunction is quite different from disjunction, proof-nets will not accept the confusion between

[3]This is indeed the vicious circle (11.1).

them. The ancient, semantical, explanation will insist on the *truth* of the sequent established by the net, which brings us back to the tarskian blind spot. But could it be a matter of *topology* of graphs, conjunction and disjunction having distinct topologies? Thus, one seeks here the procedural difference between conjunction and disjunction as a topological nuance.

11.3.2 Switches. If $\mathfrak{R}$ is a proof-structure, if L is a $⅋$-link, we call a choice left/right, corresponding to the following graphs a *switch*, in which we keep only one of the two edges, the left one or the right one[4]:

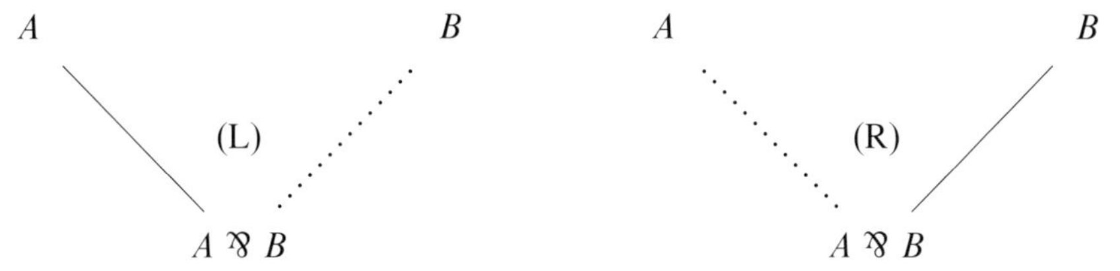

The natural justification of switches is that $A ⅋ B$ means both $A \multimapinv {\sim}B$ and ${\sim}A \multimap B$. To set a switch, this is therefore to choose a particular writing in the « natural deduction » style. By pushing this analogy, let us remember that the writing under the form of natural deduction is *arborescent*; and also that, *topologically* speaking, trees are connected/acyclic graphs. We are thus prepared for the next result, simple, but fundamental:

Proposition 16. *If $\mathfrak{R}$ is a net, then for any* switching $\mathcal{I}$ *of its « $⅋$ »-links, the resulting graph is a* **tree**, *i.e., is connected and acyclic.*

Proof. By induction on proofs in sequent calculus. The identity induces an axiom link, which is a tree. The links tensor and cut reglue two trees into another tree. Finally, the $⅋$-link adds $A ⅋ B$ to a tree containing A, B; since the new vertex is linked only to one of A and B, the graph remains a tree. □

11.3.3 Sequentialisation theorem

Definition 31 (Correctness). A proof-structure $\mathfrak{R}$ is *correct* iff for any switching $\mathcal{I}$ of its $⅋$-links, the resulting graph $\mathfrak{R}_{\mathcal{I}}$ is a *tree*.

We shall prove the converse of Proposition 16:

Theorem 33 (Sequentialisation). *A correct proof-structure is a net.*

---

[4]The electric analogy suggests the word « commutator », but this conflicts with algebra.

The proof will be the object of the next subsections; we first need a discussion.

There is a change of paradigm, one is no longer quite inside syntax, at least something happened. Usually, the writing of rules is done sequentially, which means that the writing guarantees syntactical correctness: if one prefers, a usual proof can be checked *locally*. On the other hand, one pays for this by the manipulation of *global* expressions, sequent or distant discharges in natural deduction. To sum up, since *nobody is perfect*, one must choose between:

global syntax + local correctness / local syntax + global correctness

Nets propose a purely local syntax and the price to pay is therefore a global correctness. Thus, the net

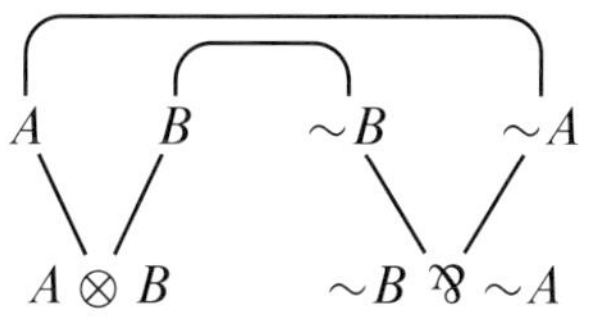

can only globally be correct. Let us experiment: if I replace the $\parr$-link with a $\otimes$-link, I get a cycle. If I had replaced the $\otimes$-link with a $\parr$-link, I would have got two connected components. If I perform both replacements, the net becomes correct again: correctness is therefore global.

Something rather impressive is the proof of cut-elimination from correctness. Since the vicious circle is incorrect, it suffices to prove that cut-elimination preserves correctness. This reduces to the case of a reduction (11.2) (page 224). One selects switches in the reduced structure and one removes the two links « cut ». The graph splits into a certain number of connected components, each of them containing at least one of the formulas A, B, $\sim A$, $\sim C$:

- A is alone in its component:
  - ► It does not contain B: there would be a cycle in the original net.
  - ► It cannot contain $\sim B$: switch $\parr$ on « left » in the original net.
  - ► Symmetrically, it cannot contain $\sim A$.
- Similarly B is alone in its component.
- Whatever way we switch $\parr$ in the original net, it only yields a direct access to $\sim A$ (or $\sim B$), the other not being accessible through A or B: hence $\sim A$ and $\sim B$ are in the same component.

Finally the components are those of A, of B and of $\sim A$, $\sim B$ and the graph obtained by regluing these three graphs by means of two cuts is still a tree.

11.3.4 Empires. The sequentialisation theorem immediately reduces to its particular case without cuts: replace any cut between $A, \sim A$ with a $\otimes$ of conclusion $A \otimes \sim A$: this is the Tortoise (Section 3.A.2). This helps in simplifying the proof below: but everything adapts, *mutatis mutandis*, to the case with cuts.

Let $A \in \mathfrak{R}$ be a formula in a correct net. We select a switching $\mathcal{I}$ and in case the graph $\mathfrak{R}_{\mathcal{I}}$ contains an edge linking A (as a premise) to the conclusion of a link, we sever it: we « cut the bridge below A ». The graph thus obtained has at most two connected components: let $\mathfrak{R}_{\mathcal{I}}(A)$ be the component of A.

Definition 32 (Empires). The *empire* of A, noted eA, is the intersection of the $\mathfrak{R}_{\mathcal{I}}(A)$, when $\mathcal{I}$ varies through all switchings of $\mathfrak{R}$.

In other words, eA is what one will *inevitably* encounter when starting from A « upwards ». Empires enjoy three essential properties: *imperialism*, *principal choice* and *simultaneous empires*:

Imperialism. It is very difficult to step out from an empire: in general, if $B \in eA$ and if B is premise/conclusion of a link L, then all other premises/conclusions of L are in eA. A first exception is $B = A$ and is a premise of L. Otherwise, we see that the links « axiom » and $\otimes$ are *imperialistic*. There remains the $\parr$-link:

- If $B = C \parr D \in eA$ and if, say, $C \notin eA$, there is a switching $\mathcal{I}$ such that C is not « upwards accessible » from A; it is therefore « downwards accessible » from A. The link L didn't contribute to this connexion, since B is « upwards accessible » from A. I can therefore modify $\mathcal{I}$ by switching L to « left »: I thus got a nice cycle.

- If $B \in eA$ is a premise of $\parr$, say $B \parr C$, it is possible that $B \parr C \notin eA$. But, if $C \in eA$, no matter what switching we set for L, $B \parr C$ will be linked to a point (B or C) of eA, hence will be in eA.

From this discussion we will remember that:

- eA is upwards closed, we cannot step out of by climbing.

- The conclusion of a $\parr$-link is in eA iff both premises are in it (and are $\neq A$).

The downmost points of eA form its *border* ∂A; they are of three kinds:

- The « main gate » A.

- Conclusions.

- Premises of $\parr$-links not shared, i.e., s.t. the other premise is not in eA.

Principal choice. There is at least one switching $\mathcal{I}$ such that $eA = \Re_{\mathcal{I}}(A)$.

In case A is a premise – say, right – of a « ⅋ », we set it to « right ». We set the other switches so as to never step out from eA: if one of the two premises of a ⅋-link is not in eA, we tie it to the conclusion.

Since the previous constraints are rather light, there are many principal choices. Algorithmically speaking, it is easy to find a principal switching: we visit eA starting from A and following the links. If we arrive at a ⅋ « from below », then the premises are still in the empire and both choices « L,R » are good. If we arrive at it « from above » for the first time, we set the switch so as not to go down: we turn back. If, later on, we come back through the other premise, we will be committed to « go down », but without exiting the empire.

Simultaneous empires. Assume that $B \notin eA$:

- If $A \notin eB$, then $eA \cap eB = \emptyset$.
- If $A \in eB$, then $eA \subset eB$.

This can be proved by choosing a switching – principal for eB and, if possible, for eA. In case of conflict, let us favour eB. We start with B « upwards ». Observe that we can enter an empire only « from below », only through a link L a premise of which is in eA, but whose conclusion is not:

- If this premise is A, this is because $A \in eB$; we switch (if it makes sense) L so as to move towards A. What comes « above A » is included in eB and contains eA.
- Otherwise, it is a ⅋-link and our switching chooses the other premise, the one not in eA.

We enter eA only when that cannot be avoided because $A \in eB$; in any case this is done through A and $eA \subset eB$. Otherwise we never enter and $eA \cap eB = \emptyset$.

11.3.5 Proof of the theorem. The theorem is proved by induction on the number of links.

- If there is only one link, it is an axiom and we are done.
- If there are more than one link, they cannot all be axioms: otherwise, the structure would be disconnected. Hence there is at least one logical link and we can concentrate upon the *terminal* logical links, those whose conclusion is a conclusion of the net, i.e., is not in turn a premise.
  - ► If a terminal link is a ⅋: remove this link, to get a correct structure, with one less link. It is sequentialisable by hypothesis and the same is true of the full structure.

► Otherwise the terminal links are $\otimes$-links. We must show that one of these links *splits*, which means that, if we remove this link, the structure has two connected components. From the splitting link, we recover by induction hypothesis two sequentialisations to which we apply a rule « Tensor ».

Consider, for all terminal links $A_i \otimes B_i$, the sets eA_i, eB_i. We choose among them one which is *maximal* w.r.t. inclusion, let us call it eA_0. I contend that the formulas of ∂A_0 other than A_0 are conclusions of $\mathfrak{R}$: if $D \in \partial A$ is not a conclusion, D must be a premise of a $⅋$-link whose conclusion (say, $C ⅋ D$) is not in eA_0. Below D stands the premise of a terminal $\otimes$-link, say B_1. $C \in eA_0 \cap eB_1$ and $B_1 \notin eA_0$ force $eA_0 \subset eB_1$, contradicting maximality.

Observe the difference between $⅋$, invertible and $\otimes$, positive.

The difficulty is that, in a net whose terminal links are tensors, not all of them split. Typically, in

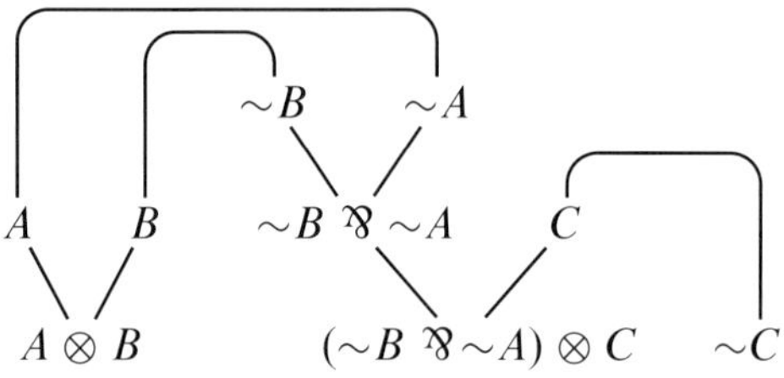

only the rightmost tensor splits. Moreover, there are in general several splitting tensors: this accounts for the maximality argument. In the above example, the empire eA contains, besides A, the sole $\sim A$. There is a « border conflict » between eA and eB, on the link $⅋$: each empire contains one of the premises of $⅋$. This is natural, one must be able to go to the conclusion of the $⅋$ by passing through A or by passing through B. But one does not know how to descend (leftwise or rightwise); hence from A, B, one must be able to rejoin both $\sim A$ and $\sim B$; since A always leads to $\sim A$, B always must lead to $\sim B$.

We saw how to compute an empire by exploring the graph and setting the switches as we go along. We can extend the method to find the splitting $\otimes$. We select A_0 and if ∂A_0 contains, besides the main gate, only conclusions, we are done. Otherwise, a border point is a premise of a non-shared $⅋$ above some B_1 (resp. A_1); we proceed with A_1 (resp. B_1). This forces us to explore another part of the graph (since $eA_1 \cap eA_0 = \emptyset$); the splitting link is thus found by exploring the graph only once. This has been really improved by Guerrini [57], who showed that correctness can be checked in linear time.

In the presence of cut, the main result would have been: if there is more than one link, but no terminal $⅋$-link, there is a splitting $\otimes$ or « cut » link.

11.3.6 An anecdote. I followed, in its main lines, the original proof of [37]. This being said, the original correctness was formulated in terms of *trips*, i.e., travels through the graph passing twice at each point. The formulation retained here is due to Danos & Regnier [19]. It dispenses with a lot of tedious details; but it does not abolish the original criterion, which is the source of *geometry of interaction* and which remains irreplaceable in the case of *non-commutative logic* (Chapter 18).

The original presentation was restructured into *imperialism, principal choice, simultaneous empires* to provide a setting in which to accommodate various extensions (quantifiers, additives).

Once in a while, I like to indulge in an informative anecdote concerning the genesis of the proof. The criterion was found by the end of 1985; then I remained more than six months making circles around the « splitting tensor ». One nice day of August 1986, I woke up in a camp in Siena and *I had got the proof*: I therefore sat down and wrote a manuscript of 10 pages. One month later, while recopying this with my typewriter, I discovered that one of my lemmas about imperialism was wrong: no importance, I made another one! This illustrates the fact, neglected by the formalist ideology, that a proof is not the mere putting side by side of logical rules, it is a *global perception*: since I had found the concept of *empire*, I had my theorem and the faulty lemma was hardly more than a misprint.

11.A More on multiplicatives

11.A.1 η in nets. There is a « net » version of η-expansion:

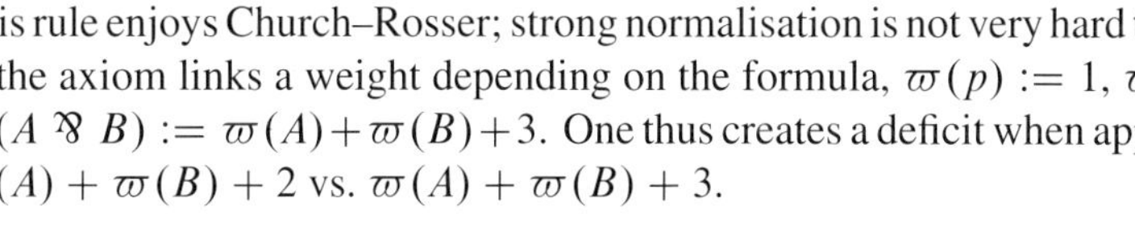

(11.5)

This rule enjoys Church–Rosser; strong normalisation is not very hard to prove: give to the axiom links a weight depending on the formula, $\varpi(p) := 1$, $\varpi(A \otimes B) = \varpi(A \parr B) := \varpi(A) + \varpi(B) + 3$. One thus creates a deficit when applying (11.5): $\varpi(A) + \varpi(B) + 2$ vs. $\varpi(A) + \varpi(B) + 3$.

11.A.2 Euler–Poincaré. Remember that it is a matter of topology of graphs: in what follows, with G a graph, **s**, **a**, **c**, **z** denote:

s: the number of vertices of G,

a: the number of edges of G,

c: the number of connected components of G,

z: the number of cycles G; one should only count the *primitive* cycles.

We write the familiar equation

$$\mathbf{c}(G) - \mathbf{z}(G) = \mathbf{s}(G) - \mathbf{a}(G) \tag{11.6}$$

which is easily proved: if G is edge-free, $\mathbf{c} = \mathbf{s}$ and $\mathbf{z} = \mathbf{a} = 0$. If one adds an edge, either it links two previously disconnected parts: $\mathbf{c}$ decreases by 1 while $\mathbf{z}$ stays the same; or it links two already connected parts: in which case $\mathbf{c}$ stays the same while $\mathbf{z}$ increases by 1.

The *Euler–Poincaré characteristic* $\mathbf{s}(G) - \mathbf{a}(G)$ is therefore equal to 1 for a tree; but it does not quite characterise trees. One can toy with and compute this number for a proof-structure.

Proposition 17. *If* $\mathfrak{R}$ *is a proof-structure, then* $\mathbf{s}(\mathfrak{R}_{\mathcal{I}}) - \mathbf{a}(\mathfrak{R}_{\mathcal{I}}) = \sharp(\text{axioms}) - \sharp(\otimes)$.

Proof. By induction on the number of logical links: for a structure only made of axioms, $\mathbf{s} - \mathbf{a} = \sharp(\text{axioms})$. A $\otimes$-link adds a vertex and two edges, which preserves the equality. *Idem* for the $⅋$, which adds a vertex and only *one* edge because of the switch. □

Of course, if we allow them, cuts must be counted like tensors:

$$\mathbf{s}(\mathfrak{R}_{\mathcal{I}}) - \mathbf{a}(\mathfrak{R}_{\mathcal{I}}) = \sharp(\text{axioms}) - (\sharp(\otimes) + \sharp(\text{cuts})) \tag{11.7}$$

Corollary 17.1. *If* $\mathfrak{R}$ *is a net,* $\sharp(\text{axioms}) - \sharp(\otimes) = 1$.

This necessary condition is not sufficient. We will immediately give a counterexample, but let us explain first where it comes from: given three formulas A, B, C, one can combine them into $A ⅋ (B \otimes C)$ or $(A ⅋ B) \otimes C$, which does not alter the *characteristic*, while the formulas are not equivalent. Indeed, let us write the proof-structure for $A ⅋ (B \otimes C) \vdash (A ⅋ B) \otimes C$:

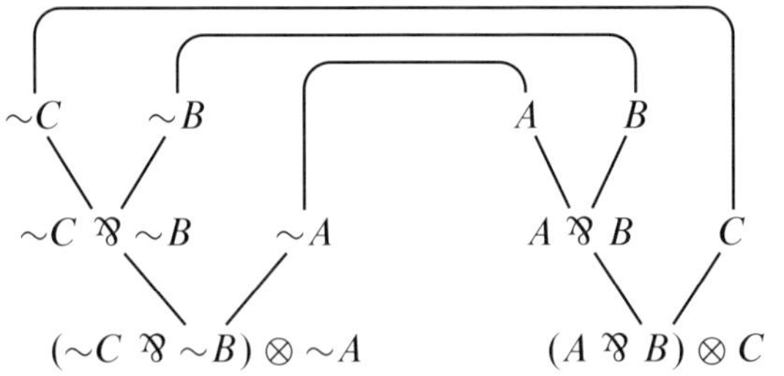

(11.8)

The characteristic is 1: if we set both switches to « R », we get only one component (hence no cycle); if we set both of them to « L », we get two components (thus one cycle).

Which shows by the way the necessity of considering *all* possible switchings.

11.A.3 Neutral elements. Euler–Poincaré suggest the following semantics:

$$\begin{aligned}
[p] &:= x, \\
[\sim p] &:= 1 - x, \\
[\perp] &:= 0, \\
[\mathbf{1}] &:= 1, \\
[A \otimes B] &:= [A] + [B] - 1, \\
[A ⅋ B] &:= [A] + [B],
\end{aligned} \tag{11.9}$$

where variables x, y, z,... are associated to the atoms p, q, r,... ; a theorem will get the value 1. It is indeed a simple instance of phase semantics, the monoid is $\mathbb{Z}$, shifted by 1 (composition law $a + b - 1$, neutral 1). The pole is $\{0\}$ and the facts are therefore the $\{n\}$, as well as $\emptyset, \mathbb{Z}$.

Of course, this semantics is not complete, since it accepts the non-provable $\perp \otimes (\mathbf{1} ⅋ \mathbf{1})$ as easily as the provable $(\perp \otimes \mathbf{1}) ⅋ \mathbf{1}$. This being said, this incomplete semantics, which contradicts classical logic, is not without charm. If topological considerations in the style « Euler–Poincaré » had led to a convincing procedural elaboration, it would have been accepted with great interest, since it would have solved the question of the extension of multiplicative proof-nets to neutral elements. Incidently, let us mention among other incomplete semantics the *classical truth tables*. The combination of both semantics is not complete either; this is more generally the case of any algebraic semantics with finitely many *distinguished* (i.e., « true ») values: in the multiplicative fragment built upon the sole neutrals, there are infinitely non-equivalent *provable* formulas.

Now, an irritating question: the category-theoretic interpretation, taken *by the book*, imposes the writing down of an axiom for $\mathbf{1}$ (link without a premise, with a single conclusion) and something of the like for $\perp$, with a nuance that should differentiate nets from mere structures. The problem is that, if a multiplicative formula is built upon the sole neutral elements, one can write only one diagram, corresponding to its decomposition, for instance

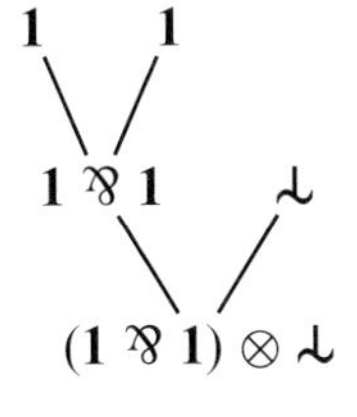

which contains zero *bits* of information. In other terms, the *correctness* of such a net is equivalent to its provability. And one can state the following « theorem »:

Theorem 34 (Not rigourous, but convincing!). *There is no correctness criterion for neutrals.*

Proof. Everything rests upon what is a correctness criterion. Let us say that there is nothing like « for all switchings... ». Indeed, such a criterion would make correctness coNP (of the same kind as classical propositional provability « for any choice of truth values... »). But Lincoln & Winkler [76] have shown that the multiplicative fragment built upon the sole neutrals is NP-complete. It is of course rather improbable that NP = coNP; and frankly impossible that this unlikely equality results from a correctness criterion! □

Obviously, this is everything but a theorem; but it is anyway very strong, since it indicates a dead end. Other ways out could be:

- Relaxation of structural rules, in favour of, for instance: weakening, or « mix », see *infra*.
- Abandonment of the logical constants $\mathbf{1}, \perp$. There are procedural reasons for that (Section 12.3.3).

Anyway, we just met a problem linked to the weakening rule: we do not know how to cope with $\perp$ (and soon will find the same problem with the weakening of an underlined formula $\underline{A}$). A solution consists in « attaching » the « weakened » formula, as we please:

- To an axiom, which amounts to writing down weakened axioms.
- To an arbitrary formula of the net.
- To a non-empty set of formulas of the net. In this case, one will need a switching to select one of the corresponding edges.

11.A.4 The « mix » rule. This rule does not preserve the characteristic. It corresponds to the implication $A \otimes B \multimap A \parr B$, which induces a proof-structure of characteristic 2, hence with two connected components. In favour of the rule, its sequent calculus expressions:

$$\frac{\vdash \Gamma \quad \vdash \Delta}{\vdash \Gamma, \Delta}$$

as well as a natural phase semantics: $x, y \in \perp \Rightarrow xy \in \perp$. One sees that, if $x \in X, x' \in \sim X, y \in Y, y' \in \sim Y$, then $xx', yy' \in \perp$ hence $(xy)(x'y') \in \perp$, which shows that $X \cdot Y \subset \sim(\sim X \cdot \sim Y)$, i.e. that $X \otimes Y \subset X \parr Y$. Which is the sign of a cut-elimination (Section 10.1.7). In terms of nets, « Mix » admits a correctness criterion due to Fleury and Rétoré: : « $\Re_I$ acyclic » [27].

Mix is natural in various category-theoretic settings: see, e.g., [13], [14]. But this rule is not completely convincing:

- One should perhaps add the 0-ary case, i.e., the axiom $\bot$, one would then contradict classical logic and also weakening[5].
- Weakening, which plays on similar grounds, seems more convincing.

This discussion (by the way, note that I am pushing very prudent views) is typical of the present state of logic. Between « mix » and weakening, none of them, both of them, layer -1 will not decide. Layer -2 neither, since we shall see (Section 12.2.3), that the hypothesis of *polarisation* destroys the heavy objections against weakening. One finds oneself naked at layer -3: procedurality must decide. But it has not yet spoken: no striking phenomenon has been observed for or against either rule.

11.A.5 Interaction nets. The *interaction nets* of Lafont [69], [71] are an interesting paradigm of parallel computation, in the style of *graph reduction*, inspired from proof-nets. The main emphasis is on *acyclicity*.

11.A.6 Natural deduction for the Lambek calculus. One can write « intuitionistic » multiplicative logic in the style « natural deduction ». To avoid the horrible elimination rule of the tensor, one writes a rule with two conclusions:

$$\frac{A \otimes B}{A \quad B} \tag{11.10}$$

And we interpret a proof of $\Gamma \vdash A$ as a deduction of A under the hypotheses Γ. This sort of net in « natural deduction style »is handled without problem: the rules $\multimap E$ and $\otimes I$ are treated as tensor links; while $\multimap I$ and $\otimes E$ are treated in the spirit of « $\parr$ »; we need a switch linking:

- $A \multimap B$ with B or to the discharged premise A.
- $A \otimes B$ with one of the two conclusions A, B.

Correctness is what one can imagine, the structure must be connected and acyclic for any switching.

Let us proceed with cyclic logic, hence with Lambek's calculus which is its *teleological* version. The rules of implication split, thus:

$$\frac{\begin{array}{c}[A]\\ \vdots\\ B\end{array}}{B \circ\!\!- A}\;(\circ\!\!-I) \qquad\qquad \frac{\begin{array}{c}\vdots\\ B \circ\!\!- A\end{array} \quad \begin{array}{c}\vdots\\ A\end{array}}{B}\;(\circ\!\!-E)$$

[5]The idea of a « 0-ary case » is an idea taken from categories and which has its own limits; think of the doubts as to the existence of « real » multiplicative neutrals (Section 12.3.3).

Non-commutativity is expressed through *planarity* constraints. The hypotheses are written in reverse order, i.e., from right to left; $\multimap I$ and $\circ\!\!- I$ are interpreted by means of an edge linking the discharged premise A and the conclusion. In the case of $\multimap I$, the edge is coming « from the left », in the other case, « from the right ». Crossings are forbidden; it is similarly forbidden to « imprison » a hypothesis or the conclusion. Thus:

$$\dfrac{\dfrac{[A] \quad A \multimap B}{B}}{B \circ\!\!- A}$$

is incorrect, since the edge connecting $[A]$ to the right side of $B \circ\!\!- A$ will imprison either $A \multimap B$, or the conclusion.

11.B Syllogistic

11.B.1 Scholastics. The expression *scholastics*, which originally applied to Middle Age teaching, is now derogatory. However, the works of medieval logicians on Aristotelian syllogisms is neither stupid nor repetitive. Of course, after more than 500 years of less and less inspired routine, one has some doubts.

One finds these – at random: in the Sorbonne – people who monotonously recite truth according to St Tarski or St Kripke; those modern scholasticians are representative of a certain *sclerosis* of the philosophical approach to logic. But this was not the case of the ancient scholastics.

11.B.2 Barbara. *Barbara* is the most noted syllogism:

$$\dfrac{\forall x(Ax \Rightarrow Bx) \quad \forall x(Bx \Rightarrow Cx)}{\forall x(Ax \Rightarrow Cx)}\ \textit{Barbara}$$

If all A are B, if all B are C, then all A are C. We already had the occasion of exposing the incredibly debilitating interpretation of Łukasiewicz: transitivity of inclusion (Section 7.1.2).

Category theory already proposes a much more stimulating reading; nets enable one to pass to a procedural reading:

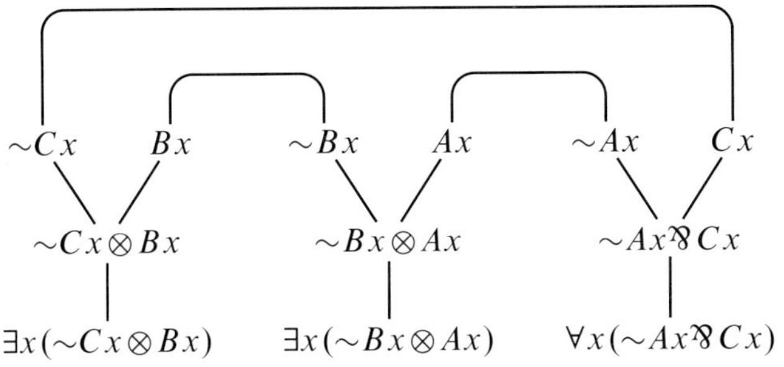

The combination of syllogisms actually corresponds to cut-elimination.

11.B.3 Ancient scholastics. Middle Age logicians classified syllogisms by means of acronyms. The vowels *aeio* designate types of formulas, i.e., *a* for « universal affirmative », *e* for « existential affirmative », *i* for « universal negative », *o* for *existential negative*: this is why *Barbara* includes *a* three times. Consonants memorise processes to pass from one to another. Those who judge this activity stupid and prefer transitivity of inclusion are overlooking procedurality. The futuristic technique of nets helps us to see nuances completely opaque to the uncouth set theoretic reduction of syllogistics.

Abrusci [3] wrote syllogisms in cyclic logic; this formulation creates three classes, according to the number of crossings of the associated nets: 0, 1, or 2. This very recent classification corresponds to the old taxomomy, the *figures* of Aristotle. Thus, only the first figure (in particular, *Barbara*) can be written without crossing. By the way, remark that Aristotle's syllogisms are all equivalent as *inference rules*. What makes them different is their writing as *implications*: at that point, crossings do matter.

11.B.4 The respect of the past. It is difficult to speak of the past, ancient, or even recent. One finds all possible attitudes:

The frozen respect: the attitude of editors of facsimiles. They understand nothing, so they transmit everything, including the misprints.

Rereading: more delicate, but necessary. It is important not to make the past say what it cannot have said. But not to despise it either: these people had their own reading grids, forever lost.

As to rereading in terms of nets, one must emphasise the fact that Aristotle was completely foreign to the debate underlying the opposition between classical, intuitionistic and linear logics. It is therefore out of the question to make him a linear logician *ante literam*. This being said, the reading « nets » has an advantage over the set theoretic reading (independently of the fact that it is much more refined): (cyclic) linear logic is more neutral than the other systems.

In the style « condescension », the set theoretic interpretation has sometimes gone a bit too far. Like everyone else, I had to suffer lectures of not quite inspired colleagues: when they have nothing « new » on the distinction free/bound variable, they contend with syllogisms interpreted « Polish-style »; this is how I first heard about incorrect syllogisms. Remember that they are based upon « if all A are B, then one A is B »; incorrect syllogisms are those containing exactly one of the letters $i\,o$: for instance *Barbari* is incorrect, while *Darii* or *Ferio* are correct. The « Polish » interpretation of these mistakes is rather insipid, tolerant: Aristotle neglected the fact that A could be empty; but one can forgive him, since he didn't know set-theory: it is the leniency one may have for doddering oldsters.

Look at this in terms of nets and forget quantifications: it says $A \multimap B \vdash A \otimes B$, in other terms, $\vdash {\sim}B \otimes A, A \otimes B$, i.e., pure gibberish. There is a cycle; indeed, this is not even an *incorrect* proof-structure:

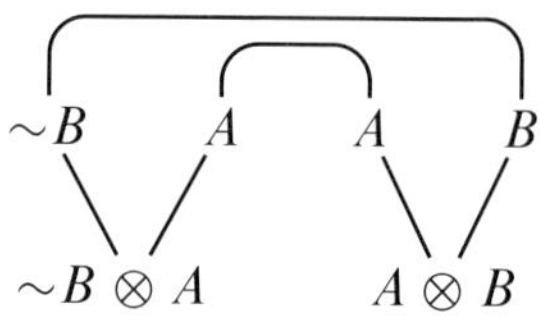

And some feel entitled to say that, after all, Aristotle wasn't that dumb… No, no and no! Aristotle syllogistic was inspired, but these syllogisms are an uncouth mistake: one sees on the drawing that they are good for nothing.

11.C General nets

11.C.1 Boxes. Boxes are an expeditious, but efficient, way of inserting sequents inside a net. A box with *conclusions* $A_1, \ldots, A_n$ $(n \neq 0)$ is a generalised axiom link whose conclusions are $A_1, \ldots, A_n$. One uses drawings of the style:

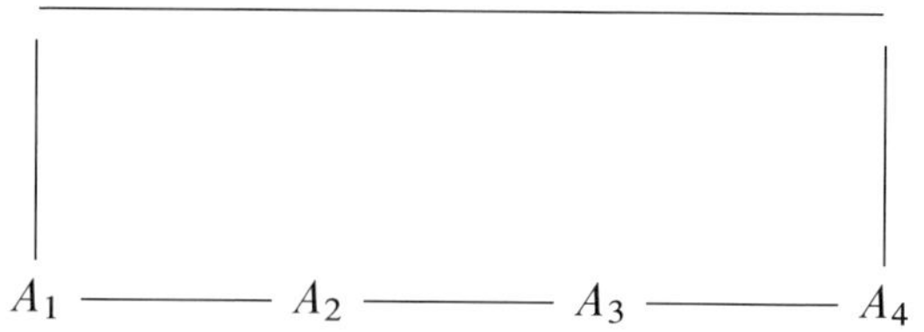

which implies that there might be something inside the box, but this does not matter at this point.

To define correctness, it suffices to associate a graph to a box: indeed, no matter which tree connects the conclusions, for instance, the graph with edges $A_1 - A_2, \ldots, A_{n-1} - A_n$. The sequentialisation theorem extends: a net is correct iff it comes from a sequent calculus proof extended by means of the extra axiom $\vdash A_1, \ldots, A_n$. This is without surprise, the box being – *topologically* speaking – only an insignificant variant of the axiom link.

Boxes will help us to palliate the limitations of nets. The present state of technology allows one to dispense with boxes for quantifiers and provides an almost satisfactory solution in the additive case. Boxes are nevertheless more or less useful in the following cases:

Weakening: they must be « attached »; among various solutions: weakened axioms (boxes) $A, \sim A$, ⅄, $?B, ?C$.

$\top$: one bluntly writes a box of conclusions $\Gamma, \top$.

&: if one does not like additive nets *(infra)*, one can introduce a box of conclusions $\Gamma, A \,\&\, B$. « In the box », one must put two nets of respective conclusions Γ, A and Γ, B. Which shows the interest of the box technology: one has sort of « taken a picture », frozen the deduction at a specific time: that of a specific & rule. Before the rule, « in the box » and after the rule, in the net, one is as atemporal as possible.

!: it is the only connective which clearly deserves a box[6] (for the others, boxes are rather a stammer of logic). The connective « ! » is violently non-linear and this non-linearity expresses itself through an aggressive sequentialisation.

11.C.2 Exponential nets. One allows the underlining of arbitrary *conclusions* of links; which permits us to give to the introduction of « ? » a pleasant aspect:

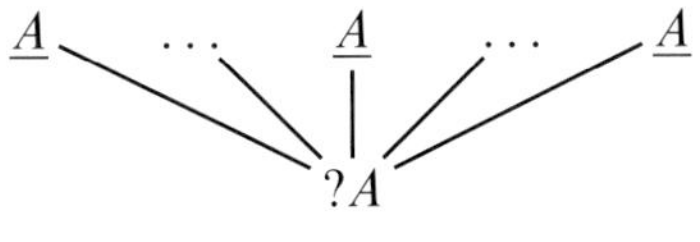

an elegant reconciliation of the triptych dereliction/contraction/weakening. This link is handled like an n-ary ⅋, i.e., the switch selects one premise. The 0-ary case, that of weakening, is problematic: nothing to select. One must in that case « attach » the conclusion to another formula, for instance by means of weakened axioms.

[6]It is a pity that the symbol $\Box$ was prostituted to do-it-yourself modalities: it fitted perfectly the operationality of « ! ».

The !-box, of conclusions $\underline{\Gamma}, !A$ « contains » a net of conclusions $\Gamma, !A$. The reduction rules will normally « open » this box (Section 15.2). But a much more exciting possibility appears: one does not open the box, one enters it. This is the vision of light logics which will be the subject of Chapter 16.

11.C.3 Nets with quantifiers. The only really satisfactory extension of proof-nets, unfortunately to the less exciting part of logic; I discuss here the first-order case.

Nets. The notion of *proof-structure* extends with two new unary links:

$$\begin{array}{c} A \\ | \\ \forall x A \end{array} \qquad\qquad \begin{array}{c} A[t/x] \\ | \\ \exists x A \end{array}$$

An obvious translation defines proof-nets in the multiplicative/quantifiers setting. It is important, in this precise issue, to be extremely pedantic as to the choice of the bound variables of universal quantifiers: they must be *pairwise* distinct.

The question of correctness poses a new problem: the sequent calculus rule

$$\frac{\vdash \Gamma, A}{\vdash \Gamma, \forall x A} \tag{11.11}$$

proposes a « photo » of a moment when one could introduce the $\forall$, this photo bringing the evidence that there was then a context Γ free of x. This context does not interest us, this is why the net forgets it; but we must find in what way the structure

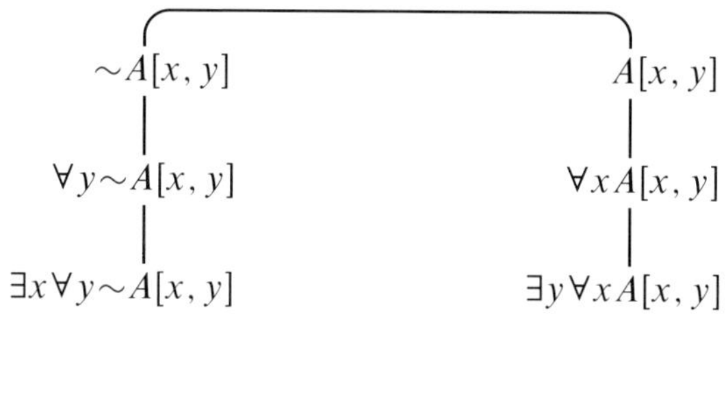

(11.12)

which expresses the logical atrocity $\forall x \exists y A[x, y] \vdash \exists y \forall x A[x, y]$, might be incorrect. The explanation lies in the story of dependencies « y depends on x », that one must translate into topological terms; in this case Herbrand would have written $A[x, f(x)] \vdash A[g(y), y]$ and concluded that

$$\begin{aligned} x &= g(y), \\ y &= f(x) \end{aligned} \tag{11.13}$$

has no solution. We have no right to function symbols, not to speak of equations, but we can give an equivalent to the *loop* of equation (11.13), by means of a *cycle*.

Correctness. For each link $\forall$, we can set a switch, linking the conclusion $\forall x A$ to another formula, as we please:

- The premise A (which need not contain x): this is the *main* switching.
- Any formula of the net where x occurs free: this sort of switching is styled a *jump*.

We must check that any switching actually yields a tree. Typically, when translating the rule (11.11), we already dispose of all formulas to which we can « jump » from $\forall x A$. And, no matter what formula of the net with conclusions Γ, A we jump to, the resulting graph will be a tree.

The criterion refuses (11.12): make $\forall y$ « jump » to $\forall x A[x, y]$ which contains y free and $\forall x$ jump to $\forall y {\sim} A[x, y]$ which contains x free: this yields a cycle.

Normalisation. Before proving sequentialisation, let us show that correctness is preserved by normalisation. The obvious reduction

(11.14)

requires the replacement of *all* free occurrences of x with t, which only makes sense when x does not occur in t. Correctness excludes this possibility: indeed, if x were occurring in t, one could set a « jump » from $\forall x A$ to $A[t/x]$, thus producing a cycle.

The preservation of the criterion offers no particular difficulty: if x occurs in C and y occurs in t, the substitution of t for x introduces a new jump from $\forall y B$ to $C[t/x]$. Which can be reduced to two jumps in the original net, one from $\forall x A$ to C, the other from $\forall y B$ to ${\sim} A[t/x]$. $\forall x A$ can be topologically assimilated to

$\sim A[t/x]$ and if one removes the cut on $A[t/x]$:

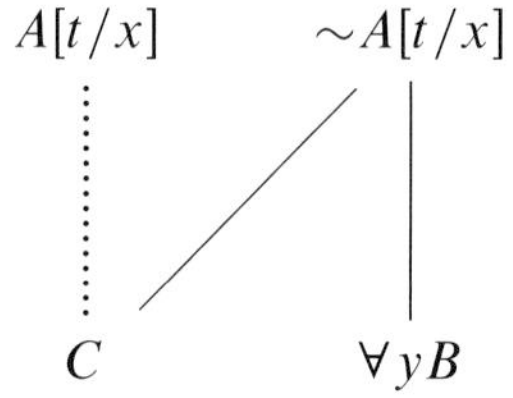

is a tree (the dotted line indicates that, since $C \in eA$ in the original net, $A[t/x]$ and C are already connected). The topology of the graph is not altered if one reintroduces the cut on $A[t/x]$ and directly jumps from $\forall yB$ to C:

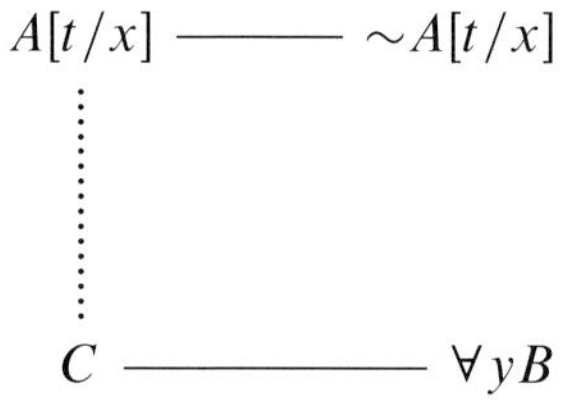

Sequentialisation. The proof of sequentialisation is very simple. We work indeed with boxed nets. One gets rid of $\forall$-links as follows: if a link admits $\forall xA$ as a conclusion, one considers $e\forall xA$. It is immediate that this substructure whose border is of the form $\Gamma, \forall xA$ is a net. One can thus replace it with a box: « in the box » one will put the net eA. One can proceed, in and outside the box, up to the total disappearance of the $\forall$ links. And, finally, sequentialise a multiplicative net (with $\exists$ which poses no problem) with boxes, then proceed with the « contents » of the boxes, etc.

There is however something to check. When I replace the empire $e\forall xA$ with a box, I must make sure that an *eigenvariable* y of some $\forall yB$ of the box does not occur freely outside the box. For this, one extends the property of *imperialism* to $\forall$-links:

$\forall yB \in eC$ *iff* B *and all possible « jumps » of* $\forall yB$ *belong to* eC.

In particular, with $C := \forall xA$, all *eigenvariables* of $\forall$-links internal to $e\forall xA$ (e.g., x) occur freely only inside $e\forall xA$.

An open problem. The additional logical principles

$$\forall X(A \otimes B) \ \vdash \ \forall XA \otimes \forall XB, \tag{11.15}$$

$$\forall X(A \oplus B) \ \vdash \ \forall XA \oplus \forall XB \tag{11.16}$$

are not provable (form $NP \vdash PN$); (11.16) is even classically inconsistent, since $\forall X(X \vee \neg X)$. But they are justified at second order by a locative interpretation (Section 14.3.3).

Independently of their justification, one would like to see a correctness criterion for multiplicative/quantifiers nets, accepting (11.15), i.e.,:

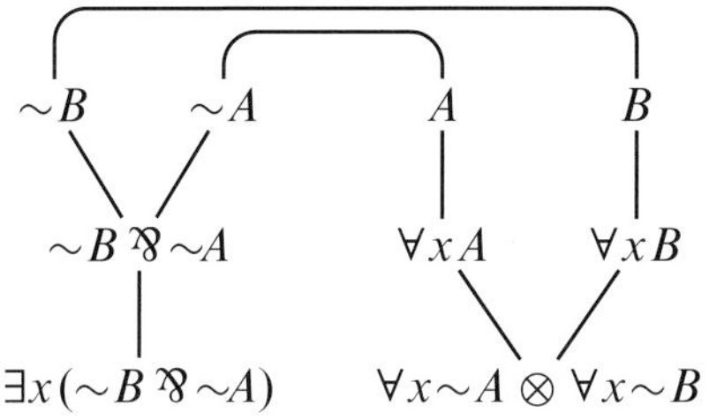

which is not even a proof-structure (one should rename the variables, but then the $\exists$-link no longer works).

11.C.4 Additive nets

Slices. We cannot bluntly « put together » two nets of conclusions Γ, A and Γ, B: indeed, since the context Γ is shared, there will be a phenomenon of *superposition* and an irreversible loss of information.

We thus prefer to keep the two « partial » nets, in which « & » is represented by a unary link, i.e., of the form:

(11.17)

We introduced the boolean *eigenvariable* p to tell which premise of the « With » has been kept: $p = 1$ for left, $p = 0$ for right.

The case of « $\oplus$ »is much easier, we write the unary links:

(11.18)

If each boolean *eigenvariable* takes a value 1 or 0, we get a sort of multiplicative net involving additional unary links, & and $\oplus$. Such a net is called a *slice*. We will represent an additive net by the set of its slices, when we make the boolean *eigenvariables* vary over all possible values.

Structures. A *proof-structure* consists in the data of finitely many formulas and links such that every link has univocal premises/conclusions among the formulas; in particular a &-link has two premises. Conversely, a formula can be a premise of only one link; but it can be a conclusion of several links. Each &-link has its *eigenvariable*; each link and each formula receives a *weight* in the boolean algebra generated by the *eigenvariables*. The weights are subject to constraints:

- $\varpi(A) = 1$ if A is a conclusion.
- $\varpi(A) = \sum \varpi(L)$, where L varies among the links of conclusion A.
- $\varpi(A) = p \cdot \varpi(L)$, $\varpi(B) = \neg p \cdot \varpi(L)$ if L is a &-link of premises A, B and *eigenvariable* p. In particular $\varpi(L) = \varpi(A) + \varpi(B)$. (Here occurs an uninteresting technicality: if ϖ is any weight in the structure, then $\varpi \cdot \neg\varphi(L)$ must not depend on p.)
- $\varpi(A) = \varpi(L)$ when A is a premise of a link L other than a &.

It is easy to translate sequent calculus proofs: the structures associated with the premises of a & rule are reglued by means of a &-link L. This link has a weight $\varpi(L) = \varpi(A \mathbin{\&} B) = 1$, while the other formulas of the structure get the weights: $\varpi(C) := p \cdot \varpi'(C) + \neg p \cdot \varpi''(C)$. Which poses the question of the *identity* between formulas (Section 15.C.4).

The criterion. As usual, we set switches for the negative links. We begin by giving a value, 1 or 0, to each boolean *eigenvariable* associated to the &-links; the formulas and links whose weight is 1 form a slice, a sort of multiplicative net, in which we set switches:

- Choices L/R for the ⅋-links.
- Jumps for quantifiers (in case).
- Jumps for &-links, to formulas depending on its *eigenvariable*.

In a &-link, of conclusion $A \mathbin{\&} B$, where one made – say – the left choice, $p = 1$, some formulas depend on this choice: they are those, which like the premise A, would disappear from the slice if one were making $p = 0$ and, more generally, those conclusions of a link that would change if p were passing from 1 to 0.

The condition *connex and acyclic* is obviously necessary. It is also stable by normalisation, while making normalisation possible: a not quite difficult variation on the theme of quantifiers.

The problem. This being said, sequentialisation fails; a counterexample may be found in [61]. The reason is simple: imitating the treatment of quantifiers, one meets a problem with the notion of *dependency*, which is not honest. One finds again our old acquaintance, the « parallel or » (Section 8.2.4): if a formula has weight $p \cup q$, it depends neither on p, nor on q for $p = q = 1$; but, if I make $p = 0$, it starts depending on q! In particular, if p is the *eigenvariable* of the link of conclusion $A \mathbin{\&} B$, I am not sure that the border $\partial(A \mathbin{\&} B)$ is the same for $p = 0$ and $p = 1$: which impedes me from reconstituting a &-box.

On the other hand, this is the case when weights are monomials: one thus gets « honest » dependencies and one can introduce a box in place of each &-link. The solution I retained in [48] consists in writing weights as explicit sums of monomials, for instance in the previous case, one can write: $p + \neg p \cdot q$ or $p \cdot \neg q + q$ or $p \cdot \neg q + p \cdot q + \neg p \cdot q$, which obviously induces distinct sequentialisations. This fall-back solution is nevertheless good enough to get rid of boxes. Hughes and van Glabbeek [61] got a sequentialisation for « real » nets, but their criterion is not the last word on the subject.

11.C.5 The « state of the art ». If one admits that the rule « ! » really deserves a box, there remain only a few problems in the theory of nets: these problems deal with the additives, since one has the choice between a slightly artificial criterion [61], or a slightly *ad hoc* modification of nets [48]; there is also the 0-ary additive case, namely the constant $\top$. Obviously, one would like a « definitive » solution. This one could come from the exploitation of *polarisation*, see next chapter. At the present time, polarisation essentially induced structures more sequentialised than nets, in the style of *ludics* (Chapters 13–14). But this might be a *péché de jeunesse*. There remain also problems linked to weakening; the solution consisting in « attaching » the weakened formulas to axioms – or to other formulas – is also only a fall-back. Polarisation, which by the way makes weakening plausible as a general structural rule, can perhaps bring some light to this matter.

Sequentialisation is not a dogma, it is a tool which enabled one to find the *procedural* contents of nets; we shall come back to this in Chapter 18. But, on the whole, one has nothing against the idea of non-sequentialisable nets, *as long as one can manipulate them*: the ultimate meaning of logic is this ability to manipulate. And sequential decomposition is not the panacea: one will find in Section 18.A a definition of the coherent interpretation based on the sole correctness. By the way, there are good criteria which only work for cut-free nets (Section 18.B.2); since they are stable under reduction, this is enough for us. Asking for more is part of technical perfectionism: never forget that a very difficult question may be without much interest.

Part IV

Polarised interpretations

Chapter 12
A hypothesis: polarisation

12.1 Faithfulness of coherent spaces

12.1.1 Polarisation. The distinction between positive and negative connectives is not novel, at least in the intuitionistic world. But it was rather of a *pragmatic* nature: it corresponded to observed phenomena, by no means with a *great divide*. It is in the years 1990 that *polarity* took its real place, through:

- The work of Andreoli on *focalisation* (Section 10.A).
- My (slightly posterior) work on classical logic (Section 12.A).

But these works, although giving much importance to polarisation, did not bestow a *status* upon it. I personally much hesitated to do so; what eventually convinced me is that I saw no other possibility to ensure *category-theoretic faithfulness*. Among the assets of *polarisation*: it yields clear and satisfactory answers, such as those given by ludics (Chapters 13–14). Among its liabilities: it makes interpretations heavier, it increases sequentialisation: for instance, one will regress w.r.t. proof-nets. At the present hour (2011) one cannot say that the debate as to polarisation is settled. But this remains a major option anyway.

12.1.2 The faithfulness problem. The question of the *faithfulness* of layer -2 is clearly more delicate than the same question at layer -1, which is plain *completeness*. The original question was to find *concrete* category-theoretic interpretations in which the only morphisms are those obtained *logically*: this is called *full completeness*. This completeness of layer -2 would imply the completeness of layer -1: indeed, say that $A \vdash B$ when there is a morphism in $A \vdash B$; then there is a proof of $A \vdash B$. In particular, since usual completeness is limited, by Gödel's theorem, to Π^1 formulas, one should not be « more royalist than the King »: one must restrict *full completeness* to closed Π^1 formulas (Section 14.A.2).

But it is rather difficult to conciliate two constraints, that of consistency – which *stricto sensu* makes certain category-theoretic objects « empty » – and faithfulness, which requires much differentiation, thus rather « full » objects. *Full completeness* turns out to be an immature ideal that one must make evolve: by *category-theoretic faithfulness*, I mean a bijection between proofs and « good » morphisms. The exact meaning of the term « good » remains willingly fuzzy; this could mean for instance « winning »… nay « plain » if « full completeness » were working.

The discussion which follows is carried out in the setting of coherent spaces. The same problems occur, *mutatis mutandis*, no matter what other *concrete* category-theoretic interpretation we choose.

12.1.3 Totality. The main contribution of Scott was the introduction of *partial* elements. These elements are used to interpret those computations which diverge, i.e., yield no results. Obviously this does not happen in logic, since strong normalisation guarantees termination. Faithfulness compels us to exclude partial elements, thus retaining the sole *total* ones. But what is a total element? Let us try to answer in a precise case: that of a *clique* in a coherent space.

Totality expresses *horror vacui*: the empty space is, indeed, the paragon of non-totality. Taking a coherent space such as the pole $\curlywedge$, with its two cliques $\emptyset$ and $\{\curlywedge\}$, one will therefore say that the sole $\{\curlywedge\}$ is total. Totality on a type « $\multimap$ » will be defined by

$$C \sqsubset X \multimap Y \text{ total} \iff \forall a \sqsubset X \text{ total} \quad (C)a \text{ total}. \tag{12.1}$$

Which yields, remembering that $\sim X \simeq X \multimap \curlywedge$:

$$a \sqsubset X \iff \forall b \sqsubset \sim X \text{ total} \quad \sharp(a \cap b) = 1. \tag{12.2}$$

One easily checks that the rules of perfect logic do preserve totality. In the case of &, one must define

$$a + b \sqsubset X \,\&\, Y \text{ total} \iff a \text{ total and } b \text{ total}. \tag{12.3}$$

Totalitarian coherent spaces now come with their notion of totality; there can be several of them over the same underlying space.

One should not confuse totality with *maximality*. In practice, total cliques are often maximal and one can legitimately consider that a non-maximal total clique is the sign of a faulty definition. The other direction is more complex: it is indeed easy to find *natural* spaces in which maximality does not entail totality. Thus the « horseshoe » $\{a, b, c, d\}$ with three strict coherences $a \frown b \frown c \frown d$ admits the maximal clique $\{b, c\}$ which does not meet the maximal anticlique $\{a, d\}$; thus both cannot be total in their respective totalitarian spaces $X, \sim X$.

Maximality is not even a notion of totality in the previous sense. Say that an anticlique b is *comaximal* when it meets all maximal cliques: there might be no comaximal anticlique. Take, for instance, the « pentagon » $\{a, b, c, d, e\}$ with five strict coherences $a \frown b \frown c \frown d \frown e \frown a$, hence five maximal cliques (with two points) has no comaximal anticlique: it should have at least three points… but the negation of our pentagon is the pentagon $\{a, c, e, b, d\}$ which has no cliques with three points. In this case, the bipolar of maximality contains all cliques!

12.1.4 Additive problems. Additive neutrals are differentiated by totality: indeed, $\top$ has a total clique, $\emptyset$; while $\mathbf{0}$ has none. On an empty web, those are the only two possible choices.

If X is a totalitarian space, then $X \multimap \mathbf{0}$ and $\sim(X \multimap \mathbf{0})$ have an empty web; one is thus equal to $\top$, the other to $\mathbf{0}$. And $\sim(X \multimap \mathbf{0}) \oplus (X \multimap \mathbf{0})$ is equal to $\top$, hence has a total clique. Full completeness would compel us to add the axiom $\forall X(\sim(X \multimap \mathbf{0}) \oplus (X \multimap \mathbf{0}))$ of which we immediately see that it contradicts the disjunction property: if A is any formula, $\sim(A \multimap \mathbf{0}) \oplus (A \multimap \mathbf{0})$ will be provable, although we cannot tell which side is provable.

Full completeness suggests that we relinquish the disjunction property, i.e., the category-theoretic spirit. One can keep both, by deciding all formulas, i.e., by becoming « non-monotonic »; this is even worse. This is a typical case where the logic of the idea tends to destroy the idea[1].

Morality: there cannot be any empty space, we must « fill » $\mathbf{0}$.

12.1.5 Multiplicative problems. Totality does not tell the difference between $\bot$ and its negation $\mathbf{1}$: in both spaces, the only total clique is the singleton $\{\bot\}$. To get a differentiation, one must analyse the atom x such that $a \cap b = \{x\}$; and suppose that x carries a sort of *result* of the duel between the two spaces, thus designating a « winner », or perhaps no winner at all, but *not two* winners whatsoever.

To totality, one adds a division of the web $|X|$ in three parts: the points where x wins, those where x loses and the undecided. A clique $a \sqsubset X$ is *winning* when included in the winning part of the web.

Winning is an *introspective* notion, while totality is *extraspective* (Section 13.7). Indeed, totality refers to the *result* (the intersection, of which one retains the sole cardinal), while winning refers to the *way* in which the result has been obtained. Introspection enables one to distinguish, among cliques, some which are « more cliques than others »; but this notion – contrary to totality, extraspective – needs the others to live: there can be winners only if there are losers. As soon as one speaks of winning, one relinquishes the naïve idea of *full completeness* to enter the more flexible paradigm of *category-theoretic* faithfulness (Section 12.1.2).

Anyway, we are still far from being done: it is enough to consider the multiplicative formulas built upon $\mathbf{1}$ and $\bot$. There is only one total clique, hence winning would reduce to provability. But this question is NP-complete (Section 11.A.3): this means that there is no simple definition of winning on neutrals – or that the spaces are too small.

[1] Just like the – legitimate – hatred of the French Army pushed certain pacifists – Déat and many others, e.g., Lucien Rebatet (*Les Décombres* 1942) – to support... Hitler.

12.1.6 The Gustave function. This function is due to Berry [12], nicknamed « Gustave » . It takes three boolean arguments:

$$\begin{aligned}\varphi(\mathbf{v},\mathbf{f},z) &= \mathbf{v},\\ \varphi(x,\mathbf{v},\mathbf{f}) &= \mathbf{v},\\ \varphi(\mathbf{f},y,\mathbf{v}) &= \mathbf{v},\\ \varphi(\mathbf{v},\mathbf{v},\mathbf{v}) &= \mathbf{v},\\ \varphi(\mathbf{f},\mathbf{f},\mathbf{f}) &= \mathbf{v}.\end{aligned} \tag{12.4}$$

What counts is not the output (I deliberately chose the same), but the way to get it. Since it sends booleans to booleans, this function is indeed total; but its ternary symmetry impedes any sequentialisation.

I translate this in linear logic: consider the ternary « $⅋$ » (written as « , ») $\vdash (A \oplus (B \mathbin{\&} C)), (A \oplus (B \mathbin{\&} C)), (A \oplus (B \mathbin{\&} C))$, in which one has a clique made of five pieces:

$$\begin{aligned} a_1 &\sqsubset \; \vdash B, C, A,\\ a_2 &\sqsubset \; \vdash A, B, C,\\ a_3 &\sqsubset \; \vdash C, A, B,\\ a_4 &\sqsubset \; \vdash B, B, B,\\ a_2 &\sqsubset \; \vdash C, C, C.\end{aligned} \tag{12.5}$$

If each part is total in the respective subspace, then the union will be total in $\vdash (A \oplus (B \mathbin{\&} C)), (A \oplus (B \mathbin{\&} C)), (A \oplus (B \mathbin{\&} C))$.

What does not work in « Gustave » is that there is an abstract possibility of computing, but *but no first step*. The first step should be something like: given a « $⅋$ » of three « $\oplus$ », one of these $\oplus$ simplifies (into its left or right part). Obviously this *partial* information (which « $\oplus$ » simplifies) should be represented in the clique. Once more we observe that the interpretations are « too small ». See also Ehrhard's *hypercoherences* (Section 12.C).

This example corresponds to a $⅋$ of $\oplus$, in turn combined with &: there are alternations of polarities. Incidently, the previous counterexamples to faithfulness were also linked to alternations of polarity.

12.2 A prototype

12.2.1 Summary of the discussion. The previous discussion raised the following points:

- *Horror vacui* (Section 12.1.4); all spaces and especially **0**, must be non-empty, which contradicts *full completeness*.

- One recovers a notion of *faithfulness*, by requiring an additional condition, *winning* (Section 12.1.5).
- Which is not enough; one must also add additional atoms representing the « first rule » (Section 12.1.6).

All this put together will drag us far astray from coherent spaces, so far that we shall be forced to start again from scratch: see Chapter 13. We shall make a few steps in this way with *polarised coherent spaces*, a transient notion that will help us to introduce a major hidden actor of logic: the *daimon*.

12.2.2 The *daimon*. One can easily justify the demon, or better, *daimon*, the inhabitant of the formula **0** by the need to be non-empty, but above all by a *procedural* evidence coming from proof-search:

When I meet **0**, *I give up.*

Giving up has nothing to do with stalling: it is a positive attitude.

The *daimon* (✠) is the unique inhabitant of this space, once empty. It is total, moreover *losing* (since a failure). One is led to *polarise* coherent spaces.

Incidently, remember *reducibility* (Section 6.2.4). A basic point is that a reducibility candidate is never empty: it contains the variables. Should one restrict to closed terms, everything would collapse. In some sense, the *daimon* corrects this anomaly by producing the missing « closed term ».

12.2.3 Polarised coherent spaces. A positive (resp. negative) coherent space is of the form $\{✠\} \oplus X$ (resp. $\{✠\} \mathbin{\&} X$). These polarised spaces are totalitarian spaces with a notion of *winning*, i.e., certain points of the web are winning, others are losing; negation exchanges winners and losers. The distinguished point ✠ is losing in positive spaces, winning in negative spaces. $\{✠\}$ is total in positive spaces; hence, in order to meet $\{✠\}$ (12.2), a total negative clique must contain it. Which is not enough to make it winning, since ✠ is presumably not the only point of our negative clique.

0, now the positive space $\{✠\}$, remains neutral; this is because the *daimon* receives a special treatment. Thus $(\{✠\} \oplus X) \otimes (\{✠\} \oplus Y) := \{✠\} \oplus (X \otimes Y)$, $(\{✠\} \oplus X) \oplus (\{✠\} \oplus Y) := \{✠\} \oplus (X \oplus Y)$. Should one perform a positive operation on spaces which are not positive, one first « positivises » them by adding a *daimon*. In this way, the constructions made around the neutrals will involve bigger and bigger spaces, following polarity changes.

Weakening – the rejection of which essentially brings complications – is rehabilitated by the *daimon*. Indeed, a constant function from $\{✠\} \mathbin{\&} X$ into $\{✠\} \mathbin{\&} Y$: $\varphi(a) := b_0$ becomes (observing that total cliques must contain ✠) $\varphi(a) := b_0$ if $✠ \in a$, $\varphi(a) := \emptyset$ otherwise. This does not yet prove weakening, it only makes it *plausible*.

12.3 Objections to polarisation

12.3.1 Literals. If all compound formulas admit a polarity, it is not the case for atoms and their negations, the *literals*. Is there enough ground to refuse polarity?

I don't think so, because an *atom* is the *unknown formula*: since one does not know what it means, it is difficult to attribute to it any procedurality. One can see it as a (implicitly) universally quantified second-order variable.

One is forced to attribute a polarity to everything: atoms will be positive and their negations negative. This creates small problems since one likes to *decompose up to the change of polarity*: but how to decompose an atom? In *ludics* the problem has been « fixed » by allowing no atoms, everything decomposing up to infinity. Which nevertheless introduces technical complications out of proportion with the importance of the point.

12.3.2 Category-theoretic viewpoint. Polarisation is completely foreign to the category-theoretic viewpoint; however, one could try to distinguish between:

Positive: those are the *inductive* limits (a.k.a. colimits), i.e., « on the right ».

Negative: those are *projective* limits (a.k.a. limits), i.e., « on the left ».

Categories do not separate them, because of trivial (co)limits. For instance, since $A \otimes \mathbf{1} \simeq A$, one cannot say that a « $\otimes$ » is positive.

But, between us, do you think that the definitions have been made from the 0- and 1-ary cases? More likely, these cases have been added later for – a very understandable – commodity.

One can dissipate an objection, that of the opposite category. By swapping the sense of arrows, one obtains a category where projective and inductive have been swapped. This should be a strong argument against my explanation of polarities; but not at all! Take for instance the principle $PN \vdash NP$, which is expressed in categories under the form of various morphisms $\varinjlim \varprojlim \mapsto \varprojlim \varinjlim$. The passage to the opposite category truly swaps $\varinjlim \varprojlim$ and $\varprojlim \varinjlim$, but also replaces $\mapsto$ with $\hookleftarrow$, thus $\varprojlim \varinjlim \hookleftarrow \varinjlim \varprojlim$: one has replaced – say – $\otimes \,\&\, \vdash \,\&\, \otimes$ with $\oplus \parr \vdash \parr \oplus$. Finally, the swapping of arrows is nothing but linear negation.

12.3.3 The geometry of neutrals. To understand the nature of the contention with category-theory, let us look at the multiplicative neutrals and try to understand the absence of multiplicative nets in that case.

Observe that the tensor is handled at layer -3 in a *locative* spirit. This means that A, B occupy well-defined *locations* and that, to form $A \otimes B$, one only needs to link them to the conclusion. There is therefore no real difference between $A \otimes B$ and $B \otimes A$, in other words:

*The tensor product is **literally** commutative.*

In contradiction to the category-theoretic viewpoint: a tensor product *cannot* be literally commutative (but it could be literally associative). Which means that the reading of the third underground can by no way be *reduced* to its category-theoretic dimension – however important this dimension might be.

A neutral element in the same « -3 » spirit should be literally neutral, i.e., occupy *no* space: two distinct locations would indeed induce two literally distinct elements. This non-localisation of the neutrals impedes any drawing of links: those require a definite location. Note that this becomes a problem only if one changes the polarities: as long as **1** is tensorised, $\perp$ cotensorised, there is not the slightest need for « neutral » links to draw satisfactory nets.

The problem begins with the tensorisation of $\perp$: in $\perp \otimes \perp, \Gamma, \Delta$ one $\perp$ belongs to Γ, the other to Δ. One needs to indicate it by « attaching » the $\perp$. Which means that, by tensorisation, the $\perp$ « found a location » somewhere, that one can now hook links on them: that's exactly the change of polarity.

We shall soon see (Section 12.6.2) that locativity has an unexpected consequence, the disappearance of the formula $\mathbf{1} \oplus \mathbf{1}$! Finally, layer -3 suggests that multiplicative neutrals do not exist – in contrast to additive neutrals. This contradicts the category-theoretic viewpoint which wants neutrals everywhere: but category theory has no *jus primae noctis* on logic.

12.4 Logic and games

Polarisation is linked to the idea of game: « I play » is positive (active), « you play » is negative (passive). Following a touching tradition, the players are sometimes named (H)éloïse and Abélard[2] (the symbols $\exists, \forall$ are, like ⅋, topsy-turvied, here the letters E, A); Héloïse is positive, Abélard is negative.

12.4.1 « Historical » interpretations

Gentzen. It is in Gentzen's first consistency proof [32] that one can find the first interpretation of a formula of logic – or rather arithmetic – by a game between a *player* `Me` and an *opponent* `You`. Thus:

$\forall x A$: `You` begins to play by giving a value n to x: in this way he challenges `Me` to show that $A[\bar{n}/x]$.

$\exists x A$: `Me` begins to play by proposing a value n for x; he can later change his mind and propose a second value for x: `Me` indeed enunciates the sequent

[2]The famous lovers were also used, unwillingly, as advertisement of the *Père Lachaise*: they were (re)inhumated with pomp and circumstance in 1817 to « launch » this fashionable graveyard.

$\vdash A[\bar{n}/x], \exists x A$. But he will not be allowed to « stall » and continuously change his mind.

If changing one's mind in the existential case is not allowed, one immediately sees that a winning strategy is nothing but a truth « computation »: to win in $\exists x A$, I will look for n such that $A[n]$ is true. The notion of winning strategy contains the notion of truth; by allowing changes of mind, one can restrict to recursive, effective, strategies, which is not the case of a brutal truth « computation » – precisely non-computable. But a truth calculus – even a sophisticated one – is inadmissible in a consistency proof. This being said, it is the very idea of a consistency proof which is inadmissible!

One can also, legitimately, contend that foundations do not necessarily mean consistency and that shedding a new light on the *nature* of mathematical objects is the real goal of foundations. In this perspective of « non-reductive foundations », Gentzen's contribution is an authentic breakthrough, especially when replaced in its historical context:

Logic is a sort of game; proofs are winning strategies.

Gödel. The interpretation called *Dialectica* [56] is the contribution of Gödel to the subject (Section 6.C.1). This is a game as well, which takes the form of the interaction $a[x, y]$ between a strategy x^S for `Me` and a counter-strategy y^T for `You`; `Me` wins when $a[x, y] = 0$. $\exists x^S \forall y^T a = 0$ thus enunciates the existence of a winning strategy for `Me`.

I already mentioned the poor global structure of the interpretation. In [21], de Paiva slightly ameliorates this interpretation by means of a linear layer. Whatsoever, *Dialectica* is only a half-success, which also means a half-failure.

Lorenzen. Lorenzen is presumably the only logician of those times to have fully assumed the *dialectic* dimension of logic.

His School must have been in a cantilevered position w.r.t. the milieu:

- Difficult being an intuitionist in Germany, of hilbertian – thus formalist – culture.

- Difficulty to « sell » a dialectic vision of intuitionism, at a time – after WWII – when Brouwer was no longer very creative; and Heyting had not the necessary independence of thought.

- No conscience of a layer below -1; thus Felscher formulated his « theorem » under the form « if there is a winning strategy, then A is *provable* », where one would expect: « ..., then it comes from a *proof* ».

Lorenzen's contribution to the ideas of interaction, game, … is hard to assess. The achievements of his School, [79], [78], [26], are of an unsurpassed bureaucratic mediocrity: *grosso modo*, `Me` plays the right rules, `You` plays the left rules, more or less *literally*. Rather than painful paraphrases that bring nothing new, one will prefer the calculus itself.

Here too, one must question the relation to past; what to make of precursors who « fumbled everything » and who durably discredited the viewpoint they were trying to push? That's delicate; in any case, one should not try to « fix Lorenzen » at any price by changing something here and there: such *rehabilitations* show that their authors are more competent than the « rehabilitated »; or, at least, have the distance he was so cruelly in want of. It is therefore of no use; *Matthew* VIII.20: *leave the dead with the dead.*

Rather than trying at any price, in the name of an « eternal return », of an essentialism in the style « 2001 » – negating any scientific progress – to spruce failures up, it is more interesting to analyse the difference between these stutters and the (relative) modern successes. This is essentially a matter of dialectics analysis/synthesis (Section 12.6.5).

12.4.2 Recent interpretations. During the past twenty years many game-based interpretations have appeared; the main reference on the topic is the games of Nickau–Hyland–Ong [84], [62]. *Geometry of interaction* (Chapter 19) although not a game, induced game theoretic by-products, see for instance [2]; and, of course, *ludics*, which is not quite a game: we shall soon come to this point.

12.4.3 My contention. Games allow model to enter the arena, hence, instead of « proof/counter-model », one gets « proof/counter-proof ». This is a brutal jump from layer -1 to layer -3. It is important not to work haphazardly:

- The notion of game, with its checkerboard, its alternating moves, etc., does not belong to « good » mathematics. One will thus prefer a less exotic approach, the closest possible to the heart of mathematics: this is what we shall try for instance in *geometry of interaction*. And confine the aspect « game » to the domain of metaphor – this being said, an excellent metaphor.

- The Achilles' heel of the game is its *rule*, which will depend on the formula to prove. One exposes oneself to an analysis of the rule in terms of games… and where are we heading this way? Presumably in the same meanders where the functional interpretation got stuck long ago, see the discussion of Section 5.A: the meanders of river Meta.

Nevertheless, games remain the best metaphor of layer -3, this is why I will not deprive myself of it. But, once more, let us not confuse the metaphor with what it illustrates.

12.5 Proofs and tests

12.5.1 Rereading the functional interpretation. We shall revisit the functional interpretation in a *monist* way. Instead of saying that a proof of A is a θ *such that...*, we will introduce *tests*, homogeneous to θ and will say that a proof of A is a θ which « passes » certain tests. This replacement is not innocent: indeed, it is not *a priori* obvious that the conditions imposed upon θ can be reduced to a family of tests; typically, one could require that θ is a polytime algorithm, which cannot be directly « tested ». One will thus arrive at the outline of a duality proofs/tests; since it is a discussion in the style *Pons Asinorum*, I do not go through all cases. With the notations of Section 5.1.2:

Conjunction: if I want to test $\theta = (\theta_1, \theta_2)$, I must test θ_1 as a proof of A and θ_2 as a proof of B, which can be done by means of two series of tests: tests for A, tests for B. Hence:

A test for $A \wedge B$ is a pair (i, τ), where either $i = 1$ and τ is a test for A, or $i = 2$ and τ is a test for B.

It is fundamental to remark that the notion of test is considerably subtler than the notion of counter-model. Indeed, a counter-model will refute A or B, without telling which one, because of this ambiguous case where both are false. It is in this sort of detail that one sees that semantics is about formulas, not about proofs.

Disjunction: to test a proof $\theta = (i, \theta_1)$, I must prepare two tests τ_1, τ_2, one in case $i = 1$, one in case $i = 2$. In other terms:

A test for $A \vee B$ is a pair (τ_1, τ_2): a test τ_1 for A, a test τ_2 for B.

Observe how much conjunction and disjunction are symmetrical.

Implication: a test for the function θ consists in the data of an *argument* θ' – a proof of A – and a test for B, i.e. for $\theta(\theta')$:

A test of $A \Rightarrow B$ is the pair (θ', τ) of a proof θ' of A and a test τ for B.

Using the temporary notation A^t to speak of tests for A:

$$\begin{aligned} (A \wedge B)^t &= A^t \vee B^t, \\ (A \vee B)^t &= A^t \wedge B^t, \\ (A \Rightarrow B)^t &= A \wedge B^t. \end{aligned} \tag{12.6}$$

Which suggests an identification between A^t and $\neg A$, based on:

Test for $A \simeq$ Proof of $\neg A$

12.5.2 Linear version. Of course, this does not work well; either we keep logic as it is and one is eventually led to an unsatisfactory *Dialectica*-like approach or we change the logical setting. Indeed, it is not the same « $\wedge$ » which is used in the cases of implication (both components θ, τ participate in the test) and disjunction (only one of them is activated). Which corresponds to the distinction between $\otimes$ (both) and & (one of them).

Pushing this analysis, we arrive at a more robust linear version:

$$\begin{aligned} (A \,\&\, B)^t &= A^t \oplus B^t, \\ (A \oplus B)^t &= A^t \,\&\, B^t, \\ (A \multimap B)^t &= A \otimes B^t, \end{aligned} \tag{12.7}$$

and to the identification, that we will take literally

$$\textit{Test for } A \;\simeq\; \textit{Proof of} \sim\! A.$$

12.5.3 Proofs and paraproofs. Let us introduce a neologism; instead of a test for $\sim\! A$, let us rather speak of a *paraproof*[3] of A. We just observed the great similarity between proofs and *paraproofs*, but the notions cannot coincide: there cannot be proofs of both A and $\sim\! A$, while one expects that there will be paraproofs of any formula A (so as to be able to test $\sim\! A$).

This being said, this analogy is not fortuitous and one easily gets convinced that the logical rules of formation of proofs do apply to paraproofs. In particular:

Any proof is a paraproof.

Which suggests that paraproofs are *generalisations* of formal proofs obtained by means of additional logical rules: rules which are « incorrect » from the traditional logic standpoint – from layer -1 –, but perfectly acceptable from a procedural standpoint.

Using a game-style metaphor, one would get:

$$\begin{aligned} \text{Formula } A &= \text{Rule of the game} \\ \text{Paraproof } \tau &= \text{Strategy} \\ \text{Proof } \theta &= \text{Winning strategy} \\ \text{Negation } &= \text{Swapping the players} \end{aligned} \tag{12.8}$$

The cut between the proof θ of A and the paraproof τ of $\sim\! A$ consists in performing the test τ on θ. Of course, in order to make this work well, one must be able to perform a test on a test – since the real proofs are to be recruited among tests.

[3] In French, *épreuve*, expression suggested by Pierre Livet.

Which essentially amounts to extending cut-elimination to the logical system with paraproofs.

Paraproofs are obtained by adding a new principle, the *daimon*, which I just introduced in Section 12.2.2. The *daimon* is basically the possibility to *prove* anything, which requires a *polarised* version of logic.

12.6 Hypersequentialised logic

12.6.1 Motivations. I will propose a variation on multiplicative/additive logic. This version will fully integrate polarity, under its two dual aspects, inversion/focalisation (Section 10.A). Furthermore, the system will have a *daimon*.

To get a naturally focalised system, one must use *synthetic*, i.e., arbitrary multiplicative/additive combinations, but of a single polarity (Section 10.A.5). For instance $\Phi(A, B, C) := A \oplus (B \otimes C)$. This is indeed a positive cluster, which can only be followed by a negative cluster; then, how can we impede a positive cluster from following another positive cluster? Simply by combining connectives with negation, thus: $\Psi(A, B, C) := {\sim}A \oplus ({\sim}B \otimes {\sim}C)$, which produces a positive formula from negated positive constituents, with, by the way, a change of polarity. One only sees positive formulas: negatives occur in disguise on the left side of sequents.

12.6.2 Connectives

Definition 33 (Synthetic connectives). An n-ary *synthetic connective* is a subset $\Phi \subset \wp(\{1, \ldots, n\})$.

Some examples:

- $\Phi = \emptyset$ corresponds to the constant $\mathbf{0}$.
- $\Phi = \{\emptyset\}$ corresponds to the constant $\mathbf{1}$.
- $\Phi = \{\{1\}\}$ corresponds to negation $\Phi(A_1) = {\sim}A_1$. This is not quite a negation; indeed, all formulas are positive, hence ${\sim}A_1$ is negative. Something is therefore done in order to « positivise » it: there is a *change of polarity*, noted $\downarrow$[4]. Indeed, one is dealing with $\Phi(A_1) = \downarrow {\sim}A_1$.
- $\Phi = \{\{1, 2\}\}$ corresponds to the negation of $⅋$, $\Phi(A_1, A_2) = {\sim}A_1 \otimes {\sim}A_2$.
- $\Phi = \{\{1\}, \{2\}\}$ is the negation of &, $\Phi(A_1, A_2) = {\sim}A_1 \oplus {\sim}A_2$.
- $\Phi = \{\{1, 2\}, \{1, 3\}\}$ is $\Phi(A_1, A_2, A_3) = ({\sim}A_1 \otimes {\sim}A_2) \oplus ({\sim}A_1 \otimes {\sim}A_3)$, or, if one prefers ${\sim}A_1 \otimes ({\sim}A_2 \oplus {\sim}A_3)$, etc.

[4]The downwards arrow is reminiscent of the exponential « ! », the typical operation making one pass from negative to positive.

- In general, $\Phi = \{\{a_1, \ldots, a_k\}\}$ corresponds to the negation of a k-ary ⅋; $\Phi = \{\varphi_1, \ldots, \varphi_p\}$ corresponds to the $\oplus$ of the p connectives $\{\varphi_1\}, \ldots, \{\varphi_p\}$.

In a metaphor in the style « game », positive means « I start » and negation is the swapping of players, hence to be positive again, one must introduce a *dummy move*. Since this operation is not idempotent, this is not scot-free: two dummy moves only yield an approximation of the original game – there are possibilities of early surrender linked to the *daimon* which slightly changes the possibilities. At chess, we should remember the « Anderssen opening » `a2-a3`, supposed to block the terrifying Morphy, by playing « defense »; this almost dummy move which gives the initiative to black, impedes the symmetric form of the Lopez – hence is not quite innocent. Nevertheless, Anderssen was defeated!

We see that the notion of a synthetic connective contains in itself, under the form of equalities, the canonical isomorphisms. Why is it an equality and not an isomorphism, as category-theory would require[5]? This is due to the *locative* character, already observed for multiplicative nets (Section 12.3.3): in $\Phi(A_1, A_2)$, the indices 1, 2 are not graphical conveniences, they are intangible locations. This is better understood with a ternary example: instead of $\Phi(A_1, A_2, A_3)$, I can write $\Phi(A_3, A_1, A_2)$; indeed, if $A_1 = a$, $A_2 = b$, $A_3 = c$, both mean « Φ with a at place 1, b at place 2, c at place 3 ». We shall by the way see that, in ludics, all binary functions do enjoy $f(a, b) = f(b, a)$, not because of something weird, but because *the location matters*[6]. Such phenomena are typical of layer -3 and, by the way, correspond to the procedurality of the computer. The real question is to determine whether or not to « quotient » w.r.t. locations and come back to a more traditional category-theoretic universe, or keep layer -3 as it is.

A strange feature, still locative: we will not recover the formula $\mathbf{1} \oplus \mathbf{1}$. Indeed, it is the « $\oplus$ » of two connectives represented by the empty set; but $\{\emptyset, \emptyset\}$ makes no sense. Those are the tiny differences between layer -3 and layer -2. $\mathbf{1} \oplus \mathbf{1}$ is sometimes used to represent booleans; we shall be forced to use something else, for instance $\forall X(X \otimes X \multimap X \otimes X)$.

We define the language **HS**: starting with atoms $P, Q, R, \ldots$, we can recursively build formulas by means of any synthetic connective.

This language is limited to positive formulas, but one can more or less translate multiplicative/additive propositions. For instance, $\sim P \otimes (Q \,\&\, R)$ is obtained in two steps, first $\sim Q \oplus \sim R$, then $\sim P \otimes \sim(\sim Q \oplus \sim R)$. One observes problems of translation only with $\mathbf{1} \oplus \mathbf{1}$, which no longer makes sense; there are also problems with atoms, thus one cannot write $P \otimes Q$, $P \oplus Q$, but this is only because one cannot further decompose P, Q.

Up to minor details, the language we obtained represents – under a very in-

---

[5] Remember that a tensor product, while it can be literally associative, can be commutative only up to isomorphism.

[6] Hence *Locus Solum* [51]: the location only.

convenient, but heavily structured, form – a version of *perfect* propositional logic. Its peculiarity lies in its *hypersequentialisation*, which induces the best sequent calculus ever obtained: it indeed enjoys Church–Rosser!

12.6.3 Sequent calculus. A sequent of **HS** is of the form $A \vdash \Delta$ (*negative* sequent) or $\vdash \Delta$ (*positive* sequent). One sees that negative formulas, which have been banished from syntax, are indeed present under the form of sequents $A \vdash$. Sequents are formulated up to right permutations, which extricates from exchange.

Identity/Negation

$$\frac{}{A \vdash A}\ (\mathit{identity}) \qquad\qquad \frac{\Gamma \vdash A, \Pi \quad A \vdash \Delta}{\Gamma \vdash \Delta, \Pi}\ (\mathit{cut})$$

Logic. Let $\Phi \subset \wp(\{1, \ldots, n\})$ be a synthetic connective. There is a right rule for each $\varphi \in \Phi$ and a single left rule with one premise for each $\varphi \in \Phi$. In what follows, we suppose that $\varphi = \{i_1, \ldots, i_k\}$:

$$\frac{A_{i_1} \vdash \Delta_{i_1} \quad \ldots \quad A_{i_k} \vdash \Delta_{i_k}}{\vdash \Delta_{i_1}, \ldots, \Delta_{i_k}, \Phi(A_1, \ldots, A_n)}\ (\vdash \varphi) \qquad \frac{\ldots \quad \vdash \Delta, A_{i_1}, \ldots, A_{i_k} \quad \ldots}{\Phi(A_1, \ldots, A_n) \vdash \Delta}\ (\Phi \vdash)$$

Daimon

$$\frac{}{\vdash \Gamma}\ (\maltese)$$

12.6.4 Normalisation in HS. The most beautiful formulation of normalisation consists in writing the system **HS** in the style « nets ». One uses the links axiom and cut, everything else using boxes. Due to the absence of ⅋-links, these nets are *literally* connected and acyclic. I will later speak of the *empire* $e{\sim}A$ of the premise of a cut: this is just the connected component of $\sim A$ obtained by « removing » the cut.

Let us examine the various types of cuts – other than the cut with an identity link, that we already know:

Key case. A cut between a box $(\vdash \Phi)$ and a box $(\Phi \vdash)$ normalises as follows: the box $(\vdash \Phi)$ corresponds to a rule $(\vdash \varphi)$, applied to nets $\mathfrak{R}_1, \ldots, \mathfrak{R}_k$ with respective conclusions $\sim A_{i_1}, \Delta_{i_1} \ \ldots \ \sim A_{i_k}, \Delta_{i_k}$; the box $(\Phi \vdash)$ contains, among others, a premise for φ, i.e., a net $\mathfrak{R}$ of conclusions $\Delta, A_{i_1}, \ldots, A_{i_k}$. One reglues $\mathfrak{R}, \mathfrak{R}_1, \ldots, \mathfrak{R}_k$ by means of k cut links on $A_{i_1}, \ldots, A_{i_k}$.

Demoniac case. In case of a cut on A with a *daimon* of conclusions Γ, A, we seek the empire (i.e., the connected component) $e{\sim}A$ of ${\sim}A$, whose border is ${\sim}A, \Delta$. We replace it with a *daimon* of conclusions Γ, Δ.

It could be the case that $\Gamma, \Delta = \emptyset$. We would thus get a net without conclusions, but still the *daimon*.

Commutations. The trouble with boxes is commutations. Take a cut-link between A (positive) and ${\sim}A$. The empire of ${\sim}A$ is thus a net of border ${\sim}A, \Delta$:

- If A is conclusion of a $(\vdash \psi)$ (whose conclusion introduces $B \neq A$); one of the premises of the rule contains A. Replace this premise (which is indeed a net) with the same net « cut » with the empire of ${\sim}A$; then apply $(\vdash \psi)$. We get a new box, whose conclusions differ from the original one: A has been replaced with Δ.
- If A is a conclusion of a $(\Psi \vdash)$; all premises of the rule do contain A. Replace each premise (which is indeed a net) with the same net « cut » with the empire of ${\sim}A$; then apply $(\Psi \vdash)$. We get a new box, whose conclusions differ from the original one: A has been replaced with Δ.

We must check that the empires verify Church–Rosser and this is very easy: the empires are trivial.

Theorem 35 (Strong normalisation). *The sequent calculus* **HS** *enjoys strong normalisation.*

Proof. We associate a weight to nets, an integer which strictly decreases with normalisation: in what follows, the symbols $\mathcal{B}$ and $\mathfrak{R}$ respectively denote boxes (including axiom and *daimon*) and nets:

$$\varpi(\mathcal{B}) := 1 \quad (\maltese)$$

$$\varpi(\mathcal{B}) := 2 \quad \text{(axiom link)}$$

$$\varpi(\mathcal{B}) := 1 + \sum_{1}^{k} \varpi(\mathfrak{R}_i) \quad \text{(rule } (\vdash \varphi))$$

$$\varpi(\mathcal{B}) := 1 + \sum_{\varphi \in \Phi} \varpi(\mathfrak{R}_\varphi) \quad \text{(rule } (\Phi \vdash))$$

$$\varpi(\mathfrak{R}) := \prod_{1}^{n} \varpi(\mathcal{B}_i) \text{ if } \mathfrak{R} \text{ is made of } n \text{ boxes linked by cuts.}$$

□

12.6.5 Digression: analysis and synthesis. *Ludics* (Chapters 13–14), based upon an analysis of **HS**, does not differ, in its basic purpose, from the work of Felscher [26], which is also a paraphrase of sequent calculus. However, the result is rather different – incomparably better –, so why? A truly interesting question, which poses the question of the relation between *analysis* and *synthesis*.

One is *a priori* free to analyse anything in no matter which way. For instance to cut a man into slices and to declare that those are his primary constituents. But there is a major difference between a dumb analysis like this and the analysis of water by Lavoisier which has the quality of *synthesis*: one can reconstitute water from hydrogen and oxygen, while synthesis « à la Frankenstein » does not work.

Which means that an analysis is necessarily *oriented* in view of a synthesis. Coming back to Felscher, one cannot say that he made a completely dumb analysis: he took into account the necessary isomorphism between $(A \wedge B) \Rightarrow C$ and $A \Rightarrow (B \Rightarrow C)$, but forgot the even more basic one between $(A \wedge B) \wedge C$ and $A \wedge (B \wedge C)$, which conditions the necessary associativity, i.e., *compositionality*, of the interpretation. It is obvious that Felscher, late follower of Lorenzen, hadn't heard about, or rather didn't understand, « Curry–Howard ».

Coming back to *ludics*, what we shall see is the result of several to-and-froes analysis/synthesis; by the way, several of those occurred inside these notes. For instance, the remark that one is obliged to extend the notion of formal proof, so as to admit tests – essentially the *daimon* – to avoid the anomalies linked to the empty set. And, by the way, the synthesis made from the analysis of logic in terms of *designs*, *behaviours*, stresses the importance of a structuration around several *analytical* results (Section 13.6): new, more refined, analyses must therefore be expected.

Finally, analysis and synthesis interact like existence and essence, like the hen and the egg. One can oppose them only if one keeps in mind their deep organic unity.

12.6.6 Towards designs. Very little is missing to try a synthesis:

- The first remark is to imagine a *wild* situation: what happens if one selects the « wrong » logical rule? In other terms, if, by accident, one faces a cut between $(\vdash \Phi)$ and $(\Psi \vdash)$. If one prefers, if one relinquishes the « superego » which compels us to follow logical rules: one instead follows the laws of nature, one accepts the rules which are *geometrically* possible. Say that $(\vdash \Phi)$ is indeed a $(\vdash \varphi)$ for some $\varphi \in \Phi$:

  - If $\varphi \in \Psi$, then one normalises without the slightest problem, without even noticing the mistake.
  - Otherwise the process diverges. This means that one cannot proceed with normalisation.

  Which means that, taking into account diverging normalisation, one has *extended* the notion of normal form.

- One sees that only φ matters (and not the Φ such that $\varphi \in \Phi$).
- There is no reason to force Ψ to be finite, the negative rule could indeed have infinitely many premises, since only one matters for normalisation.
- Since one is ill at ease with atoms, one dumps them! Axiom links will disappear!

All this lead to *dessins*, (almost) the right notion.

12.A Classical polarity

12.A.1 Relative structural rules. Classical polarity comes from the following remark:

Theorem 36 (Relative structural rules). *In linear logic, the structural rules of weakening and contraction on* ***negative*** *formulas are* ***provable*** *from the same rules applied to their constituents.*

Proof. Using the mixed calculus of Section 9.4.5, let A be negative; then it admits a unique rule of the form:

$$\frac{\vdash \Gamma, \Delta_1 \quad \dots \quad \vdash \Gamma, \Delta_n}{\vdash \Gamma, A}$$

where the Δ_i are made of immediate subformulas of A, possibly underlined. It is moreover *invertible*, which means that, given a proof of $\vdash \Gamma, A$, one can get proofs of the $\vdash \Gamma, \Delta_i$ (Section 10.A.2). By hypothesis, weakening and contraction are available for the Δ_i, hence:

Weakening: one proves it as follows:

$$\frac{\dfrac{\vdash \Gamma}{\vdash \Gamma, \Delta_1} \quad \dots \quad \dfrac{\vdash \Gamma}{\vdash \Gamma, \Delta_n}}{\vdash \Gamma, A}$$

Contraction: if $\vdash \Gamma, A, A$ is provable, then, by invertibility, the same is true of the $\vdash \Gamma, \Delta_i, \Delta_i$ and one writes:

$$\frac{\dfrac{\vdash \Gamma, \Delta_1, \Delta_1}{\vdash \Gamma, \Delta_1} \quad \dots \quad \dfrac{\vdash \Gamma, \Delta_n, \Delta_n}{\vdash \Gamma, \Delta_n}}{\vdash \Gamma, A}$$

□

By the way, note that invertibility also yields the sequents $\vdash \Gamma, \Delta_i, \Delta_j$ which are destroyed by contraction when $i \neq j$. Concretely, the double conditional

$$\begin{aligned} f(\iota_l x, \iota_l x') &= t[x, x'], \\ f(\iota_l x, \iota_r y') &= u[x, y'], \\ f(\iota_r y, \iota_l x') &= v[y, x'], \\ f(\iota_r y, \iota_r y') &= w[y, y'] \end{aligned} \tag{12.9}$$

contracts into

$$\begin{aligned} g(\iota_l x) &= t[x, x], \\ g(\iota_r y) &= w[y, y]. \end{aligned} \tag{12.10}$$

12.A.2 Connectives. If we want to translate classical logic, we should rather choose a formulation which is coherent from the viewpoint of linear logic; hence structural rules come freely on one side and there is no need to formulate them. Which avoids many complications such as the tiresome « commutative cuts » of Section 3.4.3. Note that the connective $\wedge$ admits two formulations, a positive one, $\otimes$, a negative one, &; in the same way, there are two formulations, **1** and $\top$, of the neutral element « true ». Which induces by duality an alternative in the formulation of the connective $\vee$ and its neutral element « false ». On the other hand, quantifiers are univocal, $\forall$ is negative, $\exists$ is positive[7].

How to choose? Indeed no choice is really satisfactory. The idea will be to attribute *polarities* to formulas and, according to these polarities, to use one formulation or the other; with, in case of heterogeneous components, a necessary change of polarity for one component, a change operated by means of exponentials. This is indeed very close to what we did for Gödel's translation; to be quite exact, one should rather say that the translation of Section 7.A.6 appears, in the original paper [45], as a by-product of what we present here. In what follows, P, Q, R stand for positive, L, M, N for negative formulas.

Conjunction

$\wedge$	M	Q
L	$L \,\&\, M$	$!L \otimes Q$
P	$P \otimes !M$	$P \otimes Q$

(12.11)

It is immediate that conjunction is associative, commutative, with neutral element $\top$, all of this in the sense of category-theoretic isomorphisms. For associativity,

[7]The idea of a « multiplicative quantifier », a sort of parametric « ! » remained as the state of a phantasm.

the crucial case is $P \otimes !(M \ \& \ N) \simeq (P \otimes !M) \otimes !N$, which comes from $!(M \ \& \ N) \simeq !M \otimes !N$.

Disjunction

$$\begin{array}{c|cc} \vee & M & Q \\ \hline L & L \parr M & L \parr ?Q \\ P & ?P \parr M & P \oplus Q \end{array} \qquad (12.12)$$

Which is also associative, commutative, with neutral element **0**: it suffices to remark that this table is the image of the previous one *modulo* De Morgan.

The distributivity of $\otimes$ over $\oplus$ implies that $\wedge$ distributes over positive disjunctions and $\vee$ distributes over negative conjunctions (still in the sense of layer -2). Full distribution of $\wedge$ over $\vee$ and of $\vee$ over $\wedge$ is of course hopeless.

Quantifiers translate, depending on the case, by: $\forall x L$, $\forall x ?P$, $\exists x !L$, $\exists x P$.

This interpretation induces a sequent calculus **LC** (Section 15.4).

12.B Intuitionistic logic

12.B.1 Intuitionistic polarities. As long as we stay within the *negative* fragment of intuitionistic logic, everything works well. We use the following dictionary:

$$\begin{aligned} A \wedge B &:= A \ \& \ B, \\ A \Rightarrow B &:= !A \multimap B, \\ \forall x A &:= \forall x A, \\ \text{true} &:= \top. \end{aligned}$$

And this is really perfect. Problems appear with positive connectives:

$$\begin{aligned} A \vee B &:= !A \oplus !B, \\ \exists x A &:= \exists x !A, \\ \text{false} &:= \mathbf{0}. \end{aligned}$$

Indeed the « $\vee$ » thus obtained is non-associative: compare $!A \oplus !(!B \oplus !C)$ and $!(!A \oplus !B) \oplus !C$[8]. We must do something, i.e., declare *polarities*, yielding

$$\begin{array}{c|cc} \vee & M & Q \\ \hline L & !L \oplus !M & !L \oplus Q \\ P & P \oplus !M & P \oplus Q \end{array} \qquad (12.13)$$

[8] It is comical that linear logic is, for a wide part, issued from this interpretation: non-associativity didn't shock me at that time!

resolutely associative, on which the polarised conjunction given by the classical table (12.11) distributes.

The problem comes rather from implication: indeed the table

$$\begin{array}{c|cc} \Rightarrow & M & Q \\ \hline L & !L \multimap M & !L \multimap Q \\ P & P \multimap M & P \multimap Q \end{array} \tag{12.14}$$

does not ensure $A \Rightarrow (P \wedge Q) \simeq (A \Rightarrow P) \wedge (A \Rightarrow Q)$. These are the limits of the *unified logic* I proposed in [46] and which aims at finding a calculus of which classical, intuitionistic and linear logics are *fragments*. The idea is seducing and successful as to the linear and classical fragments; as to the intuitionistic fragment, it is *così così*.

12.B.2 Translations in F. The second-order translations do invert polarities. Thus[9]:

Tenseur: $\forall X\ (A \multimap ((B \multimap X) \multimap X))$ yields a positive version of « $\otimes$ ».

Plus: $\forall X\ (!(A \multimap X) \multimap (!(B \multimap X)) \multimap X)$ yields a negative version of « $\oplus$ ».

With: $\exists X\ (!(X \multimap A) \otimes !(X \multimap B) \otimes X)$ yields a positive version of « & ».

This inversion of polarities is due to deep reasons – so deep that they remain completely obscure.

12.C Hypercoherences

In the *hypercoherences* of Ehrhard [23], coherence is no longer restricted to pairs: it applies to finite sets of cardinal > 1. A clique is a subset of the web of which all subsets of cardinal > 1 are coherent. A family of cliques $a_1, \ldots, a_n$ $(n > 1)$ is *coherent* when for all $x_1 \in a_1, \ldots, x_n \in a_n$, the set $\{x_1, \ldots, x_n\}$ is coherent. a, b, c can be coherent while its subfamily a, b is incoherent, see *infra*. *Strong stability* is the preservation of intersections of coherent families:

$$F(a_1 \cap \cdots \cap a_n) = F(a_1) \cap \cdots \cap F(a_n). \tag{12.15}$$

A finite set $\{(x_1, y_1) \ldots, (x_n, y_n)\}$ $(n > 1)$ is coherent in $X \otimes Y$ when its projections $\{x_1, \ldots, x_n\}$ (resp. $\{y_1, \ldots, y_n\}$) are either of cardinal 1 or coherent in X (resp. Y). A finite set $a + b$ is coherent in $X \oplus Y$ when it wholly lies in one of them and is coherent. In X & Y, this remains but a sufficient condition: as soon as

[9] In the last two cases, « ! » is there to allow weakening; a heavy price for not that much.

$a, b \neq \emptyset$, $a + b$ is coherent; this looks weird, but can be explained by the negation, see *infra*. Exponentials can be handled as well by hypercoherences so as to satisfy $!(A \,\&\, B) = !A \otimes !B$.

Strong stability yields a CCC which refutes the Gustave function φ (Section 12.1.6). The cliques $a_1 := \{\mathbf{v}_2, \mathbf{f}_3\}$, $a_2 := \{\mathbf{v}_3, \mathbf{f}_1\}$, $a_3 := \{\mathbf{v}_1, \mathbf{f}_2\}$ are coherent: $x_1 \in a_1$, $x_2 \in a_2$, $x_3 \in a_3$ are not in the same component of the source of φ, the ternary « with » $(\{\mathbf{v}_1\} \oplus \{\mathbf{f}_1\}) \,\&\, (\{\mathbf{v}_2\} \oplus \{\mathbf{f}_2\}) \,\&\, (\{\mathbf{v}_3\} \oplus \{\mathbf{f}_3\})$, hence:

$$\varphi(a_1 \cap a_2 \cap a_3) = \varphi(\emptyset) = \emptyset \neq \{\mathbf{v}\} = \varphi(a_1) \cap \varphi(a_2) \cap \varphi(a_2) \tag{12.16}$$

contradicting (12.15).

At the level of coherence, negation is rendered by the complement: in analogy to $x \frown_X y$ iff $x \not\asymp_{\sim X} y$ (when $x \neq y$) a is coherent w.r.t. X iff a is not coherent w.r.t. $\sim X$, which explains the strange definition of « With ». As a consequence, this – by the way, quite brilliant – approach cannot be « desessentialised », in contrast to coherent spaces (Section 9.1.5).

Chapter 13
Designs and behaviours

The discussion of Section 12.6 leads us to *designs*; they exist in two versions, almost equivalent, *dessins* and *desseins*, the latter being the real thing.

13.1 Designs-*dessins*

13.1.1 Locations, biases, etc.

Definition 34 (Loci). A *bias*, notation $i, j, k \ldots$, is a natural number. A *ramification*, notation $I, J, K, \ldots$, is a finite set of biases. A *directory* is an arbitrary set of ramifications. A *locus*, or location, address, notation $\sigma, \tau, \upsilon, \xi \ldots$ is a finite sequence $\langle i_1, \ldots, i_n \rangle$ of biases. The *parity* of a *locus* is the parity of its length n.

Thus $\langle 3, 3, 8 \rangle$ is odd while its immediate *sublocus* $\langle 3, 3, 8, 0 \rangle$ is even. We adopt the usual conventions for concatenation, in particular $\sigma * i$ instead of $\sigma * \langle i \rangle$; sometimes even σi to save space. The *sublocus* $\sigma * \tau$ of σ is *strict* when $\tau \neq \langle \rangle$, *immediate* when $\tau = \langle i \rangle$. Two incomparable *loci* are *disjoint*, i.e., they have no common *sublocus*. We introduce the notation $\xi * I := \{\xi * i; i \in I\}$.

W.r.t. **HS**, these locative *artifacts* can be explained as follows:

- We want to prove $\vdash A$ (resp. its negation $A \vdash$) that we « place » by convention in *locus* $\langle \rangle$.
- If $A = \Phi[A_1, \ldots, A_n]$, we place the immediate subformulas A_i in the immediate *subloci* $\langle i \rangle$. More generally, the subformulas of A will be placed in distinct *loci*. Even *loci* correspond to positive subformulas on the right (resp. on the left) and *vice-versa* for odd *loci*.
- A ramification is one of those finite sets φ used for a rule $(\vdash \varphi)$.
- A rule $(\Phi \vdash)$ therefore uses a set of ramifications, i.e., a *directory*.

An important point is that these locations are *a priori* devoid of any logical significance. For instance, the fact that A is located in $\langle \rangle$ tells me *strictly* nothing about A. Locations have nothing to do with, say, « Gödel numbers »; by the way, *loci* would not accept to be (mis)treated in the spirit of coding.

13.1.2 Pitchforks

Definition 35 (Pitchforks). A *pitchfork* is an expression $\Xi \vdash \Lambda$ such that:

Incomparability: Ξ and Λ are two sets of *loci*, pairwise disjoint; in particular, any *locus* of Λ is disjoint from any *locus* of Ξ.

Handle and tines: Ξ contains at most one *locus*, its *handle*; the *loci* of Λ are called *tines*.

Each fork receives a *polarity*:

Positive: a pitchfork without handle (a « comb »).

Negative: a pitchfork $\xi \vdash \Lambda$ with a handle.

A pitchfork is *atomic* if it has only one *locus*, i.e., is of the form $\vdash \xi$ or $\xi \vdash$.

Pitchforks are « delogicalised » sequents, handled with the usual conventions of sequent calculus: thus, $\xi \vdash \Gamma, \Delta, \lambda$ means $\{\xi\} \vdash \Gamma \cup \Delta \cup \{\lambda\}$, which implies that the four sets $\{\xi\}, \Gamma, \Delta$ and $\{\lambda\}$ are disjoint.

13.1.3 Paritarism. This condition, of no theoretical import, is often verified in practice:

Paritarism: the *loci* of Λ have the same parity, opposite to the parity of the handle (if any).

Thus, the pitchforks occurring in a *design* whose basis is atomic will be paritary; this is because rules preserve paritarism. A paritary pitchfork receives a parity, that of the right side and/or the opposite of the parity of the handle. The empty pitchfork thus admits both parities.

Metaphorically, one can see parities as two players (essentially isomorphic), `Even` and `Odd`. *Actions* correspond to *moves* of the players: a proper action (Section 13.2.2) of even focus will thus be a move of `Even`. In the paritary case, the focus of the action will be on the right (tine, same parity) or on the left (handle, different parity). Hence polarity corresponds to *relative* parity, i.e., to the distinction `Me`/`You`:

Positive: same parity, I (`Me`) play.

Negative: different parity, you (`You`) play.

The practical interest of paritarism leaves no room for doubt; a parity check is enough to correct most mistakes one can make when building designs.

Finally let us examine the pros and cons of paritarism:

Pro: if I analyse a cut-free proof of A, it yields a paritary design: parity is then nothing but *signature* (Section 3.3.3).

Con: this does not extend to the case of cut, typically, the cut between two paritary *faxes* (Section 13.1.5) of bases $\xi \vdash \xi'$ and $\xi' \vdash \xi''$ will normalise into a fax on non-paritary base $\xi \vdash \xi''$. To forbid such cuts would complicate everything; indeed, three paritary faxes (the third of base $\xi'' \vdash \xi'''$) yield again a paritary base. How could we formulate *associativity* (Section 13.6.3) in such a setting? Too complicated.

13.1.4 *Dessins*

Definition 36 (Designs-dessins). A *dessin* $\mathfrak{D}$ is a proof-tree made of pitchforks. The terminal pitchfork is called a conclusion (or *base*) of $\mathfrak{D}$. Each pitchfork of the design is a conclusion of a unique *rule* chosen among the following three categories:

Daimon:

$$\frac{}{\vdash \Lambda}\,\maltese \tag{13.1}$$

Positive rule: I is a ramification, for $i \in I$ the Λ_i are pairwise disjoint and included in Λ: one can apply the rule (finite, one premise for each $i \in I$):

$$\frac{\ldots \xi * i \vdash \Lambda_i \ldots}{\vdash \Lambda, \xi}\,(\vdash \xi, I) \tag{13.2}$$

Negative rule: $\mathcal{N}$ is a directory, for $I \in \mathcal{N}$, $\Lambda_I \subset \Lambda$: one can apply the rule (possibly infinite, one premise for each $I \in \mathcal{N}$):

$$\frac{\ldots \vdash \Lambda_I, \xi * I \ldots}{\xi \vdash \Lambda}\,(\xi \vdash \mathcal{N}) \tag{13.3}$$

ξ is the *focus* of the rules $(\vdash \xi, I)$ and $(\xi \vdash \mathcal{N})$. Since I is a ramification and $\mathcal{N}$ is a directory, the symbol $\vdash$ conveys information only if $I = \mathcal{N} = \emptyset$ and one will usually omit it: thus, writing (ξ, I) or $(\xi, \mathcal{N})$.

W.r.t. **HS**, the first difference is the possibility of an infinite *directory*. There is another one, more important, concerning *weakening*. A strict linear discipline would indeed require $\Lambda = \bigcup \Lambda_i$ in the positive case, $\Lambda = \Lambda_I$ in the negative case. Why relinquish this restriction now? This is due to *procedurality*: I discovered that the global structure is better off *with* weakening; in particular, one of the analytical theorems, *separation* (Section 13.6.2), supposes weakening. This is by the way a typical backlash of synthesis on analysis!

These proof-trees are really wild ones, not the ornamental type: no soothing hypothesis, finiteness, recursiveness, well-foundedness, etc. is made. Hence, designs will be badly infinite! This being said, designs are of an *expansive* nature, which can be seen on negative rules. In terms of *desseins* (*infra*), the *inclusion* $\mathfrak{D} \subset \mathfrak{E}$ means that $\mathfrak{E}$ has been obtained by « adding » more premises in the negative rules of $\mathfrak{D}$, by « augmenting » directories. In particular, every design is the (directed) union of the finite designs obtained by restricting directories to finite subdirectories, most of them empty. Of course, normalisation will commute to this approximation by directed unions; which by the way corresponds to the *expansive* procedurality of cut-elimination.

13.1.5 A few designs

Daimon.

$$\frac{}{\vdash \Lambda}\maltese \tag{13.4}$$

This *dessin* of positive base is called *Daimon* and noted $\mathfrak{Dai}$.

Figure 13.1. $\mathfrak{Dai}$.

When $\vdash \Sigma, \xi$ is positive, the rules $\maltese$ and $(\vdash \xi, \emptyset)$ have quite the same premises and one must distinguish them. The rule $\maltese$, which has no focus, is an *improper* positive rule. One also speaks of an improper design in the case of $\mathfrak{Dai}$. By the way, observe that there is only one design with an empty base $\vdash$, precisely $\mathfrak{Dai}$.

Fax. One recursively defines a design of base $\xi \vdash \xi'$, the *Fax* $\mathfrak{Fax}_{\xi,\xi'}$.

$$\dfrac{\cdots \qquad \dfrac{\dfrac{\vdots\ \mathfrak{Fax}_{\xi' * i, \xi * i}}{\ldots \xi' * i \vdash \xi * i \ldots}}{\vdash \xi', \xi * I}(\xi', I) \qquad \cdots}{\xi \vdash \xi'}(\xi, \wp_f(\mathbb{N})) \tag{13.5}$$

The *fax* relates two disjoint « occurrences » of the same formula A, one in ξ, one in ξ', or rather two isomorphic formulas A and A' which differ by their

Figure 13.2. $\mathfrak{Fax}_{\xi,\xi'}$.

location. These *loci* are usually of opposite parities, a restriction not preserved by *normalisation* (Section 13.1.3).

This name *fax* reflects its procedurality, that of an *extension cord*, exchanging information in both senses, at a distance: the fax is a *delocation*.

Metaphorically, the *fax* reproduces the *copycat strategy*: he has the presumption to play simultaneously against two famous players, A (Anderssen, positive, with the whites, although he would prefer the blacks) and $\sim A'$ (Morphy, negative, with the blacks). When A has moved (which corresponds to the choice of a ramification I in the rule $(\xi, \wp_f(\mathbb{N}))$), the fax plays the same I against $\sim A'$ (which corresponds to the rule (ξ', I)). When $\sim A'$ answers, the fax *transmits* the answer to A, etc. Without knowing chess, this strategy is enough for the fax not to lose both plays; indeed, he wins against Anderssen! Note that black and white essentially correspond to parities. The only weak point of this – excellent – metaphor is that it does not emphasise *delocation*.

The fax is an η-expanded (Section 7.4.2) identity axiom, based upon the idea that there is no atom, that formulas can be decomposed indefinitely. Its directory is maximum (all ramifications) because A is *unknown*. Should I know more as to A, say $A = \Phi(A_3, A_4, A_7)$ with $\Phi = \{\{3,7\},\{4,7\}\}$, the following *pseudo-fax* would achieve the same task:

$$\dfrac{\dfrac{\dfrac{\vdots\,\mathfrak{Fax}_{\xi'3,\xi3}}{\xi'3 \vdash \xi3} \quad \dfrac{\vdots\,\mathfrak{Fax}_{\xi'7,\xi7}}{\xi'7 \vdash \xi7}}{\vdash \xi', \xi3, \xi7}\,(\xi',\{3,7\}) \qquad \dfrac{\dfrac{\vdots\,\mathfrak{Fax}_{\xi'4,\xi4}}{\xi'4 \vdash \xi4} \quad \dfrac{\vdots\,\mathfrak{Fax}_{\xi'7,\xi7}}{\xi'7 \vdash \xi7}}{\vdash \xi', \xi4, \xi7}\,(\xi',\{4,7\})}{\xi \vdash \xi'}\,(\xi,\{\{3,7\},\{4,7\}\}) \tag{13.6}$$

This design differs from the « real » fax only by its last rule: the finite directory $\{\{3,7\},\{4,7\}\}$ replaces the *full* directory $\wp_f(\mathbb{N})$. Most premises have been pruned back; in terms of *desseins*, see *infra*, the pseudo-fax is included in the fax. « Syntactically », both fax and pseudo-fax are η-expansions of the identity. The difference is that the fax is generic, while the pseudo-fax is more specialised; see the discussion of subtyping and *incarnation* (Section 14.1.5).

Definition 37 (Sub-dessins). If the pitchfork $\Xi \vdash \Lambda$ occurs in the design $\mathfrak{D}$, the subtree $\mathfrak{E}$ induced by $\mathfrak{D}$ « above » $\Xi \vdash \Lambda$ is a design of base $\Xi \vdash \Lambda$, called a « subdesign » of $\mathfrak{D}$.

Thus, the designs $\mathfrak{Fax}_{\xi'*i,\xi*i}$ are subdesigns of $\mathfrak{Fax}_{\xi,\xi'}$. This has nothing to do with subsets of $\mathfrak{D}$ (inclusion), not to speak of *precedence* (Section 13.6.2).

13.2 Designs-*desseins*

The homonymy between *dessin* (graphics, picture) and *dessein* (plan, plot[1]) minimises the differences between the original version (*dessins*) and its correct reformulation (*desseins*).

13.2.1 Introduction to *desseins*. An obvious problem with *dessins*, is that the label $(\vdash \xi, I)$ yields no information as to the *splitting* of the context. One could choose to make it explicit, but the dialectics analysis/synthesis impedes this choice: typically one would lose the *separation* theorem of Section 13.6.2.

Let us take an example:

$$\dfrac{\dfrac{\dfrac{\dfrac{}{\sigma 2 \vdash \xi 30}\,(\sigma 2,\emptyset)}{\vdash \xi 30, \sigma}\,(\sigma,\{2\}) \qquad \dfrac{}{\vdash \xi 30, \xi 35, \sigma}\,\maltese}{\xi 3 \vdash \sigma}\,(\xi 3,\{\{0\},\{0,5\}\}) \qquad \dfrac{\dfrac{}{\vdash \tau}\,\maltese}{\xi 7 \vdash \tau}\,(\xi 7,\{\emptyset\})}{\vdash \xi, \sigma, \tau}\,(\xi,\{3,7\}) \tag{13.7}$$

The first rule « gives » σ to 3 and τ to 7. But could one choose differently? Of course, since a rule $(\vdash \sigma, \{2\})$ occurs in a branch corresponding to 3, one can by no means give σ to 7; but what about τ, which has been given to 7 and stays idle? Indeed the *dessins*

$$\dfrac{\dfrac{\dfrac{\dfrac{}{\sigma 2 \vdash \xi 30, \tau}\,(\sigma 2,\emptyset)}{\vdash \xi 30, \sigma, \tau}\,(\sigma,\{2\}) \qquad \dfrac{}{\vdash \xi 30, \xi 35, \sigma, \tau}\,\maltese}{\xi 3 \vdash \sigma, \tau}\,(\xi 3,\{\{0\},\{0,5\}\}) \qquad \dfrac{\dfrac{}{\vdash}\,\maltese}{\xi 7 \vdash}\,(\xi 7,\{\emptyset\})}{\vdash \xi, \sigma, \tau}\,(\xi,\{3,7\}) \tag{13.8}$$

[1] « Le savant fou ruminait de noirs desseins. »

and

$$
\dfrac{\dfrac{\dfrac{\dfrac{}{\sigma 2 \vdash \xi 30}\,{\scriptstyle (\sigma 2,\emptyset)}}{\vdash \xi 30, \sigma}\,{\scriptstyle (\sigma,\{2\})} \qquad \dfrac{}{\vdash \xi 30, \xi 35, \sigma}\,{\scriptstyle \maltese}}{\xi 3 \vdash \sigma}\,{\scriptstyle (\xi 3,\{\{0\},\{0,5\}\})} \qquad \dfrac{\dfrac{}{\vdash}\,{\scriptstyle \maltese}}{\xi 7 \vdash}\,{\scriptstyle (\xi 7,\{\emptyset\})}}{\vdash \xi, \sigma, \tau}\,{\scriptstyle (\xi,\{3,7\})} \tag{13.9}
$$

are equally valid distributions of the context. So to speak, τ lies in between 3 and 7, maybe nowhere.

To sum up, the splitting of the context is a decorative (a *dessin* is a graphical artefact) diversion, which only makes sense for those *loci* used as foci (of positive rules). Ironically, remark that the use of a *locus* as focus *destroys* it, makes it disappear: this is *location post mortem*[2].

Rereading *dessins modulo* this ambiguity yield *desseins*. The *dessin* of example (13.7) becomes:

$$
\dfrac{\dfrac{(\sigma,\{2\})}{(\xi 3,\{0\})} \qquad \dfrac{\maltese}{(\xi 3,\{0,5\})} \quad \dfrac{\maltese}{(\xi 7,\emptyset)}}{(\xi,\{3,7\})} \tag{13.10}
$$

In the tree, one replaces any negative pitchfork $\xi \vdash \Lambda$ conclusion of a rule $(\xi, \mathcal{N})$ with several copies, one for each $I \in \mathcal{N}$, renamed (ξ, I): this essentially makes the negative branchings *go down* by one notch. Each positive pitchfork is renamed after the rule of which it is a conclusion.

The interpretation of the two variants (13.8) and (13.9) yields the same result (13.10) which is their common *dessein*. We now rewrite the fax (13.5):

$$
\cdots \qquad \dfrac{\dfrac{\cdots \qquad \dfrac{\dfrac{\begin{array}{c}\vdots \\ \ldots\ (\xi ij, K)\ \ldots\end{array}}{(\xi i, J)}\,{\scriptstyle j\in J, K\in\wp_f(\mathbb{N})}}{(\xi' i, J)} \qquad \cdots}{(\xi', I)}\,{\scriptstyle i\in I, J\in\wp_f(\mathbb{N})}}{(\xi, I)} \qquad \cdots \quad {\scriptstyle I\in\wp_f(\mathbb{N})} \tag{13.11}
$$

[2]In the camp of Guntánamo, the prisoners lost everything, including their identity. Some managed to recover it (June 2006), but they had to commit suicide to achieve that effect.

This many-rooted *dessein* is not a tree, it is a *forest*. The repetitiveness of the fax is even more conspicuous than on the *dessin*. We proceed with the *pseudo-fax* (13.6):

$$\dfrac{\dfrac{\cdots\quad \dfrac{\dfrac{\begin{array}{c}\vdots\\ \cdots\ (\xi 3j, K)\ \cdots\end{array}}{(\xi 3, J)}}{(\xi' 3, J)} \quad \dfrac{\dfrac{\begin{array}{c}\vdots\\ \cdots\ (\xi 7j, K)\ \cdots\end{array}}{(\xi 7, J)}}{(\xi' 7, J)} \quad \cdots}{(\xi', \{3,7\})}}{(\xi, \{3,7\})} \qquad \dfrac{\dfrac{\cdots\quad \dfrac{\dfrac{\begin{array}{c}\vdots\\ \cdots\ (\xi 4j, K)\ \cdots\end{array}}{(\xi 4, J)}}{(\xi' 4, J)} \quad \dfrac{\dfrac{\begin{array}{c}\vdots\\ \cdots\ (\xi 7j, K)\ \cdots\end{array}}{(\xi 7, J)}}{(\xi' 7, J)} \quad \cdots}{(\xi', \{4,7\})}}{(\xi, \{4,7\})} \tag{13.12}$$

This *dessein* really appears as a subset of the previous one, of which the sole roots $I = \{3,7\}, I = \{4,7\}$ were retained.

The irrelevance of τ is linked to weakening:

$$\dfrac{\vdash \Gamma}{\vdash \Gamma, A} \tag{13.13}$$

in which A (the *locus* of A) disappears (w.r.t. the procedural « look », from below). But it is not uniquely due to weakening, thus

$$\dfrac{\dfrac{}{\vdash \mathsf{T}, A} \qquad \dfrac{}{\vdash \mathsf{T}}}{\vdash \mathsf{T} \otimes \mathsf{T}, A} \tag{13.14}$$

without weakening; A might as well be given to the right premise:

$$\dfrac{\dfrac{}{\vdash \mathsf{T}} \qquad \dfrac{}{\vdash \mathsf{T}, A}}{\vdash \mathsf{T} \otimes \mathsf{T}, A} \tag{13.15}$$

13.2.2 *Desseins.* In the following we fix a *base*, i.e., a pitchfork $\Upsilon \vdash \Lambda$. A *proper action* (ϵ, ξ, I) consists in a *polarity* $\epsilon = \pm 1$, a *focus* ξ (i.e., a *locus*) and a ramification I. The *improper action* $(+1, \maltese)$, called *daimon* is « positive ».

In what follows, by a *proper action*, I mean an action κ whose focus ξ is a *sublocus* of a *locus* (necessarily unique) of the base: $\sigma \in \Lambda$ (resp. $\sigma \in \Upsilon$); its polarity is $+1$ if the parities of ξ, σ are the same (resp. opposite), -1 otherwise[3]. In practice, we forget the polarity of an action, and simply write (ξ, I)[4] or $\maltese$.

[3]If the base is paritary, the polarity is the *relative* parity of ξ and the base.

[4]This may become ambiguous with *dessein-nets* (Section 13.5.1), where the « same » action may occur twice, with both polarities.

Definition 38 (Chronicles). A *chronicle of base* $\Upsilon \vdash \Lambda$ is a sequence of actions $\langle \kappa_0, \ldots, \kappa_n \rangle$ such that:

Alternation: the polarity of κ_p is equal to the polarity of the base for p even, to its opposite for p odd.

Daimon: for $p < n$, κ_p is proper, $\kappa_p = (\xi_p, I_p)$.

Negative actions: a negative focus ξ_p must be chosen either in Υ (then $p = 0$ and the base is negative), or in $\xi_{p-1} * I_{p-1}$.

Positive actions: a positive focus ξ_p must be chosen either in Λ, or in a $\xi_q * I_q$, where (ξ_q, I_q) is one of the previous negative actions, i.e., $q < p$ and $p - q$ odd.

Destruction: foci are pairwise distinct, i.e., they cannot be reused.

A chronicle $\langle \kappa_0, \ldots, \kappa_n \rangle$ is *proper* or *improper* depending on whether or not κ_n is proper. A *subchronicle* of $\langle \kappa_0, \ldots, \kappa_n \rangle$ is a restriction $\langle \kappa_0, \ldots, \kappa_p \rangle$ $(p \leq n)$; a strict subchronicle is therefore always proper.

Definition 39 (Coherence). The chronicles $\mathfrak{c}$, $\mathfrak{c}'$ are *coherent* when:

Comparability: either one is a subchronicle of the other, or they first differ on negative actions, i.e., $\mathfrak{c} = \mathfrak{d} * \kappa * e$, $\mathfrak{c}' = \mathfrak{d} * \kappa' * e'$, with $\kappa \neq \kappa'$ negative.

Propagation: moreover, with the previous notations, if κ, κ' have *distinct* foci, then all remaining foci, i.e., the foci of the actions of $\mathfrak{e}$ and $\mathfrak{e}'$, are distinct.

Definition 40 (Designs-desseins). A *design* of base (or *conclusion*) $\Upsilon \vdash \Lambda$ is a set $\mathfrak{D}$ of chronicles of base $\Upsilon \vdash \Lambda$ such that:

Arborescence: $\mathfrak{D}$ is closed under restriction, in other terms it is a *forest*.

Coherence: the chronicles of $\mathfrak{D}$ are pairwise coherent.

Positivity: if $\mathfrak{c} \in \mathfrak{D}$ has no strict extension in $\mathfrak{D}$, its last action is positive.

Totality: if the base is positive, then $\mathfrak{D}$ is non-empty.

A design is *positive* or *negative* according to its base.

A few comments:

- The notion of chronicle exactly corresponds to that of branch in the forest associated to a *dessin*, see examples (13.10), (13.11), (13.12). The conditions « kill » those sequences not coming from a real *dessin*.

- *Comparability* ensures that, if $\mathfrak{c}*\kappa, \mathfrak{c}*\kappa'$ belong to the same design, with κ, κ' positive, then $\kappa = \kappa'$. In particular, a positive design – non-empty by *totality* – has a well-defined first action. Metaphorically, if a design is a strategy, positive actions are « my » answers, thus univocal.

- *Propagation* is a subtler property. Let us come back to example (13.10): above the first action $(\xi, \{3, 7\})$, a ternary branching occurs, with a choice between three negative actions $(\xi * 3, \{0\})$, $(\xi * 3, \{0, 5\})$, $(\xi * 7, \emptyset)$, say κ_i for $i = 1, 2, 3$. κ_1 and κ_2, with the same focus, are parts of the same negative rule, while κ_3 comes from another one. κ_1 and κ_3 have distinct foci since performed above two distinct premises of a logical rule *for which the context splits*: coming back to the original *dessin* (13.7), one sees that the action on σ has been performed on the extreme left, i.e., above κ_1; if we were allowing an action of focus σ above κ_3, we could no longer *dispatch* the context between 3 and 7.

- *Positivity* can be understood as follows: take a chronicle $\mathfrak{c}$ ending with a negative action; it is thus a premise of a negative rule and necessarily a conclusion of a positive rule, which corresponds to the « next action ». The same argument fails in the positive case, since the last action of $\mathfrak{c}$ may be a *daimon*, or the premise of a negative rule $(\xi, \emptyset)$. Metaphorically, positivity means that designs are sorts of strategies and that any move of `You` (negative action) stirs up a unique answer (positive action).

- *Totality* – which states the existence of a (unique) first action in a design – is only a technical variation on positivity: there must be a first move if `Me` begins.

13.2.3 From *desseins* to *dessins*. It remains to associate a *dessin* to a *dessein*; we shall thus have two irons in the fire, *dessins* for all and sundry, *desseins* when one must be rigourous:

(i) In a first step, I construct a *wrong dessin*, wrong because the context is systematically recopied (in positive rules, $\Lambda_i = \Lambda$, in negative rules, $\Lambda_I = \Lambda$). This offers no difficulty.

(ii) In each pitchfork $\Upsilon \vdash \Lambda$ of the wrong *dessin*, I remove all *loci* not used « above » $\Upsilon \vdash \Lambda$. One loses $\Lambda_I = \Lambda$, moreover, in the positive case, the Λ_i are now disjoint.

This construction is *recessive*, hence not computable. We have anyway obtained the existence of a *minimum dessin*, in the sense of the most parsimonious dispatching of contexts. But one should not identify a *dessein* with its minimum *dessin*, which is in no way canonical: for instance normalisation can destroy minimality. Finally, observe that one cannot expect a « weakening-free » maintenance of contexts; typically the design of base $\vdash \xi, \xi'$ reduced to the first action $(\xi, \emptyset)$ does not know how to « put itself out » of the context ξ'.

13.3 Partial designs

This new category is an excessive honour for the adjunction of a single object to designs; but this is a very important one.

Definition 41 (Partial designs). Same definition as *desseins*, but without *totality*.

Seen as partial designs, plain designs will therefore be styled *total*.

The unique real partial *dessein* is positive: it is the empty set, a.k.a. the *pseudo-design* or *Faith*: $\mathfrak{Fid}$. The name has been chosen to symbolise its procedurality: it is a paragon of expansivity, since it is what one hopes for and which never comes.

Figure 13.3. $\mathfrak{Fid}$.

It is typically what occurs in a *diverging* normalisation, like those induced by antinomies (by the way, the λ-term coming from Russell's antinomy is often denoted by Ω). If the result does not materialise, the faith that we kept at each moment in the termination of the computation[5] remains.

One can represent it by a « rule », which means « there is no first rule »:

$$\frac{}{\vdash \Lambda}\;\Omega \tag{13.16}$$

Procedurally speaking, this would rather be « the first rule didn't *yet* materialise ».

A natural generalisation would be to allow more partiality by forgetting *positivity*: an arbitrary positive pitchfork, not only the base, would have the right of having no rule above it. It is a fake generalisation, but a very instructive one. Indeed, such a pitchfork would be the premise of index I of a negative rule $(\xi, \mathcal{N})$ and the same effect is achieved by pruning back the premise, i.e., by replacing $\mathcal{N}$ with $\mathcal{N} \setminus \{I\}$.

[5] In Rome, it is the expected arrival of the bus 64.

One can use this generalisation the other way around: introducing another improper positive action, Ω, such that $\mathfrak{Fid}$ is the design of first action Ω, I can now decide that negative branchings are *full* (i.e., with directories $\mathcal{N} = \wp_f(\mathbb{N})$): one must add new premises and above each of them, the new action Ω. In this presentation, the status of negative rules is simplified (they are really invertible: above $\xi \vdash \Lambda$ the rule is always $(\xi, \wp_f(\mathbb{N}))$); and *totality* means that Ω cannot be the first action of a design.

This variant has no practical interest, but it suggests a certain symmetry between the two improper actions. Indeed, *precedence* between designs (Section 13.6.2) will be summarised by

$$\Omega \preceq (\xi, I) \preceq \maltese \tag{13.17}$$

This symmetry should not make us forget the procedural gap between the two improper actions. As we shall see it, Ω is naturally generated by the *divergence* of the normalisation process, while $\maltese$ is termination *subito*, a sudden death. When seeking the first action – preferably proper – of a positive design through an expansive method like normalisation, the answers Ω and $\maltese$ are both anomalies, but of a different nature:

Too late! $\maltese$ says « you will not make it »; but at least I know it.

Please wait! Ω corresponds to the case where the answer never comes; but how to know this? Not only do I get nothing, but I don't even know that.

Thinking about expansivity, undecidability results, one sees that Ω, seen as an « action » or as a « rule », is non-effective.

One can see Ω either statically: as the *reification* of the absence of information (this is set-theory), or dynamically: one thus imagines that a design is in a process of « growth », that it is expanding. For instance, during normalisation, one would systematically seek positive rules, (ξ, I) or $\maltese$, above a positive pitchfork; as long as one gets nothing, one writes Ω – stalled; later on, if the information materialises, one will *upgrade* Ω into (ξ, I) or $\maltese$. This is the meaning of the inclusion between designs, the *stable order* of Section 13.6.6:

$$\Omega \subset (\xi, I), \maltese \tag{13.18}$$

13.4 Nets and normalisation: *dessins*

13.4.1 Cuts. Nets familiarised us with cut-links. Here we shall make a slightly different choice, cut as *coincidence*. Instead of saying that a cut links together two distant *loci* (as does the fax), we say that it is the direct, physical, contact between two pitchforks, through a shared *locus*. This corresponds to a procedurality of the kind « plugging of sockets ». Which does not forbid distant plugging, through, precisely, an extension cord, i.e., a fax. By the way, remark that, a cut being a disputed *locus*, it is natural that a conflict occurs: a conflict solved by *normalisation*.

13.4.2 *Dessins*-nets

Definition 42 (Dessins-nets). A *net of* dessins (or *dessin*-net) is a non-empty finite set $\mathfrak{R} = \{\mathfrak{D}_0, \dots, \mathfrak{D}_n\}$ of *dessins* of respective bases $\Xi_p \vdash \Lambda_p$ such that:

Disjunction: the *loci* of the bases are pairwise disjoint or equal.

Cuts: each *locus* appears in at most two bases, once as a handle, once as a tine. Such a shared *locus* is called a *cut*.

Tree: the graph whose vertices are the bases $\Xi_p \vdash \Lambda_p$ and whose edges are the cuts is connected and acyclic.

The particular case $n = 0$ (no cut) corresponds to the usual *dessins*.

By Euler–Poincaré, $\sharp(components) - \sharp(cycles) = \sharp(vertices) - \sharp(edges)$; the last condition can be rewritten as « connected with n cuts », or « acyclic with n cuts ». Since n handles are *consumed* by cuts, it remains at most one « free » one; one can form a pitchfork with the uncut *loci*, the *conclusion* or *base* of the *dessin*-net; if its base is empty, the *dessin*-net is said to be *closed*. The unique *dessin* $\mathfrak{D}_i$ whose base is either positive, or negative with as handle the uncut one, is the *main dessin* of the *dessin*-net; its base is the *main pitchfork*, its main rule is the *main rule* of $\mathfrak{R}$.

$\mathfrak{R}$ is *paritary* when made of paritary *dessins* and if, moreover, its base is paritary – which is not automatic. Take two paritary *dessins* of bases $\xi \vdash \sigma$ and $\sigma \vdash \tau$, which form a non-paritary net of base $\xi \vdash \tau$. Paritarism being a way to detect mistakes, one will construct, when possible, paritary nets.

Since the conditions only mention the bases of $\mathfrak{D}_0, \dots, \mathfrak{D}_n$, certain deviances are possible:

- Replace *dessin* $\mathfrak{D}_0$ with a *dessin*-net $\mathfrak{R}_0 = \{\mathfrak{E}_0, \dots, \mathfrak{E}_m\}$ of the same basis. Isn't it very close to the net $\{\mathfrak{E}_0, \dots, \mathfrak{E}_m, \mathfrak{D}_1, \dots, \mathfrak{D}_n\}$?
- Allow the *dessins* of $\mathfrak{R}$ to be partial; concretely this means allowing $\mathfrak{Fid}$, which will be – since positive – the main *dessin* of such a *partial* net.

The combination of these two possibilities yields *associativity*, one of the analytical theorems (Section 13.6.3).

13.4.3 Normalisation: *dessins*. *Normalisation* is a strictly deterministic procedure which replaces a *dessin*-net $\mathfrak{R}$ with a *dessin* of the same base, its *normal form* $[\![\mathfrak{R}]\!]$; it may *diverge*, i.e., yield no result, or, equivalently, « give back » the partial *dessin* $\mathfrak{Fid}$.

This locally finite procedure propagates expansively from below. This version « *dessins* » remains very close to normalisation in **HS** (Section 12.6.6).

I begin with the closed case – i.e., a net of empty base –, by far the simplest and most important one.

Closed case

Definition 43 (Normalisation). Let $\mathfrak{R}$ be a closed *dessin*-net; then the main *dessin* – say $\mathfrak{D}$ – is positive, with main rule κ; three subcases:

Daimon: κ is the *daimon* ✠[6]. Then the net normalises into the only *dessin* of base $\vdash$, the *daimon*: $[\![\mathfrak{R}]\!] = \mathfrak{Dai}$. This is the only terminating case for a closed net.

Immediate failure: κ is (ξ, I); hence ξ is a cut which appears as the *handle* of a *dessin* $\mathfrak{E}$, the *adjoint dessin* of the net, whose last rule is necessarily of the form $(\xi, \mathcal{N})$. If $I \notin \mathcal{N}$, then normalisation fails.

Conversion: same as before, but $I \in \mathcal{N}$. For $i \in I$, let $\mathfrak{D}_i$ be the sub-*dessin* of $\mathfrak{D}$ whose conclusion is the premise of index i, i.e., $(\xi * i \vdash \ldots)$ of (ξ, I) and let $\mathfrak{E}'$ be the sub-*dessin* $\mathfrak{E}$ indexed by I $(\vdash \xi * I, \ldots)$ of the rule $(\xi, \mathcal{N})$. Define $\mathfrak{S}$ by replacing $\mathfrak{D}, \mathfrak{E}$ by the $\mathfrak{D}_i, \mathfrak{E}'$; $\mathfrak{S}$ need not be connected: let $\mathfrak{S}'$ be the connected component of $\mathfrak{E}'$ in $\mathfrak{S}$. Then $[\![\mathfrak{R}]\!] := [\![\mathfrak{S}']\!]$.

In the case « conversion », the replacement of $\mathfrak{S}$ with $\mathfrak{S}'$ comes from the fact that our rules contain « weakenings »: some *loci* occurring in the conclusions of the main *dessin* and its adjoint may disappear, hence connectivity disappears as well; this is why one restricts to the connected component of the new main *dessin*, $\mathfrak{E}'$. This kind of petty erasure problem disappears in the version *desseins*.

The normal form, when it exists, can only be the *daimon* $\mathfrak{Dai}$. But normalisation can also diverge, either because of a failure, or because of an infinite sequence of conversions. One decides to denote by $[\![\mathfrak{R}]\!] = \mathfrak{Fid}$ the result of a diverging normalisation. This suggests extending normalisation to partial nets; there is a new case to consider:

Faith: if the main « *dessin* » $\mathfrak{D}$ is $\mathfrak{Fid}$, the normal form is $[\![\mathfrak{R}]\!] = \mathfrak{Fid}$.

Although this is a useful convention, we should not forget that there is no way to determine whether or not a normal form is total. Remember the lesson of the incompleteness theorem: there is no way to know that one cannot know.

In an expansive approach, we could see the writing $\mathfrak{Fid}$ as a temporary way of saying that we didn't yet get the last rule. Which may change, after an undetermined number of conversions. One must also think – this is the meaning of associativity – that the adjoint *dessin* can in turn come from normalisation, which means that it is « expanding », thus that immediate failure can be eventually be replaced with a conversion or a demoniac case.

---

[6] If the net is normal, then $n = 0$ and the net is already a *daimon*.

Open case. The open case induces two unproblematic commutations:

Positive commutation: the net is positive, of main rule (ξ, I), but ξ is not a cut. If $\mathfrak{D}_i$ is defined as in the case of conversion (Definition 43) one defines $\mathfrak{R}'$ by replacing $\mathfrak{D}$ with the $\mathfrak{D}_i$. $\mathfrak{R}'$ splits into several components and $\mathfrak{D}_i$ lies in a component $\mathfrak{R}_i$, which is a net, with a normal form $\mathfrak{E}_i$[7]. The normal form of $\mathfrak{R}$ is the *dessin* with first rule (ξ, I) with, above the premise of index i, the sub-*dessin* $\mathfrak{E}_i$, i.e.:

$$[\![\mathfrak{R}]\!] = \frac{\ldots \quad [\![\mathfrak{R}_i]\!] \quad \ldots}{\vdash \Lambda, \xi}\,(\xi, I) \tag{13.19}$$

Negative commutation: the net is negative and its main *dessin* $\mathfrak{D}$ has the main rule $(\xi, \mathcal{N})$. For $I \in \mathcal{N}$, let $\mathfrak{D}_I$ be the sub-*dessin* $\mathfrak{D}$ with base the premise of index I of the first rule and let us replace $\mathfrak{D}$ with $\mathfrak{D}_I$ in $\mathfrak{R}$; let $\mathfrak{R}_I$ be the connected component of $\mathfrak{D}_I$ (again weakening!). Let $\mathcal{N}'$ be the subset of $\mathcal{N}$ made of the I such that $\mathfrak{R}_I$ has a normal form $\mathfrak{E}_I$. The normal form $\mathfrak{R}$ is the *dessin* of first rule $(\xi, \mathcal{N}')$, which proceeds, above the premise of index I, as $\mathfrak{E}_I$, i.e.:

$$[\![\mathfrak{R}]\!] = \frac{\ldots \quad [\![\mathfrak{R}_I]\!] \quad \ldots}{\xi \vdash \Lambda}\,(\xi, \mathcal{N}') \tag{13.20}$$

In other words, positive commutation recopies the first rule, then proceeds independently above each premise. The same idea holds for negative commutation, except that some premises may disappear. Since negative commutation is the only possibility for a negative net, they all normalise; in the worst case, one will get the empty directory $\mathcal{N}' = \emptyset$ for its first rule, i.e., a *skunk* (Section 13.6.2).

The replacement of $\mathcal{N}$ with $\mathcal{N}'$ in the negative commutation must be understood *dynamically*: $\mathcal{N}'$ is expanding (as soon as we get a first rule for $\mathfrak{E}_I$, we know that $I \in \mathcal{N}'$). But, on the whole, did we know $\mathcal{N}$ that well? We already noted that $\mathcal{N}$ can be the result of an expansive process; the missing premises are those that will never materialise. What appeared in our definition as an immediate failure is indeed properly infinite: one waits for a premise that will materialise the next day, the next year. Indeed, Ω is not emptiness, it is the infinite loop.

Of course, normalisation still extends to the partial case. With the convention of *full* negative branchings, negative commutation simplifies, since $\mathfrak{E}_I$ is defined for all $I \in \wp_f(\mathbb{N})$ and, of course, may be equal to $\mathfrak{Fid}$; in this presentation, the case « immediate failure » disappears, replaced with the case « faith ».

[7]For negative nets, normalisations always converge, see *infra*.

13.4.4 Examples

Daimon. Let us introduces the *negative daimon*:

$$\frac{\ldots \quad \dfrac{}{\vdash \xi * I, \Lambda}\maltese \quad \ldots}{\xi \vdash \Lambda}(\xi, \wp_f(\mathbb{N})) \tag{13.21}$$

Any cut with a *daimon* normalises into a *daimon*. Every cut between a *negative*

Figure 13.4. $\mathfrak{Dai}^-$.

daimon $\xi \vdash \Lambda$ and a *dessin* $\mathfrak{D}$ of base $\Upsilon \vdash \Sigma$ normalises:

- If $\Upsilon = \upsilon$ and $\upsilon \in \Lambda$, the normal form is a negative *daimon*.
- If $\xi \in \Sigma$ and $\Upsilon = \emptyset$, then the normal form is a positive *daimon* if $\mathfrak{D}$ is a *daimon* or if $\mathfrak{D}$ ends with a rule of focus ξ. If $\mathfrak{D}$ ends with (σ, I) with $\sigma \neq \xi$, then the normal form ends with (σ, I) with negative *daimons* above each premise:

$$\frac{\ldots \quad \dfrac{\dfrac{}{\vdash \sigma * i * I, \Lambda, \Sigma_i \setminus \xi}\maltese \quad \ldots}{\sigma * i \vdash \Lambda, \Sigma_i \setminus \xi}(\sigma * i, \wp_f(\mathbb{N}))}{\vdash \Lambda, \Sigma \setminus \xi}(\sigma, I) \tag{13.22}$$

- If $\xi \in \Sigma$ and $\Upsilon = \upsilon$, let $(\upsilon, \mathcal{N})$ be the first rule of $\mathfrak{D}$; then the first rule of the normal form is $(\upsilon, \mathcal{N})$ and its premises $\vdash \upsilon * J, \Lambda, \Sigma_J \setminus \xi$ are either proved by *daimons* or sub-*dessins* in the style of (13.22).

Fax. Take the fax $\mathfrak{Fax}$ of base $\xi \vdash \xi'$; then:

- A cut with $\mathfrak{D}$ of base $\vdash \xi$ ending with (ξ, I) normalises into a *dessin* of base $\vdash \xi'$ ending with (ξ', I).

- A cut with $\mathfrak{E}$ of base $\xi' \vdash$ ending with $(\xi', \mathcal{N})$ normalises into a *dessin* of base $\xi \vdash$ ending with $(\xi, \mathcal{N})$.

Which shows that, in the first case, the normal form is indeed the *delocation* $\rho(\mathfrak{D})$ of $\mathfrak{D}$, i.e., the *dessin* obtained by systematically replacing ξ with ξ'. In the second case, the normal form is the delocation $\rho^{-1}(\mathfrak{E})$ of $\mathfrak{E}$. More generally, a cut with the fax on ξ normalises by replacing ξ with ξ' and a cut with the fax on ξ' normalises by replacing ξ' with ξ. In particular the cut between $\mathfrak{Fax}_{\xi,\xi'}$ and $\mathfrak{Fax}_{\xi',\xi''}$ normalises into $\mathfrak{Fax}_{\xi,\xi''}$. To sum up:

$$[\![\mathfrak{Fax}, \mathfrak{D}]\!] = \rho(\mathfrak{D}), \tag{13.23}$$

$$[\![\mathfrak{Fax}, \mathfrak{E}]\!] = \rho^{-1}(\mathfrak{E}), \tag{13.24}$$

$$[\![\mathfrak{Fax}, \mathfrak{D}, \mathfrak{E}]\!] = [\![\mathfrak{D}, \rho^{-1}(\mathfrak{E})]\!] = [\![\rho(\mathfrak{D}), \mathfrak{E}]\!]. \tag{13.25}$$

What happens with the pseudo-fax (13.6)? It normalises like the fax but only retains the first actions (ξ, I) (or (ξ', I)) with $I \in \{\{3,7\},\{4,7\}\}$. Concretely, a cut with a negative *dessin* whose first rule is $(\xi', \mathcal{N})$ normalises into a *dessin* of first rule $(\xi, \mathcal{N} \cap \{\{3,7\},\{4,7\}\})$; a cut with a positive design *dessin* of first rule (ξ, I) normalises (i.e., converges) when $I \in \{\{3,7\},\{4,7\}\}$, in which case the normal form is the delocation of $\mathfrak{D}$.

13.5 Nets and normalisation: *desseins*

13.5.1 *Desseins*-nets. Normalisation should be defined in terms of *desseins*, not of *dessins*. But how to dispatch the context?

Definition 44 (Desseins-nets). A *net of desseins* (or *dessein*-net) is a non-empty finite set $\mathfrak{R} = \{\mathfrak{D}_0, \ldots, \mathfrak{D}_n\}$ of *desseins* of respective bases $\Xi_p \vdash \Lambda_p$ such that:

Disjunction: the *loci* of the bases are pairwise disjoint or equal.

Cuts: a *locus* cannot occur in two handles; a *cut* is a *locus* occurring both as a handle and a tine.

Tree: for each cut ξ let us draw an edge between ξ and *one of the* pitchforks with tine ξ (this is a *switch*). For each switching the graph must be connected and acyclic.

Propagation: if ξ is a tine in both $\Xi_p \vdash \Lambda_p$ and $\Xi_q \vdash \Lambda_q$ and if actions of focus ξ are performed in $\mathfrak{D}_p, \mathfrak{D}_q$, then $p = q$.

A typical example is that of a net $\{\mathfrak{D}, \mathfrak{D}', \mathfrak{D}''\}$ made of *desseins* of bases $\xi \vdash \sigma$, $\xi' \vdash \sigma$, $\vdash \xi, \xi'$. Propagation impedes the use of σ as focus in both of $\mathfrak{D}, \mathfrak{D}'$.

The trick is that ξ will eventually be activated in at most one of the $\mathfrak{D}_p$, but one does not know which one; this is why the condition « Tree » involves a switching. Cuts will be reduced as in the case of *dessins*, except that the context is « given » to everything.

The most natural definition of *desseins*-nets, is that of a set of chronicles, i.e., $\mathfrak{R} = \mathfrak{D}_0 \cup \cdots \cup \mathfrak{D}_n$. Of course, since some *loci* are both handle and tine, a focus can be used both positively *and* negatively. Here one must be pedantic and carefully distinguish $\kappa = (\epsilon, \xi, I)$ from its *opposite* $\widetilde{\kappa} := (-\epsilon, \xi, I)$.

We shall now give a precise definition of normalisation of *desseins*-nets; there are two equivalent versions, mauls and disputes. The former is more rigourous, but the latter is so simple!

13.5.2 Slices and mauls

Definition 45 (Slices). A *slice* is a *dessein* (more generally, a net) $\mathfrak{S}$ in which the negative rules are at most *unary*: if $\mathfrak{c} * (-1, \xi, I), \mathfrak{c} * (-1, \xi, I') \in \mathfrak{S}$, then $I = I'$. A slice of a *dessein* $\mathfrak{D}$ (or a net $\mathfrak{R}$) is any slice $\mathfrak{S} \subset \mathfrak{D}$ ($\mathfrak{S} \subset \mathfrak{R}$).

In a slice-*dessein*, two incomparable chronicles necessarily differ for the first time on actions of distinct foci; hence, by *propagation*, all ulterior foci differ. Each focus occurs exactly once and *a fortiori* each action occurs at most once. In a slice-net, each action occurs at most twice, once positively, once negatively.

We can define an *arborescent* order, i.e., of type forest (i.e., whose initial segments are linear orders), between the proper actions of a slice:

$$\kappa <_{\mathfrak{S}} \kappa' \text{ iff } \mathfrak{S} \text{ contains a chronicle } \mathfrak{c} * \kappa * \mathfrak{c}' * \kappa'.$$

We identify $\mathfrak{S}$ with its proper actions, ordered by $<_{\mathfrak{S}}$: the order is indeed enough to recover *proper* chronicles; improper chronicles are obtained by totality or positivity, since, if κ is maximal and negative, we need a *daimon* just after κ.

An action κ is *hidden* (w.r.t. a slice $\mathfrak{S}$) if it is proper and its focus ξ is *sublocus* of a cut. Otherwise, it is *visible*.

Definition 46 (Balance). A finite slice $\mathfrak{S}$ is *balanced* when for any *hidden* action κ:

$$\kappa \in \mathfrak{S} \Rightarrow \widetilde{\kappa} \in \mathfrak{S}.$$

Definition 47 (Mauls). The *maul* of a balanced slice $\mathfrak{S}$ is obtained by identifying each hidden action with its opposite, notation $(\pm 1, \xi, I)$ (one speaks of a *neutral* action). This quotients the order $<_{\mathfrak{S}}$ into $\ll_{\mathfrak{S}}$.

Proposition 18. *If $\mathfrak{S}$ is balanced, then $\ll_{\mathfrak{S}}$ is an arborescent order.*

Proof. The proof given in [51] is very technical. As a result internal to ludics, this is well; but one can obtain *legible*, almost as good, results by means of disputes (*infra*). □

Although interesting, the next results have a slightly ingrate ratio of quality/price:

Theorem 37 (Protoslices). (i) *If* $\mathfrak{S}$ *is balanced, then* $[\![\mathfrak{S}]\!]$ *is made of the* ***visible*** *actions of* $\mathfrak{S}$*, with the order induced by* $\ll_{\mathfrak{S}}$.

(ii) *Conversely, if* $\mathfrak{R}$ *is a net and* $\mathfrak{S}$ *is a finite slice of* $[\![\mathfrak{R}]\!]$*, there is a* ***unique*** *balanced slice* $\mathfrak{T} \subset \mathfrak{R}$*, the* ***protoslice of*** $\mathfrak{S}$ ***along*** $\mathfrak{R}$*, such that* $\mathfrak{S} = [\![\mathfrak{T}]\!]$.

Proof. As before, we drop it: the case of disputes will be enough for us. □

Corollary 37.1. *A positive net converges iff it contains at least one balanced slice.*

Corollary 37.2. *A closed net converges iff it contains at least one balanced slice; this slice* $\mathfrak{S}$ *is unique and is linearily ordered by* $\ll_{\mathfrak{S}}$.

Disputes (*infra*) (re)prove this nice result. One could define disputes (those ending with a positive action) as follows:

Definition 48 (Disputes). A *dispute* of $\mathfrak{R}$ is a balanced slice $\mathfrak{S} \subset \mathfrak{R}$ such that $\ll_{\mathfrak{S}}$ is a total order.

13.5.3 Disputes. We will now give a direct definition of normalisation by means of disputes that we will by the way (re)define.

To a *dessein*-net $\mathfrak{R}$ one associates sequences of actions, called *disputes*. These sequences are obtained by building a slice $\mathfrak{S}_n$ such that $\ll_{\mathfrak{S}_n}$ is total. When we start, all directories are empty; to pass from $\mathfrak{S}_n$ to $\mathfrak{S}_{n+1}$, one will often « open » a unary directory.

Start: two cases, depending on the base:

> **Positive:** there is, among all chronicles of $\mathfrak{R}$, a unique sequence $\langle\kappa\rangle$ of length 1. This is by definition the unique dispute of length 1 of $\mathfrak{R}$.
>
> **Negative:** let ξ be the main handle and let $\langle(\xi, I)\rangle$ be a chronicle of $\mathfrak{R}$; it induces a dispute of the same $\mathfrak{R}$: we just opened a unary directory.

Continuation: assume that $\mathfrak{c} * (\xi, I)$ is a chronicle; three cases occur:

> (ξ, I) **visible and negative:** then, by *positivity*, there is in the slice a *unique* action κ which extends (ξ, I). Our dispute extends into $\mathfrak{c} * (\xi, I) * \kappa$; indeed, its unique immediate extension.
>
> (ξ, I) **visible and positive:** one chooses, if possible, a negative action « extending » κ and our dispute extends into $\mathfrak{c} * (\xi, I) * \kappa$; κ is visible. Observe that there need not be unicity, and that one « opens » a new unary directory.

(ξ, I) hidden: if (ξ, I), hidden, occurs at the end of a dispute, it is necessarily as a positive action. There is at most one possibility to « open » a unary directory containing the action $(-1, \xi, I)$. This action extends, by positivity, in a unique way, into a positive action κ. My chronicle thus extends into $\mathfrak{c} * (\xi, I) * \kappa$.

A dispute is therefore a sequence of visible and hidden actions. The *tunnels*, sequences of hidden actions, can be of arbitrary length, in particular infinite. A tunnel must begin with the start (then the base is positive) or just after a visible negative action. If one ever exits the tunnel, it is through a visible positive action – hence, in the paritary case, the tunnel is of even length.

In any case, we now have our definition:

Definition 49 (Normal form). Let $\mathfrak{R}$ be a *dessein*-net; its *normal form* $[\![\mathfrak{R}]\!]$ is obtained:

(i) By taking all disputes $\mathfrak{c}$ whose last action is *visible*.

(ii) Retaining the sole visible actions, in the same order.

One should:

(i) Prove that this definition actually yields a *dessein*: easy and tedious.

(ii) Prove the equivalence with the same definition for *dessins*. This is deeply useless, since *dessins* are only a graphical diversion; but this is true. By the way, remark that visible actions correspond to the demoniac case and to the two commutations. The tunnels are sequences of conversions.

The only important thing to remark is a property of *unicity*:

Proposition 19. *If $\mathfrak{c} \in [\![\mathfrak{R}]\!]$, then there is a unique chronicle $\mathfrak{d}$ of $\mathfrak{R}$ whose visible actions are, following the order, those of $\mathfrak{c}$ and ending with a visible action.*

Proof. Unicity comes from the fact that the only element of freedom in the building of a dispute is the extension of a *visible* positive action with a negative action – or the choice of a first negative action – which is visible, hence which remains present in $\mathfrak{c}$. □

To link this with the previous subsection, observe that, if $\mathfrak{c}$ ends with a positive action, then the $\mathfrak{d}$ of Proposition 19 is the *protoslice* of $\mathfrak{c}$.

13.5.4 Disputes and coherent spaces

Definition 50. Suppose that $\mathfrak{D}$, $\mathfrak{E}$ of bases $\vdash \xi$ and $\xi \vdash$ are such that $[\![\mathfrak{D}, \mathfrak{E}]\!] = \mathfrak{Dai}$; then the protoslice of $\langle \maltese \rangle$ along $\{\mathfrak{D}, \mathfrak{E}\}$ (noted $[\mathfrak{D} \rightleftharpoons \mathfrak{E}]$) is the *dispute* between $\mathfrak{D}$, $\mathfrak{E}$. It is the sequence $\langle \kappa_0, \ldots, \kappa_{n-1}, \maltese \rangle$ of the $n - 1$ conversions performed, followed with a final *daimon*.

In ludics *ante litteram*, a design $\mathfrak{D} \in \mathbf{G}$ was identified with the set $Dsp_{\mathbf{G}}(\mathfrak{D}) = \{[\mathfrak{D} \rightleftharpoons \mathfrak{E}]; \mathfrak{E} \in \sim\mathbf{G}\}$; this perfectly legitimate approach was following the pattern of Section 12.2.3: coherent spaces as *cliques of disputes*. Relinquished at the time (1999) for want of legibility, this approach seems wrong for a deeper reason, although controversial: is it necessary that objects are built from « atomic » constituents, designs as sets of disputes and behaviours as sets of designs? Are the stars made of atoms, in the same way the galaxies are made of stars? This suits our attitude of analytical humans, but what does nature think of analyticity? It does not give a damn about it; the expression might be crude, but is nature polite?

Finally, coherent spaces are only one of the aspects of logic, surely more exciting than T_0 « topologies », but to put anyway on the same shelving, slightly withdrawn.

13.6 Analytical theorems

These theorems occupy the very center of ludics; they are *analytical*, since they deal with the basic objects, before synthesis, *designs*. There are four of them:

Separation: the analogue of *Böhm's theorem*, (Theorem 43 *infra*); the very origin of the replacement « *dessin* $\rightsquigarrow$ *dessein* » .

Associativity: the analogue of Church–Rosser, the key to layer -2.

Monotonicity: the old inheritance of Scott domains.

Stability: the contribution of coherent spaces.

These properties are indeed much more important than the objects (designs) which support them; having fulfilled their task, designs could eventually be replaced with something else enjoying the same analytical theorems.

13.6.1 Duality

Definition 51 (Duality). Let $\mathfrak{D}$ be a design of base $\Xi \vdash \Lambda$ and let $\mathfrak{E}_\sigma$ be designs of respective bases $\vdash \sigma$ (if $\sigma \in \Xi$) and $\sigma \vdash$ (if $\sigma \in \Lambda$). The notation $\ll \mathfrak{D} \,|\, (\mathfrak{E}_\sigma) \gg$ stands for the normal form $[\![\mathfrak{D}, (\mathfrak{E}_\sigma)]\!]$ ($\mathfrak{Dai}$ or $\mathfrak{Fid}$ in case of divergence) of the design-net $\{\mathfrak{D}, \ldots, \mathfrak{E}_\sigma, \ldots\}$ (a sort of « bilinear form »). $\mathfrak{D}$ and the $(\mathfrak{E}_\sigma)$ are *polar* when the normal form is total, i.e., when $\ll \mathfrak{D} \,|\, (\mathfrak{E}_\sigma) \gg = \mathfrak{Dai}$, notation $\mathfrak{D} \perp (\mathfrak{E}_\sigma)$.

We took the viewpoint of the base $\Xi \vdash \Lambda$; the bases $\sigma \vdash$ $(\sigma \in \Xi)$ and $\vdash \sigma$ $(\sigma \in \Lambda)$ (or $\vdash \sigma$) are called *anti-bases* or *counter-bases*. We will speak of an *anti-design* or a *counter-design* to mean of a design whose conclusion is one of the counter-bases.

By far the most important case is that of an atomic base $\xi \vdash$ or $\vdash \xi$: then there is exactly one $\mathfrak{E}_\sigma$; we then use the simplified notations $\ll \mathfrak{D} \,|\, \mathfrak{E} \gg$ and $\mathfrak{D} \perp \mathfrak{E}$. In this case, base and counter-base play symmetric roles.

We might as well allow the pseudo-design in our definition. It is polar to nobody, since $[\![\mathfrak{Fid}, (\mathfrak{E}_\sigma)]\!] = \mathfrak{Fid}$.

13.6.2 Separation

Definition 52 (Precedence). The set of designs of base $\Xi \vdash \Lambda$ is equipped with the topology generated by the closed sets $\sim(\mathfrak{E}_\sigma)$. *Precedence* is the preorder:

$$\mathfrak{D} \preceq \mathfrak{D}' :\Longleftrightarrow \sim\{\mathfrak{D}\} \subset \sim\{\mathfrak{D}'\}$$

Since the closure of a point is its bipolar $\sim\sim\{\mathfrak{D}\}$, precedence might as well be defined by $\mathfrak{D} \preceq \mathfrak{D}' \Leftrightarrow \mathfrak{D}' \in \sim\sim\{\mathfrak{D}\}$. This topology will be T_0 exactly when $\preceq$ is a partial order.

Theorem 38 (Separation). *$\preceq$ is a partial order, i.e., the topology is T_0. Indeed $\mathfrak{D} \preceq \mathfrak{D}'$ iff $\mathfrak{D}$ is* ***more defined*** *than $\mathfrak{D}'$, i.e., if each chronicle $\mathfrak{c} \in \mathfrak{D} \setminus \mathfrak{D}'$ can be written $\mathfrak{c}' * \mathfrak{d}$ for a certain $\mathfrak{c}'$ such that $\mathfrak{c}' * \maltese \in \mathfrak{D}'$.*

Proof. The difficult part of the theorem is necessity. Indeed, this is more tedious than really difficult: I therefore proceed assuming – which only changes tiny details – that the base is atomic, say, $\xi \vdash$ and I will work on two examples that summarise the real proof [51]. Suppose that $\mathfrak{D} \preceq \mathfrak{D}'$ and let $\mathfrak{c} \in \mathfrak{D}$. We consider the cases:

(i) $\mathfrak{c} = \langle(\xi, \{1,3\}), (\xi 3, \{7\}), (\xi 37, \emptyset), (\xi 4, \{5\})\rangle$, and

(ii) $\mathfrak{c} = \langle(\xi, \{1,3\}), (\xi 3, \{7\}), (\xi 37, \emptyset), (\xi 4, \{5\}), \maltese\rangle$.

We construct a counter-design (in fact, a slice) $\mathfrak{Opp}_\mathfrak{c}$, with the opposites of the proper actions of $\mathfrak{c}$ and by adding or removing the *daimon*:

(i) $\langle(\xi, \{1,3\}), (\xi 3, \{7\}), (\xi 37, \emptyset)\rangle$, $\langle(\xi, \{1,3\}), \maltese\rangle$ (and subchronicles), and

(ii) $\langle(\xi, \{1,3\}), (\xi 3, \{7\}), (\xi 37, \emptyset)\rangle$, $\langle(\xi, \{1,3\}), (\xi 4, \{5\})\rangle$ (and subchronicles).

$\mathfrak{Opp}_\mathfrak{c}$ is such that $\mathfrak{D} \perp \mathfrak{Opp}_\mathfrak{c}$, hence $\mathfrak{D}' \perp \mathfrak{Opp}_\mathfrak{c}$. Which is possible only if $\mathfrak{c} \in \mathfrak{D}'$ or if a strict restriction of $\mathfrak{c}$ to which a *daimon* has been added, i.e., $\langle(\xi, \{1,3\}), \maltese\rangle$, or $\langle(\xi, \{1,3\}), (\xi 3, \{7\}), (\xi 37, \emptyset), \maltese)\rangle$ belongs to $\mathfrak{D}'$.

On the other hand, the condition is obviously sufficient: if

$$[\mathfrak{D} \rightleftharpoons \mathfrak{E}] = \langle \kappa_0, \ldots, \kappa_{n-1}, \maltese \rangle$$

and if $\mathfrak{D}$ is more defined than $\mathfrak{D}'$, then

$$[\mathfrak{D}' \rightleftharpoons \mathfrak{E}] = \langle \kappa_0, \ldots, \kappa_{n'-1}, \maltese \rangle$$

for an $n' \preceq n$. Finally it is obvious that « more defined than » is antisymmetric, hence this relation is a partial order. $\square$

In terms of *dessins*, $\mathfrak{D} \preceq \mathfrak{D}'$ means that $\mathfrak{D}'$ is « larger and shorter », that it has been obtained from $\mathfrak{D}$ by means of a broadening of negative rules and a replacement of positive rules by *daimons*. But how come that $\mathfrak{D}$ is more defined? This is clear with the *daimon*, an *opportunistic* action, less informative, less risky than a proper action. In the case of negative rules, think that $\mathfrak{D}$, with its more restricted directories, takes more risks, knows better what it wants or does not want. Thus, the *negative daimon* conveys no sort of information.

$\mathfrak{D} \preceq \mathfrak{D}'$ can be decomposed in two steps: first shorten the branches (which yields $\mathfrak{D}''$), then broaden the negative rules. Here one reaches the limits of the style « *dessin* »: one can produce common *dessins* for $\mathfrak{D}$, $\mathfrak{D}''$ and for $\mathfrak{D}''$, $\mathfrak{D}'$, but not always for $\mathfrak{D}$, $\mathfrak{D}'$, which means that the context does not split in the same way in both.

Precedence is summarised by the equation

$$\Omega \preceq (\xi, I) \preceq \maltese \tag{13.26}$$

which means that one can always replace Ω (absent a premise of a negative rule) with a « real » premise and that a proper action can be replaced with a *daimon*.

By the way, which are the maximal/minimal elements in terms of precedence? The *daimons* $\mathfrak{Dai}$, $\mathfrak{Dai}^-$ are obviously maximal; more opportunist than me, you die! These *yes men* are polar to everybody.

When the base is negative, there is a design smallest w.r.t. $\preceq$, the *Skunk*, indeed the empty design: $\mathfrak{Skunk} := \emptyset$.

$$\frac{}{\xi \vdash \Lambda}\,(\xi, \emptyset) \tag{13.27}$$

Figure 13.5. $\mathfrak{Skunk}$.

The minimal designs of base $\vdash \Lambda$ are the *positive skunks* of base $\mathfrak{Skunk}_{(\lambda, I)}$ (with $\lambda \in \Lambda$):

$$\cfrac{\cfrac{}{\lambda * i \vdash}\,(\lambda * i, \emptyset)}{\vdash \Lambda}\,(\lambda, I) \tag{13.28}$$

Figure 13.6. $\mathfrak{Skunk}^{+}$.

If the discussion had been extended to partial designs, one would have found that $\mathfrak{Fid}$ is minimum. It is paradoxical to think that the partial element is the most defined. Indeed, he is so demanding that he didn't (yet) make it!

On the base $\vdash \Lambda$, the designs which are maximal among those distinct from the *daimon* are the *Ramifications* $\mathfrak{Ram}_{(\lambda,I)}$:

$$\dfrac{\cdots \qquad \dfrac{\cdots \quad \dfrac{}{\vdash \lambda * i * J}\maltese \quad \cdots}{\lambda * i \vdash}\,(\lambda * i, \wp_f(\mathbb{N})) \qquad \cdots}{\vdash \Lambda}\,(\lambda, I) \tag{13.29}$$

Figure 13.7. $\mathfrak{Ram}^{+}$.

13.6.3 Associativity. Given a net of nets $\mathfrak{S} = \{\mathfrak{R}_0, \ldots, \mathfrak{R}_n\}$, one gets the same result, whether one first normalises each $\mathfrak{R}_p$ then the net $\mathfrak{S}' = \{\mathfrak{D}_0, \ldots, \mathfrak{D}_n\}$ of their normal forms or directly $\mathfrak{S}'' = \mathfrak{R}_0 \cup \cdots \cup \mathfrak{R}_n$:

Theorem 39 (Associativity). *Let $\{\mathfrak{R}_0, \ldots, \mathfrak{R}_n\}$ be a net of nets, then*

$$[\![\mathfrak{R}_0 \cup \cdots \cup \mathfrak{R}_n]\!] = [\![[\![\mathfrak{R}_0]\!], \ldots, [\![\mathfrak{R}_n]\!]]\!]. \tag{13.30}$$

Moreover – and this is the most important point –, the equation persists for partial nets.

Proof. Immediate from disputes. □

Technically speaking, this is hardly more than the Church–Rosser theorem for **HS**; one could by the way establish it by means of a Church–Rosser theorem in terms of *dessins*.

13.6.4 The closure principle. By combining the two previous theorems, one gets a principle of proof by *adjunction*: « one can restrict to closed nets ». Thus, if $\mathfrak{D}, \mathfrak{E}$ are designs of respective bases $\xi \vdash \lambda$ and $\vdash \xi$, the normal form $[\![\mathfrak{D}, \mathfrak{E}]\!]$ is the *unique* $\mathfrak{D}'$ of base $\vdash \lambda$ such that for all $\mathfrak{F}$ of base $\lambda \vdash$,

$$[\![\mathfrak{D}', \mathfrak{F}]\!] = [\![\mathfrak{D}, \mathfrak{E}, \mathfrak{F}]\!]. \tag{13.31}$$

The normal form of a net $\mathfrak{S}$ is determined by those of all its closures, i.e., the completions of $\mathfrak{S}$ into a closed net. The interest of the thing is obvious: thus, in an approach of style *dessin*, one can « skirt » commutations.

Theorem 40 (Closure Principle). *Let $\mathfrak{R}$ be a net of base $\Xi \vdash \Lambda$. The normal form $[\![\mathfrak{R}]\!]$ of $\mathfrak{R}$ is the unique $\mathfrak{D}$ such that for all counter-designs $(\mathfrak{E}_\sigma)$, $\mathfrak{D} \perp (\mathfrak{E}_\sigma)$ iff $[\![\mathfrak{R} \cup \cdots \cup \mathfrak{E}_\sigma \cup \cdots]\!] = \mathfrak{Dai}$, i.e., if the normal form of $\mathfrak{R} \cup \ldots \cup \mathfrak{E}_\sigma \cup \ldots$ converges.*

Proof. $\mathfrak{R}$ verifies the condition by associativity; unicity comes from separation. □

An example: if $\mathfrak{F}$ is a design of base $\xi \vdash \xi'$ such that for all $\mathfrak{E}$ of base $\vdash \xi$,

$$[\![\mathfrak{F}, \mathfrak{E}]\!] = \rho(\mathfrak{E}) \tag{13.32}$$

with ρ the *delocation* introduced by equation (13.23), then $\mathfrak{F} = \mathfrak{Fax}_{\xi,\xi'}$.

The closure principle is at work in multiplicative constructions (Section 14.2) where it permits definitions by adjunction which thus turn out (which is non-trivial) to be associative.

The closure principle is obviously inspired from the definition of the adjoint in Hilbert spaces:

$$\langle u(x) \mid y \rangle = \langle x \mid u^*(y) \rangle. \tag{13.33}$$

In ludics the same holds, but – as a consequence of locativity – $u = u^*$: if $\mathfrak{U}, \mathfrak{X}, \mathfrak{Y}$ are of respective bases $\xi \vdash \xi'$, $\vdash \xi$ and $\xi' \vdash$ then

$$\ll [\![\mathfrak{X}, \mathfrak{U}]\!] \mid \mathfrak{Y} \gg = \ll \mathfrak{X} \mid [\![\mathfrak{U}, \mathfrak{Y}]\!] \gg = [\![\mathfrak{X}, \mathfrak{U}, \mathfrak{Y}]\!]. \tag{13.34}$$

13.6.5 Monotonicity

Theorem 41 (Monotonicity). *Normalisation is monotonic w.r.t. the order* $\preceq$: *if* $\mathfrak{D}_0 \preceq \mathfrak{E}_0, \ldots, \mathfrak{D}_n \preceq \mathfrak{E}_n$, *then* $[\![\mathfrak{D}_0, \ldots, \mathfrak{D}_n]\!] \preceq [\![\mathfrak{E}_0, \ldots, \mathfrak{E}_n]\!]$.

Proof. This is typical by-product of the closure principle: let $\mathfrak{R} :=\{\mathfrak{D}_0, \ldots, \mathfrak{D}_n\}$, $\mathfrak{S} :=\{\mathfrak{E}_0, \ldots, \mathfrak{E}_n\}$.

(i) We first establish the property for an empty base; this is more or less the definition of $\preceq$.

(ii) Let us now take a non-empty base, say, $\xi \vdash$. If $\mathfrak{F}$ is of base $\vdash \xi$, then $[\![\mathfrak{R} \cup \mathfrak{F}]\!] \preceq [\![\mathfrak{S} \cup \mathfrak{F}]\!]$ by the closed case, hence $[\![[\![\mathfrak{R}]\!] \cup \mathfrak{F}]\!] \preceq [\![[\![\mathfrak{S}]\!] \cup \mathfrak{F}]\!]$ by associativity. We conclude that $[\![\mathfrak{R}]\!] \preceq [\![\mathfrak{S}]\!]$.

□

As with other analytical theorems, monotonicity persists for partial nets.

13.6.6 Stability

Theorem 42 (Stability). *Normalisation commutes with compatible intersections: if* K *is non-empty and* $\mathfrak{R}_k \subset \mathfrak{R}$ *for all* k, *then*

$$[\![\bigcap_{k\in K} \mathfrak{R}_k]\!] = \bigcap_{k\in K} [\![\mathfrak{R}_k]\!] \tag{13.35}$$

Proof. The inclusion $[\![\bigcap_k \mathfrak{R}_k]\!] \subset \bigcap_k [\![\mathfrak{R}_k]\!]$ is immediate. Conversely, if $\mathfrak{c} \in \bigcap_k [\![\mathfrak{R}_k]\!]$, then $\mathfrak{c}$ comes from a unique dispute $\mathfrak{r}_k \subset \mathfrak{R}_k$; but $\mathfrak{r}_k \subset \mathfrak{R}$ and, by unicity of the dispute $\mathfrak{r}$ of $\mathfrak{R}$ which yields $\mathfrak{c}$, $\mathfrak{r}_k = \mathfrak{r}_{k'} = \mathfrak{r}$ for all $k, k' \in K$: then $\mathfrak{r} \subset \bigcap_k \mathfrak{R}_k$ and $\mathfrak{c} \in [\![\bigcap_k \mathfrak{R}_k]\!]$. □

A typical example of stability is given by:

$$\ll \mathfrak{D} \,|\, \bigcap_k \mathfrak{E}_k \gg = \bigcap_k \ll \mathfrak{D} \,|\, \mathfrak{E}_k \gg \tag{13.36}$$

which will be used to define *incarnation* (Section 13.8.2).

« Double » stability:

$$\ll \mathfrak{D}_1 \cap \mathfrak{D}_2 \,|\, \mathfrak{E}_1 \cap \mathfrak{E}_2 \gg = \ll \mathfrak{D}_1 \,|\, \mathfrak{E}_1 \gg \cap \ll \mathfrak{D}_2 \,|\, \mathfrak{E}_2 \gg \tag{13.37}$$

cannot be directly deduced from the « unary » case (13.36).

13.7 Introspective vs. extraspective

13.7.1 The « intensional ». We find in logic many definitions in the style

$$f \preceq g \quad \text{iff} \quad f(a) \preceq g(a) \quad \text{for all } a. \tag{13.38}$$

Which is often called *extensional order*, by reference to the extensionality axiom of set-theory. And that one might as well call « pointwise order », by reference to a well-established style of mathematical definition.

A tenacious logical prejudice wants good definitions to be « extensional ». And since it is not tenable, one has created a duplicate of the word – a sort of bad bank – « intensional »[8], which means strictly nothing, since its only the negation of the former. The vocabulary works as follows: *extensional* for « good definition », *intensional* for « no matter what ». For instance, one distinguishes a left-handed cup from a right-handed one and does not know how to get out of this mess; one thus styles it « intensional ».

Extensional is painful mainly because of the insistence of logicians in calling differently what already exists – and better; *intensional* obviously opens the door to irresponsibility. But the opposition between these terms is more than a convenience: for instance, in mathematics, besides simple (pointwise) convergence stands uniform convergence; in ludics, besides the pointwise order $\preceq$ stands the stable order $\sqsubset$. But can one style uniform convergence, stability, *intensional*? Immediately, one will think that it is rubbish! This is why I propose to change the terms, to avoid the *newspeak*, Orwellian phenomenon, around « intensional ». I thus propose to use non-prostituted words, like *introspective* and *extraspective*; one could also choose invisible/visible.

13.7.2 The extraspective. The extraspective is everything dealing with inputs/outputs: one gives arguments and one observes the result. As Poincaré would say, one puts the pig on the conveyor belt and one recuperates the sausages. Are extraspective separation, associativity and their by-products, monotonicity, closure; and also *totality* (Section 12.1.3). The extraspective viewpoint is prominent in the noted:

Theorem 43 (Böhm's theorem). *If t and u are closed normal λ-terms which differ* modulo η, *then one can find $v_1, \dots, v_n$ such that $(\dots(t)v_1\dots)v_n \rightsquigarrow x$ and $(\dots(u)v_1\dots)v_n \rightsquigarrow y$.*

Proof. See, e.g., [9]. □

Some will thus say that λ-calculus without « η » is intensional. Rubbish!

---

[8] The spelling « inten*s*ional » tries in vain to attenuate the afflictive subjectivism conveyed by the standard spelling « intentional » .

13.7.3 The introspective. Stability is a typical example of introspection: one does not observe the result, but the *way*. It is not the final chronicle that matters, it is the dispute that generated it. This idea of « way » is compatible with the rather indefinite sense of « intensional », but is more precise, slightly too precise for the taste of « intensionalists ».

Among introspective ideas that I mentioned so far, is the extension of Curry–Howard to classical logic, an extension that can only be understood by introspection: it is no longer a matter of inputs/outputs, but of computing procedures (control instructions), hence of *style*, of way.

Among « intensional » scoria refused by « introspective », there is everything which is superfluous, but that one cannot dump. For instance, the writing of additive weights as sums of monomials in my version of additive nets (Section 11.C.4), is a « control » often redundant and rather arbitrary: two nets with different monomials can « behave in the same way ». It is where one would speak of « intensional net »; I prefer to speak of a half-baked definition.

The notion of formal system is obviously extraspective, since this is the very nature of deduction. On the other hand, Rosser's variant (Section 2.D.3), is not extraspective, since the « way » influences deduction. More generally, all paraconsistent systems, « labelled deductive systems » which are precisely... non-deductive, play against extraspection. If something deserves to be styled « intensional », this is surely those... doohickeys. In other words, « introspective » does not quite recover « intensional »!

13.7.4 Winning. One more beautiful introspective definition:

Definition 53 (Winning). A design is winning when it does not use the *daimon*.

On the basis of what we observed so far, a winning design thus looks very much like a proof.

The introspective character of winning is easy to understand: suppose that $\mathfrak{D} \perp \mathfrak{E}$; this means that $[\![\mathfrak{D}, \mathfrak{E}]\!] = \mathfrak{D}\mathfrak{ai}$. A *daimon* has been produced, but we don't know by whom. Going back to the dispute $[\mathfrak{D} \rightleftharpoons \mathfrak{E}]$, we see that the final *daimon* comes from one of them: we can thus find a *winner* to their confrontation.

We can say it in a more synthetic way:

Proposition 20. *If $\mathfrak{D} \perp \mathfrak{E}$, then one of them is not winning.*

This takes its full sense when we reconstruct logic (Chapter 14). If the *truth* of the behaviour **G** is defined as the existence of a winning design in **G**, then **G** and $\sim$**G** cannot be both true (Section 14.A.2).

Which is essential, the only thing one needs to reconstruct an acceptable layer -1, i.e., a *consistent* truth definition. Finally, consistency appears as an introspective notion!

13.8 Behaviours

13.8.1 Definition and examples

Definition 54 (Behaviours). A *behaviour* is a set $\mathbf{G}$ of designs of a given base equal to its bipolar. A behaviour is *positive* or *negative* according to its base.

Example 1. (i) The set of all designs of a given base form a behaviour, the *Skunk*, equal to $\sim\emptyset$. One uses the notations $\mathbf{T}^{\epsilon}$ where ϵ is the polarity of the base and, more simply, $\mathbf{T}$ when the base is negative.

(ii) The set $\{\mathfrak{Dai}^{\epsilon}\}$ is a behaviour, the *Daimon*, indeed the smallest behaviour of a given base, equal to $\sim\sim\emptyset$. One uses the notation $\mathbf{0}^{\epsilon}$ and, more simply, $\mathbf{0}$ when the base is positive.

(iii) In general, when $\mathbf{E}$ is a set of counter-designs, $\sim\mathbf{E}$ is a behaviour; every behaviour is of this form (with $\mathbf{E} = \sim\mathbf{G}$).

Example 2. There is a smallest (styled *principal*) behaviour containing a given design $\mathfrak{D}$, indeed $\sim\sim\{\mathfrak{D}\}$. By separation:

$$\sim\sim\{\mathfrak{D}\} = \{\mathfrak{D}'; \mathfrak{D} \preceq \mathfrak{D}'\}. \tag{13.39}$$

Since *daimons* are polar to all designs, a behaviour is necessarily non-empty; it contains the *daimon*, positive or negative. The problems related to emptiness (Section 12.1.4) disappear for good.

Behaviours verify certain closure properties:

Theorem 44 (Closure). *If $\mathfrak{D} \preceq \mathfrak{E}$ and $\mathfrak{D} \in \mathbf{G}$ then $\mathfrak{E} \in \mathbf{G}$.*

If K is non-empty, if $\mathfrak{D}_k \subset \mathfrak{D} \in \mathbf{G}$ for all $k \in K$, then $\bigcap_k \mathfrak{D}_k \in \mathbf{G}$.

Proof. Immediate from monotonicity and stability. □

13.8.2 Incarnation. If $\mathfrak{D} \in \mathbf{G}$ and $\mathfrak{D} \subset \mathfrak{E}$, then $\mathfrak{E} \in \mathbf{G}$, but for « bad » reasons: none of the new chronicles in $\mathfrak{E}$ is needed to guarantee belonging to $\mathbf{G}$. So to speak, as members of $\mathbf{G}$, $\mathfrak{D}, \mathfrak{E}$ are « equivalent » and $\mathbf{G}$ is naturally equipped with an equivalence, the symmetric and transitive closure of inclusion. One can distinguish a design $|\mathfrak{D}|_{\mathbf{G}}$ in each class, to the effect that $\mathfrak{D} \simeq \mathfrak{E} \Leftrightarrow |\mathfrak{D}|_{\mathbf{G}} = |\mathfrak{E}|_{\mathbf{G}}$.

Theorem 45 (Incarnation). *If $\mathfrak{E} \in \mathbf{G}$ there is a smallest $\mathfrak{D} \subset \mathfrak{E}$ such that $\mathfrak{D} \in \mathbf{G}$.*

Proof. The set of designs of $\mathbf{G}$ included in $\mathfrak{E}$ is non-empty. By Theorem 44, the intersection $\mathfrak{D}$ of this family is still in $\mathbf{G}$. □

Definition 55 (Incarnation). The design $\mathfrak{D}$ of Theorem 45 is the *incarnation* of $\mathfrak{E}$, noted $|\mathfrak{E}|$, or $|\mathfrak{E}|_{\mathbf{G}}$ in case of doubt:

$$|\mathfrak{E}|_{\mathbf{G}} = \bigcap\{\mathfrak{E}'; \mathfrak{E}' \subset \mathfrak{E} \text{ and } \mathfrak{E}' \in \mathbf{G}\}. \tag{13.40}$$

A design $\mathfrak{D} \in \mathbf{G}$ is *incarnated* or *material* when $\mathfrak{D} = |\mathfrak{D}|$. One defines the incarnation $|\mathbf{G}|$ of $\mathbf{G}$ as the set of its material designs.

The incarnation of $\mathfrak{E}$ is the part of $\mathfrak{E}$ which can be interactively recognised through cuts with designs of $\sim\mathbf{G}$. As a matter of example, one will easily understand that the pseudo-fax of (13.6) can be an incarnation of the fax.

Incarnation is contravariant:

$$\mathbf{G} \subset \mathbf{H} \Rightarrow \mathfrak{E}_{\mathbf{H}} \subset \mathfrak{E}_{\mathbf{G}}. \tag{13.41}$$

The incarnation of $\mathfrak{E}$ is maximum when $\mathbf{G}$ is the principal behaviour $\sim\sim\{\mathfrak{E}\}$ containing $\mathfrak{E}$; in this case, $|\mathfrak{E}| = \mathfrak{E}$, as proved by the separation theorem. The incarnation is minimum when $\mathbf{G}$ is the biggest behaviour $\mathbf{T}^{\epsilon}$. Thus, on a negative base,

$$|\mathfrak{E}|_{\mathbf{T}} = \mathfrak{Skunk}, \tag{13.42}$$

i.e., the incarnation is empty. As a behaviour, the negative skunk is the intersection of the empty family; at the level of incarnation, it only contains the empty design, i.e., remembering that $\{\emptyset\}$ is the empty product:

$$\Big|\bigcap_{\emptyset}\Big| = \prod_{\emptyset}. \tag{13.43}$$

Which anticipates the *mystery of incarnation*. The negative design « skunk » is – as the name suggests – asocial: $\mathfrak{Skunk} \perp \mathfrak{E}$ only if $\mathfrak{E} = \maltese$. As a counterpart, the house of the skunk is very welcoming: $\sim\sim\{\mathfrak{Skunk}\} = \mathbf{T}$, i.e., everybody. But, grab much, gain little: the elements of $\mathbf{T}$ are here only very symbolically since their incarnation, their *useful part*, is $\mathfrak{Skunk}$, i.e., is empty.

Incarnation, a typical introspective notion, is linked to the dialectics existence/essence. Indeed one can see behaviours as essence (type) and designs as existence. But when a design begins to contribute to an essence, it « degenerates », part of its functions become *atrophied*[9]. The incarnation is the design as it should be, if it were only perceived as representative of an essence.

13.8.3 Behaviours and games. My main contention with games (Section 12.4.3) is the presence of a superego at layer -1, rules of the game, referee, etc. Ludics speaks – as the word suggests – of games, but games without referees, of *games by consensus*.

[9]Analogy taken from biology; thanks to Giuseppe Longo.

Let us examine, from that viewpoint, a closed net $[\![\mathfrak{D}, \mathfrak{E}]\!]$ made of two designs of atomic bases.

Players: they are named `Even` and `Odd`. Once a viewpoint has been taken, for instance, that of $\mathfrak{D}$, one is called `Me` (positive), the other `You` (negative). A positive game is a game where `Me` begins, a negative game is a game where `You` begins.

Plays: a play is the sequence of conversions used in the converging normalisation of $\ll \mathfrak{D} \,|\, \mathfrak{E} \gg$, i.e., a *dispute*. Once a viewpoint has been taken, the actions of `Me` are positive, those of `You` negative and their parity alternates following an obvious pattern: `Me` always plays even (resp. always odd), `You` always plays odd (resp. always even).

Strategies: $\mathfrak{D}$ plays the role of a strategy for `Me` and $\mathfrak{E}$ plays the role of a strategy for `You`.

Winner: to lose is to play the *daimon*, in other terms, to give up. Thus, the other wins.

So far, so good… But there is a hitch: in a real game, the idea is that both players have roughly balanced chances: but `Me` can choose for $\mathfrak{D}$ – of positive base $\vdash \xi$ – the design

$$\mathfrak{Bomb}^+ := \frac{}{\vdash \xi}\,(\xi, \emptyset). \tag{13.44}$$

Figure 13.8. $\mathfrak{Bomb}^+$.

Which wins, since `You` can only react with a *daimon*, i.e., by giving up: the only chronicle extending $\langle(-1, \xi, \emptyset)\rangle$ is $\langle(-1, \xi, \emptyset), \maltese\rangle$! In other words, `Me` possesses the « atomic weapon » and one hardly sees who would like to play such a dumb game, where the first move wins!

But this is because we are dealing with general designs; one will restrict them and everything will work well.

Rule: in a behaviour, `Me` and `You` *behave* according to two sets **G**, $\sim$**G** of designs. The condition of polarity can be perceived as a constraint: the designs of **G** must be polar to those of $\sim$**G**: $\sim$**G** is the « rule of the game **G** », while **G** is the « rule of the game $\sim$**G** ».

This sort of restriction is of the same nature as the re-definition of « proofs » by means of *set of tests* (Section 12.5). Very few games fall into the category just described;[10] but this sort of restriction has enormous advantages.

How can `You` impede `Me` from using the *atomic weapon* $(\xi, \emptyset)$? It suffices that `You` puts in $\sim$**G** the negative design $\mathfrak{E} = \mathfrak{Dir}_{\wp_f(\mathbb{N})\setminus\{\emptyset\}}$:

$$\mathfrak{E} := \frac{\cdots \quad \dfrac{}{\vdash \xi * I}\maltese \quad \cdots}{\xi \vdash}(\xi, \wp_f(\mathbb{N}) \setminus \{\emptyset\})\,. \tag{13.45}$$

$\ll \mathfrak{Bomb}^+ \mid \mathfrak{E} \gg = \mathfrak{Fid}$, i.e., $\mathfrak{Bomb}^+ \not\perp \mathfrak{E}$: in other terms $\mathfrak{E}$ « impugns » the action $(\xi, \emptyset)$, there is a *dissensus*.

The design (13.45) is not a real strategy for `You`, since `You` has no chance of winning[11]. Indeed, `You` plays for a draw, since he cannot win. The idea of the counter-design is that, if `Me` plays the atomic weapon, there is no agreement; if `Me` doesn't play the atomic weapon, then `Me` wins against the design $\mathfrak{E}$ of `You`. Which does not bring much to `Me`, since `You` is not forced to play such a pessimistic design; in terms of military strategy, $\mathfrak{E}$ is a *deterrence*.

To see a design as a strategy is abusive: in **G**, two designs with the same incarnation induce the same disputes, i.e., the same plays. One must thus see the incarnation $|\mathfrak{D}|_{\mathbf{G}}$ as the strategy *induced* by $\mathfrak{D}$ in the « game » **G**.

If this explanation by consensus is rather satisfactory, it does not tell *how* such a consensus is reached. There is incompatibility between $\mathfrak{Bomb}^+$ and $\mathfrak{E}$, but how do we reach an argument? *Mystère et boule de gomme...* The only thing one knows is that, in real life, infinite disputes – e.g., domestic quarrels – are always avoided in the same way: one of the two contenders (the wisest) throws in the towel, i.e., « plays the *daimon* »: « *You are too dumb, I make for the door* ».

13.8.4 Behaviours and syntax/semantics. If a design is a sort of proof, a counter-design is a sort of counter-model. Hence the duality between **G** and $\sim$**G** can be read as syntax vs. semantics, if I decide to put the syntax on the side of **G**, the semantics on the side of $\sim$**G**. Which is natural, since the side of `Me` is the side of the subject!

[10] Or rather, logicians are so befogged with the idea of reproducing the obsolete distinctions of layer -1 that they never thought of a game by consensus

[11] Unless `Me` is so defeatist that he begins with giving up, by playing the *daimon*.

The duality links a potential proof of **G** with a potential refutation of **G**. Although we abolished the most uncouth differences between syntax and semantics, a practical and enormous distinction remains: syntax is usually *given* by a set of rules, which generates a semantics, which in turn enables one to pose the problem of *completeness*.

Definition 56 (Internal completeness). An *ethics* is a set $\mathbf{E}$ of designs of a given base $\Xi \vdash \Upsilon$; if $\mathbf{G} = \sim\sim\mathbf{E}$, one says that $\mathbf{E}$ *is an ethics for* $\mathbf{G}$. An ethics is *complete* when it contains the incarnation of its bipolar, i.e., when $|\sim\sim\mathbf{E}| \subset \mathbf{E}$.

In general, a behaviour $\mathbf{G}$ is presented by an ethics, i.e., $\mathbf{G} = \sim\sim\mathbf{E}$. The typical example is the set $\mathbf{E}$ of designs obtained from the *syntactical* cut-free proofs of a given formula A. The counter-models of A will be replaced with $\sim\mathbf{E}$, hence $\sim\sim\mathbf{E}$ corresponds to what is validated by the counter-models of A; finally completeness for A is the fact that the bipolar brings nothing new.

Sometimes – especially in the negative case –, $\sim\sim\mathbf{E} = \mathbf{E}$ is impeded by stupid reasons: this is why incompleteness is only up to incarnation. A typical example is given by the behaviour T – which corresponds to the constant $\top$ of logic – and which, although made of *all* designs, admits the finite ethics $\{\mathfrak{Skunk}\}$! This is indeed a completeness theorem in the usual sense, since $\mathfrak{Skunk}$ interprets the only cut-free proof of $\top$, an axiom.

13.9 An example: the shift

A « connective » is any way of building behaviours. Each « good » connective has its own internal form of completeness, which will thus induce completeness results in the more traditional sense. Thus, the internal completeness of additive disjunction is nothing but the *disjunction property*, the internal completeness of the additive conjunction corresponds to the mystery of incarnation (Section 14.1).

Here, we are interested in a seemingly minor connective, the change of polarity, a.k.a. *shift*; it is indeed one of the *major* novelties of ludics: it swaps polarities by means of a dummy action.

Definition 57 (Shift). Let $\mathfrak{c}$ be a chronicle of base $\vdash \Lambda, \xi * i$ (resp. $\xi * i \vdash \Lambda$); the *shift* $\updownarrow \mathfrak{c}$ of $\mathfrak{c}$ is the chronicle $(\xi, \{i\}) * \mathfrak{c}$ of base $\xi \vdash \Lambda$ (resp. $\vdash \xi, \Lambda$).

If $\mathfrak{D}$ is a design of base $\vdash \Lambda, \xi * i$ (resp. $\xi * i \vdash \Lambda$), the *shift* of $\mathfrak{D}$ is the design $\updownarrow \mathfrak{D} = \{\updownarrow \mathfrak{c}; \mathfrak{c} \in \mathfrak{D}\} \cup \{\langle(\xi, \{i\})\rangle\}$ de $\xi \vdash \Lambda$ (resp. $\vdash \xi, \Lambda$).

If $\mathbf{G}$ is a behaviour of base $\vdash \xi * i$ (resp. $\xi * i \vdash$), the *shift* of $\mathbf{G}$ is the behaviour $\updownarrow \mathbf{G} = \sim\sim\{\updownarrow \mathfrak{D}; \mathfrak{D} \in \mathbf{G}\}$ of base $\xi \vdash$ (resp. $\vdash \xi$).

It is more limpid to replace $\updownarrow$ with $\downarrow$ (resp. $\uparrow$) when the polarity of the shift is positive (resp. negative). And remember that $\downarrow$ is reminiscent of !, an eminently positive operation!

Proposition 21. *If* $\mathbf{G}$ *is negative, then* $\downarrow \mathbf{G} = \{\downarrow \mathfrak{D};\, \mathfrak{D} \in \mathbf{G}\} \cup \{\mathfrak{Dai}\}$.
If $\mathbf{G}$ *is positive, then* $\{\uparrow \mathfrak{D};\, \mathfrak{D} \in \mathbf{G}\}$ *is a complete ethics for* $\uparrow \mathbf{G}$.
Moreover, $\updownarrow \sim\mathbf{G} = \sim(\updownarrow \mathbf{G})$.

This immediate proposition establishes the *completeness* of the shift by exhibiting complete ethics. In the positive case, one must add the *daimon* to the shift of the designs of **G**. In particular, the shift is not involutive, even up to isomorphism. In the negative case, one only gets a complete ethics: one can always add « barren » chronicles not beginning with $(\xi, \{i\})$.

The shift is the operation which permits us to define connectives in case the polarities do not match; thus, when one of the behaviours **G**, **H** is negative, one will define $\mathbf{G} \otimes \mathbf{H}$ by $\mathbf{G} \otimes \downarrow \mathbf{H}$, $\downarrow \mathbf{G} \otimes \mathbf{H}$, or $\downarrow \mathbf{G} \otimes \downarrow \mathbf{H}$.

Chapter 14
Ludics: the reconstruction

It is now time to *reconstruct* logic *ex nihilo* (or almost). This reconstruction, limited to *perfect* logic (second-order multiplicative/additive propositional calculus), splits in two well-defined steps:

Reconstruction: of connectives; we will prove their *internal* completeness.

Faithfulness: *external* completeness that we will not follow in detail.

Reconstruction is by far the most important step: upon its quality depends our feeling of progress – or not – in our understanding of logic. Faithfulness is slightly less exciting:

- It takes its inspiration from an interesting, although immature, idea, « full completeness » (Section 12.1.2). It is therefore a deliberate transfer of paradigms from layer -1 to layer -2, here, to layer -3. This is of undeniable interest, but only in the sense of checking that we did the right thing: remember that we are no longer in a duality syntax/semantics.
- Indeed, if everything, syntax, semantics, lives at layer -3, faithfulness should be quite immediate: nothing but a by-product of reconstruction. The technical difficulty in arriving at the end of faithfulness shows that something is still missing, especially concerning literals. The problem is not so much to *prove* faithfulness as it is to prove it in a simple, natural, *convincing*, way.

In our reconstruction steps, the base will be atomic, $\vdash \langle\rangle$ or $\langle\rangle \vdash$; *modulo* delocation (Section 14.1.2 below), this is perfectly general.

14.1 Additives

14.1.1 Locative additives

Directories. The positive designs $\mathfrak{Ram}_{(\langle\rangle,I)}$ have already been introduced ((13.29), Section 13.6.2). I now introduce the negative designs $\mathfrak{Dir}_{\mathcal{N}}$ « directory »:

$$\mathfrak{Dir}_{\mathcal{N}} := \quad \frac{\cdots \quad \dfrac{}{\vdash I}\maltese \quad \cdots}{\langle\rangle \vdash}(\langle\rangle, \mathcal{N})\cdot \tag{14.1}$$

Figure 14.1. $\mathfrak{Dir}^-$.

The negative *daimon*, the skunk, respectively correspond to $\mathcal{N} = \wp(\mathbb{N})$, $\mathcal{N} = \emptyset$; we have already met $\mathfrak{Dir}_{\mathbb{N}\setminus\{\emptyset\}}$ in Section 13.8.3 and we will soon introduce $\mathfrak{Dir}_{\{\emptyset\}}$ (Section 14.2.2).

Definition 58 (Directory). If $\mathbf{G}$ is a positive behaviour, its *directory* is the set $\P\mathbf{G} := \{I\,;\mathfrak{Ram}_{(\langle\rangle,I)} \in \mathbf{G}\}$.

If $\mathbf{G}$ is a negative behaviour, the incarnation $|\mathfrak{Dai}^-|_{\mathbf{G}}$ of the negative *daimon* is of the form $\mathfrak{Dir}_{\mathcal{N}}$ for a certain $\mathcal{N}$, the *directory of* $\mathbf{G}$: $|\mathfrak{Dai}^-|_{\mathbf{G}} = \mathfrak{Dir}_{\P\mathbf{G}}$.

Proposition 22. *If* $\mathbf{G}$ *is positive, then* $\P\mathbf{G}$ *consists of the* I *such that* $(\langle\rangle, I)$ *is the first action of a design of* $\mathfrak{D} \in \mathbf{G}$. *If* $\mathbf{G}$ *is negative, then* $\P\mathbf{G}$ *is the directory of the first rule* $(\langle\rangle, \mathcal{N})$ *of no matter which* ***material*** *design* $\mathfrak{D} \in \mathbf{G}$. *Moreover,* $\P{\sim}\mathbf{G} = \P\mathbf{G}$.

Proof. If $(\langle\rangle, I)$ is the first action of $\mathfrak{D} \in \mathbf{G}$, positive, then $\mathfrak{D} \preceq \mathfrak{Ram}_{(\langle\rangle,I)}$ hence $\mathfrak{Ram}_{(\langle\rangle,I)} \in \mathbf{G}$. The incarnation of $\mathfrak{Dai}^-$ in ${\sim}\mathbf{G}$ is of the form $\mathfrak{Dir}_{\mathcal{N}}$; it is immediate that $I \in \mathcal{N}$ iff $(\langle\rangle, I)$ is the first action of a design $\mathfrak{D} \in \mathbf{G}$. If $\mathbf{G}$ is negative, observe that, in general

$$\mathfrak{F} \preceq \mathfrak{G} \Rightarrow |\mathfrak{F}| \preceq |\mathfrak{G}| \tag{14.2}$$

and conclude that all material designs have the same first rule. The last property is immediate. □

The connective « Inter »

Definition 59 (Inter). Let $\mathbf{G}_k$ be a family of behaviours of the same base; one defines $\bigcap_k \mathbf{G}_k$ ($\boxed{\cap}^{-}$, $\boxed{\cap}^{+}$ if one wants to indicate the polarity) as the intersection of the $\mathbf{G}_k$.

The definition makes sense because an intersection of polars can be expressed as the polar of a union.

Proposition 23. *The connective $\bigcap$ is **literally** commutative and associative. Its neutral element, the empty intersection, is the **Skunk** T^ϵ; it also admits an absorber, the full intersection, i.e., the smallest behaviour: the **Daimon** $\mathbf{0}^\epsilon$.*

The most important point is the word « literally »: we are really at layer -3, i.e., not within categories (where commutativity cannot be *literal*).

The connective « Union »

Definition 60 (Union). Let $\mathbf{G}_k$ be a family of behaviours of the same base; one defines $\biguplus\!\!\!\ast_k\, \mathbf{G}_k$ (i.e., depending on the polarity, $\biguplus\!\!\!-_k\, \mathbf{G}_k$, $\biguplus\!\!\!+_k\, \mathbf{G}_k$) as $\sim\sim(\bigcup_k \mathbf{G}_k)$.

Proposition 24. *The connective $\uplus\!\!\!\ast$ is **literally** commutative and associative, with $\mathbf{0}^\epsilon$ as neutral element and T^ϵ as absorber.*

Internal completeness fails in the case of $\uplus\!\!\!\ast$: there is no way to remove the bipolar so as to get a « reasonable » *ethics*. In a certain sense, $\uplus\!\!\!\ast$ is the only incomplete connective (Section 14.3.2).

Intersection and incarnation

Theorem 46.

$$|\mathfrak{D}|_{\bigcap_k \mathbf{G}_k} = \bigcup_k |\mathfrak{D}|_{\mathbf{G}_k}. \tag{14.3}$$

Proof. Let $\mathfrak{D}' = |\mathfrak{D}|_{\bigcap_k \mathbf{G}_k}$ $\mathfrak{D}'' = \bigcup_k |\mathfrak{D}|_{\mathbf{G}_k}$. The inclusion $\mathfrak{D}'' \subset \mathfrak{D}'$ is immediate (contravariance of incarnation). Conversely, observe that $\mathfrak{D}''$ is polar to $\bigcup_k \sim\mathbf{G}_k$, hence belongs to $\bigcap_k \mathbf{G}_k$. Since $\mathfrak{D}'$ is material, this forces equality. □

There is nothing of the like for the dual connective *union*.

Locative additives and directory. « ¶ » is covariant in the positive case; furthermore:

Proposition 25.

$$\P \bigoplus_k \mathbf{G}_k = \bigcap_k \P\mathbf{G}_k, \tag{14.4}$$

$$\P \biguplus_k \mathbf{G}_k = \bigcup_k \P\mathbf{G}_k. \tag{14.5}$$

In the negative case, the directory is contravariant and:

Corollary 25.1.

$$\P \bigcap\!\!\!\!-_{k} \mathbf{G}_k = \bigcup_{k} \P\mathbf{G}_k, \tag{14.6}$$

$$\P \biguplus\!\!\!\!-_{k} \mathbf{G}_k = \bigcap_{k} \P\mathbf{G}_k. \tag{14.7}$$

14.1.2 Additives

Plus and With

Definition 61 (Disjunction, connectedness). Two behaviours **G**, **H** of the same polarity are *disjoint* when their directories are disjoint. A behaviour **G** is *connected* when its directory $\P\mathbf{G}$ is a singleton $\{I\}$, in which case I is called the *ramification* of the behaviour.

We use the notations $\oplus$, $\&$, instead of $\uplus$, $\cap$ to say that the operations have been applied to disjoint behaviours.

If $\P\mathbf{G}$, $\P\mathbf{H}$ are *alien*, i.e., made of distinct biases, e.g., $\P\mathbf{G} \subset 3\mathbb{N}$, $\P\mathbf{G} \subset 3\mathbb{N}+1$, then $\P\mathbf{G}$ and $\P\mathbf{H}$ can only have $\emptyset$ in common; which makes them most likely disjoint (Section 14.2.5).

Proposition 26. **G** *and* **H**, *positive, are disjoint iff* $\mathbf{G} \cap \mathbf{H} = \mathbf{0}$ $(= \{\mathfrak{Dai}\})$.

G *and* **H**, *negative, are disjoint iff* $|\mathfrak{D}|_{\mathbf{G}} \cap |\mathfrak{E}|_{\mathbf{H}} = \emptyset$ *for all* $\mathfrak{D} \in \mathbf{G}$, $\mathfrak{E} \in \mathbf{H}$.

Proof. Obvious; for instance, in the negative case $|\mathfrak{D}|_{\mathbf{G}} \preceq \mathfrak{Dir}_{\P\mathbf{G}}$. □

A dilemma: locative vs. spiritual. The connectives $\oplus$, $\&$ are partial, in contrast to the *spiritual* logical tradition. By this I mean the idea that logical operations can be performed independently of any spatio-temporal constraint and are therefore total[1]. At layer -2, spirituality is expressed through « everything is up to isomorphism »; at layer -1, by the absence of any organic link between a formula and its semantics. The spiritual approach is hidden behind the notion of *occurrence*, the « same thing » in two distinct places, a form of *bilocation*, as in Fátima. We have the choice between two alternatives:

(i) Keep things as they are; the connective remains partial, with exceptional properties, e.g., equalities instead of isomorphisms.

[1]One must witness the indignation of certain scholars when they hear about locativity; they lapse into gibberish, e.g., epistemic « logic », but this is spiritual, syntax/semantics/meta. This indignation reminds me of the code of honour of the *made men* of the *mafia*.

(ii) Define total operations, corresponding to the familiar connectives. We must thus fix *delocations*, for instance

$$\begin{aligned}\varphi(i * \sigma) &= 3i * \sigma,\\ \psi(i * \sigma) &= (3i + 1) * \sigma,\end{aligned} \tag{14.8}$$

and redefine $\mathbf{G} \oplus \mathbf{H}$ as $\varphi(\mathbf{G}) \uplus \psi(\mathbf{H})$, etc. We no longer get equalities, but only canonical isomorphisms[2]. Thus the disjunction property (Section 14.1.4) becomes $\mathbf{G} \oplus \mathbf{H} = \varphi(\mathbf{G}) \cup \psi(\mathbf{H})$.

I shall make here the choice of literal, but partial, operations: an equality is much better than an isomorphism, canonical or not.

Delocations. By the way, let us speak of delocations: this is a translation of the famous story of the *Hilbert hotel*: « Not enough room, you are kidding ! Move room $\sharp n$ to $\sharp 3n$ (resp. to $\sharp 3n + 1$)... »: two hotels find places for all in a single one and there is even room left (the $\sharp 3n + 2$).

Definition 62 (Delocation). A *delocation* from *locus* ξ to *locus* ξ' is an injective function θ from the *subloci* of ξ to the *subloci* of ξ' such that:

- $\theta(\xi) = \xi'$.
- For each σ there is an injective function θ_σ from biases to biases such that: $\theta(\sigma * i) = \theta(\sigma) * \theta_\sigma(i)$.

A delocation is *positive* when ξ, ξ' have the same parity, *negative* otherwise.

If $\mathfrak{c} = \langle \ldots, (\sigma_p, I_p), \ldots \rangle$ is a proper chronicle of base $\vdash \xi$ (resp. $\xi \vdash$), one defines $\theta(\mathfrak{c})$ of base $\vdash \xi'$ (resp. $\xi' \vdash$) by $\theta(\mathfrak{c}) := \langle \ldots, (\theta(\sigma_p), \theta_{\sigma_p}(I_p)), \ldots \rangle$. In the improper case, one defines $\theta(\mathfrak{d} * \maltese) := \theta(\mathfrak{d}) * \maltese$.

If $\mathfrak{D}$ is a design of base $\vdash \xi$ (resp. $\xi \vdash$) one defines $\theta(\mathfrak{D}) = \{\theta(\mathfrak{c}); \mathfrak{c} \in \mathfrak{D}\}$, a design of base $\vdash \xi'$ (resp. $\xi' \vdash$).

Definition 63 (Delocation: behaviours). If $\mathbf{G}$ is a behaviour of base $\vdash \xi$ (resp. $\xi \vdash$) and θ is a delocation from ξ to ξ', one defines the behaviour $\theta(\mathbf{G})$ of base $\vdash \xi'$ (resp. $\xi' \vdash$) by

$$\theta(\mathbf{G}) := \sim\sim\{\theta(\mathfrak{D}); \mathfrak{D} \in \mathbf{G}\}. \tag{14.9}$$

The image under θ of the behaviour T is not T, unless θ is surjective. But it contains the unique material design of $\theta(\mathsf{T}) = \mathsf{T}$, i.e., the skunk; in general:

Proposition 27. $|\theta(\mathbf{G})| = \{\theta(\mathfrak{D})\,;\ \mathfrak{D} \in |\mathbf{G}|\}$.

Corollary 27.1. *The image of* $\mathbf{G}$ *under* θ *is a complete ethics for* $\theta(\mathbf{G})$.

If θ is not always defined, one speaks of a *partial* delocation. One must then make sure that $\theta(\mathfrak{D})$ is at least defined for *material* designs.

[2]One stumbles on a tenacious problem with the empty ramification (Section 14.2.5).

14.1.3 Internal completeness. We shall construct complete ethics for the connectives & and $\oplus$.

14.1.4 The mystery of incarnation. The result, stated in the binary case, holds in full generality.

Theorem 47 (Mystery of incarnation).

$$|\mathbf{G} \,\&\, \mathbf{H}| = |\mathbf{G}| \times |\mathbf{H}| \tag{14.10}$$

Proof. If $\mathfrak{D} \in \mathbf{G} \,\&\, \mathbf{H}$ is material, the two incarnations $\mathfrak{E} = |\mathfrak{D}|_{\mathbf{G}}$ and $\mathfrak{F} = |\mathfrak{D}|_{\mathbf{H}}$ are included in $\mathfrak{D}$. We conclude that $\mathfrak{E} \cup \mathfrak{F} \subset \mathfrak{D}$.

Conversely, if $\mathfrak{E}, \mathfrak{F}$ are respectively material in **G** and **H**, Proposition 26 shows that they are disjoint, hence their union is a design $\mathfrak{D}$ belonging to **G** and **H**, i.e., to **G** & **H**. If $\mathfrak{D}$ were not incarnated in **G** & **H**, then $\mathfrak{E}' \cup \mathfrak{F}' \subsetneq \mathfrak{D}$ for appropriate $\mathfrak{E}', \mathfrak{F}'$ and at least one of $\mathfrak{E}, \mathfrak{F}$ would not be incarnated.

To sum up, the material designs of **G** & **H** are exactly the unions of a material design of **G** and a material design of **H**, this decomposition being unique. □

There is something fishy about equation (14.10): the left-hand side is literally commutative, etc., while the cartesian product on the right is notoriously commutative, etc., only up to isomorphism. Indeed, among all isomorphic definitions of the product, I picked up the sole natural in that case:

$$X \uplus\!\!\!\!\times\, Y := \{x \cup y; x \in X, y \in Y\}. \tag{14.11}$$

This *locative* product is literally commutative, associative, with $\{\emptyset\}$ as neutral element. When all $x \in X$ are disjoint from all $y \in Y$, one can use the notation $X \times Y$. It is therefore a partial operation, which is to the set-theoretic cartesian product what our & is to the spiritual connective. A delocated, spiritual, version of the mystery of incarnation would be

$$|\mathbf{G} \,\&\, \mathbf{H}| \simeq |\mathbf{G}| \times |\mathbf{H}|. \tag{14.12}$$

Which is only an isomorphism; one must be rather fluent in category theory to express in what sense this isomorphism is natural. No problem with (14.10), the isomorphism being a degenerated version of an underlying equality!

Mystery of ideology: I never succeeded in explaining the mystery of incarnation to category-theorists: for them an equality is an undetected isomorphism[3]. However, to explain subtyping in category-theoretic terms is as desperate as explaining « revision » – i.e., the possibility of changing axioms – on the basis of classical logic, or the theory of evolution of species on the basis of Genesis.

[3] In the days of Pharaoh Akhenaten, it was forbidden to represent a man as other than his true likeliness, i.e., prognathic.

The locative product is to the cartesian product what union is to disjoint sum. And, by the way, as a cartesian product distributes – up to isomorphism – over a disjoint sum, the locative product distributes – literally – over the union:

$$X \boxtimes (Y \cup Z) = (X \boxtimes Y) \cup (X \boxtimes Z). \tag{14.13}$$

Observe also that $\wp(X \cup Y) = \wp(X) \boxtimes \wp(Y)$, *idem* for $\wp_{\text{fin}}$, which is reminiscent of the properties of the exponential « ! ».

An analogue of the mystery of incarnation in mathematics is the Chinese remainder theorem (Section 2.C.3), which relates a product to an intersection, under « locative » hypotheses: if p, q are mutually prime, then $p\mathbb{Z} \cap q\mathbb{Z} = pq\mathbb{Z}$.

The disjunction property. This is the dual of the mystery of incarnation.

Theorem 48 (Disjunction property). *If the indexing set is non-empty,*

$$\bigoplus_k \mathbf{G}_k = \bigcup_k \mathbf{G}_k. \tag{14.14}$$

Proof. The binary case is enough: if $\mathfrak{D} \in {\sim}{\sim}(\mathbf{G} \cup \mathbf{H}) \setminus (\mathbf{G} \cup \mathbf{H})$, then $\mathfrak{D}$ is neither polar to a $\mathfrak{E} \in {\sim}\mathbf{G}$ nor to a $\mathfrak{F} \in {\sim}\mathbf{H}$, that one can suppose both material. By the mystery of incarnation (Theorem 47) $\mathfrak{E} \cup \mathfrak{F}$ is a design in ${\sim}\mathbf{G} \cap {\sim}\mathbf{H}$, although not polar to $\mathfrak{D}$. □

We thus found a complete ethics for « Plus ». In particular, $\mathbf{G} \oplus \mathbf{H} = \mathbf{G} \cup \mathbf{H}$, is nothing but the old property « A cut-free proof of $A \oplus B$ is a proof of A or a proof of B ». This disjunction is not exclusive, since $\mathbf{G} \cap \mathbf{H} = \{\mathfrak{Dai}\}$; however, since the *daimon* is losing, it remains exclusive at the level of *winning* designs.

The presence of the *daimon* in any positive behaviour explains the restriction « non-empty indexing set ». The fact that two disjoint behaviours have anyway the *daimon* in common is obviously reminiscent of the case of the empty clique in the coherent space $X \oplus Y$ (Section 9.2.2).

Additive decomposition

Theorem 49 (Additive decomposition). *Every positive behaviour can be written in a unique way as a* $\bigoplus$ *of connected behaviours:*

$$\mathbf{G} = \bigoplus_{I \in \P\mathbf{G}} \mathbf{G}_I. \tag{14.15}$$

Proof. $\mathbf{G}_I$ is made of the $\mathfrak{D} \in \mathbf{G}$ beginning with $(\langle\rangle, I)$ as well as the *daimon*. Since $\mathbf{G} = \bigcup \mathbf{G}_I$ is a behaviour, ${\sim}{\sim}\mathbf{G}_I \subset \mathbf{G}$ and $\P\mathbf{G}_I = \{I\}$ forces $\mathbf{G}_I$ to be a behaviour. □

Corollary 49.1. *Every negative behaviour can be written in a unique way as a & of connected behaviours.*

$$\mathbf{G} = \mathop{\&}_{I \in \P\mathbf{G}} \mathbf{G}_I \tag{14.16}$$

14.1.5 Subtyping

Subtyping and incarnation. The *mystery of incarnation* relates the two readings of the additive conjunction, intersection and product: this concerns crucial questions, *subtyping* and *inheritance*, which are mistreated at layer -2.

$\mathbf{G} \mathbin{\&} \mathbf{H} \subset \mathbf{H}$ means that each object « of type $\mathbf{G} \mathbin{\&} \mathbf{H}$ » is also « of type $\mathbf{H}$ ». One is in the spirit of « records »: a component for $\mathbf{G}$, a component for $\mathbf{H}$ and perhaps others without interest, typically if $\mathfrak{D}$ belongs to $\mathbf{G} \mathbin{\&} \mathbf{H} \mathbin{\&} \mathbf{K}$. Thus, what is incarnation? This is the part of a design relative to a behaviour. Hence $|\mathfrak{D}|_{\mathbf{G}\&\mathbf{H}}$ only retains the part of $\mathfrak{D}$ relevant to $\mathbf{G}$, $\mathbf{H}$, while $|\mathfrak{D}|_{\mathbf{H}}$ forgets the information relative to $\mathbf{G}$. Since this information is disjoint, independent, one gets the result.

The « mystery » is made possible by the coexistence of two notions:

Existentialist: the official definition, for which an object is an object, hence if it is of type $\mathbf{G} \mathbin{\&} \mathbf{H}$ it is also of type $\mathbf{H}$. In this context, subtyping corresponds to inclusion.

Essentialist: the old category-theoretic, typed, version for which a pair $(a, b) \in A \mathbin{\&} B$ is not of type A: one indeed works with *material* designs. Subtyping still makes sense provided one destroys those parts which become useless, i.e., compute the incarnation in a supertype: what is known as a *coercion*.

Additive decomposition corresponds to a sort of generalised « record style »: each ramification I denotes a *field*, maybe absent. For a material design of $\mathbf{G} \mathbin{\&} \mathbf{H}$, there is a component for each field $I \in \P\mathbf{G} \cup \P\mathbf{H}$; the coercion from $\mathbf{G} \mathbin{\&} \mathbf{H}$ to $\mathbf{H}$ corresponds to the erasure of the I of $\P\mathbf{G}$. Such a coercion can be implemented, provided our behaviours have been delocated in disjoint *loci* $\xi \vdash$ (for $\mathbf{G} \mathbin{\&} \mathbf{H}$) and $\xi' \vdash$ (for $\mathbf{H}$), by a partial fax, induced by the delocation $\theta(\xi * \sigma) = \xi' * \sigma$. For instance, the pseudo-fax of example (13.6) corresponds to $\P\mathbf{H} = \{\{3, 7\}, \{4, 7\}\}$.

Incarnation and records. Imagine the following record:

$$\texttt{coord:}\ (3, 4) \qquad \texttt{colour:}\ \textit{green} \qquad \texttt{shape:}\ \textit{circle} \tag{14.17}$$

The fields `coord`, `colour`, `shape` are respectively encoded by the biases 2, 3, 8: they become negative behaviours, respectively included in: $\sim\!\sim(\mathfrak{Ram}_{(\langle\rangle,\{2\})})$, $\sim\!\sim(\mathfrak{Ram}_{(\langle\rangle,\{3\})})$ and $\sim\!\sim(\mathfrak{Ram}_{(\langle\rangle,\{8\})})$, (*infra*). Planar coordinates (m, n) are rendered by $\{2m, 2n + 1\}$, here $(3, 4)$ by $\{6, 9\}$; colours by numbers, e.g., *green* by 8; the shape *circle* is rendered by the bias 0. Our record becomes the negative design:

$$\dfrac{\dfrac{\dfrac{}{26\vdash}{\scriptstyle(26,\emptyset)}\qquad \dfrac{}{29\vdash}{\scriptstyle(29,\emptyset)}}{\vdash 2}{\scriptstyle(2,\{6,9\})} \qquad \dfrac{\dfrac{}{38\vdash}{\scriptstyle(38,\emptyset)}}{\vdash 3}{\scriptstyle(3,\{8\})} \qquad \dfrac{\dfrac{}{80\vdash}{\scriptstyle(80,\emptyset)}}{\vdash 8}{\scriptstyle(8,\{0\})}}{\langle\rangle\vdash}{\scriptstyle(\langle\rangle,\{\{2\},\{3\},\{8\}\})}$$

- The first branching (negative) is a list of fields (questions) 2, 3, 8.
- The next branchings (positive) yield answers – the values of the fields.
- Above, the design is the simplest possible, i.e., $\mathfrak{Skunk}$. In other terms, $\texttt{coord} := \uparrow \oplus_{m,n} (\downarrow \mathbf{T} \otimes \downarrow \mathbf{T})$, $\texttt{coulour} := \uparrow \oplus_n \downarrow \mathbf{T}$ and $\texttt{shape} := \uparrow \oplus_n \downarrow \mathbf{T}$.

Say that we are not interested in shapes, that only coordinates and colours matter: our design is of type $\texttt{coord} \,\&\, \texttt{colour}$. Shape is useless, one could replace the design with its incarnation:

$$\dfrac{\dfrac{\dfrac{}{26\vdash}{\scriptstyle(26,\emptyset)}\qquad \dfrac{}{29\vdash}{\scriptstyle(29,\emptyset)}}{\vdash 2}{\scriptstyle(2,\{6,9\})} \qquad \dfrac{\dfrac{}{38\vdash}{\scriptstyle(38,\emptyset)}}{\vdash 3}{\scriptstyle(3,\{8\})}}{\langle\rangle\vdash}{\scriptstyle(\langle\rangle,\{\{2\},\{3\}\})}$$

which corresponds to the truncated record

$$\texttt{coord:}\ (3,4); \quad \texttt{colour:}\ \textit{green} \tag{14.18}$$

Furthermore, if one forgets colour, one gets the incarnation

$$\dfrac{\dfrac{\dfrac{}{26\vdash}{\scriptstyle(26,\emptyset)}\qquad \dfrac{}{29\vdash}{\scriptstyle(29,\emptyset)}}{\vdash 2}{\scriptstyle(2,\{6,9\})}}{\langle\rangle\vdash}{\scriptstyle(\langle\rangle,\{\{2\}\})}$$

i.e., the record

$$\texttt{coord:}\ (3,4) \tag{14.19}$$

of type $\texttt{coord}$, while retaining only coordinates yields the incarnation

$$\dfrac{\dfrac{\dfrac{}{38\vdash}{\scriptstyle(38,\emptyset)}}{\vdash 3}{\scriptstyle(3,\{8\})}}{\langle\rangle\vdash}{\scriptstyle(\langle\rangle,\{\{3\}\})}$$

i.e., the record

$$\texttt{colour:}\ \textit{green} \tag{14.20}$$

of type `colour`. Up to incarnation, a record of type `coord` & `colour` is the pair (i.e., the disjoint union) of two records of types `coord` and `colour`.

The next example is a sort of fax which takes a record and which – without looking at the other fields – will repaint it in *black*, coded by 0. Of course, source and target must be delocated in disjoint ξ' and ξ.

$$\dfrac{\ldots \quad \dfrac{\begin{array}{c}\vdots\ \mathfrak{Fax}_{\xi' i,\xi i}\\ \ldots \xi' i \vdash \xi i \ldots\end{array}}{\vdash \xi', \xi I}\,(\xi',I) \qquad (I \neq \{3\}) \ldots \quad \dfrac{\dfrac{}{\xi 30 \vdash}\,(\xi 30,\emptyset)}{\vdash \xi', \xi 3}\,(\xi 3,\{0\})}{\xi \vdash \xi'}\,(\xi,\wp_f(\mathcal{N}))$$

Modulo normalisation, this design transforms any record $\xi' \vdash$ into the same record $\xi \vdash$, but painted in black, the colour coded $c \in \mathbb{N}$ becoming 0, i.e., black. The part « fax » recopies everything outside the field $\{3\}$, thus `coord,shape` but also the other fields if any.

As to planar coordinates, let us mention the alternative solution consisting of two fields, 0 for x-coord, 1 for y-coord,

$$\texttt{x-coord:}\ 3 \qquad \texttt{y-coord:}\ 4 \tag{14.21}$$

rendered by

$$\dfrac{\dfrac{\dfrac{\dfrac{}{\vdash 031}\,(031,\emptyset)}{03 \vdash}\,(03,\emptyset)}{\vdash 0}\,(0,\{3\}) \qquad \dfrac{\dfrac{\dfrac{}{\vdash 141}\,(141,\emptyset)}{14 \vdash}\,(14,\emptyset)}{\vdash 1}\,(1,\{4\})}{\langle\rangle \vdash}\,(\langle\rangle,\{\{0\},\{1\}\})$$

admitting coercions into records containing only `x-coord` or `y-coord`.

Discussion. I think that these examples should encourage us to revisit the theory of subtyping with *locative* glasses. It becomes obvious that categories are not adapted to records. They can provide us with ordered pairs (a, b), hence a projection, left or right, a notion deeply non-associative: a record by no ways decomposes into a left and a right component. The locative viewpoint offers much more interesting projections, e.g., « retain the sole ramifications of the form $\{i\}$ » or « retain those made of even biases ». Those are *locative* projections, with all the advantages of a literal associativity; if $\pi_{\mathcal{N}}$ denotes the projection of a negative design on the directory $\mathcal{N}$,

$$\pi_{\mathcal{N}}(\mathfrak{D}) := \{(\langle\rangle, I) * \mathrm{c};\ (\langle\rangle, I) * \mathrm{c} \in \mathfrak{D}, I \in \mathcal{N}\}, \tag{14.22}$$

it is immediate that $\pi_{\mathcal{M}} \circ \pi_{\mathcal{N}} = \pi_{\mathcal{M}\cap\mathcal{N}}$.

This methodological discussion (to be continued with a discussion on quantifiers, which deal with « intersection types ») should lead to pose *differently* – not necessarily under the form of completeness, of which one quickly sees the absolute limitations – the question of a *useful* syntax for subtyping. And one should not confine this discussion to the setting of ludics: I think that this applies *mutatis mutandis* to pure λ-calculus, provided one does what is never done: be careful with locations, i.e., with names of variables.

14.2 Multiplicatives

14.2.1 The adjunction

Definition 64 (Tensor product of designs). Let $\mathfrak{A}$, $\mathfrak{B}$ be positive designs; we define their *tensor product* $\mathfrak{A} \otimes \mathfrak{B}$:

- If $\mathfrak{A}$ or $\mathfrak{B}$ is a *daimon* then $\mathfrak{A} \otimes \mathfrak{B} = \mathfrak{Dai}$.
- Otherwise $\mathfrak{A}$, $\mathfrak{B}$ have respective first actions $(\langle\rangle, I)$ and $(\langle\rangle, J)$. If $I \cap J \neq \emptyset$, then $\mathfrak{A} \otimes \mathfrak{B} = \mathfrak{Dai}$. Otherwise, let us replace in $\mathfrak{A}$, $\mathfrak{B}$ the first action $(\langle\rangle, I)$ or $(\langle\rangle, J)$ with $(\langle\rangle, I \cup J)$, so as to get $\mathfrak{A}'$, $\mathfrak{B}'$. We finally define: $\mathfrak{A} \otimes \mathfrak{B} := \mathfrak{A}' \cup \mathfrak{B}'$.

Theorem 50 (Adjunction). *Let $\mathfrak{F}, \mathfrak{A}, \mathfrak{B}$ be designs, $\mathfrak{F}$ negative, $\mathfrak{A}$, $\mathfrak{B}$ positive. There is a* ***unique*** *negative design $(\mathfrak{F})\mathfrak{A}$ (not depending on $\mathfrak{B}$) such that*

$$\ll \mathfrak{F} \mid \mathfrak{A} \otimes \mathfrak{B} \gg \, = \, \ll (\mathfrak{F})\mathfrak{A} \mid \mathfrak{B} \gg . \tag{14.23}$$

Proof. If $\mathfrak{A} = \mathfrak{Dai}$, it is enough to take $(\mathfrak{F})\mathfrak{A} = \mathfrak{Dai}^-$. Otherwise, $\mathfrak{A}$ has a first action $(\langle\rangle, I)$, hence, for each $i \in I$, a subdesign $\mathfrak{A}_i$ of base $i \vdash$. With the convention of full negative directories (Section 13.3), $\mathfrak{F}$ admits a subdesign (maybe partial) $\mathfrak{F}_J$ of conclusion $\vdash J$ for each ramification J. For K a ramification, one defines $(\mathfrak{F})\mathfrak{A}_K$, with two cases:

$I \cap K = \emptyset$: one forms a net between $\mathfrak{F}_{I\cup K}$ and the $\mathfrak{A}_i$, of conclusion $\vdash K$. $(\mathfrak{F})\mathfrak{A}_K$ is by definition the normal form of this net.

$I \cap K \neq \emptyset$: $(\mathfrak{F})\mathfrak{A}_K := \mathfrak{Dai}$.

Unicity comes from separation. □

Proposition 28. *$\otimes$ is commutative, associative, with neutral element $\mathfrak{Bomb}^+$ (Section 13.8.3).*

Corollary 28.1. *$\mathfrak{F} \perp \mathfrak{A}$ iff the chronicle $\langle(\langle\rangle, \emptyset)\rangle$ belongs to $(\mathfrak{F})\mathfrak{A}$.*

Corollary 28.2.

$$((\mathfrak{F})\mathfrak{A})\mathfrak{B} = ((\mathfrak{F})\mathfrak{B})\mathfrak{A} = (\mathfrak{F})(\mathfrak{A} \otimes \mathfrak{B}) = (\mathfrak{F})(\mathfrak{B} \otimes \mathfrak{A}). \tag{14.24}$$

In ludics, the application of a function to its arguments does not depend on the order; this is because everybody already was given a location. In spiritual logic, e.g. in the category-theoretic world, $f(a)$ means that a has been put in contact with f by means of a delocation, hence $f(a)(b)$, $f(b)(a)$ are fundamentally distinct, i.e., non isomorphic. Here, each argument comes with its place, hence no ambiguity. Which we already remarked while discussing **IIS** in Section 12.6.2.

14.2.2 Multiplicative connectives

Definition 65 ($\otimes$, $\invamp$). If **G**, **H** are positive, we define

$$\mathbf{G} \otimes \mathbf{H} := \sim\sim\{\mathfrak{A} \otimes \mathfrak{B}; \mathfrak{A} \in \mathbf{G}, \mathfrak{B} \in \mathbf{H}\}. \tag{14.25}$$

If **G**, **H** are negative, we define

$$\mathbf{G} \invamp \mathbf{H} := \sim\{\mathfrak{A} \otimes \mathfrak{B}; \mathfrak{A} \in \sim\mathbf{G}, \mathfrak{B} \in \sim\mathbf{H}\}. \tag{14.26}$$

Proposition 29. *If* **G**, **H** *are negative, then*

$$\mathfrak{F} \in \mathbf{G} \invamp \mathbf{H} \iff \forall\mathfrak{A}(\mathfrak{A} \in \sim\mathbf{G} \Rightarrow (\mathfrak{F})\mathfrak{A} \in \mathbf{H}), \tag{14.27}$$

$$\mathfrak{F} \in \mathbf{G} \invamp \mathbf{H} \iff \forall\mathfrak{B}(\mathfrak{B} \in \sim\mathbf{H} \Rightarrow (\mathfrak{F})\mathfrak{B} \in \mathbf{G}). \tag{14.28}$$

Proof. Immediate consequence of (14.24). □

This essential proposition enables us to define the « Par » by means of linear implication, from $\sim$**G** into **H**, or from $\sim$**H** into **G**. *Both* ways are needed to prove the next theorem:

Theorem 51 (Associativity, distributivity $\otimes$, $\invamp$). *The connective $\otimes$ is literally commutative, associative, with neutral element* **1** *and absorber* **0**; *it distributes over the locative union* $\uplus$.

The connective $\invamp$ is literally commutative, associative, with neutral element $\bot$ *and absorber* $\top$; *it distributes over the locative intersection* $\Cap$.

Proof. The two parts of the theorem are strictly equivalent; this being said, try to prove directly the positive part: no way, because of the bipolars! So, let us prove the part concerning $\invamp$. The idea is that, between the equivalent formulations (14.27) and (14.28), at least one of them is available – i.e., does not use the bipolar. Thus, if I want to prove distributivity on the right, I will express $\invamp$ by means of functions from left to right, sending $\sim$**G** into **H**, or **K**, or **H** $\cap$ **K**. In the same way, in order to speak of **G** $\invamp$ (**H** $\invamp$ **K**), I will choose successively an argument in $\sim$**G**, then one in $\sim$**K**, which yields a result in **H**; the same that one would have got with (**G** $\invamp$ **H**) $\invamp$ **K**, where the arguments would have been chosen first in $\sim$**K**, next in $\sim$**G**, with the same output. □

The neutral of « ⊗ » is reduced to the sole $\mathfrak{Bomb}^+$, see (13.44). The neutral of « ⅋ » has the sole material design $\mathfrak{Dir}_{\{\emptyset\}}$, a.k.a. *negative bomb*:

$$\mathfrak{Bomb}^- := \dfrac{\dfrac{}{\vdash}\,\maltese}{\langle\rangle\vdash}\,(\langle\rangle,\{\emptyset\}) \tag{14.29}$$

Figure 14.2. $\mathfrak{Bomb}^-$.

14.2.3 Multiplicatives and directory

Proposition 30.

$$\P(\mathbf{G}\otimes\mathbf{H}) = \{I\cup J; I\in\P\mathbf{G}, J\in\P\mathbf{H}, I\cap J=\emptyset\}. \tag{14.30}$$

14.2.4 Internal completeness

Negative case. The direct characterisation of $\mathbf{G}⅋\mathbf{H}$ by means of (14.27) and (14.28) can be seen as an internal completeness. It has been obtained without hypothesis, which will no longer be the case with $\mathbf{G}\otimes\mathbf{H}$.

The projection lemma

Definition 66 (Reservoirs). A *reservoir* is a set of biases; the *reservoir* of $\mathbf{G}$ is defined as $\S\mathbf{G} := \bigcup\P\mathbf{G}$.

Let $\mathbb{X}\subset\mathbb{N}$ be a reservoir. Every positive design $\mathfrak{A}$ of first action $(\langle\rangle, K)$ is uniquely written as the tensor product $\mathfrak{D}\otimes\mathfrak{B}$ of a design $\mathfrak{D}$ beginning with $(\langle\rangle, K\cap\mathbb{X})$ and a design $\mathfrak{B}$ beginning with $(\langle\rangle, K\setminus\mathbb{X})$.

Definition 67 (Projection). The design $\mathfrak{D}$, denoted by $\mathfrak{A} \upharpoonright \mathbb{X}$, is the *projection* of $\mathfrak{A}$ on $\mathbb{X}$; we also define $\mathfrak{Dai} \upharpoonright \mathbb{X} = \mathfrak{Dai}$. If $\mathbf{E}$ is an ethics, the set $\{\mathfrak{A} \upharpoonright \mathbb{X}; \mathfrak{A} \in \mathbf{E}\}$ is called the *projection* of $\mathbf{E}$ on $\mathbb{X}$ and noted $\mathbf{E} \upharpoonright \mathbb{X}$.

Theorem 52 (Projection). *If* $\mathbf{E}$ *is* connected, *projection commutes to the bipolar:* $\sim\sim\mathbf{E} \upharpoonright \mathbb{X} = \sim\sim(\mathbf{E} \upharpoonright \mathbb{X})$.

Proof. Let K be the *ramification* of $\mathbf{E}$; and let $I := K \cap \mathbb{X}$:

$\sim\sim\mathbf{E} \upharpoonright I \subset \sim\sim(\mathbf{E} \upharpoonright I)$**:** let $\mathfrak{F} \in \sim(\mathbf{E} \upharpoonright I)$; if $\mathfrak{F}$ is material, we can replace its first action $(\langle\rangle, I)$ with $(\langle\rangle, K)$, yielding $\mathfrak{F}'$ (this is a « weakening »). If $\mathfrak{D} = \mathfrak{A} \upharpoonright I$, it is clear that $\ll \mathfrak{A} | \mathfrak{F}' \gg = \ll \mathfrak{D} | \mathfrak{F} \gg$ and if we let $\mathfrak{A}$ run through $\mathbf{E}$ that $\mathfrak{F}' \in \sim\mathbf{E}$. We redo the same with the weaker hypothesis $\mathfrak{A} \in \sim\sim\mathbf{E}$ and we conclude that $\mathfrak{D} \perp \mathfrak{F}$, i.e., that $\mathfrak{D} \in \sim\sim(\mathbf{E} \upharpoonright I)$.

$\sim\sim(\mathbf{E} \upharpoonright I) \subset \sim\sim\mathbf{E} \upharpoonright I$**:** let $\mathfrak{F}' \in \sim\mathbf{E}$ and $\mathfrak{F} = (\mathfrak{F}')\mathfrak{B}$, with $\mathfrak{B} := \mathfrak{Ram}_{(\langle\rangle, K \setminus I)}$. If $\mathfrak{A} \in \mathbf{E}$ and $\mathfrak{D} = \mathfrak{A} \upharpoonright I$, then $\mathfrak{A} \preceq \mathfrak{D} \otimes \mathfrak{B}$ hence $\mathfrak{F}' \perp \mathfrak{D} \otimes \mathfrak{B}$, which yields by adjunction $\mathfrak{F} \perp \mathfrak{D}$ and we conclude that $\mathfrak{F} \in \sim(\mathbf{E} \upharpoonright I)$. If $\mathfrak{D} \in \sim\sim(\mathbf{E} \upharpoonright I)$ we can now conclude that $\mathfrak{D} \otimes \mathfrak{B} \perp \mathfrak{F}'$, hence $\mathfrak{D} \otimes \mathfrak{B} \in \sim\sim\mathbf{E}$ and $(\mathfrak{D} \otimes \mathfrak{B}) \upharpoonright I = \mathfrak{D}$ finally yields $\mathfrak{D} \in \sim\sim\mathbf{E} \upharpoonright I$. □

Extraneousness

Definition 68 (Extraneousness). $\mathbf{G}$ and $\mathbf{H}$ of the same polarity are *alien* when $\S\mathbf{G} \cap \S\mathbf{H} = \emptyset$.

If $\mathbf{G}, \mathbf{H}$ are positive let $\mathbf{G} \circledcirc \mathbf{H} := \{\mathfrak{A} \otimes \mathfrak{B}; \mathfrak{A} \in \mathbf{G}, \mathfrak{B} \in \mathbf{H}\}$:

Theorem 53 (Extraneousness). *If* $\mathbf{G}, \mathbf{H}$ *are alien, then* $\mathbf{G} \circledcirc \mathbf{H}$ *is a complete ethics for* $\mathbf{G} \otimes \mathbf{H}$.

Proof. One writes $\mathbf{G} \circledcirc \mathbf{H} = \bigcup_K (\bigcup_{I \cup J = K} \mathbf{G}_I \circledcirc \mathbf{H}_J)$. By the disjunction property, one is reduced to showing that $\bigcup_{I \cup J = K} \mathbf{G}_I \circledcirc \mathbf{H}_J$ is a behaviour. But, if $\mathbb{X}$, $\mathbb{Y}$ are the respective reservoirs of $\mathbf{G}$, $\mathbf{H}$, $\bigcup_{I \cup J = K} \mathbf{G}_I \circledcirc \mathbf{H}_J = \mathbf{G}_{K \cap \mathbb{X}} \circledcirc \mathbf{H}_{K \cap \mathbb{Y}}$ and the theorem reduces to the connected case.

Let us thus assume $\mathbf{G}, \mathbf{H}$ connected, of respective ramifications I and J, with $I \cap J = \emptyset$. If $\mathfrak{A} = \mathfrak{D} \otimes \mathfrak{B} \in \mathbf{G} \otimes \mathbf{H}$, then $\mathfrak{D} \in \mathbf{G} = \sim\sim\mathbf{G}$ and $\mathfrak{B} \in \mathbf{H} = \sim\sim\mathbf{H}$ by the projection lemma, hence $\mathfrak{A} = \mathfrak{D} \otimes \mathfrak{B} \in \mathbf{G} \circledcirc \mathbf{H}$. □

14.2.5 The constant 1. Observe that one can be alien without being disjoint: due to the empty ramification, $\S\mathbf{G} \cap \S\mathbf{H} = \emptyset$ does not imply $\P\mathbf{G} \cap \P\mathbf{H} = \emptyset$. One meets a problem with $\mathbf{1}$, which is self-alien but not self-disjoint. There is no way to delocate it, for instance, $\varphi(\mathbf{1}) \oplus \psi(\mathbf{1})$ makes no sense, which limits the interest of the definition of Section 14.1.2.

As $\mathbf{1} \oplus \mathbf{1}$ no longer works, one can fall back on $\varphi(\downarrow \mathbf{T}) \oplus \psi(\downarrow \mathbf{T})$, which works well, especially in the presence of weakening, a natural property in ludics.

To sum up, layer -3 has no real multiplicative unit, as we already remarked in Section 12.3.3: categories are too optimistic on that issue. Ten years after the conception of ludics, I consider that the empty ramification has no room here. Everything works really better without it!

14.3 Quantifiers

Technically speaking, quantifiers have already been introduced with the notations $\bigcap_k$, $\uplus_k$. But the departure spiritual/locative is different.

First, a delicate point: the word « quantifier » – say universal – corresponds to two opposite approaches that tradition can hardly separate:

First-order: a big conjunction, possibly uniform in a sense to make precise: this is the viewpoint of model theory and also of German-style proof-theory – where quantification is handled by an infinite rule, the ω-rule:

$$\frac{\ldots \vdash \Gamma, A[\bar{n}/x] \ldots}{\vdash \Gamma, \forall x A} \tag{14.31}$$

This sort of translation (rereading of Gentzen by Schütte) is problematic (infinite syntax) (Section 3.C). Just remark that it is actually about *numerical* quantification.

Second-order: an intersection: this is the viewpoint of the « forgetful » interpretation of system **F** (Section 6.1.2), the one I shall follow here.

The first-order approach is spiritual – i.e., supposes many delocations and is defined only up to isomorphism –, while the second-order approach is locative. Since spiritualism is a hegemonic tendency of logic, second-order has been treated as a poor relative of first-order... Thus, how to understand that one is complete, the other is not? The word « quantifier » is one more Orwellian trap; but in that case, the amalgamation is too ancient to allow an efficient reaction at the level of terminology.

In what follows, one restricts to the locative, i.e., second-order, quantifier.

14.3.1 Universal quantification

Definition 69 (Universal). Let $\mathbf{G}_d$ be a family of behaviours of the *same* polarity, indexed by a set $\mathbb{D}$. We define $\forall d \in \mathbb{D}\ \mathbf{G}_d$ as the intersection $\bigcap_{d \in \mathbb{D}} \mathbf{G}_d$.

The cardinal of $\mathbb{D}$ is arbitrary, for instance 0 (which yields $\mathbf{T}^{\epsilon}$), 2 (an « intersection type »), $\aleph_0$ (first-order quantification handled locatively), and $2^{2^{\aleph_0}}$. Indeed, in the second-order case, $A[X]$ is interpreted by a behaviour $\mathbf{H_X}$ depending on a parameter $\mathbf{X}$ which interprets the variable X: $\mathbf{X}$ is an *arbitrary* behaviour of base $\vdash \langle\rangle$. If A is a formula depending on X, its interpretation $\mathbf{H_X}$ will use the logical connectives defined in this chapter. The interpretation of $\forall X A[X]$ will thus be the intersection of the $\mathbf{H_X}$, where $\mathbf{X}$ ranges among *all* behaviours of base $\vdash \langle\rangle$, a set of enormous cardinality, $2^{2^{\aleph_0}}$.

14.3.2 Existential quantification

Definition 70 (Existential). Let $\mathbf{G}_d$ be a family of behaviours of the *same* polarity, indexed by a set $\mathbb{D}$. One defines $\exists d \in \mathbb{D}\; \mathbf{G}_d$ as the bipolar of $\bigcup_{d \in \mathbb{D}} \mathbf{G}_d$:

$$\exists d \in \mathbb{D}\; \mathbf{G}_d := \sim\sim\Big(\bigcup_{d \in \mathbb{D}} \mathbf{G}_d\Big). \tag{14.32}$$

The unfortunate existentials will have a hard life: there is no complete ethics in this case. When the indexing set $\mathbb{D}$ is something like the set of all behaviours – as in second-order –, Cantor's theorem will impede any form of completeness. Which is consistent with the fact that second-order existential quantification yields Π^0_1 formulas, incomplete by Gödel's theorem. The circle closes when one remembers what Gödel owes to Cantor!

We have just described an incompleteness of enumerative nature. But the problem begins when $\mathbb{D}$ has two elements, i.e., with « intersection types », see equation (14.53) below.

If you are not *a priori* convinced of the incompleteness of $\exists$, observe that, for any ethics $\mathbf{E}$,

$$\sim\sim\mathbf{E} = \exists \mathfrak{E} \in \mathbf{E}\; \{\mathfrak{E}'; \mathfrak{E} \preceq \mathfrak{E}'\}. \tag{14.33}$$

In other words, the bipolar is an existential quantification, hence, if incompleteness exists, this must be ascribed to existential quantification.

14.3.3 Prenex forms

Commutation theorems

Theorem 54 (Commutation). *$\forall d$ commutes to all connectives but $\exists$; thus, ludics admits prenex forms.*

Essentially, $\forall$ commutes with any complete connective. The general idea is to use the completeness of the ethics $\mathbf{E}_d$ to replace $\forall d \sim\sim\mathbf{E}_d$ with $\bigcap_d \mathbf{E}_d$: everything is almost immediate. The most important cases are $\forall/\oplus$ and $\forall/\otimes$ (dually $\exists/\&$ and $\exists/\invamp$). The theorem enunciates unary commutations, typically

$$\forall d(\mathbf{G}_d \oplus \mathbf{H}) = (\forall d \mathbf{G}_d) \oplus \mathbf{H}, \tag{14.34}$$

but one often gets binary commutations – so strong that the commutation $\forall/\oplus$ (14.37) even contradicts classical logic!

I take an example, that of the commutation $\forall/\oplus$. Here, the connective $\oplus$ has its spiritual meaning:

$$\mathbf{G} \oplus \mathbf{H} = \sim\sim(\varphi(\mathbf{G}) \cup \psi(\mathbf{H})) \tag{14.35}$$

with the delocations φ, ψ of Section 14.1.2. Hence, in $\mathbf{G}_d \oplus \mathbf{H}_d$, the delocations do not depend on d. By the way, one could not do differently, since $\mathbb{D}$ can be of enormous cardinality, say $2^{2^{\aleph_0}}$ and one will not find enough pairwise disjoint delocations φ_d, ψ_d.

This is nothing but good old *realisability*, (Section 6.C.3): the realiser of a disjunction is of the form $(1, r)$ or $(2, r)$, where the numbers $1, 2$ indicate fixed delocations of r; the implication $\forall X(A[X] \vee B[X]) \Rightarrow (\forall X A[X]) \vee (\forall X B[X])$ is thus realised by the identity function. Around 1970, this type of remark was part of the weirdnesses of realisability, hardly separable from the mistakes linked to – say – the bad behaviour of negation.

Technically speaking, the result is based on a subtle interplay between locative and spiritual aspects:

- The disjunction property, which removes the bipolar in (14.35).
- The disjointedness of the reservoirs $\mathbb{X} = \varphi(\mathbb{N})$ and $\mathbb{Y} = \psi(\mathbb{N})$, to the effect that the two cases are well distinguished in $\mathbb{X} \cup \mathbb{Y}$.

For questions of legibility, I will not use delocations. Which means that I suppose that, for all $d \in \mathbb{D}$ $\S\mathbf{G}_d \subset \mathbb{X}, \S\mathbf{H}_d \subset \mathbb{Y}$; I shall not mention the indexing set $\mathbb{D}$, unless necessary.

Commutations of $\forall$

$$\forall/\downarrow: \quad \forall d \downarrow \mathbf{G}_d = \downarrow \forall d\, \mathbf{G}_d \tag{14.36}$$
$$\forall/\oplus: \quad \forall d\, (\mathbf{G}_d \oplus \mathbf{H}_d) = (\forall d\, \mathbf{G}_d) \oplus (\forall d\, \mathbf{H}_d) \tag{14.37}$$
$$\forall/\otimes: \quad \forall d\, (\mathbf{G}_d \otimes \mathbf{H}_d) = (\forall d\, \mathbf{G}_d) \otimes (\forall d\, \mathbf{H}_d) \tag{14.38}$$
$$\forall/\uparrow: \quad \forall d \uparrow \mathbf{G}_d = \uparrow \forall d\, \mathbf{G}_d \tag{14.39}$$
$$\forall/\&: \quad \forall d\, (\mathbf{G}_d \,\&\, \mathbf{H}_d) = (\forall d\, \mathbf{G}_d) \,\&\, (\forall d\, \mathbf{H}_d) \tag{14.40}$$
$$\forall/\parr: \quad \forall d\, (\mathbf{G}_d \parr \mathbf{H}) = (\forall d\, \mathbf{G}_d) \parr \mathbf{H} \tag{14.41}$$
$$\forall/\forall: \quad \forall d \in \mathbb{D}\, \forall e \in \mathbb{E}\, \mathbf{G}_{d,e} = \forall e \in \mathbb{E}\, \forall d \in \mathbb{D}\, \mathbf{G}_{d,e} \tag{14.42}$$

The commutations (14.39)–(14.42) are just invertibility. On the other hand, the first three are novel:

(14.36): immediate; observe that $\forall$ is positive on the left, negative.

(14.37): if $e \in \mathbb{D}$ and $\mathfrak{D} \in \forall d(\mathbf{G}_d \oplus \mathbf{H}_d)$ is proper, $\mathfrak{D} \in \mathbf{G}_e \oplus \mathbf{H}_e$ and, by the disjunction property, belongs to $\mathbf{G}_e$ or to $\mathbf{H}_e$. This remains true for $e' \in \mathbb{D}$ and the locative hypotheses force $\mathfrak{D}$ to « stay on the same side ».

(14.38): if $\mathfrak{D} \in \forall d(\mathbf{G}_d \otimes \mathbf{H}_d)$ then $\mathbf{G}_d \otimes \mathbf{H}_d = \mathbf{G}_d \copyright \mathbf{H}_d$ and $\mathfrak{D} \upharpoonright \mathbb{X}$ belongs to $\forall d \mathbf{G}_d$ for all d, hence to $\forall d \mathbf{G}_d$, etc.

The equation (14.37) is violently anti-classical: classically $\forall X(X \vee \neg X)$, while $\forall X X$ and $\forall X \neg X$ are false: $\forall$ commutes more « politely » with the multiplicative disjunction $⅋$!

Commutations of ∃

$$\exists/\uparrow : \quad \exists d \uparrow \mathbf{G}_d = \uparrow \exists d\, \mathbf{G}_d \tag{14.43}$$
$$\exists/\& : \quad (\exists d\, \mathbf{G}_d)\ \&\ (\exists d\, \mathbf{H}_d) = \exists d\, (\mathbf{G}_d\ \&\ \mathbf{H}_d) \tag{14.44}$$
$$\exists/⅋ : \quad \exists d\, (\mathbf{G}_d ⅋ \mathbf{H}_d) = (\exists d\, \mathbf{G}_d) ⅋ (\exists d\, \mathbf{H}_d) \tag{14.45}$$
$$\exists/\downarrow : \quad \exists d \downarrow \mathbf{G}_d = \downarrow \exists d\, \mathbf{G}_d \tag{14.46}$$
$$\exists/\oplus : \quad \exists d\, (\mathbf{G}_d \oplus \mathbf{H}_d) = (\exists d\, \mathbf{G}_d) \oplus (\exists d\, \mathbf{H}_d) \tag{14.47}$$
$$\exists/\otimes : \quad \exists d\, (\mathbf{G}_d \otimes \mathbf{H}) = (\exists d\, \mathbf{G}_d) \otimes \mathbf{H} \tag{14.48}$$
$$\exists/\exists : \quad \exists d \in \mathbb{D}\, \exists e \in \mathbb{E}\, \mathbf{G}_{d,e} = \exists (d,e) \in \mathbb{D} \times \mathbb{E}\, \mathbf{G}_{d,e} \tag{14.49}$$

14.3.4 Discussion. Finally, a quantifier commutes with everything but its dual:

$$\forall d \in \mathbb{D}\, \exists e \in \mathbb{E}\, \mathbf{G}_{d,e} = \exists f \in \mathbb{E}^{\mathbb{D}}\, \forall d \in \mathbb{D}\, \mathbf{G}_{d,f(d)} \tag{14.50}$$

is completely wrong, even for $\mathbb{D}$ finite: since $\exists$ is very badly incomplete, hence – in set-theoretic terms – one is not with $\forall\exists$, but with $\forall{\sim}{\sim}\exists$.

Among the most extravagant principles that I just stated:

$$\exists d \forall e(\varphi(\mathbf{G}_d) \multimap \uparrow \varphi'(\mathbf{G}_e)) \tag{14.51}$$

obtained by commutation from

$$(\forall d\ \varphi(\mathbf{G}_d) \multimap \forall e \uparrow \varphi'(\mathbf{G}_e)). \tag{14.52}$$

Which could be written in the setting of « intersection types » as

$$(\varphi(\mathbf{G}) \multimap \uparrow \varphi'(\mathbf{G} \cap \mathbf{H})) \uplus (\varphi(\mathbf{H}) \multimap \uparrow \varphi'(\mathbf{G} \cap \mathbf{H})) \tag{14.53}$$

a very surprising equation.

Locative existentials do not enjoy the existence property, but this has no importance. Indeed, the *useful* existence property concerns numerical quantification, which is not locative: a numerical existence can be written $\exists x(x \in \mathbb{N} \otimes \ldots)$.

Whether the quantifier $\exists x$ is spiritual or locative, this hardly matters; what will make the decision is the formula $x \in \mathbb{N}$: it is enough to have completeness for $\exists x\, x \in \mathbb{N}$. I have a tendency to think that first-order quantification is – in the constructive world – a mistake: its only use is in the numerical case and its justification outside this setting is only a servile imitation of the classical case.

Prenex forms cannot persist in the context of exponentials; one would indeed get $\forall d \neg\neg \mathbf{G}_d = \neg\neg \forall d \mathbf{G}_d$. But, since we have $\forall X \neg\neg(X \oplus \neg X)$, one would get by commutation (14.37), $\neg\neg(\forall X\ X \oplus \forall X\ \neg X)$, a real contradiction.

Finally, I would like to mention again the following problem (Section 11.C.3): extend proof-nets with quantifiers so as to take into account the prenex forms, especially (14.38) and (14.45). More generally, « take seriously » locative quantifiers. They have no « semantical » meaning at layer -1, but they have surprising properties; and since they are part of a globally deductive setting, these astonishing properties may have consequences in the traditional « semantic » sense. But how to use them? I confess that I don't know[4].

14.A Faithfulness

14.A.1 Interpretation of perfect logic. Second-order perfect logic, without the constant **1** (and its dual $\perp$), but polarised, i.e., with the adjunction of the shifts $\updownarrow$, is naturally interpreted in ludics. One must select adequate locations and infinite reservoirs:

- We begin with the formula A which interests us. We locate it in $\vdash \langle\rangle$ or in $\langle\rangle \vdash$, according to its polarity; with the reservoir $\mathbb{N}$.
- If B is located in $\vdash \xi$ (resp. $\xi \vdash$), with the reservoir $\mathbb{X}$ and if $B = C \oplus D$, $C \otimes D$ (resp. $B = C \,\&\, D, C \parr D$), then C and D are located in the same place; each of them receives an infinite « half » of $\mathbb{X}$.
- If B is located in $\vdash \xi$ (resp. $\xi \vdash$), with the reservoir $\mathbb{X}$ and if $B = \forall X C$, $\exists X C$, then C gets the same location; and the same $\mathbb{X}$.
- If B is located in $\vdash \xi$ (resp. $\xi \vdash$), with the reservoir $\mathbb{X}$ and if $B = \downarrow C$ (resp. $B = \uparrow C$), then C is located in $\xi * a \vdash$ (resp. $\vdash \xi * a$), where a is the smallest element of $\mathbb{X}$; with the reservoir $\mathbb{N}$.

An « occurrence » of the variable X gets a location $\vdash \xi$ (or $\xi \vdash$) and a reservoir $\mathbb{X}$, enumerated by a monotonic function f; let φ be the delocation:

$$\begin{aligned} \varphi(\langle\rangle) &:= \xi, \\ \varphi(\langle i \rangle * s) &:= \xi * \langle f(i) \rangle * s. \end{aligned} \tag{14.54}$$

[4] A few years later after writing those lines, I have some doubts as to these commutations, which rest too much on ludics and seem more problematic in GoI. It might be the case that some of them ($\forall/\otimes$) are mere accidents; I would remain more optimistic as to $\forall/\oplus$.

The « occurrence » X is then interpreted by $\varphi(\mathbf{X})$ or $\varphi(\sim\mathbf{X})$ (according to the case), where $\mathbf{X}$ is an arbitrary behaviour of base $\vdash \langle\rangle$.

From the interpretation of literals, our reconstruction, as well as the shift of Section 13.9, enables us to associate a behaviour $\mathbf{B}$ to any subformula B of A. Moreover, if A is closed, $\mathbf{A}$ depends upon no parameter $\mathbf{X}$: such parameters disappear with quantifications.

14.A.2 The result. « Full completeness » conjectures can be stated as follows: if a belongs to the interpretation $\mathbf{A}$ of A, then a comes from a proof of A. Let us try to formulate this in our case; we dissect the naïve formulation:

If $\mathfrak{D} \in \mathbf{A}$, then $\mathfrak{D}$ comes from a (cut-free) proof of A.

Free variables: The first reef lies within the free variables of A; in general, the authors of full completeness results ask A to be a first order and the object $a \in \mathbf{A}$ to be *parametric*, i.e., to verify some uniformity condition, see for instance [58]. Remembering the analysis done in Section 2.A.3, one will rather replace A with its universal closure, which enables one to get rid of free variables.

Daimon: One should pay attention to the *daimon*: either prove results limited to *winning* designs, or – which is more natural – prove a more general result w.r.t. a syntax with a *daimon*, that one easily specialises afterwards.

Logical complexity: One must then remark that this completeness cannot hold beyond Π^1. Indeed, define the *truth* of a closed formula A:

Definition 71 (Truth). A is *true* iff $\mathbf{A}$ contains a winning design.

One already observed (Section 13.7.4), that A and $\sim A$ cannot be both true. What interests us here, is the following corollary of our supposed full completeness, obtained by forgetting both design and proof:

If A is true, then A is provable.

This is completeness of layer -1, forever limited to the Π^1 setting. Prenex form results enable one to bring quantifiers back *in front of* A: the general Π^1 case reduces to the universally quantified first-order formulas. Equivalences in the style of (14.37), which are not Π^1, are true without being provable.

Incarnation: If ludics were a category-theoretic setting, we would be done; here, we must take into account the *subtyping* implicit in the notion of incarnation. Take the negative behaviour $\mathbf{T}$; it interprets the constant $\top$, which has only one cut-free proof, the axiom $\vdash \top$. This behaviour contains $2^{\aleph_0}$ designs, many of them winning.

They are not represented in syntax and, by the way, no « reasonable » syntax could account for them. On the other hand, if one restricts to the sole material design of **T**, the *skunk* $\mathfrak{Skunk}$, completeness holds, since the axiom is precisely interpreted by $\mathfrak{Skunk}$.

Finally, one reaches the following statement:

Theorem 55 (Faithfulness). *If A is closed and Π^1, if $\mathfrak{D} \in \mathbf{A}$ is winning and material, then $\mathfrak{D}$ is the interpretation of a proof of A.*

Proof. The good news is that the result has been proved in the last chapters of [51]. The bad news is that the proof is very technical – whereas it should be no more than an exploitation of internal completeness. One must complicate the interpretation, behaviours becoming *bihaviours*, see next section. □

The proof requests the introduction of *sequents of behaviours*. For instance, if $\mathbf{G}, \mathbf{H}$ are positive behaviours of disjoint respective bases ξ, ξ', the behaviour-sequent $\mathbf{G} \vdash \mathbf{H}$, of base $\xi \vdash \xi'$ is defined by

$$\mathfrak{F} \in \mathbf{G} \vdash \mathbf{H} \iff \forall \mathfrak{D} \in \mathbf{G} \quad [\![\mathfrak{F}, \mathfrak{D}]\!] \in \mathbf{H}. \tag{14.55}$$

We would easily arrive at « proving » an incarnated design by means of internal completeness results... if all of this was not stumbling on literals.

14.B Bihaviours

14.B.1 The problem of literals. The typical problem occurs with a base of the form $X \vdash X \oplus X$; we suppose that the base is $\xi \vdash \xi'$ and that the reservoirs associated to the three « occurrences » of X are respectively $\mathbb{N}, 3\mathbb{N}, 3\mathbb{N}+1$. Here, our hypothesis will be that the design (winning and material) $\mathfrak{F}$ is such that, for each behaviour $\mathbf{X}$ of base $\vdash \langle\rangle$ – respectively delocated in $\mathbf{X}', \mathbf{X}'', \mathbf{X}'''$ – we have $\mathfrak{F} \in \mathbf{X}' \vdash \mathbf{X}'' \oplus \mathbf{X}'''$. We would like to prove that $\mathfrak{F}$ comes from a fax $\mathfrak{F}'' \in \mathbf{X}' \vdash \mathbf{X}''$ or a fax $\mathfrak{F}''' \in \mathbf{X}' \vdash \mathbf{X}'''$, which correspond to the two proofs of $X \vdash X \oplus X$.

If $\mathbf{X}$ is such that $\S\mathbf{X} = \{I\}$, then $[\![\mathfrak{F}, \mathfrak{Ram}_I]\!] \in \mathbf{X}'' \oplus \mathbf{X}'''$. Thus, the chronicles of length 2 of $\mathfrak{F}$ are of the form $\langle(\xi, I), (\xi', 3I)\rangle$ or $\langle(\xi, I), (\xi', 3I+1)\rangle$. But there is no reason why the departure $3I/3I+1$ should be the same for all values of I; nothing opposes the choice $\langle(\xi, I), (\xi', 3I)\rangle$ for, say, $\sharp(I)$ even and $\langle(\xi, I), (\xi', 3I+1)\rangle$ for $\sharp(I)$ odd.

14.B.2 Uniformity, external version. Let us equip behaviours with a partial equivalence relation (PER), i.e., a symmetric and transitive relation. We will require a condition of *uniformity*, consisting in saying that equivalence is preserved by $\mathfrak{F}$.

The equivalence on a « $\oplus$ » is defined in such a way that – apart from the *daimon* – nothing from $\mathbf{X}''$ is equivalent with nothing of $\mathbf{X}'''$ in $\mathbf{X}'' \oplus \mathbf{X}'''$; hence, if the PER on **X** is such that $\mathfrak{Ram}_I$, $\mathfrak{Ram}_J$ are equivalent, then $\mathfrak{F}$ can only be uniform by answering in the same way: $3I, 3J$ or $3I + 1, 3J + 1$.

Which is only half satisfactory, since this uniformity is the comeback *in extremis* of the external view, of arbitrariness, of essence. Moreover, at the occasion of a difficult technical question, but slightly sordid. Is essence reduced to hide behind obscure considerations?

14.B.3 Bihaviours. One limits the losses with *bihaviours*. This technical extension does not deserve a detailed description:

- One considers the *partial* designs of **G**. By this one means a design (partial) included in a design of **G**.
- A PER on partial designs of **G** induces a *polar* PER on partial designs of $\sim$**G**:

$$\mathfrak{F} \sim \mathfrak{F}' \;:\Longleftrightarrow\; \forall \mathfrak{D}, \mathfrak{D}' \quad \mathfrak{D} \sim \mathfrak{D}' \Rightarrow [\![\mathfrak{F}, \mathfrak{D}]\!] = [\![\mathfrak{F}', \mathfrak{D}']\!]. \tag{14.56}$$

A *bihaviour* is a behaviour together with a PER equal to its bipolar. One requires that $\mathfrak{Dai}$ is only equivalent to itself, the same for $\mathfrak{Fid}$ (positive case) and that $\mathfrak{Skunk} \sim \mathfrak{Skunk}$, $\mathfrak{Dai}^- \sim \mathfrak{Dai}^-$ (negative case).

Definition 72 (Uniformity). If **G** is a bihaviour, a design $\mathfrak{D} \in$ **G** is *uniform* when it is equivalent to itself, i.e., when $\mathfrak{D} \sim_{\mathbf{G}} \mathfrak{D}$.

One redoes the reconstruction with bihaviours. One must, in each case, define an appropriate PER. For instance, in the case of a « Par »:

$$\mathfrak{F} \sim \mathfrak{F}' \;:\Longleftrightarrow\; \forall \mathfrak{D}, \mathfrak{D}' \quad \mathfrak{D} \sim \mathfrak{D}' \Rightarrow \mathfrak{F}(\mathfrak{D}) \sim \mathfrak{F}'(\mathfrak{D}'). \tag{14.57}$$

Which should be compared with equation (14.56). One understands that the new notion will enable us to prove Theorem 55 under the more correct form:

Theorem 56 (Faithfulness). *If A is closed and Π^1, if $\mathfrak{D} \in$ **A** is winning, incarnated and uniform, then $\mathfrak{D}$ is the interpretation of a proof of A.*

This being said, the proof remains technical, with beautiful constructions of bihaviours. But can technical ingenuity compensate for a deficiency of the global architecture? I don't think so.

14.B.4 Discussion. One has the impression that uniformity is a superfluous property, that a more refined analysis, new ideas as to designs, could help us to relinquish. The intuition is that something is wrong in the analysis of $\mathfrak{F} \in \mathbf{X}' \vdash \mathbf{X}'' \oplus \mathbf{X}'''$: *this decomposition of* $\mathbf{X}'$ *should not be allowed.* If it is sure that the destiny of X is to be replaced with a behaviour, one anyway commits a crime against potentiality in this case: one should not « actualise ». But how can we say that X « is more than all possible behaviours », how to avoid this list of all possible I without losing separation, without entering « intensional » fiddling? One will find a hint in Section 17.5.4.

There are however other reasons for uniformity, let us mention:

Spiritual quantification: the quantifier $\forall x$ appears as a product (non-locative version), of which one can consider the *uniform* version, which is the sole complete one (Section 6.A.2). This requires a quantification over domains which reintroduces uniformity, see [28].

Non-uniform exponentials: this is essentially for questions of *separation* that one may think of modifying the exponential so as to make it look closer to an infinite tensor product (Section 15.B). Here too, we should use uniformity.

This being said, the progress in science is first that of *questions*. Thus, spiritual first-order quantification offers a limited interest; one will not justify uniformity on such limited grounds. Exponentials are a more serious argument; but, while we shall soon – in the next chapters – call into question exponentials, shouldn't we by the way call also into question the idea of faithfulness of imperfect logic?

To sum up, I don't know. Although more bearable than uniformities « falling from the sky », bihaviours are slightly bad taste. They must eventually be either eliminated or fully integrated, so that they no longer look like this artificial trick which destroys a harmony – rather convincing otherwise.

14.C Parsimony

This tediously proven faithfulness is obtained at the price of weakening. One can enunciate a property of *parsimony* corresponding to the idea of *destroying* all locations. But this interactive notion is not faithful to the rejection of weakening and one must replace it with *exactness*, a blunt, essentialist, request of non-weakening: see the paper by Faggian [25].

If one follows the procedural viewpoint which is that of these lectures, one must thus note, that, no matter how strong are the arguments against weakening – *grosso modo* material implication –, its prohibition is at the price of contortions which justify it *a contrario*.

That's it, the dialogue between analysis and synthesis, existence and essence: one layer of analysis (-2) takes us away from weakening, another layer (-3) brings us back to it!

14.D Epilogue (?)

If one tries to give a provisional epilogue to these lectures – before entering the zone of turbulence of the last chapters –, one could say that the procedural viewpoint recognises the immense interest of the logical inheritance, while being wary of it, since it is transmitted through an ideological, essentialist filter.

It is a general problem, of which we can find multiple instances in everyday life. Thus, in the laboriorious childhood of my great-grandmother Octavie, the grocery where she was working offered two qualities of petrol, styled « à un sou » and « à deux sous[5] », coming in fact from the same barrel hidden under the counter. Which would only be yet one more variation on the opposition sense/denotation, if the wealthy she-clients did not plebiscit the « twopence », supposedly « less smoky »: one thus sees that *essence* comes, not from petrol, but from its price. Take another example, the cooking of *pasta*; some contend that one must salt water only *after* ebullition; my objections against this medieval taboo received a blunt « This is like that ! ». Reiterating my objection, I met the imposing figure of *Zia Ermenegilda*: « You will not teach my aunt how to cook *tonnarelli* ! ». Which illustrates the blasphematory character of any discussion of the dogma... of the Holy Pasta. Here, essence is located in the *metapasta* of aunt Ermegilda, who must now cook in Heaven, still salting after ebullition[6].

One must relinquish these interdictions, related to habits, ignorance, superstition, nay fanaticism when mixed up with religion. It is the idea of Ockham's razor: get rid of useless hypotheses, assimilated to a beard. This being said, one hardly meets cases as simple as those of the petrol or the *pasta*. Thus, in the logical world, one faces a much more serious problem, since one has difficulties in separating such and such restrictions, because of combined effects. Moreover, various prejudices – one knows a bit too well what one is looking for – impede a neutral, « objective », revisitation of the restrictions bequeathed by tradition. But here, beware of the sophism: one is indeed seeking a *trivial* interpretation of logic, trivial since – everything having found its real place – it seems that nothing else can be done. By moments, in this chapter, one had this impression of self-evidence, of triviality. But the search for a trivial synthesis is *the least trivial* task that I know in logic. One should not « see the hands »: for instance, in the case of the Lorenzen group, the hands were so conspicuous that they were hiding the poor ideas that these people could have!

The short story by Borges « Pierre Ménard, author of the Quixotte », should not be reduced to a mockery of the snobbery of certain literary *milieux*, preferably

[5] Respectively, one and two pence.

[6] Magazines specialised in cooking must sustain the dogma: for instance I recently found the theory that « salt slows down the increase in temperature of water »! Rather reading that than being blind... But this unhealthy need to justify *at any price* the worst baloney, to which obscure part of the human soul does it refer?

French. Ménard rewrites the Quixotte, line after line, in an internal way. The idea being to reach a certain self-evidence; the difference between Ménard rewriting Cervantes and the project of this book, is the possibility of writing something else – a possibility not considered by Borges.

At the point I reached, the natural evidence produced would at least require one to revisit the atoms. It seems also that this evidence arbitrated in favour of weakening and in defavour of the multiplicative unit **1**. One seeks a trivial synthesis of logic; *even if one must change logic to do so!*

A first step in this direction has been the discovery of linear logic; it is indeed, more a reformulation than a revolution, articulated around the opposition perfect/imperfect. This procedural (and not semantical) articulation opens the way for the real revolution: the putting into question of imperfection, of the world of the « always ».

Chapter 15
Orthodox exponentials

It is hard to ignore the strength of the *perenniality* sustaining classical logic, which is to be found again, just so, in intuitionistic and linear logics; in the latter, perenniality is but an attribute of specific connectives, the exponentials. This chapter is sort of a farewell to perenniality, to a stable, nay frozen, world.

15.1 The perennial perenniality

15.1.1 Perenniality in perenniality. If exponentials are treated only so late, this is because here lies the actual blind spot of logic, the mother of all opacities. Imperfection is the non-completed, the non-finished or infinite, i.e., *in-finished*, in other terms, *perenniality*.

Perenniality is a strange thing. It indeed concerns nothing sublunar, no life or kingdoms, not to speak of no solar or galactic systems: *sic transit*. It is from the other world: thus one « perennialises » the dead through remembrance. As a matter of perenniality, the physical world only offers but vague approximations summarised by the image of this glass of water that one can indefinitely dip from the sea. It is a modest perenniality, which does not allow one to nibble a sea from the sea – at least not indefinitely.

On the other hand, our creation, this world of ideas to which our logical laws supposedly apply, is perennial beyond all material experience. Not at all in the sense of this cliché once heard from a philosopher: « Indeed, infinity is but the possibility of adding one more point »; he was still at the stage of the sea and the glass of water. He didn't know that infinity contains an infinity of infinities and even an infinity of such infinities, that one cannot disentangle without entanglement in circles, without biting one's tail. This difficulty is linked to our occidental culture and the Thomism which sustains it; Thomism is indeed a theory, not only of deity, but of the impunity of deity. What is akin to the perenniality of perenniality, the infinity of infinity.

Can one « measure » the infinite, distinguish several sizes of infinity? This depends upon the finesse of the grain. Set-theory classes sets according to their cardinalities, but this is very rough. We would rather distinguish the simple reuse from the reuse of reuse, etc.; one is soon faced with a hierarchy of nested reuses, indexed by, say, ordinals. These ordinals bring us back to infinity; like Dupond and Dupont in the desert[1], we just found again our own track. Instead of making

[1]Hergé: *Tintin au pays de l'or noir*, pp. 29–30, 1950.

loops and meta-loops like the Dupondt, we might as well acknowledge that we met a blind spot, hardly surpassable.

There is indeed no convincing way of « unscrewing » the infinity of classical logic – of the exponentials –: it is a hard, compact, opaque kernel, which only accepts genuflexion. Which will give even more value to the results – although experimental – of Chapter 16.

15.1.2 Perenniality of perenniality. Prior to its formal expression, perenniality is already in our minds. One can verify it with one example, the axiomatisation of *revision*. One is dealing with something simple, how a formal actor – a machine – updates its data. Perenniality enunciates a principle of intangibility, if not of the machine, at least of the photo of the machine. Revision thus becomes no more than a matter of cinematography: one changes the photo 24 times per second. This is the very base of Kripke models (Section 4.E): how to recreate life by animated images, in the same way that Lloyola directed the Jesuit *perinde ac cadaver*. This yields various logics: modal, temporal and above all *temporary* logics.

Nevertheless, the *perfect* fragment of linear logic seems well-adapted to revision and the attempts in this direction are rather convincing, see for instance the work of Vauzeilles & al. [81]. But this stumbles against a specious objection: this is not classical, in other terms, not perennial. To found revision upon perenniality, there is some bad faith here! This bad faith rests upon a certain gregarious common sense according to which logic is classical or is not; which has exactly the same value as the everyday experience telling us that the Sun rotates around the Earth, in accordance with the Book of Joshua. The non-monotonic logics[2] are the monstrous metastases of this postulate. By the way, if one wants to revise while staying classical, there is no alternative. Except that the final « product » is everything but logic, classical or not... This is where sectarianism leads: to the very opposite of the original goal. This is the eternal story of the monkey who found a nut in a hole: his closed fist impedes his liberation; eventually he loses both freedom and the nut. The same with the guy who chooses the wrong queue and which *rentabilises* his mistake by staying in the same line: the more time he wastes, the more reasons he finds to stay there[3]!

The lie is not only logical: the pseudo-classical formalisation of a revision of the type operated by `ftp` is *procedurally* faulty. Either, by means of « photos » of the globality of the system, one enforces a strict synchronisation of *all* actors, whether or not they are related; this ensures a total paralysis. Or one neglects the signals of termination, by declaring that a revision is over when it no longer modifies

[2]They have of names: circumscription, default reasoning (swell name!), ... Beyond their superficial differences, they agree on one basic point: the refusal (or the ignorance) of incompleteness.

[3]According to G. W. Bush, the defence of democracy rested upon the dungeons of the CIA, i.e., a return to the Middle-Ages. It is thus natural that the defence of classical logic rests upon the relinquishment of deduction!

anything: but how to know that without waiting « up to the end »? One must then imagine machines operating after the Last Judgment, i.e., according to a *transfinite* temporality[4].

15.2 Exponential nets: normalisation

It is time to come back to nets with exponential boxes, introduced in Section 11.C.2. By themselves, they are not that impressive: everything is hidden in their *normalisation*.

In what follows we reduce a cut-link $!A/?{\sim}A$ with premises given by:

- An exponential box $\mathfrak{B}$ with conclusions $\underline{\Delta}$, $!A$; in the box there is a net $\mathfrak{R}$ of conclusions $\underline{\Delta}, A$.
- A link « ? » of conclusion $?{\sim}A$.

For the sake of simplicity, we will assume that $\Delta = B$ and will restrict to three particular cases:

(i) The link « ? » is 0-ary: we erase everything, the cut-link, the link « ? » and the box $\mathfrak{B}$. One $\underline{B}$ disappears; it was a premise of an $n + 1$-ary « ? », it is now a premise of an n-ary « ? ». If $n = 0$, we must « attach » the conclusion $?B$; but where? At the very place where $?{\sim}A$ was « attached ».

(ii) The link « ? » has two premises, $\underline{{\sim}A}$, $\underline{{\sim}A}$, which are not conclusions of exponential boxes: we « open » $\mathfrak{B}$ and make two copies of $\mathfrak{R}$; which yields two « occurrences » of A that we « cut » with the two ${\sim}A$; this is possible, since the $\underline{{\sim}A}$ are not conclusions of exponential boxes, so that the underlinings can be removed. There are also two $\underline{B}$'s, instead of a single one; it was a premise of an n-ary « ? », they will be premises of an $n + 1$-ary « ? ».

(iii) The link « ? » has a single premise, $\underline{{\sim}A}$, which is the conclusion of an exponential box; in the box, we find again the same $\underline{{\sim}A}$, which may be in turn the conclusion of an exponential box, etc. Finally, opening d nested boxes, we obtain an $\underline{A}$ which is not the conclusion of an exponential box: let $\mathfrak{S}$ be the net in the dth nested bow, which contains among other conclusions $\underline{{\sim}A}$, but that we can « un-underline ». We « dig » $\mathfrak{R}$ at depth d and operate a cut at this depth between A and ${\sim}A$. At depth 0, the context $\underline{B}$ takes the place of the underlined formula $\underline{{\sim}A}$.

[4] Some have actually constructed « models » of the *closed world assumption* on the basis of transfinite iteration. Hence the following perfidious comment: « PROLOG is not bugged, one does not wait long enough, period »!

This definition has all possible virtues; Church–Rosser, strong normalisation, etc. But at this point, this is no longer the real problem. One would like to know what is so badly infinite in the normalisation process, what is this thing that Germanic proof-theory used to measure by means of ordinal *Panzerdivisionen*. The second step produces only multiplications; it is the third step, which, by *burying* boxes inside boxes, produces monstrous sizes. This burying – which remains of fixed depth in simply typed calculi – becomes of variable depth when one passes to system **F**. In the beginning, it can be measured with ordinals, $\epsilon_0, \Gamma_0, \ldots$, but soon enough, infinity loses any sense of measure. In any case, let us remember for the next chapter that:

Infinity hides inside the nesting of boxes.

15.3 Categories and classical logic

15.3.1 A thorny question. We already met, in an order opposite to their genesis, two *polarised* interpretations of classical logic: in Section 7.A.6, a Gödel-style interpretation; in Section 12.A, a linear interpretation. What appeared then is a *non-degenerated* categorical interpretation, involving beautiful isomorphisms: commutativity, associativity and a reasonable amount of distributivity.

One thus tries to translate the sequent calculus **LK** as follows: a sequent $\vdash \Gamma$ splits as $\vdash \Gamma', \Gamma''$, where the Γ' are negative and the Γ'' are positive. I will interpret my sequent in linear logic by $\vdash \Gamma', ?\Gamma''$, the « ? » before the Γ'' corresponding to the fact that they have neither contraction nor weakening so that one must « provide » the Γ'' with those rules.

To my greatest surprise, when I developed this idea by the end of 1990, I was not able to interpret **LK**. For instance, cut-elimination does not enjoy Church–Rosser, i.e., is not associative. Indeed, if A, B are positive, there is no way to interpret the double cut:

$$\frac{\vdash \Gamma, A \qquad \vdash \neg A, \neg B \qquad \vdash B, \Delta}{\vdash \Gamma, \Delta} \tag{15.1}$$

or rather, there are too many of them! Supposing Γ, Δ negative, this could be translated as:

$$\frac{\dfrac{\vdash \Gamma, ?A}{\vdash \Gamma, !?A} \qquad \dfrac{\dfrac{\vdash {\sim}A, {\sim}B}{\vdash !{\sim}A, {\sim}B}}{\vdash ?!{\sim}A, !{\sim}B} \qquad \vdash ?B, \Delta}{\vdash \Gamma, \Delta} \tag{15.2}$$

or the mirror version, which favours A:

$$\dfrac{\vdash \Gamma, ?A \qquad \dfrac{\dfrac{\vdash {\sim} A, {\sim} B}{\vdash {\sim} A, !{\sim} B}}{\vdash !{\sim} A, ?!{\sim} B} \qquad \dfrac{\vdash ?B, \Delta}{\vdash !?B, \Delta}}{\vdash \Gamma, \Delta} \tag{15.3}$$

These versions are violently antagonistic. Indeed, suppose that the last respective rules of the given proofs of $\vdash \Gamma, A$ and $\vdash B, \Delta$ are weakenings on A and on B: then (15.2) will destroy the left part, while (15.3) will destroy the right part. One finds again an old dilemma, linked to the double weakening in classical logic; it is an effective version of the category-theoretic degeneracy of Section 7.A.4 due to Lafont, the impossibility of normalising:

$$\dfrac{\dfrac{\vdash \Gamma}{\vdash \Gamma, A} \qquad \dfrac{\vdash \Delta}{\vdash \neg A, \Delta}}{\vdash \Gamma, \Delta} \tag{15.4}$$

Polarisation fixes the problem – this is why linear logic can now envisage weakening. Classical polarisation should do the same... By no means! The example just given reproduces the dilemma with formulas A, B of the same polarity – positive, which is not by chance.

15.3.2 What to do? The problem is posed identically for the conjunction of positive formulas[5]. This is easy to understand: the double cut (15.1) can be translated into a cut between $A \wedge B$ and $\neg(A \vee B)$.

We could, just for a minute, imagine that positive conjunction could have two rules, one favouring A, the other favouring B. This does not work: the ternary case would call for six rules, corresponding to all possible orderings of the constituents; but the bracketing $A \wedge (B \wedge C)$ yields the orderings ABC, ACB, BCA, CBA, while $(A \wedge B) \wedge C$ yields ABC, BAC, CAB, CBA.

I even thought of a crazier solution: to give no rule, just saying that the conclusion of the conjunction rule is non-deterministic. I even thought to make here a link with the quantum world. This was not the case and this for a simple reason: at that time (1990), I had only in mind set-theoretic notions, graphs, coherent spaces; but the quantum does not belong in this very world. Like all logicians, I was badly underestimating the radicality of the « anti-set-theoretism » underlying the quantum.

[5]And dually the disjunction of negative formulas: the restriction to negatives is thus no solution.

15.3.3 Comonoids. The idea is as follows: try to individualise proofs that are insensitive to the problem. In other words, for which the two protocols (15.2) and (15.3) yield the same result.

For this we come back to the category-theoretic explanation of exponentials. Since it is – as expected – illegible, we will content ourselves with a concrete description in terms of coherent spaces.

Definition 73 (Comonoids). A (commutative) *comonoid* is a coherent space $\mathcal{P}$, together with a stable linear map $+$ from $\mathcal{P}$ to $\mathcal{P} \otimes \mathcal{P}$ and an anti-clique $\mathbf{1}_{\mathcal{P}} \sqsubset \sim\mathcal{P}$, enjoying the dual forms of commutativity, associativity and neutrality.

Which can be written by means of diagrams. But one can more simply introduce the notation, for $x, y, z \in |\mathcal{P}|$: $x \rightsquigarrow y, z$ to say that $(y, z) \in +\{z\}$.

Commutativity: if $x \rightsquigarrow y, z$, then $x \rightsquigarrow z, y$.

Associativity: if $x \rightsquigarrow y, z$ and $z \rightsquigarrow t, u$, then there is a v such that $x \rightsquigarrow v, u$ and $v \rightsquigarrow y, t$.

Neutrality: if $x \in |\mathcal{P}|$, then $x \rightsquigarrow x, y$ for a certain $y \in \mathbf{1}_{\mathcal{P}}$. Conversely, if $x \rightsquigarrow z, y$ and $y \in \mathbf{1}_{\mathcal{P}}$, then $x = z$.

The typical comonoid is $!X$, with:

- $a \rightsquigarrow b, c$ iff $a = b \cup c$.
- $\mathbf{1}_{!X} = \{\emptyset\}$.

If $\mathcal{P}, \mathcal{Q}$ are comonoids, the same is true of $\mathcal{P} \otimes \mathcal{Q}$:

- $(a, a') \rightsquigarrow (b, b'), (c, c')$ iff $a \rightsquigarrow b, c$ and $a' \rightsquigarrow b', c'$.
- $\mathbf{1}_{\mathcal{P} \otimes \mathcal{Q}} = \mathbf{1}_{\mathcal{P}} \times \mathbf{1}_{\mathcal{Q}}$.

And of $\mathcal{P} \oplus \mathcal{Q}$:

- $(a, i) \rightsquigarrow (b, j), (c, k)$ iff $i = j = k$ and $a \rightsquigarrow b, c$.
- $\mathbf{1}_{\mathcal{P} \oplus \mathcal{Q}} = \mathbf{1}_{\mathcal{P}} \cup \mathbf{1}_{\mathcal{Q}}$.

The notion of comonoid exactly matches the idea of a space equipped with a way of performing weakenings and contractions.

15.3.4 Central morphisms

Definition 74 (Central morphisms). A stable linear map φ from the comonoid $\mathcal{P}$ to the comonoid θ is *central* when it commutes to the sum and the neutral element: $\mathbf{1}_{\mathcal{Q}} \circ \varphi = \curlywedge_{\mathcal{P}}$, as well as the commutative diagram:

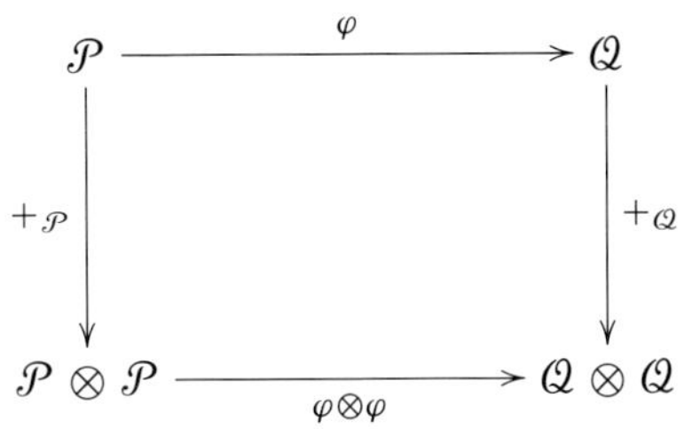

(15.5)

We proceed with presenting $!X$ as the solution of a universal problem: given X, find a comonoid $!X$ and a linear map δ from $!X$ into X such that for any comonoid $\mathcal{P}$ and any linear function f from $\mathcal{P}$ to X, there is a unique *central* morphism $!f$ from $\mathcal{P}$ to $!X$ rendering commutative the diagram:

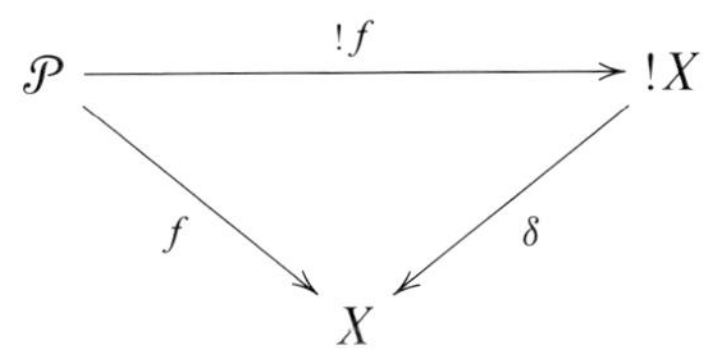

(15.6)

We recognise dereliction (δ) and promotion.

This offers as usual not much interest… except that, in this very case, this is wrong! Indeed, this is not $!X$ which is universal, but a variant obtained from finite multisets. We modify whatever needed to make it work.

The interest of the thing lies outside these threadbare ideas. Going back to the counterexamples (15.2) and (15.3) of Section 15.3.1, we see that the two protocols yield the same output when one of the two functions from $\sim\Gamma$ to A or from $\sim\Delta$ to B is central. This is by the way the origin of the terminology: a central morphism « commutes » with cut.

From that point, we find our solution: beyond linear maps, one must individualise those which are central. This will lead to the calculus **LC** of the next section. One should check the categorical soundness of **LC**, but this is trivial; what was not trivial was to reach this point.

15.4 The system LC

15.4.1 Stoups. The solution consists in changing the structure of sequents by introducing a special zone, the *stoup*. A sequent thus is written $\vdash \Gamma; \Pi$, where

Π occupies the stoup. Indeed, Π consists in at most one formula, positive. The interpretation of a proof of $\vdash \Gamma', \Gamma''; P$ (Γ' negative, Γ'' positive) is a *central* morphism from the tensor product of the $\sim\Gamma', !\sim\Gamma''$ into P.

The stoup is indeed the very origin of the *handles* of ludics. But, while ludics tends to write everything positively, here everything is rather negative; see Section 15.C.6 for a discussion.

15.4.2 The calculus LC

Identity

$$\frac{}{\vdash \neg P; P}\ (identity) \qquad \frac{\vdash \Gamma; P \quad \vdash \neg P, \Delta; \Pi}{\vdash \Gamma, \Delta; \Pi}\ (p\text{-}cut)$$

$$\frac{\vdash \Gamma, N; \quad \vdash \neg N, \Delta; \Pi}{\vdash \Gamma, \Delta; \Pi}\ (n\text{-}cut)$$

Structure

$$\frac{\vdash \Gamma; \Pi}{\vdash \tau(\Gamma); \Pi}\ (X) \qquad \frac{\vdash \Gamma; P}{\vdash \Gamma, P;}\ (D)$$

$$\frac{\vdash \Gamma; \Pi}{\vdash A, \Gamma; \Pi}\ (\vdash W) \qquad \frac{\vdash A, A, \Gamma; \Pi}{\vdash A, \Gamma; \Pi}\ (\vdash C)$$

Logic

$$\frac{}{\vdash ; \mathbf{v}} \qquad \frac{}{\vdash \Gamma, \neg \mathbf{f}; \Pi}$$

$$\frac{\vdash \Gamma; P \quad \vdash \Delta; Q}{\vdash \Gamma, \Delta; P \wedge Q} \qquad \frac{\vdash \Gamma, M, N; \Pi}{\vdash \Gamma, M \vee N; \Pi}$$

$$\frac{\vdash \Gamma; P \quad \vdash \Delta, N;}{\vdash \Gamma, \Delta; P \wedge N} \qquad \frac{\vdash \Gamma, M, Q; \Pi}{\vdash \Gamma, M \vee Q; \Pi}$$

$$\frac{\vdash \Gamma, M; \quad \vdash \Delta; Q}{\vdash \Gamma, \Delta; M \wedge Q} \qquad \frac{\vdash \Gamma, P, N; \Pi}{\vdash \Gamma, P \vee N; \Pi}$$

$$\frac{\vdash \Gamma, M ; \Pi \qquad \vdash \Gamma, N ; \Pi}{\vdash \Gamma, M \wedge N ; \Pi} \qquad \frac{\vdash \Gamma ; P}{\vdash \Gamma ; P \vee Q} \qquad \frac{\vdash \Gamma ; Q}{\vdash \Gamma ; P \vee Q}$$

$$\frac{\vdash \Gamma, M ; \Pi}{\vdash \Gamma, \forall x M ; \Pi} \qquad \frac{\vdash \Gamma ; P[t/x]}{\vdash \Gamma ; \exists x P}$$

$$\frac{\vdash \Gamma, P ; \Pi}{\vdash \Gamma, \forall x P ; \Pi} \qquad \frac{\vdash \Gamma, M[t/x] ;}{\vdash \Gamma ; \exists x M}$$

15.4.3 Properties. It goes without saying that **LC** enjoys cut-elimination: for instance, translate everything in linear logic. The most important fact is the following:

Definition 75 (Hereditarily positive). A formula is *hereditarily positive* iff it is constructed from positive atoms by means of the positive operations $\wedge$, $\vee$, $\exists$.

Theorem 57 (Existence and disjunction). *The hereditarily positive fragment of* **LC** *enjoys the existence and disjunction properties.*

Proof. A cut-free proof of $\vdash P;$ comes from $\vdash ; P$ or $\vdash P; P$ by a rule (D). We recursively climb up the rules and never find an empty stoup: since the formulas outside the stoup are of no use, we are reduced to the sole case of $\vdash ; P$. □

Since there are few hereditarily positive formulas, the technical interest of this result is limited. But one could imagine a formulation embodying arithmetic, with hereditarily positive operations such as bounded universal quantifications (treated like a sort of finite « $\otimes$ ») and numerical existential quantification, which would make the hereditarily positive quite correspond to the Σ^0_1.

This result remains the ultimate argument against the rule « Mix » of Section 11.A.4. Indeed, one could prove $\vdash P, Q;$ (with P, Q hereditarily positive) by « mixing » a proof of $\vdash P;$ with a proof of $\vdash Q;$. Then, farewell, disjunction property, at least under its exclusive form!

The system **LC** is the source of interesting developments: the fragments **LKQ** and **LKT** which respectively correspond to positive and negative polarities, see [18], the same **LKT** being more known under a functional presentation, the $\lambda\mu$*-calculus* (Section 15.D); but the most satisfactory and general syntax for **LC** remains the *polarised nets* (Section 15.C).

15.A Exponentials and analytic functions

I shall shortly develop coherent Banach spaces, mainly for the explanation of exponentials in terms of analytic functions. We shall find them again in Chapter 17, but only in finite dimension.

15.A.1 Coherent Banach spaces

Definition 76 (Coherent Banach spaces). A *coherent Banach space* (CBS for short) $(E, \sim E, \langle\cdot \mid \cdot\rangle)$ consists of two complex Banach spaces E, $\sim E$, together with a bilinear form such that

$$\forall x \in E \quad \|x\| = \sup \{|\langle x \mid y\rangle|; y \in \sim E, \|y\| \leqslant 1\}, \tag{15.7}$$

$$\forall y \in \sim E \quad \|y\| = \sup \{|\langle x \mid y\rangle|; x \in E, \|x\| \leqslant 1\}. \tag{15.8}$$

In other words, each of the Banach spaces $E, \sim E$ can be identified with a subspace of the dual of the other; this is why one will not indicate the bilinear form $\langle\cdot \mid \cdot\rangle$. The most typical case is the pair ℓ^∞, ℓ^1: ℓ^∞ is the dual of ℓ^1, whereas the dual of ℓ^∞ contains, besides ℓ^1, the ultrafilters. The fact that Banach spaces are seldom *reflexive*, i.e., equal to their bidual, leads us to « give in advance » the dual, so as to get an involutive negation. The typical reflexive Banach spaces are *Hilbert spaces* (Section 17.A.1) such as ℓ^2.

In a CBS, the norm plays the part of coherence. In particular, cliques become vectors of norm ≤ 1.

Additives. On the direct sum $E \oplus F$, several norms are available, in particular

$$\|x \oplus y\|_{E \oplus F} = \|x\|_E + \|y\|_F, \tag{15.9}$$

$$\|x \oplus y\|_{E \& F} = \sup(\|x\|_E, \|y\|_F). \tag{15.10}$$

Which is enough to define the « Plus » of two CBS $(E, \sim E)$ and $(F, \sim F)$ as $(E \oplus F, \sim E \,\&\, \sim F)$. Observe that negative constructions involve suprema while positive constructions involve sums.

Multiplicatives. Here constructions are slightly more complicated:

$E \parr F$: it is made of all bilinear forms $b(\cdot, \cdot)$ on $\sim E, \sim F$, which can be seen either as linear maps from $\sim E$ into F or linear maps from $\sim F$ into E:

$$\forall x' \in \sim E \ \exists y \in F \ \forall y' \in \sim F \quad b(x', y') = \langle y \mid y'\rangle, \tag{15.11}$$

$$\forall y' \in \sim F \ \exists x \in E \ \forall x' \in \sim E \quad b(x', y') = \langle x \mid x'\rangle, \tag{15.12}$$

equipped with the norm

$$\|b\| = \sup \{|b(x, y)|; \|x\|, \|y\| \leqslant 1\}.$$

$E \otimes F$**:** the *algebraic* tensor product $E \odot F$ is equipped with the norm

$$\|a\|_{E \odot F} := \inf \{\sum_i \|x_i\|_E \cdot \|y_i\|_F\},$$

the infimum being taken w.r.t. all decompositions $a = \sum_i x_i \cdot y_i$. $E \otimes F$ is the norm-completion of $E \odot F$.

Which enables one to define the « Tensor » of two CBS $(E, \sim E)$ and $(F, \sim F)$, as $(E \otimes F, \sim E \parr \sim F)$.

Stable functions. An *analytic* function from the open unit ball of E into F can be written $\varphi(x) = \varphi_0 + \varphi_1(x) + \varphi_2(x, x) + \cdots + \varphi_n(x, \ldots, x) + \cdots$, where the φ_n are symmetrical multilinear functions from the E^n into F. Such a function will be *stable* when:

- $\varphi_n \in (E \otimes \cdots \otimes E) \multimap F$ for all n.
- The norm $\|\varphi)\| := \sup \{\|\varphi(x)\|; x \in E, \|x\| < 1\}$ is finite.

This definition works well, in particular, the stable maps from E into $\mathbb{C}$ enable one to define $?{\sim}E$, which implies that there is a dual definition of $!E$, see *infra*. On the other hand, observe that a « clique » in $E \Rightarrow F$ is an analytic map sending the *open* unit ball of E into the *closed* unit ball of F. Since there is no reasonable way to extend an analytic function to the boundary, one cannot compose stable maps. This non continuity is to be put side by side with the loss of control over the infinite. It sheds an interesting light on the alleged continuity à la Scott. Although the outcome of coherent Banach spaces remains – as for all concrete categorical interpretations – rather modest, one will notice the qualitative jump from the purely formal analytic functions of [40] and those of [50], where the actual questions begin to appear.

The space $!E$ is obtained as follows:

- If $x \in E$, $\|x\| < 1$, we consider the « Dirac mass » $!x$. If $\varphi \in ?{\sim}E$, i.e., is a stable map from E into $\mathbb{C}$, we pose $\langle !x \mid \varphi \rangle := \varphi(x)$. Which we extend by linearity to the space $\sharp E$ of finite linear combinations of Dirac masses. Equation (15.7) thus yields a norm on $\sharp E$.
- $!E$ is obviously the norm completion of $\sharp E$.

In $!E$ stand all uniform limits of barycentres, i.e., of Darboux sums; hence the contour integrals, the integral kernels, which is nice. For $\|x\|, \|y\| < 1 - \epsilon < 1$, one easily sees that $\|!x - !y\| \leq \|x - y\|/\epsilon$ which shows that evaluation (i.e., the Dirac mass) is truly continuous on the open ball. But, as we move towards the boundary, the points $!x$ step away one from another up to the distance 2.

In this order of ideas, let us mention the *differential λ-calculus* of Ehrhard et al. [24]; this remains experimental… and interesting for that very reason.

15.B Exponential ludics

15.B.1 Separation and uniformity. The exponential is *uniform*: it is not a matter of an infinite tensor product, but rather of a symmetrised product. Which is conspicuous when we look at the rule of contraction: one passes from $f(x, y)$ to $f(x, x)$, i.e., one has two copies, but twice the same. Uniformity occurs too in a CBS (see *supra*), where stable maps involve *symmetrical* coefficients.

In particular, in a double conditional ((12.9) and (12.10), Section 12.A) there is a loss of information, two lines out of four disappear. One can fear the definite loss of category-theoretic faithfulness, which would be, for instance, the case with coherent spaces.

Fortunately, CBS shows us the way out. Indeed, on a CBS, one must take into account convex combinations, thus with the notations of (12.9), (12.10),

$$g(\lambda\iota_l x + (1-\lambda)\iota_r y) = \lambda^2 t[x,x] + \lambda\mu(u[x,y] + v[y,x]) + \mu^2 w[y,y] \quad (15.13)$$

which are not as desperate as (12.10) and which by the way works correctly in the recent extension of ludics to exponentials due to Maurel.

15.B.2 Exponential ludics. Fundamentally, one allows the reuse of *loci* and convex combinations of positive actions to allow separation. But this is not that simple:

Loci: which poses a problem is not quite the reuse of a focus, it is, at the second degree, when two $\xi * i$ have been created by two focusings on ξ (positive actions (ξ, I) and (ξ, J) with $i \in I \cap J$), to differentiate those two $\xi * i$ as foci. Developing a remark of Curien, one can use a relative addressing – De Bruijn style – of the kind « focalise on $\xi * i * j$ created *three* steps before in the chronicle ».

Coefficients: separation becomes delicate since one cannot separate a chronicle repeating a positive action from the same one before the repetition: if κ is positive, κ' is negative, $\mathfrak{c} = \langle \kappa, \kappa', \maltese \rangle$ and $\mathfrak{d} = \langle \kappa, \kappa', \kappa, \kappa', \maltese \rangle$, will react in the same way to any counter-chronicle. Except if one admits real coefficients; thus, with $\mathfrak{e} = \langle \widetilde{\kappa'}, \lambda \cdot \widetilde{\kappa} \rangle$, one gets $[\![\mathfrak{c}, \mathfrak{e}]\!] = \lambda \cdot \maltese$, $[\![\mathfrak{c}, \mathfrak{e}]\!] = \lambda^2 \cdot \maltese$. It is indeed what comes from a CBS.

The Pandora's box: one must *separate the separators. Matthew* VII.2: *for in the way you judge you will be judged.* This poses an endless list of technical questions, see [82].

Uniformity: there are many new objects, separators, nay separators of separators. These objects were not in the syntax – otherwise it would not have been that difficult to find them. Their task being done, one must dismiss them, say that they are « no good ». Only a want of *uniformity*, i.e., the recourse to *bihaviours* (Section 14.B), can eliminate them.

One must stress the *experimental* character of this extension, for which faithfulness has not been established.

15.C Polarised linear logic

15.C.1 The language

Formulas. At the origin of **LC**, the translation of Section 12.A only uses formulas of a very peculiar form, indeed:

- The positive connectives $\otimes$, $\oplus$, $\exists$ are only applied to positive constituents: $P \otimes Q$, $P \oplus Q$, $\exists x P$, $\exists X P$; the exponential « ! » is only applied to a negative constituent: $!N$.
- Dually, the negative connectives $⅋$, $\&$, $\forall$ are only applied to negative constituents: $M ⅋ N$, $M \& N$, $\forall x M$, $\forall X M$; the exponential « ? » is only applied to a positive constituent: $?P$.

We decide, by pure convention, to give the positive polarity to atomic formulas – other than the neutrals which already have one. We thus get two classes of formulas (positive and negative) with as sole way of communication the exponentials; which is not that far from ludic polarisation in the style of **HS** (Section 12.6): the exponential « ! » replaces the shift « $\downarrow$ ». One can by the way see $\downarrow N$ as a supertype of $!N$ – dually, $\uparrow P$ as a subtype of $?P$ where contraction is illicit.

The calculus LLP. The sequent calculus **LLP** is inspired from **LC**; sequents have at most one positive formula – which corresponds to the *stoup*. Structural rules are valid for all negative formulas:

$$\frac{\vdash \Gamma, N, N}{\vdash \Gamma, N} \qquad\qquad \frac{\vdash \Gamma}{\vdash \Gamma, N}$$

The same is true of the context of the promotion rule:

$$\frac{\vdash \mathcal{N}, M}{\vdash \mathcal{N}, !M}$$

Taking into account Section 12.A, we see that the sole novelty of **LLP** w.r.t. linear logic is the principle $?{\sim}X \vdash {\sim}X$ restricted to negated atoms.

The system **LLP** is a conservative extension of linear logic – to which the principle $?{\sim}X \vdash {\sim}X$ has been added; this can be established by induction on cut-free proofs. One must pay attention to a small difficulty – reminiscent of the oddity observed by Schellinx (Section 11.1.2) –: one should prove that the restriction « at most one positive formula » propagates from conclusion to premise, which is not the case for dereliction. On the other hand, the presence of at least two positive formulas propagates from conclusion to (one of the) premise(s), except in the case of the axiom for $\top$, without premise: conservation requires replacing a proof using general axioms $\vdash \Gamma, \top$ with another one in which those axioms are restricted to contexts Γ with at most one positive formula, which can be done without problem.

15.C.2 Polarised proof-nets. In Section 12.1, we observed that changes of polarity have consequences difficult to control. In *polarised* linear logic, the conflicts linked to changes of polarity are so to speak « sequentialised » by the exponentials in charge of those changes – hence much less violent. Indeed, the omnipresence of exponential boxes simplifies everything, to the point that cut-free nets will be trivially sequentialisable: Laurent's correctness criterion only concerns cuts – just as for the *desseins* of ludics.

Let us begin with the multiplicative/exponential fragment of polarised logic. The formulas are built from constituents of the form $?X$, $!{\sim}X$, by means of the polarised versions of $\otimes$, $⅋$, $!$, $?$; the neutrals $\bot$, $\mathbf{1}$ which would bring nothing and the literals $X, {\sim}X$ which would complicate the study of contraction have been excluded. For this very peculiar fragment, the definition of proof-nets given in Section 11.C.2 still applies: it involves Axiom, Cut, Tensor and Par-links, exponential boxes as well as an n-ary « ? » link; when $n = 0$, this link – called « weakening » – induces a want of connectivity that one cannot handle satisfactorily – at least in the non-polarised case.

As usual, one handles separately the nets inside the boxes and the net built from these boxes – reduced to generalised axioms with conclusions $\underline{P_1}, \ldots, \underline{P_n}, !Q$. The correctness criterion of Laurent [74] is an « oriented graph » version of the usual one: positive « climb up », negative « climb down ». Which can also be expressed by means of a preorder relation between the formulas of a proof-structure; $\preccurlyeq$ is the reflexive/transitive closure of the following cases:

Identity link: $P \preccurlyeq {\sim}P$.

Box: $!Q \preccurlyeq \underline{P_1}, \underline{P_n}$.

Cut link: ${\sim}P \preccurlyeq P$.

Tensor link: $P \otimes Q \preccurlyeq P, Q$.

Par link: $M, N \preccurlyeq M ⅋ M$.

Link « ? »: $\underline{P} \preccurlyeq ?P$ for each « occurrence » of $\underline{P}$ which is a premise of the link.

Observe that:

- The *maximal* points of $\preccurlyeq$ are the negative conclusions of the structure and the conclusions $!A$ of exponential boxes without context.

- The *minimal* elements of the preorder $\preccurlyeq$ are of three kinds:
  (i) The positive conclusions of the structure.
  (ii) The underlined formulas which are not « auxiliary » conclusions of an exponential box: which indeed corresponds to a *dereliction*.
  (iii) The $?P$ obtained by weakening, i.e., by a 0-ary « ? »-link.

Among proof-structures, *positive trees* play a special role:

Definition 77 (Positive tree). *Positive trees* are defined recursively: an axiom link or an exponential box are positive trees; the « $\otimes$ » of two positive trees is a positive tree.

We verify that, for any « correct » – i.e., coming from **LLP** – structure:

- The preorder $\preccurlyeq$ is antisymmetric, i.e., is an order.
- There is exactly one minimal element of sort (i) or (ii), i.e., either a positive conclusion or a dereliction.

The first condition is the « oriented » version of the correctness criterion, more precisely, of *acyclicity*; observe that there are no more switchings so that the criterion is of linear algorithmic complexity. The second condition corresponds to connectivity, or rather to Euler–Poincaré: in a « Danos–Regnier » graph, each connected component contains exactly one minimal formula; the number of components is equal to the number of weakenings plus 1.

Conversely, the two previous conditions are sufficient:

Theorem 58 (Sequentialisation). *A correct polarised structure is a proof-net, i.e., comes from a proof in* **LLP**.

Proof. We argue by induction on the size of the structure to be sequentialised and consider the three following cases:

- If there is a terminal $\parr$ or ? link, we can remove this link and apply the induction hypothesis.
- Otherwise, if there are cut-links of premises $P_1, {\sim}P_1, \ldots, P_n, {\sim}P_n$, we select one such that P_i is maximal. Then, above P_i, there are only positive rules, axiom links and exponential boxes, i.e., a *positive tree*. We can remove the cut, which yields two correct structures to which the induction hypothesis applies.
- Otherwise, the structure is a positive tree (with its positive conclusion possibly underlined). Such a tree is immediately sequentialisable. □

To sum up, polarisation enables us to handle weakening by means of the Euler–Poincaré characteristic.

15.C.3 Cut-elimination. What precedes adapts, *mutatis mutandis*, to various proof-net style syntaxes for the multiplicative/exponential case and defines cut-elimination by following the principles of Section 15.2.

- One can admit the literals X, $\sim X$, provided one adds a « preformula » $\widetilde{\sim X}$ (a sort of underlined formula): one then uses an axiom link between $\widetilde{\sim X}$ and X and an n-ary link with premises $\widetilde{\sim X}$ and conclusion $\sim X$.

- This leaves a few problems at second order, since $\sim X$ can become, through substitution, $\sim P$; if $P = \,!N$, then $\widetilde{\sim X}$ naturally becomes $\underline{\sim N}$; if P is a tensor, one performs petty manipulations, neither very difficult, nor very elegant, to eliminate the illegal expression $\widetilde{\sim P}$ (Section 12.A).

The multiplicative neutrals are perfectly at ease in this setting: the axiom for $\mathbf{1}$ behaves like an exponential box and the rule for $\bot$ is indeed a 0-ary ?-link.

Certain proof-net syntaxes, closer to **LLP**, will replace the underlined formulas with arbitrary negative formulas. Which fixes the petty substitution problem just mentioned, but which, on the other hand, introduces a big confusion at the level of structural rules: an irrelevant multiplicity of writings. These nets sequentialise by means of a criterion imitated from the one of the previous section. They also enjoy a nice notion of normalisation: here appears a problem of technological transfer from exponentials to positive formulas and, principally, to find an analogue of exponential boxes. It turns out that positive trees possess all useful properties of exponential boxes:

- Every positive formula is a conclusion of a positive tree.

- A positive tree possesses a positive conclusion and all other conclusions are negative (or positive underlined, in the « standard » syntax).

- They are « local » structures, in the sense that their size is bounded by the size of the positive conclusion (if boxes are counted as of size zero).

- They are sequential structures.

One can extend the cut-elimination of **LL** to the case of **LLP** by modifying the three cases of Section 15.2:

(i) We erase the positive tree.

(ii) We duplicate the positive tree.

(iii) We bury the positive tree inside the box.

Church–Rosser and strong normalisation are established without problem.

This criterion is close to perfection; but the last word as to the ideal formulation of structural rules has not yet been said: whether they are restricted to underlined formulas or extended to all negative formulas, problems remain – substitution on one hand, redundancy of expressions on the other.

15.C.4 General case. We will go quickly through polarised proof-nets in the general case (quantifiers, additives), from the sole viewpoint of the correctness criterion.

Quantifiers. The preorder $\preccurlyeq$ extends to structures involving quantifiers – say, first order:

Existential link: $\exists x P \preccurlyeq P[t/x]$.

Universal link: $A \preccurlyeq \forall x M$, if $A = M$ or A is any *positive* formula in which x is free.

The correctness criterion of Section 15.C.2 remains necessary. Sufficiency is established as Theorem 58, except that one must be careful with the second case: one still chooses P_i maximal but must then show that the positive tree[6] above P_i contains no *eigenvariable* of a $\forall$. Indeed, if $P_i \preccurlyeq A$ and x is free in A, one can suppose A positive, hence $A \preccurlyeq \forall x M$ for an appropriate M and, since there is no terminal negative link, $\forall x M$ is subformula of a $\sim P_j$, which implies $\forall x M \preccurlyeq \sim P_j$; one has therefore $P_i \preccurlyeq A \preccurlyeq \forall x M \preccurlyeq \sim P_j \preccurlyeq P_j$: either $i = j$, contradicting antisymmetry, or $i \neq j$ and P_i is not maximal.

Additives. One defines the relation $\preccurlyeq$ on formulas (of non-empty weight) of a proof-structure by means of the additional cases[7]:

Plus link: $P \oplus Q \preccurlyeq P, Q$.

With link: $A \preccurlyeq M \,\&\, N$ when $A = M, N$ or A is positive and its weight depends on the boolean *eigenvariable* associated to the &-link.

One can easily prove sequentialisation under the hypotheses:

(i) $\preccurlyeq$ is antisymmetric.

(ii) Either there is a positive conclusion, or the sum $\sum \varpi(P)$ of the weights of all « derelicted » positive formulas equals 1.

The criterion is necessary as well, provided « one does not superpose cuts », see *infra*. Indeed, a supposed cycle in a structure would lay fully within the formulas which are « cuts » and their subformulas; if no identification between cuts occurs during the interpretation of a « & », no cycle can be created.

---

[6]Definition adapted to $\exists$-links.

[7][74] gives another correctness criterion, « slicewise », but without « jump »; this criterion also accommodates the axiom for $\top$.

Superposition. Let us take this opportunity to come back to a question not addressed in Section 11.C.4, that of the identity of formulas (and of links): when interpreting a « & », what should one identify between the two structures $\mathfrak{R}, \mathfrak{S}$?

Cuts: one must consider the cuts of $\mathfrak{R}$ and $\mathfrak{S}$ as « private »: in other words, the cut-formulas of both are up to delocation, i.e., distinct. Any other choice only leads to inextricable complications illustrating the dumbness of the notion of occurrence.

Underlinings: contraction poses a similar problem: if there are two $\underline{A}$'s in $\mathfrak{R}$ and three in $\mathfrak{S}$ can they be equal and, in case, how to identify them? There is no way: these formulas must be considered as private, thus not superposable.

Quantifiers: if $\mathfrak{R}, \mathfrak{S}$ use $\exists$-links with the same conclusion $\exists x A$ but with distinct *witnesses* $t \neq u$, the subformulas of $A[t/x]$ and those of $A[u/x]$ cannot be identified.

15.C.5 λ-calculus and proof-nets. The simply typed λ-calculus, based upon the sole connective $\Rightarrow$, can be translated, *modulo* Curry–Howard, in natural deduction, sequent calculus, thus in proof-nets. Let us make this translation explicit: if one operates brutally, a normal term will translate into a net with cuts. This problem, familiar since Section 4.4, is solved by underlining all free variables, but perhaps one of them – corresponding to a head variable, i.e., to the stoup; the term t of type A, depending upon underlined variables of types U_i and perhaps a non-underlined one of type T translates into a net whose conclusions are: A, a certain number of $\underline{\sim U_i}$ and also a $\sim T$ in case there is a non-underlined free variable.

Variable: an axiom link of conclusions $\sim A$, A (here, $T = A$).

Underlining: replacing x with $\underline{x}$ in t translates as an underlining of the conclusion $\sim T$.

Application, generic case: if t, u, where all free variables are underlined, of types $A \Rightarrow B$ and A are respectively interpreted by nets $\mathfrak{R}$ and $\mathfrak{S}$, one forms a box $!\mathfrak{S}$ whose conclusion $!A$ is tensorised with the conclusion $\sim B$ of an axiom link, thus producing a net $\mathfrak{T}$ whose non-underlined conclusions are B and $\sim B \otimes !A$, i.e., $\sim(A \Rightarrow B)$. $\sim(A \Rightarrow B)$ is « cut » with the conclusion $A \Rightarrow B$ of $\mathfrak{R}$, so as to produce the desired interpretation.

Application, « head case »: remains the case of a term $t = u[(y)v/x]$, i.e., obtained by substituting in t, interpreted by $\mathfrak{R}$, the head variable x^B with $(y^{A \Rightarrow B})v$: if v is interpreted by $\mathfrak{S}$, one tensorises the conclusion $!A$ of the box $!\mathfrak{S}$ with the conclusion $\sim B$ of $\mathfrak{R}$, which produces a conclusion $\sim B \otimes !A$, i.e., $\sim(A \Rightarrow B)$, still not underlined.

λ-abstraction: in order to abstract, one must regroup a certain number of conclusions – those corresponding to a certain variable x – say, $\underline{\sim U}$ by means of a ?-link, which yields a conclusion $?{\sim}U$; between this $?{\sim}U$ and the conclusion A, one performs a « Par », which yields a conclusion $?{\sim}U \parr A$, i.e., $U \Rightarrow A$.

The same applies to the pure λ-calculus, that one can type by means of a unique negative type Λ enjoying the fixed point equation:

$$\Lambda = \Lambda \Rightarrow \Lambda \tag{15.14}$$

The fixed point (15.14) by no means affects what concerns correctness – especially the simplified criterion of Section 15.C.2; on the other hand, normalisation need no longer converge; which we already knew.

The interpretation of λ-calculus by means of nets induces identifications, which one calls σ-equivalence, see [17], [88]:

$$((\lambda x t)u)v \sim_\sigma (\lambda x(t)v)u \quad \text{with } x \notin v, \tag{15.15}$$

$$(\lambda x \lambda y t)u \sim_\sigma \lambda y(\lambda x t)u \quad \text{with } y \notin u. \tag{15.16}$$

They are satisfied by the proof-nets translating the terms; nothing new for normal terms, since no term occurring in equations (15.15)–(15.16) is normal[8].

The translation of λ-calculus in proof-nets induces a decomposition of operations, an exponential step and a multiplicative step. If one only eliminates multiplicative cuts, then only the « duplicating machine » of exponentials remains: one thus gets intermediate objects which do not correspond to « real » λ-terms. If one wants to find an antecedent to them in λ-calculus, one must introduce an appropriate notation for virtual substitutions, i.e., *not performed.* Symbols of the kind $t[u/x]$ become part of the official syntax: one thus arrives at calculi with *explicit substitutions*, see [22].

15.C.6 Polarised nets and ludics. Ludics is « at most one negative », while **LLP** is « at most one positive »; however, the same restriction is indeed at work in both cases[9]:

(i) An exponential box of conclusions $!N, \underline{\Gamma}$ can be seen as a negative pitchfork with handle N and tines Γ.

(ii) By opening the box, N becomes accessible and invertible; the inversion of this conclusion corresponds to a negative rule of ludics.

[8] By Böhm's theorem (Section 13.7.2), the equivalence between normal terms is anyway limited to « η ».

[9] One supposes that the axiom links are atomic.

(iii) Pushing inversion to its end, one eventually reaches a net – several in the additive case – whose conclusions are all underlined, i.e., of the form $\underline{P}$: which corresponds to a positive pitchfork.

(iv) One of these conclusions (and only one) can be « un-underlined »; by doing this one induces a positive tree. A positive tree is nothing but a positive action. By pushing the decomposition of the tree, one arrives at boxes, i.e., by means of a positive action, to negative pitchforks.

Eventually, polarised proof-nets are the « net » version of ludics: simply forget contraction. *A contrario*, polarised nets are directly adapted to exponential ludics in the style of [82].

15.D The $\lambda\mu$-calculus

15.D.1 Linear negation and functionality. To start with, let us place oneself in the linear setting, more « neutral ». If one wants to give a « functional » perfume to a proof φ of a sequent $\Gamma \vdash \Delta$, the natural solution is as follows:

- For each $B \in \Delta$, one is given a multilinear function φ_B from $\Gamma, \sim(\Delta \setminus B)$ into B, the *teleologisation* of φ w.r.t. the « goal » B.
- If φ_B and φ_C $(B, C \in \Delta)$ are two teleologisations of the same φ, one must write a relation between them.

The simplest solution – especially since Δ can be empty – consists in adding a function φ_* from $\Gamma, \sim\Delta$ to $\perp$. The relations between the various φ_B thus reduce to the relations between the φ_B and φ_*.

For each $A \in \Gamma$ one introduces, as usual, a variable (also styled λ-variable) x^A of type A; for each $B \in \Delta$, one introduces a *covariable* (also styled μ-variable) α^B of *cotype* B, i.e., of type $\sim B$. φ can therefore be written under the two forms:

- For each $B \in \Delta$, a term φ_B of type B, depending on the variables $x^A (A \in \Gamma)$ and the covariables $\alpha^C (C \in \Delta \setminus B)$.
- A term φ_* of type $\perp$ depending on the variables $x^A (A \in \Gamma)$ and the covariables $\alpha^C (C \in \Delta)$.

With the obvious relation

$$\varphi_B = \lambda\alpha^B \varphi_*, \tag{15.17}$$

in other terms

$$\varphi_* = (\varphi_B)\alpha^B. \tag{15.18}$$

Which can be expressed by means of specific symbols for « coabstraction » and « coapplication ». The following principles thus take care of negation:

μ-abstraction: if t is a term of type $\curlywedgeuparrow$ and if α is a covariable of cotype B, then $\mu\alpha t$ is a term of type B.

Naming: if t is a term of type B and α is a covariable of cotype B, then $[\alpha]t$ is a term of type $\curlywedgeuparrow$.

The terms of the μ-calculi – the calculi based upon these two new primitives – have therefore as type either a formula or the pole $\curlywedgeuparrow$, which is treated as a special type, not necessarily internal to the system. There is a more or less obvious immediate reduction

$$[\alpha]\mu\beta t \rightsquigarrow t[\alpha/\beta], \tag{15.19}$$

to which one can, if one really wants it, add an equation of the kind « η »,

$$\mu\alpha[\alpha]t = t, \tag{15.20}$$

if α is not free in t, which can be oriented as a reduction rule.

To sum up, « μ » looks like a simplified version of « λ », simplified in the sense that, the only « coterms » being the $\alpha, \beta, \ldots$, there is therefore no expression of the form $[t]u$.

15.D.2 The classical case. The classical case allows contraction/weakening on variables and covariables; the functions lose their (multi-)linear character and negation becomes usual negation. If one is not cautious, one is once again confronted with the usual problems of classical logic: indeed, μ-abstraction produces a term of type $\neg\neg B$, by no way assimilable to B. Fortunately, the analysis made for **LC**, more precisely its intuitionistic presentation (Section 7.A.6) will get us out of trouble: it is enough to restrict to the case of negative polarity, i.e., to the fragment **LKT** of **LC** [18]. Indeed, $B \in \Delta$ can thus be written $\neg B_*$, so that one can see the covariables as of type B_* – a type much more economical than $\neg B$.

Bringing back the tables of Section 7.A.6, we get the following definitions:

$$(A \Rightarrow B)_* := A \wedge B_*, \tag{15.21}$$

$$(A \wedge B)_* := A_* \vee B_*, \tag{15.22}$$

$$(\forall x A)_* := \exists x A_*, \tag{15.23}$$

$$(\forall X A)_* := \exists X A_*. \tag{15.24}$$

$\neg A$ must be understood as in Section 7.A.6, i.e., as $A \Rightarrow \curlywedgeuparrow$, with an arbitrary pole $\curlywedgeuparrow$. Negation, which is primarily used to type correctly μ-abstraction, can be internalised by posing $(\neg A)_* := A$, which is all but involutive: the $\lambda\mu$-calculus does not respect the involutivity of negation. This is because we have only a part of **LC**: positive formulas, indispensable to an involutive negation, have been relinquished: they would require a *linear* maintenance of positive variables and covariables, which is delicate to carry out in detail. For the same reasons, the existential quantifier – fundamentally positive – cannot be integrated: we can replace it with $\neg\forall\neg A$, but this is a fall-back solution.

From this we can devise various μ-calculi, adapted to the fragments of **LKT**. One will restrict to the case of implication, i.e., the $\lambda\mu$-calculus of Parigot [85]. It is a simply typed calculus based upon implication and combining the primitives for implication (variables, λ-abstraction, application) with the primitives for negation (covariables, μ-abstraction, naming). The interaction of the two groups of primitives yields an additional immediate reduction:

$$(\mu\alpha t)u \rightsquigarrow \mu\beta t\left[{}^{[\beta](v)u}/{}_{[\alpha]v}\right]. \tag{15.25}$$

In (15.25), any subterm t beginning with α, hence of the form $[\alpha]v$ for a certain v, must be replaced with a $[\alpha](v)u$. This additional rule is easily justified from the equality – or rather the isomorphism – $A \Rightarrow B = \neg(A \wedge B_*)$.

The $\lambda\mu$-calculus can thus be translated back in the λ-calculus with a product type: it thus inherits all its properties, Church–Rosser, normalisation.

15.D.3 $\lambda\mu$-calculus and LC. Since the $\lambda\mu$-calculus is exactly concerned with the negative fragment **LKT** of **LC**, it no wonder that translations exist in both senses. In particular, the $\lambda\mu$-calculus can be translated in terms of polarised nets, see *infra*.

Although this calculus has a functional look, its category-theoretic significance is not that clear. Indeed, terms must be seen as morphisms and one must compose them, which mechanically brings us back to the problems of Section 15.3.1, which are problems of associativity of cut – i.e., of category-theoretic composition. The *control categories* of Selinger [93] provide a general framework embodying the concrete solution given for **LC** in Section 15.3 (comonoids, central morphisms).

15.D.4 $\lambda\mu$-calculus and polarised nets. **LLP** is the linear version of **LC**. One can thus give a translation of the $\lambda\mu$-calculus in terms of polarised proof-nets [75]:

(i) We will work with underlining of free variables (not covariables); as in Section 15.C.5, we associate a net to a term; when the term is of type $\perp$, $\perp$ is not part of the conclusions. We must also take into account the covariables $\alpha^B, \beta^C \ldots$: each of them is represented by a *unique* conclusion, thus – because of weakening – they must be listed together with the term to avoid ambiguity. By the way, *locativity* (Section 5.1.5) evacuates the false question of two covariables of the same cotype, also of a variable whose cotype would be that of the term: formulas are considered as equal or distinct whether or not one desires to apply contraction to them, in which case, the same covariable is used... fundamentally there is only one of them for each cotype[10]!

[10] In other words, the name of the covariable is the location of its cotype; two distinct covariables induce distinct locations, hence distinct cotypes; see Section 5.3.3.

(ii) The constructions of the « μ-system » are not represented; indeed, a term n of type $⅄$ is typed by a formula not present in the translation. On the other hand, the interpretation of binary constructions – the two variants of application, « generic » and « head », see p. 348 – requires contracting certain conclusions.

We get a quotient of $\lambda\mu$-terms by a « σ-equivalence » which generalises that of Section 15.C.5; besides (15.15) and (15.16), we get

$$\begin{aligned}
(\lambda x\mu\alpha[\beta]u)v &\sim_\sigma \mu\alpha[\beta](\lambda xu)v, && \alpha \notin v,\\
[\alpha'](\mu\alpha[\beta'](\mu\beta n)u)v &\sim_\sigma [\beta'](\mu\beta[\alpha'](\mu\alpha n)v)u, && \alpha \neq \beta,\ \alpha \neq \beta',\ \beta \neq \alpha',\\
& && \alpha \notin u,\ \beta \notin v,\\
[\alpha']\lambda x\mu\alpha[\beta']\lambda y\mu\beta n &\sim_\sigma [\beta']\lambda y\mu\beta[\alpha']\lambda x\mu\alpha n, && \alpha \neq \beta,\ \alpha \neq \beta',\ \beta \neq \alpha',\\
& && x \neq y,\\
[\alpha'](\mu\alpha[\beta']\lambda x\mu\beta n)u &\sim_\sigma [\beta']\lambda x\mu\beta[\alpha'](\mu\alpha n)u, && \alpha \neq \beta,\ \alpha \neq \beta',\ \beta \neq \alpha',\\
& && x \notin u,\ \beta \notin u
\end{aligned}$$

as well as

$$\begin{aligned}
[\beta]\mu\alpha n &\sim_\sigma n[{}^\beta/_\alpha],\\
\mu\alpha[\alpha]u &\sim_\sigma u, && \alpha \notin u.
\end{aligned}$$

Which corresponds to the fact that the « μ-system » is not represented – it is only a teleological subjective, version of right contraction, which only subsists.

15.D.5 What about symmetry? One can question the *real* interest of $\lambda\mu$-calculus, which is eventually a contrived way of operating contractions on the right, without recovering the sane naïveness of λ-calculus. And above all, which runs into a non-involutive negation, which is antagonistic to the proclaimed classical character of the thing: what is « classical » if not an involutive negation? $\lambda\mu$-calculus reminds us of Orwell and his *Animal Farm*: from now on, A and $\neg A$ will be handled symmetrically, but one will remain more symmetrical than the other.

The original system, **LC**, had a truly involutive negation: which requires *both* polarities. Polarised nets, not confined to the negative fragment **LKT**, are winning on all issues: the advantages of **LC** without the defects of sequent calculus. Of course this is not functional; but who decreed that classical logic should be – or rather should pretend to be – functional?

Part V

Iconoclasm

Chapter 16
Heterodox exponentials

16.1 The quarrel of images

One can contend that ludics resolves *perfect* logic, at least in principle: it provides a setting – the analytical theorems – where one can perform a not too fabricated synthesis; we are not done, but we can see the flickering light at the end of the tunnel! The question is to determine what to do with the *imperfect* part, i.e., with exponentials, which has been outlined by *exponential ludics* (Section 15.B.2). But one starts to have doubts about the endeavour. Indeed, the gap separating a perfect world – of very restricted expressivity, but harmonious – from an imperfect one – where the growth of functions can no longer be controlled – is the sign that something is going wrong. Remember those towers of exponentials whose height is a tower of exponentials... do we really believe in that? Such monsters are nevertheless the necessary consequence of the « mental image » of the logical world that nests in our minds; while the experience of perfect logic entitles one to question this badly infinite infinity, this very perennial pereniality. The question is therefore: should one respect mental images, be *iconodule*, or should one be *iconoclast*[1]? Instead of synthesising the exponentials as we know them, shouldn't we rather modify them to make them closer to the perfect world?

16.1.1 Classical absolutism. The strongest iconodule argument is *evidence*: the world is classical, because our fundamental intuitions are classical; the classical is an absolute that one cannot surpass... This conformism rests upon a long experience, upon an undeniable internal coherence; at the foundational level, it also rests upon a marked taste for essence, for revealed truth.

One must say that even constructivists are of the same opinion: thus, Martin-Löf believes in the set of integers; should we conclude that his theory of types is only a layer of constructive varnish applied on a classical and set-theoretic wall? In the same spirit, an American colleague once told me that the work of Abrusci (Section 11.B.3) on Aristotelician syllogisms [3] is pure baloney, under the fallacious argument that Aristotle had in mind... classical logic! This incredible anachronism indicates an unconscious prejudice: « Everything you just read is nice, but not that serious: when the recess is over, the children must go back to the classical ».

[1] The *quarrel of images* ravaged Byzantium during the VIIIth and IXth centuries; the iconoclast – enemies of images – emperors, e.g., Constantin Copronymos *(sic)*, destroyed all mosaics – except in the places no longer under their control, Ravenna for instance.

This pregnancy of the « classical » is almost unstoppable. Except that this is circular, that this is the very *blind spot*; you may be right, Mr. Iconodule, but admit that one cannot see anything: in a case like this, one should only trust *indirect* evidence.

Fortunately for iconoclasm, there are **LLL** and **ELL**, systems that we have referred to in passing and shall soon introduce. These experimental systems invert, to their profit, the « revealed » aspect of exponentials, by giving *light* versions of them, with a maintenance of a subtler infinity, inaccessible to set-theoretical methods. The sole existence of these systems is enough to refute the *a priori* objections, resting upon a so-called priority of the classical: while admitting a certain amount of *perenniality*, they present a less absolute, less desperately frozen, version of it.

16.1.2 Mathematics. Modifying logic – since this is the eventual goal of iconoclasm – means giving up mathematics, which can be expressed, as one knows, in set-theory. A light logic would thus lose mathematical results: « You want to destroy the mosaics ! » – say the iconodules. This is not that obvious:

- First, one should not confuse mathematics with mathematics revisited by logicians, which contains an enormous amount of infinite combinatorics. « Real » mathematics makes a more restricted use of the infinite, thus being less sensitive to its precise formulation.

- If, as in **LLL**, the function $m \leadsto 2^m$ is no longer available, this does not mean that a result $\forall m \exists n A[m, n]$ where the solution n is bounded by 2^m is irremediably lost: one could still write it $\forall m \exists n A[\log m, n]$. This is more complicated, but this removes the objection of principle.

Take a musical analogy: the equal temperament is extremely convenient; however, the tempered intervals are slightly out of tune. This does not mean that all music written for the equal temperament – but perhaps for certain exaggerations, e.g., dodecaphonism – should be dumped. In the same way, classical mathematics is only *slightly* wrong, as long as one does not enter into logicist exaggerations.

16.1.3 Sophistics. A sophist is the guy that says to his teacher – of sophistics: « of two things, either you were a bad teacher and I owe no money to you, or you educated me well and I can produce a sophism proving that I owe nothing to you ». The same kind of argument proves the impossibility of motion or the dumbness of general relativity, not to speak of quantum mechanics; if a three-dimensional variety is *embeddable* in a four-dimensional euclidian space, sophistics will conclude that the world is eventually euclidian: this is the theme of *hyperspace*, familiar to science-fiction fans.

The impressive work done around 1900 enables us to « encode everything » in set-theory. Thus, even the most delirious iconoclasm can be represented in set-theory: which is thus primal, according to a certain sophistics, as much as in the aforementioned hyperspace. But nobody takes hyperspace seriously: the euclidian space is rather seen as a convenient frame for the approximation of « true » geometry. In the same way, one can contend that set-theory has no real sense, that it is only a convenient *reification* of a reality difficult to access.

16.1.4 Iconoclast inconsistencies. The iconoclast viewpoint is delicate, since lacking in coherence; in particular, the present systems, like **LLL** and **ELL** are only experimental. But this is a *dynamic* position, with its future ahead and a captivating motivation: complexity theory. The gradual setting of an iconoclast logic should lead to pose questions in a radically different way. In particular, to find nuances, mistakes, in the prevailing foundational paradigm, which mainly rests on an uncouth approach to natural numbers.

16.2 Exponentials

16.2.1 Kronecker. Any foundational iconoclasm sooner or later stumbles on the *absoluteness* of integers. Thus, the various techniques introduced in these lectures – which mainly belong in finite combinatorics – bring us back to natural numbers. And one does not know how to « unscrew » natural numbers. Said Kronecker: « God created the integers, everything else is the deed of man ». How can one call into question this absoluteness of integers?

Non standard: non-standard models of (classical) arithmetic introduce integers « after » the « real ones ». This is not very convincing: who has ever seen, who has ever been able to compute, a non-standard integer?

Ultrafinitist: a proposition, not quite serious, by Essenin-Volpin: there would be integers only up to – say – 19. This want of earnestness is confirmed by his claim that these methods can be used to prove... the consistency of set-theory **ZF**!

Dynamic: one no longer tries to enlarge or shorten the set $\mathbb{N}$ of integers, since $\mathbb{N}$ is presumably only a *reification*. When one enunciates a theorem on integers, it is true *because it has been proved*, which in no way presupposes that $\{0, 1, 2, 3, \dots\}$ makes any sense. What is important is the *process* of construction, the *dynamics*.

16.2.2 A challenge: complexity. The dynamical viewpoint tries to rationalise the intuition that a tower of exponentials whose height is in turn a tower of exponentials is a perverse effect of the classical formalism; that one accepts as a fall-back solution, by no way a reality. This cannot be seen on such and such value of the argument – this would bring us back to the rut of ultrafinitism –, but on the wholeness of its *parametric* construction. Set-theory does not allow the distinction between a parametric construction and the *set* of its instanciations corresponding to the *values* $n = 0, 1, 2, \ldots$ of the parameter, i.e., it neglects dynamics.

Due to the emergence of algorithmic *complexity* in the last part of the XXth century, the question of dynamics can now be seen as a central one. The main achievement of this theory is the individuation of *complexity classes*, corresponding to the time or the space needed for the computation. Above all one knows P (problems *computable* in polynomial time) and NP (problems *verifiable* in polynomial time) and the famous 1 000 000 $ question:

$$\mathrm{P} \stackrel{?}{=} \mathrm{NP}\,.$$

Each complexity class possesses indirect and interchangeable characterisations, none of them being a mathematical definition in the noble sense of the term – e.g., a preservation property. Which might perhaps be related to the rather modest outcome of the area: in more than thirty years, not a sole serious separation result between complexity classes!

I propose to take complexity seriously, not as a problem of informatics, but as a *problem of logic*. Although iconoclastic, the following thesis is very exciting:

Complexity classes do correspond to various sizes of infinity.

Obviously, this no longer belongs in the Cantorian infinity, which classifies the set-theoretic, static, infinite, in terms of *cardinals*. Nor in an *intensional* infinity that would classify the infinite according to stronger or weaker systems, e.g., the *reverse mathematics* of Section 3.C.4 or various « bounded arithmetics », those bleak bureaucracies of complexity. One can surpass this infantile stage by playing on the sole *formal* trace of infinity in logic, exponentials.

16.2.3 Exponentials and integers. Moreover, we have a trump card, precisely this abrupt, sharp, modal, side of exponentials: « this is like that » . Which can be read *a contrario*: « one can modify them *ad libitum* ». We will thus explore the universe of alternative exponentials. Everything is permitted, but in order to take only the fair side of this freedom, we will impose a constraint: the modified exponentials must be of « tame » growth. No matter which, polynomial, exponential, provided the growth is tame. This is a job for the *lightened* logics, **LLL** and **ELL**: « Light Linear Logic » and « Elementary Linear Logic » first introduced in [49].

System **F** enables us to define integers à la Dedekind (Section 6.1.6):

$$\mathbf{nat} := \forall X(X \Rightarrow ((X \Rightarrow X) \Rightarrow X)),$$

written in a more legible way[2] **nat** $:= \forall X((X \Rightarrow X) \Rightarrow (X \Rightarrow X))$. This definition translates as **nat** $:= \forall X(!(!X \multimap X) \multimap (!X \multimap X))$, with a big stock of exclamation marks. The same effect can be obtained with the simplified version

$$\textbf{nat} := \forall X(!(X \multimap X) \multimap (X \multimap X)) \tag{16.1}$$

which is the type of the functional sending $f \in X \multimap X$ to $f \circ \cdots \circ f \in X \multimap X$.

Analysing integers, surpassing Kronecker's absolute, could thus be an « unscrewing » of the « ! » that struts about definition (16.1). Which is not that simple, since the possible definitions of the exponential deal with *finite* reuses, *finite* cliques, i.e., *presuppose* the integers. More generally, we see that all analyses based upon categories eventually run in circles. We will seek our happiness in layer −3, with proof-nets. Later, in operator algebras, if possible; see our last chapters.

16.2.4 The lesson of nets. Without much noise, something essential did occur with multiplicative nets. Let us indeed compare the proof of normalisation for natural deduction (Section 4.3.5) with the (erroneous) proof of normalisation for proof-*structures* of Section 11.2.6. In the first case, a parameter, the *degree*, measures in its way the logical complexity, thus the algorithmic complexity of normalisation, i.e., the height of the tower of exponentials of Section 4.C.4. In the second case, we see that normalisation is performed in linear time, since every step shrinks the size – i.e., the number of links – of the structure. What could happen is that one eventually reaches a vicious circle (Section 11.2.6), but one would then be wise enough to stop.

Let us go further and think of system **F**; there is no longer any degree, since functions grow too fast, but there is still a logical control, operated by *reducibility candidates* (Section 6.2.4). On the other hand, if I extend multiplicative logic to second order, no significant increase in the complexity of normalisation can be observed: the number of steps remains bounded by the number of links.

To sum up, in the perfect world, logic does not control complexity – while its does in the imperfect world. Could one find a weakened version of the imperfect world – *light* exponentials – in which one would observe the same phenomenon, i.e., *a priori* bounds on the complexity of normalisation, independent from the logical complexity of formulas and even from the correctness of their proofs?

16.3 Russell's antinomy

16.3.1 A paragon of complexity. What is the paragon of algorithmic or logical complexity? Imagine a very inexpressive system: from its viewpoint, there is not much difference between a tower of towers of exponentials and a diverging normalisation, between a system of high logical complexity and an inconsistent one. In

[2] Neglecting the technicalities linked to « η » (Section 7.4.2)!

other words, we can metaphorically consider non-termination, Russell's antinomy, as the worst possible complexity. Hence the idea of analysing what, in Russell's antinomy provokes non-termination: one knows that it is due to exponentials; if one can isolate the principles causing non-termination and if one can redesign viable exponentials by relinquishing these principles, one will get a logical system whose complexity is *a priori* bounded.

What follows is very unexpected by many standards: in one century, thousands of pages have been written on Russell's antinomy without any significant progress. The decomposition perfect/imperfect operated by linear logic enables one to make a breakthrough on this very issue.

16.3.2 Dissection of exponentials. Russell's antinomy can be decomposed in two steps:

(i) The construction of a fixed point for negation, say $A = !{\sim}A$.

(ii) The logical transformation of this fixed point into a contradiction.

The first step is without interest: it means that logical complexity is not bridled. Everything concentrates on the second step and on the fine grain analysis of exponentials. Here follows a list of « micro-properties » of exponentials:

$$!(A \,\&\, B) \vdash !A \otimes !B \tag{16.2}$$

$$!A \otimes !B \vdash !(A \,\&\, B) \tag{16.3}$$

$$\frac{A \vdash B}{!A \vdash !B} \tag{16.4}$$

$$!A \vdash ?A \tag{16.5}$$

$$!A \otimes !B \vdash !(A \otimes B) \tag{16.6}$$

$$!A \vdash A \tag{16.7}$$

$$!A \vdash !!A \tag{16.8}$$

The first two principles correspond to the isomorphism at the origin of the word « exponentials »: (16.2) expresses contraction and (16.3) expresses weakening. (16.4) expresses the functoriality of « ! »: it is the solution of a universal problem. (16.5) is a weak form of dereliction. These four principles constitute the base of **LLL**.

(16.6), combined with (16.4), corresponds to *multilinear* functoriality: from $\Gamma \vdash B$ conclude $!\Gamma \vdash !B$. **ELL** corresponds to the first five principles.

(16.7) is dereliction. « Burying » (16.8) corresponds to the fact that, from $!\Gamma \vdash A$ one deduces $!\Gamma \vdash\ !A$ and not simply $!!\Gamma \vdash\ !A$. Both principles are *separately* faulty: in the presence of a fixed point $A = !{\sim}A$, one gets the empty sequent $\vdash$ – which is not quite contradictory, but not cut-free provable anyway – in two ways:

(i) From (16.2) + (16.4) + (16.7).

(ii) From (16.2) + (16.4) + (16.5) + (16.8).

Indeed, $\vdash A, ?A$ is an identity axiom; one deduces $\vdash ?A, ?A$, as we please:

(i) By a dereliction (16.7) which directly yields $\vdash ?A, ?A$.

(ii) (16.4) yields $\vdash !A, ??A$, then (16.5) yields $\vdash ?A, ??A$ and (16.8) removes the extra « ? ».

From $\vdash ?A, ?A$, contraction (16.2) yields $\vdash ?A$, i.e., $\vdash {\sim}A$, which, by (16.4), entails $\vdash !{\sim}A$. One concludes by a cut between $\vdash ?A$ and $\vdash !{\sim}A$.

The principles (16.7) and (16.8) are excluded for good. The pseudo-dereliction (16.5), which occurs in the reference version [49], has not been retained here. **LLL**, with the sole (16.4) would be too weak: which explains the « spare » modality, $\S A$, a sort of castrated « ! », i.e., a « supertype » of « ! ».

16.3.3 Russell and normalisation. Let us try to understand how cut-elimination works in the two possible translations of Russell's antinomy.

The cut between $\vdash ?A$ and $\vdash !{\sim}A$ reduces to (two) cuts between $\vdash ?A, ?A$ and two copies of $\vdash !{\sim}A$. Then we proceed differently:

(i) If $\vdash ?A, ?A$ is obtained through a dereliction from $\vdash A, ?A$, we *open* one of the boxes $\vdash !{\sim}A$ which yields $\vdash {\sim}A$, which yields in turn two cuts between $\vdash A, ?A$, $\vdash {\sim}A$ and $\vdash !{\sim}A$. Since $\vdash A, ?A$ is an identity axiom, the cut between $\vdash A, ?A$ and $\vdash {\sim}A$ reduces to $\vdash ?A$, which finally yields back a cut between $\vdash ?A$ and $\vdash !{\sim}A$, the beginning of an infinite loop – of an eternal golden braid, would say the poet!

(ii) In the second case, the system reduces into two cuts between $\vdash ?A, ??A$, $\vdash !{\sim}A$ and $\vdash !!{\sim}A$, next between $\vdash !A, ??A$, $\vdash ?{\sim}A$ and $\vdash !!{\sim}A$. Which simplifies into a cut between $\vdash ??A$ and $\vdash !!{\sim}A$. One « loops » too, but by entering one notch « inside the boxes ».

We see that dereliction opens a box (depth 1) to pour it out at depth 0, while burying moves the cut from depth 0 to depth 1. The principles (16.7) and (16.8) which do not respect the depth – contrary to the others – are the cause of non-termination and, more generally, of unlimited complexity.

16.4 LLL and ELL

16.4.1 The systems. The language contains multiplicatives, first- and second-order quantifiers, as well as the exponentials !, ?, §, $\tilde{\S}$, with $\sim\S A := \tilde{\S}\sim A$, ...; §, $\tilde{\S}$ are useful only in the case of **LLL**; note the absence of additives. The exponential rules are as follows:

(i) Weakening on all negative formulas ($\tilde{\S}A$ being declared negative).

(ii) Contraction on formulas $?A$.

(iii) Two *promotion* rules:

$$\frac{\vdash B_1, \ldots, B_n, A}{\vdash \mu_1 B_1, \ldots, \mu_n B_n, \nu B} \tag{16.9}$$

Either $\nu = !$ and for $i = 1, \ldots, n$, $\mu_i = ?$; in the case of **LLL**, $n \leq 1$. Or $\nu = \S$ and for $i = 1, \ldots, n$, $\mu_i = \tilde{\S}$ or $\mu_i = ?$.

Which is not quite the original system, but rather a simplified variant due to Asperti [7]. The original system [49] is indeed complicated by the presence of *additives*; but, in the presence of weakening, they can be defined *without exponentials* by means of

$$A \oplus B := \forall X((A \multimap X) \otimes (B \multimap X) \multimap X), \tag{16.10}$$
$$A \,\&\, B := \exists X((X \multimap A) \otimes (X \multimap B) \otimes X) \tag{16.11}$$

which are simplified versions of the translations of Section 12.B.2. Weakening is acceptable in a polarised world (Sections 14.C and 15.3.1), which is the case here.

In general, there are many variants of **ELL** and especially **LLL**, without one being able to arbitrate in favour of this one or that one. Thus, the original version of **LLL** contained the extra principle (16.5), which was not retained here; it declared the « paragraph » as self-dual: $\tilde{\S} = \S$; it had no weakening and was fussing about a pedantic restriction: exactly one formula in the context of the promotion of « ! ». All these variants are based upon the respect for the depth of boxes, but we can hardly see anything beyond. So, if one is to navigate by sight, one might as well go to the simplest, i.e., Asperti's variant.

16.4.2 Light nets. We use underlinings (Section 11.C.2); but the sole underlined formulas will be conclusions of boxes, which accounts for the absence of dereliction. From a net with conclusions Γ, A[3], we can construct a box of conclusions $\underline{\Gamma}, !A$ or $\underline{\Gamma}, \S A$; in **LLL**, the boxes $\underline{\Gamma}, !A$ are restricted to the case where Γ has at most one

[3] These conclusions are not underlined, which corresponds to the absence of burying.

element. The link $\tilde{\S}$, of conclusion $\tilde{\S}A$, has at most one premise $\underline{A}$; this premise cannot be the conclusion of a link « ! ». The link ?, of conclusion $?A$, has n premises $\underline{A}$, ($n \geq 0$). The 0-ary links $\tilde{\S}$ and ? are particular cases of a more general « weakening » link, without a premise and a negative conclusion.

The *size* is measured depthwise: $s_0, s_1, s_2, \dots$. For a certain d, $s_d = 0$ and this *degree* will stay the same during normalisation. s_0 counts the links at depth 0, with a certain pounderation: each link has a weight, the number of its non-underlined conclusions; thus boxes, multiplicative and quantifier links count for 1; the same for the links $\tilde{\S}$, ? and weakening. But the axiom counts for 2, while cut is not counted.

16.4.3 Bounds for LLL. We know that everything normalises, so I will content myself with bounds on the size of the normalised net. These bounds are easily converted into bounds on the computation time.

(i) We begin by normalising at depth 0 all cuts, except exponential ones. We know that the size shrinks, so that we can keep the same bounds.

(ii) A cut between $\S A$, conclusion of a box of conclusions $\underline{\Gamma}$, $\S A$ and $\tilde{\S}{\sim}A$, obtained by a link $\tilde{\S}$ from a box of conclusions $\underline{{\sim}A}, \underline{\Delta}, \S B$ normalises by « burying » the cut at depth 1, between A and ${\sim}A$ and by reforming a box of conclusions $\underline{\Gamma}, \underline{\Delta}, \S B$; the size shrinks.

Finally, only cuts !/? remain. These cuts are handled as in the case of the « paragraph », with an essential difference, due to duplication. If I start with $?A_0$, cut with $!{\sim}A_0$, there is a multiplicative factor equal to $n-1$, where n is the arity of the link of conclusion $?A_0$. What really causes a problem is the context $\underline{A}_1$ of the box introducing $!{\sim}A_0$: this context is also multiplied by $n-1$. We see that the duplication in A_0 is transmitted just so to A_1; we can proceed in the same way if $\underline{A}_1$ is a premise of a « cut » contraction, which can lead to A_2. Indeed, if we want to count the duplications, we must take the paths $A_0, A_1, A_2, \dots, A_k$ which are maximal, i.e., which cannot be extended, neither before 0, nor after k. These paths are finite, i.e., $A_0 \neq A_k$: it suffices to choose an adequate switching. How many of them? This is very simple, as many as possible choices for A_0, i.e., less than the size s_0 of the net at depth 0. Therefore, the size does not increase at depth 0, but it is at most multiplied by a factor s_0 in the lower depths. We can redo this at depths 1, then 2, etc. We get the following bounds:

$$
\begin{aligned}
\text{Depth } 0: &\quad s_0, s_1, s_2, \dots, s_d, \\
\text{Depth } 1: &\quad s_0, s_0 s_1, s_0 s_2, \dots, s_0 s_d, \\
\text{Depth } 2: &\quad s_0, s_0 s_1, s_0^2 s_1 s_2, \dots, s_0^2 s_1 s_d, \\
&\quad \vdots \\
\text{Depth } d: &\quad s_0, s_0 s_1, s_0^2 s_1 s_2, \dots, s_0^{2^{d-1}} s_1^{2^{d-2}} \dots s_{d-2}^2 s_{d-1} s_d.
\end{aligned}
$$

We thus see that, when normalisation is over, the global size has been changed from $s = s_0 + s_1 + s_2 + \cdots + s_d$ to at most s^{2^d}. For a given d, this a polynomial, hence the polynomial time.

16.4.4 Bounds for ELL. In **ELL**, exponential boxes are of conclusions $\underline{\Gamma}, !A$, without constraint on Γ. We see that the « cleansing » of depth 0 induces a multiplication by an exponential factor of the lower depths, since there are many more sequences $A_0, A_1, A_2, \ldots, A_k$: the number of such sequences is bounded by 2^{s_0}. We see that the complexity of **ELL**, is, for a fixed d, dominated by a tower of exponentials.

16.5 Expressive power

We know that the complexity of normalisation – for a given depth – is polynomial (**LLL**) or *elementary*, i.e., bounded by a tower of exponentials (**ELL**). We shall prove the converse. The essential effort will be put on **LLL**.

16.5.1 Coding of polynomials. In **LLL**, integers are typed by

$$\mathbf{nat} := \forall X(!(X \multimap X) \multimap \S(X \multimap X)). \tag{16.12}$$

We must have an exponential « in output », so as to preserve depth. In **ELL**, we can take « ! »; this does not work with **LLL**, since we could not type integers $2, 3, \ldots$ which require (16.6). This is why we use the « paragraph » and this is by the way the sole reason for its creation.

Given A and $f \in A \multimap A$, we can form $!f \in !(A \multimap A)$. If $x \in \mathbf{nat}$, then $(\{x\}A)!f \in \S(A \multimap A)$. In other terms, one can iterate and the result is of type $\S(A \multimap A)$.

We can represent the following functions:

Sum: $m, n \rightsquigarrow m + n$ of type $\mathbf{nat}, \mathbf{nat} \vdash \mathbf{nat}$: here we basically use $\S(X \multimap X)$, $\S(X \multimap X) \vdash \S(X \multimap X)$.

Product: $m, n \rightsquigarrow n \cdot m$ of type $\mathbf{nat}, !\mathbf{nat} \vdash \S\mathbf{nat}$; we iterate addition, of type $!\mathbf{nat} \vdash !(\mathbf{nat} \multimap \mathbf{nat})$, which yields $\mathbf{nat}, !\mathbf{nat} \vdash \S(\mathbf{nat} \multimap \mathbf{nat})$, hence $\mathbf{nat}$, $!\mathbf{nat}, \S\mathbf{nat} \vdash \S\mathbf{nat}$, corresponding to $m, !n, \S p \rightsquigarrow \S(p + n \cdot m)$.

Square: we can build an object of type $\mathbf{nat} \vdash \S(\mathbf{nat} \otimes !\mathbf{nat})$ corresponding to $n \rightsquigarrow \S(n \otimes !n)$. Combining this with the product, we see that squaring can receive the type $\mathbf{nat} \multimap \S\S\mathbf{nat}$, which corresponds to $n \rightsquigarrow \S\S n^2$.

General polynomials: for instance, n^4 can be typed $\mathbf{nat} \multimap \S\S\S\S\mathbf{nat}$.

16.5.2 Coding of polynomial time. We must code integers in base 2:

$$\mathbf{nat}_2 := \forall X(!(X \multimap X) \multimap (!(X \multimap X) \multimap \S(X \multimap X))). \tag{16.13}$$

We can type polynomials in the length $|n|$ of a binary integer. For this it suffices to remark that the identification of the two arguments in a binary integer n (by contraction) yields the integer $|n|$ of type **nat**. The next step is to code a Turing machine in **LLL**:

Tape: the tape can be represented by a generalisation of $\mathbf{nat}_2$, say $\mathbf{nat}_N$, where N is the number of symbols that can be written on it.

States: a type of the form $\mathbf{bool}_S := \forall X(X \otimes \cdots \otimes X \multimap X)$, with S elements, can be used for the current state of the machine.

Reading head: the type **nat** can be used for the current position of the reading head.

So that the current position of a Turing machine can be given the type $\mathbf{Tur} := \mathbf{nat}_N \otimes \mathbf{bool}_S \otimes \mathbf{nat}$ and the machine itself the type $\mathbf{Tur} \multimap \mathbf{Tur}$. Given the input $n \in \mathbf{nat}_2$, one can type the initial position of the machine $p_n \in \mathbf{Tur}$, as well as the iteration $P(|n|)$ of the machine from the initial position p_n.

By cooking all this together, we actually get that every polynomial time algorithm can be given the type $\mathbf{nat}_2 \multimap \S\S\ldots\S\S\mathbf{nat}_2$, with a number of « paragraphs » depending on the degree of the polynomial.

16.5.3 About the « paragraph ». A cut-free proof of $\S(A \oplus B)$ ends with a « $\oplus$ » followed by a « $\S$ » that one can commute « handwise » into $\S A \oplus \S B$, but not in an internal way: one would seek in vain a proof of $\S(A \oplus B) \vdash \S A \oplus \S B$. I give a *procedural* argument; the same would also work against $!(A \oplus B) \vdash !A \oplus !B$.

In the presence of this principle (supposedly doing what one thinks), every polynomial time algorithm with a boolean output (hence of type $A \oplus B$) normally typed $\mathbf{nat}_2 \vdash \S\S\ldots\S\S(A \oplus B)$, could be retyped $\mathbf{nat}_2 \vdash \S\S\ldots\S\S A \oplus \S\S\ldots\S\S B$. But normalisation is done in such a way that depth 0 is « cleansed » in linear time. Nothing new will befall us later; now the *bit* left/right which distinguishes the two values has been taken back to depth 0. No need to proceed with normalisation, we already know the result!

16.5.4 Case of ELL. We define, without scheming,

$$\mathbf{nat} := \forall X(!(X \multimap X) \multimap !(X \multimap X)), \tag{16.14}$$

and we check that multiplication by 2 can be given the type $\mathbf{nat} \multimap \mathbf{nat}$. Which enables us to give the type $\mathbf{nat} \multimap !(\mathbf{nat} \multimap \mathbf{nat})$, hence $\mathbf{nat}, !\mathbf{nat} \vdash !\mathbf{nat}$ to the iteration of multiplication by 2, i.e., to the function $n, !p \rightsquigarrow !(p.2^n)$. We thus get the type **nat** $\vdash$!**nat** for the exponential $n \rightsquigarrow !2^n$, hence **nat** $\multimap$ $!!\ldots!!$**nat** for a tower of exponentials.

Chapter 17
Quantum coherent spaces

The experimental gait of this chapter is easily *regressive*. Thus, we relinquished *coherent spaces* – for reason of unfaithfulness, see the discussion in Section 12.1; this being said, it is a most simple technique which can sometimes directly lead to the essential: this is why we shall use them again to establish a link with the quantum world. This link, see [52], is soley valid for finite-dimensional spaces; to go further, other techniques (*Geometry of Interaction*) will be needed, but what we shall see here is worth the detour.

17.1 Logic vs. quantum

17.1.1 A missed encounter. No need to go very far to understand why the relation between logic and quantum has been this complete failure: in the same way Frege had the nerve to make fun of the revolutionary ideas of Riemann, logicians were not afraid to declare that *nature makes mistakes* and therefore tried to *reform*, to « reformat » it. Thus the notorious quantum logic – an expression of the style « popular democracy », where the role of the adjective is to negate the noun[1]. It is necessary to make it clear from the very start:

What follows has nothing to do with quantum « logic ».

17.1.2 Characteristics of the quantum. What strikes people about the quantum, is its non-determinism, hardly accepted for essentially ideological reasons, even by the great Einstein: « God does not play dice with the world ». However, in a strictly deterministic world, the theory of *chaos*, initiated by Poincaré, shows the *practical* impossibility of predicting the outcome of the national lottery – or the position of the solar system in 10^6 years: this is the famous metaphor of the butterfly. But the chaos is not too shocking, since it leaves open the possibility of an *abstract*, inhuman, determinism: if one actually *knew* how to photograph the world at instant t, one *could* deduce its position at instant $t + 10$. It is however necessary to remark that this « if one knew » is just as dubious as the idea that – beyond incompleteness – there *would be* a stable, eternal, notion of truth: *chaos* looks like a finite, effective, version of incompleteness, even in some of its readings.

[1]Another example: « labeled deductive systems »; « labeled » meaning that the system is not deductive; think also of the slogan « logic plus control ».

Nothing of the like in the quantum world: one cannot be reassured with the idea that there would be too many parameters, hardly measurable. One must admit that measurement influences the result in a very strong sense: not only one modifies it, but *creates* it. Thus, when I measure a *spin*, I measure it w.r.t. an axis, say, $\vec{Z}$ and I find an actual value $\pm 1/2$ w.r.t. this axis, while the electron under measurement had none before. This sort of behaviour entails non-determinism, which shocked many people; it induces various « hidden variables » theories, all of them worn out[2].

Deeper than non-determinism is the *imbrication* between the observer and the system under observation. Contrary to the usual physical world, the quantum does not accept a dichotomy subject/object.

17.1.3 Quantum logic. Von Neumann himself should be held « responsible » for the birth of quantum logic; but not guilty, since, in the beginning of the years 1930, it was natural to make attempts.

This was indeed an approach of layer -1, based upon a modification of the truth values. One knows that classical logic admits a *semantics* in terms of the truth values $\mathbf{v}, \mathbf{f}$ and, more generally, in terms of *boolean algebras*. Von Neumann did propose replacing boolean algebras with the lattice of closed subspaces of a Hilbert space... which yielded strictly nothing. By the way, von Neumann soon took a much more fruitful direction – known to us as *von Neumann algebras*.

Independently from its technical vacuousness, quantum logic was a mistake *a priori*. Indeed, layer -1 rests upon the duality syntax/semantics, i.e., the *schizophrenia* subject/object, in opposition to quantum mechanics which rests upon their *imbrication*. This level of reading supposes a fregean position: thus the expression « the impulsion of M » has a denotation, i.e., a *value*, which is a real number; one can make use of the swing which opposes – thus relates – the *sense* and its *denotation*. But, when I say « the *spin* of e », this has strictly no denotation, no « value », in no space, over-ornate or not.

Let us quickly conclude: if an idea is bad, one cannot *fix* it by a formalisation. This is nevertheless what quantum logicians did by introducing « orthomodular lattices », thus kicking out the only interesting datum, the Hilbert space. Since that time, quantum logic has vegetated as a theory of lattices, preferably ill-behaved, without any relation to quantum mechanics. Some quantum logicians *even didn't know the notion* of Hilbert space.

The attitude of logic w.r.t. quantum can be summarised by a sophism: one can describe the quantum world with mathematics and mathematics can be embedded in set-theory, thus in logic. It only remains to code this « doohickey ». Anything is good, preferably the most *ad hoc* possible, to mark a very fregean reprobation in the face of the mistake committed by Nature. This is reminiscent of the anecdote

[2] But in logic: the fashion of quantum computation induced a comeback to oldies, style « Gleason's theorem », in a new rear-guard attempt at justifying hidden variables and rehabilitating determinism!

reported by Herodotus (VII,35): a tempest having destroyed the fleet of ships he had set over the Hellespont, Xerxes had the sea *whipped*. In the same way, quantum logic is a « whipping » of nature, guilty of « illogicality ».

17.1.4 The logic of the gendarme. Remember Valéry and his inimitable pump: « nous autres, civilisations, savons maintenant que nous sommes mortelles ». The quantum is a sort of realisation, not of the mortality of science, but of its *subjectivity*. This is indeed the final stage of a process which dates back to Copernicus.

Which should have been the historical endeavour of logic, supposed to put the *subject* on the front stage; but this hardly tallies up. Thus, in the fregean explanation, the distinction sense/denotation (Section 7.1.1) – the morning star and the evening star referring to the same object – reduces to the impossibility of a *faithful* nomenclature. Which one finds again in the inglorious « intensionality »: two functions receive distinct names (= senses, intensions), while they have the same graph (= denotation, extension). Thus, for Frege, there is an objective reference, that one has no means to reach. His epigones split, on one side the cultivated philosophers who have heard of incompleteness and (try to) cope with it, on the other hand the half-wits of artificial intelligence who produce software for deciding everything, i.e., for computing the denotation.

To understand to what extent logic missed its target, let us examine the worst logical deliriums: instead of freeing the subject, the logical fiddlers deprived him of the limited space bestowed by Frege. Thus, epistemic « logic », supposed, more than any other, to deal with the *subject*, denies him any autonomy: in this wedding cake known as the Baghdad cuckolds (Section 2.3.3), there is an *objective* truth (which is/is not a cuckold) and evolving *subjects* (the cuckolds) of which, at each time, one can say what they know or don't know, deduction taking a compulsory character; one thus inexorably proceeds towards the final conclusion, i.e., the throat cutting of the guilty spouses. This confusion between *constatation* and *reasoning*, corresponds to no deductive reality; one finds no more evidence of it in economic behaviours, where the idea of a crystalline transparency would be rather… unwelcome, Mr. Madoff! One can, however, put it side by side with the techniques officialised by the Americans during the war in Iraq: unless exceptionally heroic, he who answers « I don't know » under the « question » has nothing to say. This police conception of the subjective uses specific cognitive tools, e.g., bathtubs.

To sum up, the logicist version of subjectivity is circumscribed to limited limitations of cognition. Whereas the quantum teaches us that the object is constituted concomitantly with the subject from whom it cannot – at least theoretically – be detached.

17.1.5 The question. And still the problem of the relation between logic and quantum remains. Here, more than ever, one sees that the question at stake is precisely that of the *choice of the question*. There is a problem, but which one indeed?

- This problem is not that of a logical explanation of the quantum, although one necessarily expects clarifications. There is a reason, the quantum formalism – wave functions, density matrices – cannot be replaced with sets, graphs, two words that are ubiquitous in the logical world.
- *A contrario*, the belief in a world of ideas – moreover, a rather set-theoretic one – ruling the universe from above is an essentialist prejudice. What if the world of ideas were not what one believes, but were rather quantum? After all, rather than teaching nature, why not try to learn from her?

Instead of trying to interpret the quantum into logic, one will interpret logic into the quantum[3].

17.1.6 Methodology. I must be precise about what I mean by « quantum ». Surely not quantum physics which maintains its own life, far from logic; indeed, I only mean the process of quantum measurement, the way in which quantum beings interact. This could go as far as trying to integrate the types of quantum *socialisation*, e.g., the *fermions* – asocial creatures, like « electrons » – or *bosons* – gregarious particles – in a re-reading of the notion of equality.

One must perhaps consider a third partner, *quantum computating*, a fashionable and interesting idea, albeit still at the stage of *science-fiction*. Quantum coherent spaces perhaps bear some relation to quantum computation and such a relation would be welcome; the papers [94] and [99] are anyway encouraging. If not, their theoretical interest lies in a break with set-theory, a break that category-theory sought without finding it. This is why the viewpoint taken here is that of the opening of the logical space to non set-theoretic techniques. After reading this chapter, it will no longer be tenable to indulge in sophisms like « ideas are language, the language being written with symbols a, b, c, etc. »; indeed, what if those symbols do not commute?

17.1.7 PCS and QCS. The quantum is based upon *superposition*, just like linear logic which comes from coherent spaces and a sort of superposition principle, remember (Section 9.1.1): if $a = \sum_i a_i$, then $F(a) = \sum_i F(a_i)$ (preservation of disjoint unions). We shall start from that, with the idea of an analogy cliques/wave functions (rather: density operators), to revitalise layer -2.

One proceeds in two steps, first a probabilistic, i.e., « commutative » generalisation, then the general case. As to usual coherent spaces, the Ariadne thread is as follows: points do constitute a distinguished basis and cliques correspond to

[3]This shift of viewpoint took me 30 years of (part-time) reflection.

subspaces which express themselves diagonally in this basis, i.e., with coefficients 0, 1. The probabilistic version allows real coefficients on the diagonal. As to the quantum version, it « steps out from the diagonal », by allowing, for instance, an identity function which cannot be described as a clique, i.e. as a subspace.

17.2 Probabilistic coherent spaces

17.2.1 Desessentialisation. We shall lazily approach the quantum, by first adding a probabilistic, i.e., non-deterministic, aspect. Here, the main reference is the « desessentialisation » of coherent spaces of Section 9.1.5. Remember that coherent spaces can be defined, given a *carrier* $|X|$, by the duality between subsets of $|X|$:

$$a \perp b :\iff \sharp(a \cap b) \leq 1, \tag{17.1}$$

and that linear implication corresponds to the adjunction

$$\sharp((F)a \;\cap\; b) = \sharp(\mathrm{Sk}(F) \;\cap\; a \times b). \tag{17.2}$$

Non-determinism essentially concerns the connective « $\oplus$ »: one will randomly choose between A and B in $A \oplus B$. One could thus get expressions $\lambda \cdot a + (1 - \lambda) \cdot b$, with $a \sqsubset A, b \sqsubset B$. Which suggests the replacement of cliques with functions taking their values in [0, 1]. The formulation (17.1) as well as the adjunction (17.2) will survive this generalisation: typically (17.1) becomes (17.3).

By the way, should I recall it? We are at layer -2 and this has – thank you my God! – nothing to do with fuzzy logic!

17.2.2 The bipolar theorem

Definition 78 (Duality). Let $|X|$ be a finite set; if $f, g \colon |X| \mapsto \mathbb{R}$, let $\langle f \mid g \rangle := \sum_{x \in |X|} f(x) \cdot g(x)$ be the *scalar product*. The positive functions f, g are *polar*, notation $f \perp g$, when

$$\langle f \mid g \rangle \leq 1. \tag{17.3}$$

The space $\mathbb{R}^+(|X|)$ of all functions from $|X|$ into $\mathbb{R}^+$ is thus equipped with a *duality*, whose pole is the segment [0, 1] (Section 7.1.1). One defines as usual the *polar* of a set $A \subset \mathbb{R}^+(|X|)$ and:

Definition 79 (Probabilistic coherent spaces). A *probabilistic coherent space* (PCS) is the pair $(|X|, X)$ of a finite *carrier* $|X|$ and a subset $X \subset \mathbb{R}^+(|X|)$ equal to its bipolar.

Theorem 59 (Bipolar). *Let X be a PCS; then*

(i) *X is non-empty; it indeed contains the null function: $0_{|X|} \in X$.*

(ii) *X is a closed convex set.*

(iii) *X is downwards stable: if $f \leq g \in X$, then $f \in X$.*

Conversely, every subset of $\mathbb{R}^+(|X|)$ *enjoying* (i)–(iii) *is a PCS.*

Proof. Any PCS satisfies (i), (ii) and (iii); for instance, if $f, g \in X$, if $h \in {\sim}X$ and $0 \leq \lambda \leq 1$, then $\langle \lambda f + (1-\lambda)g \mid h\rangle = \lambda\langle f \mid h\rangle + (1-\lambda)\langle g \mid h\rangle \leq 1$, hence X is convex. Conversely, if X satisfies (i)–(iii) and if $f \notin X$, the Hahn–Banach theorem applied to the real Banach space $\mathbb{R}(|X|)$ says that a hyperplane separates X (closed convex) from f. In other terms, one can find a linear form φ such that $\varphi(X) \leq 1$, $\varphi(f) > 1$. This linear form is induced by an element $h \in \mathbb{R}(|X|)$ such that $\varphi(g) = \langle g \mid h\rangle$; one defines $h'(x) := \sup(h(x), 0)$. Obviously, $\langle f \mid h'\rangle \geq \langle f \mid h\rangle > 1$. If $g \in X$, define $g'(x) := g(x)$ if $h(x) \geq 0$, $g'(x) := 0$ otherwise. Since $g' \leq g$, (iii) yields $g' \in X$. But $\langle g \mid h'\rangle = \langle g' \mid h\rangle \leq 1$, hence $h' \in {\sim}X$. Then $f \notin {\sim\sim}X = X$. □

Additives. One adopts here a locative viewpoint: one supposes that the carriers $|X|$ and $|Y|$ are disjoint. The additive connectives will build PCS with carrier $|X| \cup |Y|$. If $f \in \mathbb{R}^+(|X|)$, $g \in \mathbb{R}^+(|Y|)$, one defines $f \cup g \in \mathbb{R}^+(|X| \cup |Y|)$ in the obvious way, by gluing; one identifies f with $f \cup 0_{|Y|}$, g with $0_{|X|} \cup g$. The set

$$X \mathbin{\&} Y := \{f \cup g;\ f \in X, g \in Y\} \tag{17.4}$$

is the polar of ${\sim}X \cup {\sim}Y$. On the other hand, $X \cup Y$ is not a PCS; $X \oplus Y$ must be defined as ${\sim\sim}(X \cup Y)$, without hope of removing the bipolar. But Theorem 59 yields:

Proposition 31.

$$X \oplus Y = \{\lambda f \cup (1-\lambda)g;\ f \in X, g \in Y, 0 \leq \lambda \leq 1\}. \tag{17.5}$$

Proof. It is enough to remark that the right-hand side of (17.5) enjoys conditions (i)–(iii): it is therefore a PCS, moreover the smallest containing $X \cup Y$, indeed its convex envelope. □

Multiplicatives. The multiplicative connectives will produce a PCS of carrier $|X| \times |Y|$.

Definition 80 (Adjunction). If $\Phi \in \mathbb{R}^+(|X| \times |Y|)$, if $f \in \mathbb{R}^+(|X|)$, one defines $(\Phi)f \in \mathbb{R}^+(|Y|)$:

$$((\Phi)f)(y) := \sum_{x \in |X|} \Phi(x, y) \cdot f(x) = \langle \Phi(\cdot, y) \mid f\rangle. \tag{17.6}$$

Which makes sense due to the finiteness of $|X|$.

Theorem 60. *The function* $\Phi \rightsquigarrow (\Phi)\cdot$ *is a bijection between* $\mathbb{R}^+(|X| \times |Y|)$ *and the set of linear applications from the convex cone* $\mathbb{R}^+(|X|)$ *to the convex cone* $\mathbb{R}^+(|Y|)$. Φ *can be recovered from the associated linear function* $\varphi = (\Phi)\cdot$ *by means of*

$$\Phi(x, y) = \varphi(\delta_x)(y) \tag{17.7}$$

with $\delta_x(x) := 1$, $\delta_x(y) := 0$ *for* $y \neq x$.

Proof. A linear function satisfies $\varphi(\lambda f + \mu g) = \lambda\varphi(f) + \mu\varphi(g)$ for $\lambda, \mu \geq 0$ and is thus determined by its value on the δ_x, which explains (17.7). Everything is by the way more or less immediate. □

In the ground case (sets, coherent spaces) this would not work: if Φ and f are sets (characteristic functions), $(\Phi)f$ has no reason to be a set. This is why one introduced coherence together with its corollary, the unicity of the witness a in equation (8.14).

Definition 81 (Linear implication). If X, Y are PCS, one defines the PCS $X \multimap Y$ of carrier $|X| \times |Y|$ as the set of all Φ such that $(\Phi)\cdot$ sends X into Y.

Thus, the characteristic function $\Delta_{|X|}$ of the diagonal belongs to $X \multimap X$; indeed $(\Delta_{|X|})f = f$. $X \multimap Y$ is the polar of $\{f \times g; f \in X, g \in {\sim}Y\}$, it is why it is a PCS. Which enables one to introduce $X \parr Y := {\sim}X \multimap Y$ and, dually, $X \otimes Y = {\sim\sim}\{f \times g; f \in X, g \in Y\}$.

Proposition 32. $\parr$ *is commutative, associative and distributes over* &.

Proof. By introducing the obvious notation $\cdot(\Phi)$, one might as well define $X \multimap Y$ as the set of Φ such that $\cdot(\Phi)$ sends ${\sim}Y$ into ${\sim}X$. The result follows by imitation of Theorem 51 (Chapter 14). □

17.3 Quantum coherent spaces

It is only a slight exaggeration to say the quantum version corresponds to the probabilistic one when we have forgotten the basis $\{\delta_x; x \in |X|\}$. Anyway, the first thing to do is to come back to the PCS so as to draw some general considerations.

17.3.1 Methodological backlash. PCS have a vague resemblance to the coherent Banach spaces (CBS) of Section 15.A. If one forgets exponentials, one can restrict to finite-dimensional real spaces. In such a case, a CBS can be handled by means of a euclidian (i.e., finite-dimensional real Hilbert space) space E, by means of the duality

$$x \perp y \iff |\langle x \mid y\rangle| \leq 1. \tag{17.8}$$

Indeed, if X denotes the unit ball of a normed space on the same E, it is immediate that the unit ball of its dual can be identified with $\sim X$, see equation (15.8). Hence, any real finite-dimensional CBS can be described as a set equal to its bipolar in an appropriate euclidian space E. This being said, the definition by means of equation (17.8) is slightly more general than CBS; indeed, assuming $X = \sim\sim X$, $\|x\| := (\sup\{\lambda; \lambda x \in X\}^{-1})$ needs not define a norm. It is possible that $\|x\| = 0$ or worse, that $\|x\| = +\infty$. Which one usually handles by restricting to points of finite « norm »and quotienting by the points of null « norm ». This eventually amounts to modifying E, which shows that (17.8) is not quite more general than the definition of CBS. But, taking into account the *locative* viewpoint, the *subtyping* $X \subset Y$ means that, on the same vector space E, one can have more « coherent » objects, i.e., that the unit ball increases. In other words, the norm decreases, $\|\cdot\|_Y \leq \|\cdot\|_X$. It can thus become null, i.e., become a *semi-norm*; dually, it can become infinite and in this case, there is not even a name for what one gets.

PCS are not defined on euclidian spaces, but on positive cones linked to a *distinguished* basis, which is foreign to the spirit of linear algebra. This being said, positivity can be desessentialised:

Proposition 33. *$f \in \mathbb{R}^+(|X|)$ iff for all $g \in \mathbb{R}^+(|X|)$ the scalar product $\langle f \mid g\rangle$ is positive.*

Proof. Immediate. □

In particular, we shall see that the QCS – which have however an intrinsic notion of positivity, positive hermitians – call for this variability of positivity.

To sum up, the bilinear form $\langle x \mid y\rangle$ induces three dualities, whether one takes as pole $[-1, +1]$ (which yields the norm, i.e., coherence), $[0, +\infty]$ (which yields positivity), or the intersection of both, $[0, 1]$, which corresponds to PCS and what we shall keep just so.

17.3.2 The bipolar theorem strikes back. Let us go back to Theorem 59, in a more general setting. In what follows, E is a finite-dimensional Hilbert space, i.e., a euclidian space. The duality is defined by

$$x \perp y :\iff 0 \leq \langle x \mid y\rangle \leq 1. \tag{17.9}$$

The problem is the characterisation of bipolars.

Theorem 61 (Bipolar). *A subset $C \subset E$ is its own bipolar iff:*

(i) $0 \in C$.

(ii) *C is a closed convex set.*

(iii) *If $nx \in C$ for all $n \in \mathbb{N}$, then $-x \in C$.*

(iv) *If $x, y \in C$, if $\lambda, \mu \geq 0$ and $\lambda x + \mu y \in C$, then $\lambda x \in C$.*

Proof. We begin with necessity; (i) and (ii) are immediate.

(iii): if $nx \in C$ for $n \in \mathbb{N}$ and $z \in {\sim}C$, then $\langle x \mid z\rangle \in [0, 1/n]$ for $n \in \mathbb{N}$, hence $\langle -x \mid z\rangle = \langle x \mid z\rangle = 0 \in [0,1]$.

(iv) induces a sort of converse to (iii): if $x, -x \in C$, then $nx + n(-x) = 0 \in C$, hence

(iii′) if $x, -x \in C$, $nx \in C$.

Let us now pass to sufficiency and assume that C satisfies (i)–(iv); let C^+ be the cone $\bigcup_{n\in\mathbb{N}} n \cdot C$ $(= \bigcup_{\lambda\in\mathbb{R}^+} \lambda \cdot C)$. One can rewrite (iv):

$$C = C^+ \cap (C - C^+) \tag{17.10}$$

If $b \notin C$, then, by (17.10), one must consider two cases:

$b \notin C^+$: by Hahn–Banach, there is a $d \in E$ such that $\langle b \mid d\rangle < 0 \leq \langle c \mid d\rangle$ for all $c \in C$. Condition (iii) implies that $I = \{c;\ \forall n \in \mathbb{N}\ \ nc \in C\}$ is a vector space; $\langle \cdot \mid d\rangle$ thus vanishes on I and one can write $C = I \oplus C'$, with $C' = I^\perp \cap C$. Embedding E in the projective space, C' has a compact closure, whose frontier is made of the lines $\mathbb{R} \cdot a$ included in C'; but there is no such line (they have been removed and put in I): the frontier is therefore empty and C' is compact. It follows that $\langle \cdot \mid d\rangle$ is bounded on C', hence on C, and $\langle b \mid d\rangle < 0 \leq \langle c \mid d\rangle \leq \lambda$. By renormalising d one can suppose that $\lambda = 1$ and then $d \in {\sim}C$ and $b \notin {\sim\sim}C$.

$b \notin C - C^+$: the same Hahn–Banach yields $d \in \mathbb{E}$ s.t. $\langle p \mid d\rangle \leq 1 < \langle b \mid d\rangle$ for all $p \in C - C^+$. Suppose that $\langle c \mid d\rangle < 0$ for some $c \in C$; then $-nc \in C - C^+$ for $n \in \mathbb{N}$ and the values $\langle -nc \mid d\rangle$ cannot be bounded by 1. One deduces that $0 \leq \langle c \mid d\rangle \leq 1 < \langle b \mid d\rangle$ for all $c \in C$. As above, $d \in {\sim}C$ and $b \notin {\sim\sim}C$. □

17.3.3 Norm and order. With the notations of Theorem 61:

Definition 82 (Domain). The *domain* Fin_C of C is the vector space $C^+ - C^+$ generated by $C^+ := \bigcup_{n\in\mathbb{N}} n \cdot C$.

Proposition 34. $\mathrm{Fin}_C = ({\sim}C \cap (-{\sim}C))^\perp$.

Proof. If $c \in C, d \in {\sim}C \cap (-{\sim}C)$, then $\langle c \mid d\rangle = 0$, which subsists for $c \in \mathrm{Fin}_C$, hence $\mathrm{Fin}_C \subset ({\sim}C \cap (-{\sim}C))^\perp$. Conversely, if $c \notin \mathrm{Fin}_C$ there is a vector $d \in (\mathrm{Fin}_C)^\perp$ such that $\langle c \mid d\rangle \neq 0$. But $(\mathrm{Fin}_C)^\perp = C^\perp \subset {\sim}C \cap (-{\sim}C)$, hence $c \notin ({\sim}C \cap (-{\sim}C))^\perp$. □

In other words, the domain of C is the orthogonal of the « null space » of ${\sim}C$, that we shall soon define and characterise.

Definition 83. Fin_C is equipped with a semi-norm $\|\cdot\|_C$ and a preorder $\preccurlyeq_C$:

$$\|x\|_C = \sup\{|\langle x \mid d\rangle| \; ; \; d \in {\sim}C\}, \tag{17.11}$$

$$x \preccurlyeq_C y \iff \forall d \in {\sim}C \;\; \langle x \mid d\rangle \le \langle y \mid d\rangle. \tag{17.12}$$

Let $\sim_C$ be the equivalence associated with $\preccurlyeq_C$.

Proposition 35. *The kernel* $\mathbf{0}_C$ *of the semi-norm* $\|\cdot\|_C$ *is identical to the equivalence class of* 0 *modulo* $\sim_C$.

Proof. Obvious. □

In particular, $\mathrm{Fin}_C/\mathbf{0}_C$ is a partially ordered Banach space.

Proposition 36. (i) C^+ *is the set of positive elements w.r.t.* $\preccurlyeq_C$.

(ii) $\mathbf{0}_C = C^+ \cap (-C^+) = C \cap (-C)$.

(iii) *The unit ball w.r.t.* $\|\cdot\|_C$ *is* $(C - C^+) \cap (C^+ - C)$.

Proof. (i) and (iii) are respectively the cases « $b \notin C^+$ » and « $b \notin C - C^+$ » of the proof of Theorem 61. (ii) is immediate. □

A few remarks of a slightly repetitive nature:

(i) The partial order $\preccurlyeq_C$ is continuous w.r.t. the $\|.\|_C$: if $x_n \preccurlyeq_C y_n$ and $(x_n), (y_n)$ are Cauchy sequences for $\|\cdot\|_C$ with limits x, y, then $x \preccurlyeq_C y$.

(ii) If $0 \preccurlyeq_C x \preccurlyeq_C y$, then $\|x\|_C \le \|y\|_C$.

(iii) If $x \in \mathrm{Fin}_C$, then there exist $y, z \succcurlyeq_C 0$ such that $x = y - z$ and $\|y\| \le \|x\|$.

What relation exists between norm and order for C and norm and order for ${\sim}C$? Nothing new in what follows, it is just a compilation:

Equivalence

$$x \sim_C y \iff \forall x', y'(x' \sim_{\sim C} y' \Rightarrow \langle x \mid y\rangle = \langle x' \mid y'\rangle) \tag{17.13}$$

The introduction of the domain Fin_C, i.e., the fact of considering a *partial*, non-reflexive, relation, enables this symmetrical formulation.

Positivity

$$x \in C^+ \quad \Leftrightarrow \quad \forall y(y \in (\sim C)^+ \Rightarrow \langle x \mid y \rangle \geq 0) \tag{17.14}$$

The relation $\precsim_C$ is a preorder on the domain Fin_C. There is no standard terminology for such a relation, seen as a relation on E; it enjoys *weak reflexivity*:

$$x \precsim y \Rightarrow x \precsim x \wedge y \precsim y. \tag{17.15}$$

The next result generalises the decomposition of a hermitian as a difference $u = u^+ - u^-$ of two positive hermitians (Theorem 69):

Theorem 62. *If $x \in E$, one can find $x^+ \in C^+$ and $x^- \in (\sim C)^+$ such that $x = x^+ - x^-$ and $\langle x^+ \mid x^- \rangle = 0$; this decomposition is unique.*

Proof. Let x^+ be the projection of x on the convex C and let $x^- := x - x^+$. One knows that x^- is the unique y such that $\langle y \mid x - y \rangle \geq \langle y \mid z \rangle$ for all $z \in C$. This condition is easily transformed into $y \in (\sim C)^+$ and $\langle y \mid x - y \rangle = 0$. □

Semi-norm

$$\|x\|_C = \inf \{\lambda \ ; \ \forall y \in (\sim C)^+ \ |\langle x \mid y \rangle| \leq \lambda \|y\|_{\sim C}\} \tag{17.16}$$

But it is not the case that $|\langle x \mid y \rangle| \leq \|x\|_C \cdot \|y\|_{\sim C}$ for all $x \in \mathrm{Fin}_C$, $y \in \mathrm{Fin}_{\sim C}$.

Proposition 37. *If $C \subset D$, then*

$$\begin{aligned} \mathrm{Fin}_C &\subset \mathrm{Fin}_D, \\ \precsim_C &\subset \precsim_D, \\ \sim_C &\subset \sim_D, \\ \|\cdot\|_C &\geq \|\cdot\|_D. \end{aligned}$$

The last inequality takes its full sense if one modifies the notion of semi-norm so as to admit infinite values.

17.3.4 Quantum coherent spaces. Let $|X|$ be a finite-dimensional (complex) Hilbert space; one can apply what precedes to $E := \mathcal{H}(|X|)$, the space of *hermitian* operators on $|X|$; those enjoy $h = h^*$, i.e., are *self-adjoint*:

$$\langle h(x) \mid y \rangle = \langle x \mid h(y) \rangle; \tag{17.17}$$

equivalently $\langle h(x) \mid x \rangle \in \mathbb{R}$. Remember that h is *positive* when $\langle h(x) \mid x \rangle \in \mathbb{R}^+$. Among positive hermitians, all the uu^*; indeed every positive hermitian is of this very form, with u in turn positive, i.e., $u = \sqrt{h}$.

E is a real vector space, whose dimension can easily be computed: if $|X|$ is of (complex) dimension n, then $\mathcal{L}(|X|)$ has the complex dimension n^2, hence, as a real vector space, the dimension $2n^2$. Now every operator can uniquely be written as $2u = (u + u^*) + i(iu^* - iu)$, i.e. as $h + ik$, with h, k hermitian, which shows that the dimension of the real space $\mathcal{H}(|X|)$ is n^2. This space is equipped with a scalar product (definite positive bilinear form)

$$\langle h \mid k \rangle := \mathrm{tr}(hk) \tag{17.18}$$

which makes it euclidian: $\mathrm{tr}(hk) = \mathrm{tr}(kh) = \overline{\mathrm{tr}(hk)}$, $\mathrm{tr}(h^2) > 0$ for $h \neq 0$.

Two hermitians h, k are *polar* when $0 \leq \langle h \mid k \rangle \leq 1$.

Definition 84 (Quantum coherent spaces). A *quantum coherent space* (QCS) of carrier $|X|$ is a subset $X \subset \mathcal{H}(|X|)$ equal to its bipolar.

Theorem 61 characterises QCS. One finds below two canonical examples. In both cases the order relation is the usual ordering of hermitians ($h \leq k$ iff $k - h$ is positive); on the other hand, according to the case, one gets a norm of type « supremum » or « sum », consistently with the remarks of Section 15.A.1.

Negative canonical: N is made of the positive hermitians of *norm* ≤ 1. N^+ therefore corresponds to positive hermitians; on N^+, $\|\cdot\|_N$ correspond to the usual norm $\|\cdot\|_\infty$.

Positive canonical: P is made of the positive hermitians of *trace* ≤ 1. P^+ therefore corresponds to positive hermitians; on P^+, $\|\cdot\|_P$ coincides with the trace norm $\|u\|_1 = \mathrm{tr}(\sqrt{uu^*})$; in this case, $\|h\|_1 = \mathrm{tr}(h)$.

Indeed, $P = {\sim}N$: one uses $|\mathrm{tr}(uv)| \leq \|u\|_\infty \cdot \|v\|_1$ and, for $h, k \geq 0$, $\mathrm{tr}(hk) = \mathrm{tr}(\sqrt{h}k\sqrt{h}) = \mathrm{tr}((\sqrt{h}\sqrt{k})(\sqrt{h}\sqrt{k})^*) \geq 0$ and $\mathrm{tr}(uxx^*) = \langle u(x) \mid x \rangle$.

17.4 Additives

17.4.1 A few reminders on quantum physics. Some basics of quantum mechanics, limited to finite[4] dimension:

(i) The state of a system is represented by a *wave function*, i.e., a vector x of norm 1 in a Hilbert space $|X|$.

(ii) The *measurement* is represented by a hermitian Φ on $|X|$. To say that the *value* of x w.r.t. Φ is λ means that $\Phi(x) = \lambda x$. Thus, there is seldom any value. Worse, if Φ, Ψ do not commute, it is likely that they have no common *eigenvector*, i.e., x cannot have a value w.r.t. both Φ and Ψ, as in the noted *uncertainty principle*. Thus, the *Pauli matrices*, see *infra*, which measure the *spin* along the axes $\vec{X}, \vec{Y}, \vec{Z}$, do not commute: if the *spin* is $+1/2$ along axis $\vec{Z}$, it is completely undetermined along $\vec{X}$.

[4]Thanks to Thierry Paul for his stimulating comments.

(iii) The measurement process is a Procustus bed[5], which forces the system to « have a value ». In other words, once the measurement has been performed, the wave function x is replaced with an eigenvector x' of Φ. This process is non-deterministic: if $|X|$ is decomposed into a direct sum of eigenspaces of Φ: $|X| = \bigoplus_\lambda |X|_\lambda$, so that $x = \bigoplus_\lambda x_\lambda$, then x' is one of the components x_λ, renormalised (multiplied by $1/\|x_\lambda\|$) and the probability of the transition $x \rightsquigarrow x_\lambda/\|x_\lambda\|$ is $\|x_\lambda\|^2$. This process is the *reduction of the wave packet*, reduction for short.

(iv) In this approach, wave functions are defined up to a scalar of norm 1. For instance, the *spin* x of an electron is replaced with its opposite $-x$ during a rotation of angle 2π, without affecting the system.

(v) *Density matrices* (or operators), due to von Neumann, take into account the scalar indetermination of the wave function; above all, they maintain the probabilistic aspect of the measurement process. A *density operator* is a positive hermitian of trace 1. Density operators form a compact convex set whose *extremal* points are of the form xx^*, where x is a vector of norm 1, i.e., a wave function, transformed into the orthogonal projection on the subspace it generates. The measurement process replaces xx^* with $\sum_\lambda x_\lambda x_\lambda{}^*$: this density operator is a « mixture », a convex combination of the projections $x_\lambda x_\lambda{}^*/\|x_\lambda\|^2$, whose coefficients $\|x_\lambda\|^2$ correspond to the probabilities of the possible transitions. This replacement is called *preselection* and is deterministic: occurring prior to the reduction process, it reduces the quantum to non-determinism.

(vi) This formalism is iterative, i.e., one can measure a density operator, not necessarily extremal. This means that, if one writes our density operator h under the matricial form $(h_{\lambda\mu})$ w.r.t. the decomposition $|X| = \bigoplus_\lambda |X|_\lambda$ ($h_{\lambda\mu} \in \mathcal{L}(|X|_\mu, |X|_\lambda)$), then reduction « kills » the coefficients $h_{\lambda\mu}$ outside the diagonal: after measurement, h becomes $k = (k_{\lambda\mu})$, with $k_{\lambda\lambda} = h_{\lambda\lambda}, k_{\lambda\mu} = 0$ for $\lambda \neq \mu$.

(vii) The measurement process is irreversible: if $u \rightsquigarrow v$ by measurement, then $\mathrm{tr}(v^2) \leq \mathrm{tr}(u^2)$. If X is of finite dimension n, $\mathrm{tr}(u^2)$ will therefore vary, through successive measurements, from the value 1 (extremal point xx^*), up to $1/n$: $1/n \cdot I$, the « tepid » mixture, which carries no information.

Note the finesse, the creativity, of this interpretation, compared to what logicians may have produced, typically those bleak Kripke models which put side by side arbitrary parallel universes.

[5]According to Plutarch, Procustus was a sort of old-style communist, who rubbed out the differences between men « by equating them to the length of his beds »; he was thus stretching the short and shortening the tall: hence the « Procustus bed ».

17.4.2 Quantum booleans. Booleans form a system with two states, which suggests a two-dimensional space, hence 2×2 matrices. The *classical* viewpoint will soon be breathless for reasons of triviality, while the quantum viewpoint – which really speaks of the same thing, a two-state system – will turn out to be incredibly deeper.

Commutative booleans. There is little to say if one sticks to the « traditional » viewpoint:

(i) The booleans **v**, **f** are represented by $\begin{bmatrix} 1 & 0 \\ 0 & 0 \end{bmatrix}$ and $\begin{bmatrix} 0 & 0 \\ 0 & 1 \end{bmatrix}$.

(ii) A diagonal matrix $\begin{bmatrix} \lambda & 0 \\ 0 & \mu \end{bmatrix}$, of trace 1, i.e., such that $\lambda + \mu = 1$, and positive, i.e., such that $\lambda, \mu \geq 0$, represents a probabilistic boolean. With a computer-science background, one can even admit $\lambda + \mu \leq 1$, which corresponds to a *partial* probabilistic boolean, true with probability λ, false with probability μ, undefined with probability $1 - (\lambda + \mu)$.

None of this is earth-shaking. Now, remember that matrices are operators written w.r.t. a certain basis; and let us imagine that for the reason you prefer – say, a travel accident that would have tilted the gyroscopes – the basis has been « lost ». The booleans are still there, but one can no longer read them! Then this bleak bureaucracy starts to live! One starts studying hermitians in dimension 2, which leads to the space $\mathbb{R}^3$, not to speak of space-time!

Space-time. Any hermitian is written $h = 1/2\begin{bmatrix} t+z & x-iy \\ x+iy & t-z \end{bmatrix}$, i.e., $ts_0 + xs_1 + ys_2 + zs_3$, with t, x, y, z real, where the s_i are the *Pauli matrices*:

$$1/2\begin{bmatrix} 1 & 0 \\ 0 & 1 \end{bmatrix}, \quad 1/2\begin{bmatrix} 0 & 1 \\ 1 & 0 \end{bmatrix}, \quad 1/2\begin{bmatrix} 0 & -i \\ i & 0 \end{bmatrix}, \quad 1/2\begin{bmatrix} 1 & 0 \\ 0 & -1 \end{bmatrix}. \tag{17.19}$$

The *time* t is the *trace*, $t = \mathrm{tr}(h)$. The computation of the determinant yields $4 \cdot \det(h) = t^2 - (x^2 + y^2 + z^2)$, i.e., the square of the pseudo-metrics. Note that $\mathrm{tr}((ts_0 + xs_1 + ys_2 + zs_3)(t's_0 + x's_1 + y's_2 + z's_3)) = tt' + xx' + yy' + zz'$. For $1 \leq i \neq j \leq 3$, one gets the anti-commutations $s_i.s_j + s_j.s_i = 0$.

In order to characterise *positive hermitians*, remember that any hermitian is diagonalisable: *modulo* a change of basis, i.e., a unitary transformation u, $uhu^* = \begin{bmatrix} \lambda & 0 \\ 0 & \mu \end{bmatrix}$, with $\lambda, \mu \in \mathbb{R}$; h is positive iff $\lambda, \mu \geq 0$. In other words, the condition $\det(h) \geq 0$ (vectors in position « time ») characterises hermitians which are either positive or negative. Positivity requires the additional condition: $\mathrm{tr}(h) \geq 0$: which corresponds to the « cone of the future »: $t \geq \sqrt{x^2 + y^2 + z^2}$.

The most general transformation preserving positivity is of the form $h \rightsquigarrow uhu^*$, with $\det(u) = 1$, i.e., $u \in \mathrm{SL}(2)$: this is the positive Lorenz group, which preserves both the pseudo-metrics and the future. While here, the inverse of $u \in \mathrm{SL}(2)$ is given by

$$\begin{bmatrix} a & b \\ c & d \end{bmatrix}^{-1} = \begin{bmatrix} d & -b \\ -c & a \end{bmatrix}. \tag{17.20}$$

Hence, inversion extends into an involutive anti-automorphism of the algebra $\mathcal{M}_2(\mathbb{C})$ of 2×2 matrices. In terms of space-time, this anti-automorphism replaces the spatial coordinates with their opposites.

The group SO(3) of *rotations*, which only act on space, corresponds to preservation of time, i.e., of trace. Since $\mathrm{tr}(uhu^*) = \mathrm{tr}(u^*uh)$, one sees that they are induced by *unitary* u. In other words, SO(3) admits a (double) covering by SU(2), the group of unitaries of determinant 1, whose general form is $\left[\begin{smallmatrix} a & b \\ -\bar{b} & \bar{a} \end{smallmatrix}\right]$, with $a\bar{a} + b\bar{b} = 1$. The rotations of axes $\vec{X}$, $\vec{Y}$, $\vec{Z}$ and angle θ are induced by the $e^{i\theta s_k}$, respectively:

$$\begin{bmatrix} \cos\theta/2 & i\sin\theta/2 \\ i\sin\theta/2 & \cos\theta/2 \end{bmatrix}, \quad \begin{bmatrix} \cos\theta/2 & \sin\theta/2 \\ -\sin\theta/2 & \cos\theta/2 \end{bmatrix}, \quad \begin{bmatrix} e^{i\theta/2} & 0 \\ 0 & e^{-i\theta/2} \end{bmatrix}. \tag{17.21}$$

Since there are two solutions, it is obviously a « heresy » to divide an angle by 2. This is why the covering is double; this is also why a rotation of angle 2π acts on the *spin* – seen as a « wave function » – as a multiplication by -1. This corresponds to the possibility of replacing u by $-u$ in $h \rightsquigarrow uhu^*$; one cannot continuously choose between both determinations – just as one cannot continuously determine a complex square root.

Quantum booleans. « Classical » booleans are the orthogonal projections of two 1-dimensional subspaces distinguished by the matricial representation. A *quantum* boolean will simply be a 1-dimensional subspace. This approach refuses from the start any distinction between true and false: if E is a boolean, its negation is $E^{\perp}$, period! Also remark that, for foreseeable reasons of commutation, it will not be possible to construct convincing binary connectives.

It remains to determine the 1-dimensional subspaces, i.e., the matrices of orthogonal projections of rank 1. Those are of trace 1 and determinant 0, i.e., the points of space-time $ts_0 + xs_1 + ys_2 + zs_3$, such that $t = 1$ and $x^2 + y^2 + z^2 = 1$, and are therefore in 1-1 correspondence with the sphere S^2.

What we just called quantum « boolean » is known in physics as the *spin* of an electron. The measurement of the *spin* along the axis – say $\vec{Z}$ – is given by the Pauli matrix s_3, whose eigenvalues are $\pm 1/2$. The value $+1/2$ is the image of the projection $\left[\begin{smallmatrix} 1 & 0 \\ 0 & 0 \end{smallmatrix}\right]$ (*spin* up along $\vec{Z}$), the value $-1/2$ (*spin* down along $\vec{Z}$) corresponding to $\left[\begin{smallmatrix} 0 & 0 \\ 0 & 1 \end{smallmatrix}\right]$.

Probabilistic quantum booleans. The most general case is that of a convex combination of quantum booleans, which corresponds to a positive hermitian of trace 1, a « density matrix ». One can diagonalise a hermitian in an appropriate orthonormal basis; in which sense is this unique?

(i) $\begin{bmatrix} 1/2 & 0 \\ 0 & 1/2 \end{bmatrix}$ is diagonal in all orthonormal bases; no form of unicity.

(ii) Outside this case, our boolean is written as $\lambda b + (1 - \lambda)c$, where b, c are orthogonal booleans and $0 \leq \lambda < 1/2$ and this, in a unique way.

Preselection occurs when one measures a boolean, which corresponds to the measurement of a *spin*. One must specify a basis, which diagonalises the measurement operator. One writes our hermitian, quantum boolean, probabilistic or not, in this basis, namely: $\begin{bmatrix} a & b \\ \bar{b} & c \end{bmatrix}$. Once the measurement is done, it becomes $\begin{bmatrix} a & 0 \\ 0 & c \end{bmatrix}$, i.e., **v** with probability a, **f** with probability $c = 1 - a$, along the axis $\mathbf{v}/\mathbf{f}$ corresponding to our basis.

Negation. The choice of an orthonormal basis is that of two subspaces of dimension 1, i.e., two quantum booleans π and $I - \pi$, whose space-time coordinates will therefore be $(1, x, y, z)$ and $(1, -x, -y, -z)$. The vectors $\vec{A} = (x, y, z)$ and $-\vec{A}$ correspond to the two possible senses on the same axis (*spin* up, *spin* down). The symmetry w.r.t. the origin corresponds to the anti-automorphism $\begin{bmatrix} a & b \\ c & d \end{bmatrix} \rightsquigarrow \begin{bmatrix} d & -b \\ -c & a \end{bmatrix}$ of the algebra $\mathcal{M}_2(\mathbb{C})$. This transformation corresponds to *negation*. Since symmetry w.r.t. the origin is of determinant -1, it is not in SO(3) and is not induced by an element of SU(2); by the way, the elements of SU(2) induce automorphisms, not « anti ».

Binary boolean connectives. Negation, a truly involutive operation, does not involve preselection. Which is no longer the case for binary connectives:

(i) One cannot combine non-commuting 1-dimensional projections so as to produce another projection.

(ii) Common sense says that, if one cannot tell the truth from the false, it will be even more difficult to tell a conjunction from a disjunction.

Hence binary connectives are probabilistic: they yield a probabilistic boolean even when the arguments are « pure ». Moreover, they depend upon a basis and an order of evaluation, for instance

$$\begin{bmatrix} a & b \\ \bar{b} & c \end{bmatrix} \vee \begin{bmatrix} a' & b' \\ \bar{b}' & c' \end{bmatrix} := \begin{bmatrix} a + ca' & cb' \\ c\bar{b}' & cc' \end{bmatrix}.$$

The first argument is reduced in the basis: true with probability a, in which case the answer is $\begin{bmatrix} 1 & 0 \\ 0 & 0 \end{bmatrix}$, false with probability c, in which case the answer is $\begin{bmatrix} a' & b' \\ \bar{b}' & c' \end{bmatrix}$.

There is a symmetrical choice, reducing the second argument. And also the « Jivaro choice », which reduces both: $\left[\begin{smallmatrix} a+ca' & 0 \\ 0 & cc' \end{smallmatrix}\right]$, which is symmetrical, since $a + ca' = a' + ca = a + a' - aa'$.

17.4.3 Additives and quantum

Plus and With

Definition 85 (Additives). If X, Y are QCS of respective carriers $|X|$, $|Y|$, one defines $X \oplus Y$ and $X \,\&\, Y$, QCS of carrier $|X| \oplus |Y|$:

$$X \oplus Y = \{\lambda h \oplus (1-\lambda)k;\ h \in X,\ k \in Y, 0 \leq \lambda \leq 1\}, \tag{17.22}$$
$$X \,\&\, Y = \{g;\ |X|g|X| \in X,\ |Y|g|Y| \in Y\}. \tag{17.23}$$

The subspaces $|X|$, $|Y|$ have been identified with their orthogonal projections.

Proposition 38. *$\oplus$ and & are swapped by negation.*

Proof. Since $\langle h \oplus k \mid h' \oplus k' \rangle = \langle h \mid h' \rangle + \langle k \mid k' \rangle$. □

Neither $\|\cdot\|_{X \oplus Y}$, nor $\|\cdot\|_{X \& Y}$ are norms. The definition mistreats hermitians not of the form $h \oplus k$. W.r.t. a block decomposition, a hermitian on $|X| \oplus |Y|$ is written: $H = \left[\begin{smallmatrix} h & u \\ u^* & k \end{smallmatrix}\right]$, with h, k hermitian. If $u \neq 0$, H takes an infinite norm in $X \oplus Y$ (if one prefers, H is not in $\mathrm{Fin}_{X \oplus Y}$). *A contrario*, its norm w.r.t. $X \,\&\, Y$ does not depend on u: the kernel $\mathbf{0}_{X \& Y}$ contains all the $\left[\begin{smallmatrix} 0 & u \\ u^* & 0 \end{smallmatrix}\right]$.

Dimension 2. If $|X|$ is of dimension 1, then $\mathcal{H}(|X|)$ is of dimension 1 (isomorphic to $\mathbb{R}$); the two canonical QCS of Section 17.3.4 coincide, yielding a QCS $\mathbf{1}$, corresponding to the segment $[0, 1]$ of $\mathbb{R}$, ordered and normed « naturally ».

In dimension 2, $\mathcal{H}(X)$ is of dimension 4, with several natural choices:

Bool: the positive canonical. The elements of **Bool** are the positive hermitians of trace at most 1, the « partial probabilistic quantum booleans ».

∼Bool: the negative canonical. The elements of **∼Bool** are positive hermitians of norm (in the usual sense) at most 1, the « anti-booleans ».

$\mathbf{1} \oplus \mathbf{1}$: the « $\oplus$ » of two copies of $\mathbf{1}$. The QCS $\mathbf{1} \oplus \mathbf{1}$ is made of all matrices $\left[\begin{smallmatrix} a & 0 \\ 0 & c \end{smallmatrix}\right]$ such that $0 \leq a, c \leq a + c \leq 1$, i.e., of partial probabilistic booleans. It is a subset, a « subtype » of **Bool**.

$\mathbf{1} \,\&\, \mathbf{1}$: the negation of the previous. It is made of the matrices $\left[\begin{smallmatrix} a & b \\ \bar{b} & c \end{smallmatrix}\right]$ such that $0 \leq a, c \leq 1$.

Our construction of $\mathbf{1} \oplus \mathbf{1}$ depends upon a 1-dimensional subspace – corresponding to « $\mathbf{v}$ » –, here, the one encoded by $\vec{Z}$. Which means that, given a vector $\vec{A} \in S^2$, there is a QCS of the « booleans of axis $\vec{A}$ », noted $\mathbf{Bool}_{\vec{A}}$.

Proposition 39. $\mathbf{Bool} = \bigcup_{\vec{A} \in S^2} \mathbf{Bool}_{\vec{A}}$.

Proof. Of course, $\mathbf{Bool}_{\vec{A}} \subset \mathbf{Bool}$. Conversely, $h \in \mathbf{Bool}$ can be diagonalised as $\left[\begin{smallmatrix} a & 0 \\ 0 & c \end{smallmatrix}\right]$, with $0 \leq a, c \leq a + c \leq 1$, w.r.t. a certain orthonormal basis $\mathbf{e}, \mathbf{f}$. If $\vec{A} \in S^2$ corresponds to $\mathbf{e}$, then $h \in \mathbf{Bool}_{\vec{A}}$. □

Corollary 39.1. $\sim\mathbf{Bool} = \bigcap_{\vec{A} \in S^2} \sim\mathbf{Bool}_{\vec{A}}$.

Preselection: a discussion. I said that we are seeking a logical explanation of the quantum. This being said, what we did *clarifies* the question of preselection, hence of reduction.

The next section deals with multiplicatives, thus with linear implication. In particular, one will be able to transform a boolean $h \in \mathbf{Bool}$ into something else, by means of an element of a QCS $\mathbf{Bool} \multimap \ldots$, then transform the result by means of another implication... Some of these transformations will behave like negation, they will be of « wave style », others like the binary connectives, will make use of preselection, they will be of « particle style ». By the way, one will see that logical operations are of the « particle » style, with a noticeable exception, the non-*etaspanded* identity axiom, rather of style « wave ».

To what extent is preselection subjective? Let us afford an impossible hypothesis: we assume this process of transformation of a boolean to be completed, i.e., that, in this succession of implications, one was eventually able to « close the system ». Which means that everything ended with a last implication, with values in $\mathbf{1}$. If I compose all my implications, I see that this sequence of transformations, which eventually « closes the system », is nothing but an anti-boolean $k \in \sim\mathbf{Bool}$. The result is objective: $\langle h \mid k \rangle = \mathrm{tr}(hk)$. But the choice of k (the transformations, the measurements made on h) is very subjective. We are not neutral, since « on the side of k ». Since we stand on the side of k, we might as well diagonalise it in an appropriate basis $\mathbf{e}, \mathbf{f}$. Then $h = \left[\begin{smallmatrix} a & b \\ \bar{b} & c \end{smallmatrix}\right]$, $k = \left[\begin{smallmatrix} \alpha & 0 \\ 0 & \gamma \end{smallmatrix}\right]$ and so $\langle h \mid k \rangle = a\alpha + c\gamma$. If $h' = \left[\begin{smallmatrix} a & 0 \\ 0 & c \end{smallmatrix}\right]$, then $\langle h \mid k \rangle = \langle h' \mid k \rangle$, i.e., it is as if we had reduced h.

It could be the case that we know that f is a boolean in a certain basis (e.g., if f comes from the measurement of a *spin*). One selects this basis and $h = \left[\begin{smallmatrix} a' & 0 \\ 0 & c' \end{smallmatrix}\right]$, $k = \left[\begin{smallmatrix} \alpha' & \beta' \\ \bar{\beta}' & \gamma' \end{smallmatrix}\right]$ and one can then write $\langle h \mid k \rangle = a'\alpha' + c'\gamma'$. In this case, one « reduced » the observer k into $k' = \left[\begin{smallmatrix} \alpha' & 0 \\ 0 & \gamma' \end{smallmatrix}\right]$ in such a way that $\langle h \mid k \rangle = \langle h \mid k' \rangle$.

Finally, preselection itself is subjective!

17.5 Multiplicatives

17.5.1 Linear functionals

Theorem 63 (Folklore). *Let $|X|$, $|Y|$ be finite-dimensional Hilbert spaces. Then $\mathcal{L}(\mathcal{L}(|X|), \mathcal{L}(|Y|)) \simeq \mathcal{L}(|X| \otimes |Y|)$.*

Proof. The rank 1 endomorphisms: $xw^*(y) := \langle y \mid w\rangle x$ generate the complex vector space $\mathcal{L}(|X|)$. If $\varphi \in \mathcal{L}(\mathcal{L}(|X|), \mathcal{L}(|Y|))$, define $\Phi \in \mathcal{L}(|X| \otimes |Y|)$ by

$$\langle \Phi(x \otimes y) \mid w \otimes z\rangle = \langle \varphi(xw^*)(y) \mid z\rangle. \tag{17.24}$$

Conversely, given $\Phi \in \mathcal{L}(|X| \otimes |Y|)$, if $f \in \mathcal{L}(|X|)$, define $(\Phi)f \in \mathcal{L}(|Y|)$ by

$$\langle ((\Phi)f)(y) \mid z\rangle = \mathrm{tr}(\Phi \circ (f \otimes yz^*)), \tag{17.25}$$

hence $(\Phi)\cdot \in \mathcal{L}(\mathcal{L}(|X|), \mathcal{L}(|Y|))$. □

Corollary 39.2. *If $\Phi \in \mathcal{H}(|X| \otimes |Y|)$, if $f \in \mathcal{H}(|X|)$, then $(\Phi)f \in \mathcal{H}(Y)$. The function $\Phi \rightsquigarrow (\Phi)\cdot$ is a bijection between $\mathcal{H}(|X| \times |Y|)$ and the set of linear functionals from $\mathcal{H}(|X|)$ into $\mathcal{H}(|Y|)$.*

Proof. One easily sees that $(\Phi^*)f^* = ((\Phi)f)^*$, hence a hermitian Φ sends hermitians to hermitians. Conversely, if φ is a linear functional from $\mathcal{H}(|X|)$ to $\mathcal{H}(|Y|)$, then φ uniquely extends into a $\mathbb{C}$-linear functional from $\mathcal{L}(|X|)$ to $\mathcal{L}(|Y|)$: $\varphi(u) := 1/2(\varphi(u + u^*) + i\varphi(iu^* - iu))$. The $\mathbb{C}$-linear functionals thus obtained are *hermitian*, i.e., enjoy $\varphi(f^*) = \varphi(f)^*$ and are therefore in bijection with the hermitians of $\mathcal{H}(|X| \otimes |Y|)$. □

The essential property of $(\Phi)\cdot$ is summarised by the equation

$$\mathrm{tr}(((\Phi)f) \circ g) = \mathrm{tr}(\Phi \circ (f \otimes g)). \tag{17.26}$$

Which is reminiscent of *stability* (Section 8.2.6). It was a matter of *linearity* in the absence of true linear operations, since only unions and intersections were available; the way to true linearity is marked with the reformulations of Section 9.1. We eventually managed to get a *literal* linearity, involving true sums, true coefficients. Unfortunately, just like Moses passing away at the gates of Israel, the story of coherent spaces stops short: they will not survive, under this ambitious form, in infinite dimension (Section 17.6.1).

17.5.2 A few examples

Example 3. If $\sigma_E \in \mathcal{H}(E \otimes E)$ is such that $\sigma(x \otimes y) = y \otimes x$ (the « twist »), then

$$\begin{aligned}\langle(\sigma)(xw^*)(y) \mid z\rangle &= \langle\sigma(x \otimes y) \mid w \otimes z\rangle \\ &= \langle y \otimes x \mid w \otimes z\rangle \\ &= \langle y \mid w\rangle\langle x \mid z\rangle \\ &= \langle(xw^*)(y) \mid z\rangle,\end{aligned} \tag{17.27}$$

hence $(\sigma)(xw^*) = xw^*$. By linearity, $(\sigma)f = f$.

Example 4. More generally, let u be a linear map from E to F. Then $u \otimes u^*$ sends $E \otimes F$ to $F \otimes E$ and if σ_{EF} is the « twist » from $F \otimes E$ to $E \otimes F$, then $U = \sigma \circ (u \otimes u^*) \in \mathcal{H}(E \otimes F)$. It goes without saying that $(U)f = ufu^*$.

Example 5. Let $I = E + F$ be a decomposition of the identity into orthogonal projections (subspaces). Then $R = \sigma \circ (E \otimes E + F \otimes F)$ acts as follows: $(R)f = EfE + FfF$. R is the typical preselection, which « kills » the non-diagonal blocks EfF and FfE of f.

One can ask how the identity map of $E \otimes F$ is acting. One easily sees that $(I_{E\otimes F})u = \mathrm{tr}(u) \cdot I_F$. Not very inspired… But this will be used in our last example:

Example 6. If E is of dimension 2, then $(I_{E\otimes E} - \sigma_E)\begin{bmatrix} a & b \\ \bar{b} & c\end{bmatrix} = \begin{bmatrix} c & -b \\ -\bar{b} & a\end{bmatrix}$, i.e., acts as the *negation*. $I_{E\otimes E} - \sigma_E = 2\pi$, where π is the orthogonal projection on the antisymmetric subspace of $E \otimes E$, i.e., the 1-dimensional space made of the $x \otimes y - y \otimes x$.

17.5.3 Connectives

Definition 86 (Multiplicatives). If X, Y are QCS, one defines the QCS $X \multimap Y$, of carrier $|X| \otimes |Y|$, as the set of all Φ sending X into Y:

$$X \multimap Y = \{\Phi;\ \forall f \in X\ \ (\Phi)f \in Y\}. \tag{17.28}$$

$X \multimap Y$ might as well be defined by

$$X \multimap Y = \{\Phi;\ \forall g \in {\sim}Y\ \ (\Phi)g \in {\sim}X\} \tag{17.29}$$

and also as ${\sim}\{f \otimes g; f \in X, g \in {\sim}Y\}$, which shows that $X \multimap Y$ is a QCS. From this, one defines $X \parr Y = {\sim}X \multimap Y$ and $X \otimes Y = {\sim}{\sim}\{f \otimes g; f \in X, g \in Y\}$. As usual, $\parr$ is commutative, associative and distributes over & (up to isomorphism).

Multiplicatives force us to depart from the standard ordering of hermitians. For instance, suppose that X, Y are positive canonicals, e.g., $X = Y = \mathbf{Bool}$. Then $X \multimap Y$ will consider as positive any hermitian sending the (true) positive to the (true) positive. Thus, the *twist* σ which behaves as an identity map; but σ is a *symmetry*, a hermitian by no ways positive!

Hence $X \multimap Y$ is more liberal as to positivity than what one could expect. Dually, $X \otimes Y$ is more restrictive, more positive than the King! The positive cone of $X \otimes Y$ is the closure of the set of sums $\sum_i f_i \otimes g_i$, $f_i, g_i \geq 0$. Most (truly) positive hermitians of $|X| \otimes |Y|$ are not of this form, typically the zz^*, when z is not a pure tensor.

17.5.4 η and preselection. We already met η (Section 7.4.2). Can one imagine a bleaker, a more bureaucratic topic? At the era of bibliometry, when everything is measured in terms of number of pages, of publications, witness the prosthesis for insufficient PhDs: « do it again with η ». Which means 100% perspiration, but, since the outcome is guaranteed, 0% inspiration!

One would like so much to see this principle wrong! Thus one observed that ludic faithfulness stumbled *grosso modo* on this idiocy, see the discussion in Section 14.B.4. The refutations of « η »[6] are all in the same style: one adds extra points to the domain of functions to the effect that part of their graph, being of no use, becomes ambiguous. This is yet another example of intensional fiddling, whose fabricated character paradoxically pleads in favour of « η ». It is the case to say « what kills me reinforces me ».

This dear η, let us brush it the wrong way, and refute it, not necessarily for good, but in a simple and *honest* way: in the case of a « Plus » $C = A \oplus B$, one will differentiate the « native » identity from its etaspansion. The difference is very simple: the identity is $\varphi(x) = x$, it recopies an object without really caring about; if one thinks of delocation, it is a genuine wavy operation. The etaspanded identity is based upon the idea that one is either « of type A » or « of type B »: it asks the object whether it is of type A or B and no matter the answer, recopies it identically: it is a cop who asks for your identity papers before letting you pass. The *commutative*, set-theoretic, world cannot separate the real identity from its etaspansion, the « inquisitive », essentialist, identity; it leaves no real room for the potential, reduced to a list of possibilities. In the quantum world, this is different; etaspansion is a measuring process followed by a trivial reconstruction: it is therefore a plain *preselection*.

[6]Like the one by Plotkin which had however the virtue of being the first.

Assuming that A, B have dimension 1 carriers, let us consider:

The twist: the generic identity of a space $|C|$ of dimension 2, which is written

$$\sigma = \begin{bmatrix} 1 & 0 & 0 & 0 \\ 0 & 0 & 1 & 0 \\ 0 & 1 & 0 & 0 \\ 0 & 0 & 0 & 1 \end{bmatrix} \tag{17.30}$$

in *no matter which basis* $\mathbf{e} \otimes \mathbf{e}, \mathbf{e} \otimes \mathbf{f}, \mathbf{f} \otimes \mathbf{e}, \mathbf{f} \otimes \mathbf{f}$ of $|C| \otimes |C|$.

The etaspanded twist: this is the gluing of two identities, that of A and that of B. W.r.t. a well-defined basis, corresponding to the decomposition of $|C|$ in direct sum, this yields

$$\iota = \begin{bmatrix} 1 & 0 & 0 & 0 \\ 0 & 0 & 0 & 0 \\ 0 & 0 & 0 & 0 \\ 0 & 0 & 0 & 1 \end{bmatrix}. \tag{17.31}$$

These two things are necessarily distinct: $(\sigma)\begin{bmatrix} a & b \\ \bar{b} & c \end{bmatrix} = \begin{bmatrix} a & b \\ \bar{b} & c \end{bmatrix}$ is the « true » identity. On the other hand $(\iota)\begin{bmatrix} a & b \\ \bar{b} & c \end{bmatrix} = \begin{bmatrix} a & 0 \\ 0 & c \end{bmatrix}$ is only an « identity of Procustus ». It is the identity for those matrices which are already of the appropriate form $\begin{bmatrix} a & 0 \\ 0 & c \end{bmatrix}$. For those which do not fit in this setting (i.e., in this essence), it severs the coefficients outside the diagonal. Indeed, ι is only the preselection, corresponding to the measure of the *spin* along the axis $\vec{Z}$.

In logic, only the identity « enjoys » etaspansion. Which is not the case here, since the antipode can be etaspanded too:

$$\nu = \begin{bmatrix} 0 & 0 & 0 & 0 \\ 0 & 1 & -1 & 0 \\ 0 & -1 & 1 & 0 \\ 0 & 0 & 0 & 0 \end{bmatrix} \tag{17.32}$$

is such that $(\nu)\begin{bmatrix} a & b \\ \bar{b} & c \end{bmatrix} = \begin{bmatrix} c & -b \\ -\bar{b} & a \end{bmatrix}$; it can be etaspanded into

$$\nu' = \begin{bmatrix} 0 & 0 & 0 & 0 \\ 0 & 1 & 0 & 0 \\ 0 & 0 & 1 & 0 \\ 0 & 0 & 0 & 0 \end{bmatrix}. \tag{17.33}$$

Equivalently $(\nu')\begin{bmatrix} a & b \\ \bar{b} & c \end{bmatrix} = \begin{bmatrix} c & 0 \\ 0 & a \end{bmatrix}$: ν' measures the *spin* along the axis $\vec{Z}$, then swaps it.

17.5.5 Par and imbrication

The EPR paradox. Incidently, the matrix ν represents the double of a projection, the one projecting on the « EPR[7] state », named after a famous paradox imagined by Einstein to confute quantum non-determinism: it is about the space of the $x \otimes y - y \otimes x$, a space of dimension 1. Two correlated particles, simultaneously measured at a big distance are supposed to yield opposite *spins*, in apparent contradiction with relativity.

The EPR paradox is much less indigestible when one takes into account subjectivity and more specifically, *intersubjectivity*. It is indeed a matter of two measurements – highly subjective processes – performed without *concerting* possibility; which should induce *independent* results. But, in order to relate the results, one must surely establish a contact, share milestones (for instance, the axis along which the spin is measured). The correlation of results can thus be explained by this *a posteriori* concerting, which corresponds to the creation of a common subject.

Par and imbrication. From the standpoint of QCS, the « EPR state » is a « Par », akin to the communicating vessels (Figure 10.1, p. 206): if one destroys (measures) one side, one finds it again (reversed) on the other side. It is an *imbricated* state, not a pure tensor $u \otimes v$: imbrication is the distinctive mark of the connective « $\parr$ ».

In the « set-theoretic », commutative, world, classical logic stands closest to imbrication! Indeed, Herbrand's theorem (which we first introduced in Section 3.A.3) involves the disjunction

$$R[t_1, f(t_1), u_1, g(t_1, u_1)] \vee \cdots \vee R[t_n, f(t_n), u_n, g(t_n, u_n)] \qquad (3.2)$$

with several alternative choices $(t_1, u_1), \ldots, (t_n, u_n)$ (Section 3.A.3). This is a genuine *imbrication*, for instance in

$$(P(x) \Rightarrow P(f(x)) \vee (P(f(x)) \Rightarrow P(f(f(x))) \qquad (3.3)$$

one chooses between left and right according to x; a uniform choice can only be made by deciding the formula $\forall x P(x)$, which Herbrand's formulation wisely avoids doing. One sees here that the « extrication » can be done « by values » – which is innocuous – or globally, but it is then an operation of infinite nature, thus of a rather dubious status.

In a more modern perspective, imbrication is no longer a classical attribute (result of the contraction rule or rather of reasoning by contraposition, Section 4.1.3); it expresses the *concomitance* at work in the connective « Par » (Section 10.2.2). $A \parr B$ is basically an inextricable mixture between A and B; in order to separate them, one must destroy one or the other typically by means of $\sim B$, which will lead us to

[7] Einstein–Podolski–Rosen ; the paradox has been checked by Aspect.

A « alone ». One has the right to see this extrication as the primal manifestation of the subjective: $A \parr B$ provides us with an ore (A) in its gangue (B); one can extract A only by destructing the gangue.

The radicality of the quantum makes the previous metaphor inadequate: one is no longer extracting ore; what one extracts *fundamentally* (and not only in an accidental, contingent way) depends upon the extraction process.

17.6 Discussion

17.6.1 Infinite dimension. This cannot be convincingly extended to infinite dimension.

(i) The pivot of the construction is the *trace*, which becomes problematic in infinite dimension. One can define the two-sided ideal of *trace-class* operators (Section 20.D.1), but this ideal contains no invertible operator, thus nothing like the twist.

(ii) Some von Neumann algebras (those of type $\mathbf{II}_1$) do have a trace. One could try to define QCS whose carrier would be of type $\mathbf{II}_1$. Restricted to PCS, this would consist of replacing discrete carriers with the segment $[0, 1]$, equipped with Lebesgue measure, finite sums becoming integrals. This stumbles on the identity function, which still would be represented by the characteristic function of the diagonal... which is of measure zero, thus « does not exist ». This impossibility extends *a fortiori* to the non-commutative case.

(iii) Technically, the problem is caused by the tensor products of Hilbert spaces. One should interpret the identity by something lighter, namely direct sums: this is what geometry of interaction will do.

The failure of layer -2 must be related to the impossibility of a convincing topological treatment at this level. Although of an infinitely higher quality, layer -2 reproduces the basic error of Kripke models, i.e., reduces the potential to the set of all potentialities. Thus, a proof of $A \otimes B$ which consists of a proof of A and a proof of B, should be represented by a direct sum; while every concrete use combines both proofs in a rather unpredictable way, which requires a tensor product to represent all possibilities. What does not withstand the infinite limit, is thus « actualisation », i.e., the reification of possibles, a common bias of layers -1 and -2.

17.6.2 Operators vs. sets. The most impressive foundational endeavour of the turn of the XXIth century owes nothing to logic: it is the *Non-Commutative Geometry* of Connes, [15]. It is an anti-set-theoretic rereading based upon the familiar result:

A commutative operator algebra is a function space.

Typically, a commutative C^*-algebra can be written $\mathbb{C}(X)$, the algebra of continuous functions on the compact X. Connes proposes to consider non-commutative operator algebras as sorts of algebras of functions over... non-existing sets. An impressive blow against set-theoretic essentialism!

What does this change for logic, *a priori* far astray from considerations internal to geometry? I would say that this changes our ideas of finite set, of point, of graph, etc.

- The commutative, set-theoretical, world appears as a vector space equipped with a distinguished basis. All operations are organised in relation to this basis, in particular they can be represented by linear functions whose matrices are diagonal in this basis.

- The non-commutative world forgets the basis; there is still one, but it is subjective, the one where one diagonalises the chosen hermitian operator: « his » set-theory, so to speak. But, if two hermitians f and g have non-commuting « set-theories », one sees that $f + g$ has a third set-theory bearing no relation to the previous one.

Which enables us to return for the last time to the unfortunate quantum logic. Closed subspaces are enough to speak of whatever we want – in particular of the set-theory relative to a given hermitian. But this does not socialise – not to speak of the gigantic canard of « orthomodular lattices », where the Hilbert space has disappeared. Indeed, $\begin{bmatrix} 0 & 0 \\ 0 & 1 \end{bmatrix}$ and $\begin{bmatrix} 1/2 & 1/2 \\ 1/2 & 1/2 \end{bmatrix}$ have « set-theories » corresponding to the bases $\{\mathbf{e}, \mathbf{f}\}$ and $\{\sqrt{2}/2(\mathbf{e} + \mathbf{f}), \sqrt{2}/2(\mathbf{e} - \mathbf{f})\}$, but what is the « set-theory » of their sum $\begin{bmatrix} 1/2 & 1/2 \\ 1/2 & 3/2 \end{bmatrix}$? The solution does not belong in lattice theory, since the solution of the equation, $\lambda^2 - 2\lambda + 1/2 = 0$. In other words, the order structure of subspaces does not socialise with the basic quantum operation, *superposition*.

17.A Initiation to C^*-algebras

What follows is addressed to the reader who may have forgotten the basis of functional analysis or to the one that would seek his way through a classic book on the topic, e.g., [63]. This sort of culture is primal and can in no way be replaced with category-theoretic ruminations loosely inspired from linear or operator algebra, e.g., traced monoidal categories.

17.A.1 Hilbert spaces

Definition 87 (Hilbert spaces). A (complex) *Hilbert space* is a vector space equipped with a sesquilinear form enjoying $\langle x \mid x \rangle > 0$ for $x \neq 0$ and complete w.r.t. the norm $\langle x \mid x \rangle^{1/2}$.

The most popular Hilbert space is:

Definition 88 (ℓ^2). ℓ^2 is the space of square summable sequences of complex numbers, equipped with the *form*

$$\langle (a_n) \mid (b_n) \rangle := \sum_n a_n \bar{b}_n. \tag{17.34}$$

This is more or less the sole example: indeed, any Hilbert space admits a basis $(\mathbf{e}_i)_{i \in I}$ such that $\langle \mathbf{e}_i \mid \mathbf{e}_j \rangle = \delta_{ij}$, i.e., an *orthonormal* basis. Which makes it isomorphic with the space $\ell^2(I)$ of the square summable families indexed by I; remember Parseval's formula

$$\|x\|^2 = \sum_i |\langle x \mid \mathbf{e}_i \rangle|^2. \tag{17.35}$$

The cardinal $\sharp(I)$ is the *hilbertian dimension* of the space. ℓ^2 is the most frequent case, corresponding to a denumerable I; the other important case is I finite, which corresponds to this very chapter. Any *separable* Hilbert space, i.e., admitting a dense denumerable subset, is isomorphic to some $\ell^2(I)$, with I finite or denumerable, i.e., is of hilbertian dimension $\leq \aleph_0$.

Complex coefficients come from the spectral theory, see *infra*: the spectrum is non-empty (in finite dimension, the characteristic polynomial has a root). Bilinearity is thus replaced with sesquilinearity (*sesqui* = one and a half): $\langle x \mid \lambda y \rangle = \bar{\lambda} \langle x \mid y \rangle$, allowing $\langle x \mid x \rangle \in \mathbb{R}$ (equivalently $\langle x \mid y \rangle = \overline{\langle y \mid x \rangle}$). The condition $\langle x \mid x \rangle > 0$ for $x \neq 0$ yields the noted:

Theorem 64 (Cauchy–Schwarz). $|\langle x \mid y \rangle|^2 \leq \langle x \mid x \rangle \langle y \mid y \rangle$, *equality occurring only in case of colinearity.*

This being said, $\langle x \mid x \rangle \geq 0$ is enough in practice: one thence quotients by the kernel of the form; similarly, if the space is not complete... one completes it! These two operations are known as the process of separation/completion of a *pre-Hilbert* space.

One can sum Hilbert spaces: the algebraic direct sum $\mathbf{H} \oplus \mathbf{K}$ equipped with the form

$$\langle x \oplus y \mid x' \oplus y' \rangle := \langle x \mid x' \rangle + \langle y \mid y' \rangle. \tag{17.36}$$

In particular, $\|x \oplus y\|^2 = \|x\|^2 + \|y\|^2$. The tensor product is obtained by equipping the vector space generated by the formal tensors $x \otimes y$, $x \in \mathbf{H}$, $y \in \mathbf{K}$ with the sesquilinear form defined on « pure » tensors by

$$\langle x \otimes y \mid x' \otimes y' \rangle := \langle x \mid x' \rangle \cdot \langle y \mid y' \rangle. \tag{17.37}$$

The Hilbert space $\mathbf{H} \otimes \mathbf{K}$ is defined as the separation/completion of this pre-Hilbert space. Observe that the tensor product does not factorise all bilinear maps, but only some of them (styled Hilbert-Schmidt maps): the form $\langle \cdot \mid \cdot \rangle$ cannot be factorised through a linear map from $\mathbf{H} \otimes \mathbf{H}^\sharp$ into $\mathbb{C}$.

17.A.2 The dual space. A Hilbert space is a normed complex vector space, i.e., a *Banach space*. Its dual $\mathbf{H}^\sharp$ is the set of *continuous linear forms*, equipped with the norm $\|\varphi\| := \sup\{|\varphi(x)|; \|x\| \leq 1\}$. A Banach space which is its own bidual is said to be *reflexive*, which is an exceptional situation, at least in infinite dimension. Hilbert spaces are the main examples of reflexive Banach spaces. The main tool in Banach spaces is the Hahn–Banach theorem, which takes various forms, e.g.:

Theorem 65 (Hahn–Banach). *If C is a closed convex set in a Banach space and $x \notin C$, there is a continuous linear form φ and a real number r such that*

$$\Re(\varphi(C)) \leq r < \Re(\varphi(x)).$$

This theorem is already very interesting in finite dimensions: in this case, all forms are continuous.

A Banach space E can be equipped with a *weakened* topology: $x_i \to x$ iff $\varphi(x_i) \to \varphi(x)$ for all continuous linear forms on E. This topology has an extraordinary property: the unit ball of E is *weakly compact*. Indeed, the weakened topology is nothing but convergence « coefficientwise ». W.r.t. weakened topologies, there are many compacts sets.

Theorem 66 (Krein–Milman). *Any compact convex set is the closed convex envelope of its extremal border, i.e., of the set of its extremal points.*

Remember that an extremal point of the convex set K is a point $x \in K$ which cannot be written as the barycenter of two other points of K. The extremal frontier of a disk is the frontier circle, the extremal frontier of a convex polygon is made of its vertices. This theorem applies to *density operators*, i.e., to positive hermitians of trace 1 (w.r.t the weakened topology when the dimension is infinite). It simply says that any hermitian can be approximated by barycenters of extremal hermitians – i.e., by projections of rank 1. Which is implemented in a finite dimension by diagonalisation in an orthonormal basis $(\mathbf{e}_i)$ under the form $\begin{bmatrix} \lambda_1 & 0 & \dots & 0 \\ 0 & \lambda_2 & \dots & 0 \\ \dots & \dots & \dots & \dots \\ 0 & 0 & \dots & \lambda_n \end{bmatrix}$; the operator can thus be written as the barycenter of the projections associated with the $\mathbf{e}_i$ with the respective weights λ_i.

Some euclidian geometry: let us consider the triangle of vertices $0, x, y$ and the median starting with 0, in other terms the vector $(x + y)/2$. An immediate computation yields

$$\|x\|^2 + \|y\|^2 = 2(\|(x - y)/2\|^2 + \|(x + y)/2\|^2). \tag{17.38}$$

This equality enables one to majorise, in certain cases, the norm of the third edge, $x - y$. Thus, if $E \subset \mathbf{H}$ is a non-empty closed convex set, (17.38) enables one to project on the convex E: if $(x_n \in E)$ is such that $\|x_n\|$ converges to $\inf\{\|x\|; x \in E\}$, then it is indeed a Cauchy sequence.

Proposition 40. *The minimum* $\inf\{\|x\|; x \in E\}$ *is reached at a unique point of* E. *This point is also the unique* $e \in E$ *such that* $\Re\langle e \mid e - f\rangle$ *is negative for all* $f \in E$.

Which applies to the projection of an arbitrary point on a closed subspace E. Let π be the map obtained in this way: one sees that π is linear and that $\pi^2 = \pi$. The range of π is E, its kernel is $E^{\perp}$ and the projection associated with $E^{\perp}$ is $I - \pi$. These spaces are *supplementary*, i.e., each vector of $\mathbf{H}$ is uniquely written as $x = e + e', e \in E, e' \in E^{\perp}$, i.e., $\mathbf{H}$ is isomorphic to $E \oplus E^{\perp}$.

If $e \in \mathbf{H}$, $x \rightsquigarrow \langle x \mid e\rangle$ is a continuous linear form: by Cauchy–Schwarz $|\langle x \mid e\rangle| \leq \|e\|\|x\|$, equality being reached with $x = \lambda e$, hence $\|e^*\| = \|e\|$. Conversely, any linear form φ continuous on $\mathbf{H}$ is of the form e^* for a well-chosen e – necessarily unique: it suffices to consider $\{x; \varphi(x) = 1\}$ and apply projection techniques. Hence the dual $\mathbf{H}^{\sharp}$ of $\mathbf{H}$ is canonically isomorphic to $\mathbf{H}$ by means of the map $b \rightsquigarrow b^*$; but, beware: $(\lambda b)^* = \bar{\lambda} b^*$!

The linear forms b^* induce the *weak topology* on $\mathbf{H}$, hence:

Proposition 41. *The unit ball of* $\mathbf{H}$ *is weakly compact.*

Weak convergence does not imply norm convergence; however:

Proposition 42. *If* $x_i \to x$ *weakly and* $\|x_i\| \to \|x\|$, *then* $x_i \to x$ *« normwise ».*

Among the applications of weak compactness:

Proposition 43. *The image under an operator* $u \in \mathcal{B}(\mathbf{H}, \mathbf{K})$ (*infra*) *of the unit ball of* $\mathbf{H}$ *is norm-closed in* $\mathbf{H}$.

Proof. Indeed, u is weakly continuous, hence the image B' of the unit ball is weakly compact and therefore weakly closed. It remains closed in any stronger (with more closed sets) topology. □

17.A.3 The spectral theory. If $\mathbf{H}$ and $\mathbf{K}$ are Hilbert spaces, $\mathcal{B}(\mathbf{H}, \mathbf{K})$ stands for the set of all *bounded* linear maps from $\mathbf{H}$ into $\mathbf{K}$, i.e., such that the *norm*

$$\|u\| := \sup\{\|u(x)\|; x \in \mathbf{H}, \|x\| \leq 1\} \tag{17.39}$$

is finite. u thus induces a linear map $u^{\sharp}$ from the dual $\mathbf{K}^{\sharp}$ of $\mathbf{K}$ to the dual $\mathbf{H}^{\sharp}$ of $\mathbf{H}$. Now $\mathbf{K}^{\sharp}$ and $\mathbf{H}^{\sharp}$ are isomorphic to $\mathbf{K}$ and $\mathbf{H}$, in other terms $u^{\sharp}$ defines a linear map (the *adjoint*) u^* from $\mathbf{K}$ to $\mathbf{H}$; u^* is indeed defined by

$$\langle u(x) \mid y\rangle = \langle x \mid u^*(y)\rangle. \tag{17.40}$$

The most important case is that of $\mathbf{H} = \mathbf{K}$; one simply denotes by $\mathcal{B}(\mathbf{H})$ the Banach *algebra* thus obtained. The adjunction is an involution of $\mathcal{B}(\mathbf{H})$, satisfying $(\lambda.u)^* = \bar{\lambda}.u^*$, $(uv)^* = v^*u^*$, etc. and above all:

$$\|uu^*\| = \|u\|^2. \tag{17.41}$$

An involutive Banach algebra enjoying (17.41) is called a C^**-algebra*. The most natural example of a C^*-algebra is $\mathcal{B}(\mathbf{H})$. A self-adjoint (closed under adjunction) norm-closed sub-algebra of $\mathcal{B}(\mathbf{H})$ is the most general form example of a C^*-algebra: this is the contents of the GNS theorem (Section 17.A.8).

There is a more elementary example, the algebra $\mathbb{C}(X)$ of complex continuous functions on a compact X, equipped with multiplication, adjunction being conjugation, all of those pointwise, with

$$\|f\| = \sup\{|f(x)|; x \in X\}. \tag{17.42}$$

The peculiarity of $\mathbb{C}(X)$ is to be commutative; it is indeed the most general form of a commutative C^*-algebra (Section 17.A.5).

In a Banach algebra $\mathcal{C}$, one can define the *spectrum* of an element u:

Definition 89. If $u \in \mathcal{C}$, $\mathrm{Sp}(u)$ is the set of $\lambda \in \mathbb{C}$ such that $u-\lambda.I$ is not invertible.

If $\mathbf{H}$ is finite-dimensional and $u \in \mathcal{B}(\mathbf{H})$, $\mathrm{Sp}(u)$ is the set of eigenvalues of u. If X is compact and $f \in \mathcal{C}(X)$, then $\mathrm{Sp}(f)$ is the range of the function f. Some results involve $\mathrm{Sp}'(u) := \mathrm{Sp}(u) \cup \{0\}$, typically:

Proposition 44. $\mathrm{Sp}'(uv) = \mathrm{Sp}'(vu)$.

Theorem 67. $\mathrm{Sp}(u)$ *is a non-empty compact set.*

The theorem has an effective version: one can compute the « size » of the spectrum, i.e., the *spectral radius* $\varrho(u) := \inf\{r; \forall z \in \mathrm{Sp}(u)\ |z| \le r\}$ (by compactness, this value is effectively taken):

$$\varrho(u) = \lim \|u^n\|^{1/n} = \inf \|u^n\|^{1/n}. \tag{17.43}$$

In particular, the spectral radius of a nilpotent operator is 0. In a C^*-algebra, the norm of a *normal*, e.g., hermitian, operator equals its spectral radius.

Proposition 45. *Let P be a complex polynomial; then* $\mathrm{Sp}(P(u)) = P(\mathrm{Sp}(u))$.

Corollary 45.1. *The equation $uv - vu = I$ has no solution among bounded operators.*

Thus, $pq - qp = -i\hbar I$ requires *unbounded* operators (Section 19.A.3).

Proposition 46. $\mathrm{Sp}(u^*) = \overline{\mathrm{Sp}(u)}$; *if u is invertible, then* $\mathrm{Sp}(u^{-1}) = \mathrm{Sp}(u)^{-1}$.

17.A.4 Taxonomy of operators. The elements of a C^*-algebra, thus the operators on a Hilbert space, can be classified according to the relation with their own adjoint. Typically:

Normal: an operator commuting with its adjoint, $uu^* = u^*u$. Then, the C^*-algebra generated by u is commutative and u enjoys a sort of « diagonalisation ». Among normal operators one finds hermitians and unitaries.

Unitary: an operator u with inverse u^*: $uu^* = u^*u = I$. Since $\langle u(x) \mid u(y)\rangle = \langle x \mid u^*u(x)\rangle = \langle x \mid y\rangle$, unitaries correspond to the *isometries* of $\mathbf{H}$ and therefore form a group. The spectrum of a unitary is included in the circle $\mathcal{T} = \{z; |z| = 1\}$: since $\|uu^*\| = \|u\|^2$, $\|u\| = 1$, thus $\mathrm{Sp}(u) \subset \mathcal{D} := \{z; |z| \leq 1\}$; the same is true of $\mathrm{Sp}(u^*) = \mathrm{Sp}(u)^{-1}$, which shows that $\mathrm{Sp}(u) \subset \mathcal{D} \cap \mathcal{D}^{-1} = \mathcal{T}$.

Hermitian (a.k.a. self-adjoint:) an operator u equal to its adjoint, i.e., such that $\langle u(x) \mid x\rangle \in \mathbb{R}$ for all x. The spectrum of a hermitian is real and the extremal bounds of its spectrum are the reals $\sup\{\langle u(x) \mid x\rangle; \|x\| = 1\}$ and $\inf\{\langle u(x) \mid x\rangle; \|x\| = 1\}$. The most typical hermitian (indeed, every hermitian is of this form) is a sum $u + u^*$.

Symmetry: a unitary hermitian, i.e., such that $u = u^* = u^{-1}$. The spectrum is included in $\{-1, +1\}$ and one can indeed « diagonalise » u as the difference of the projections (*infra*) $(I+u)/2$ (eigenspace of $+1$) and $(I-u)/2$ (eigenspace of -1).

Projection: an idempotent hermitian (hence positive): $u = u^* = u^2$. The spectrum is included in $\{0, +1\}$ and u corresponds to the orthogonal projection $u = I_E$ on a closed subspace, the range E of u.

Positive hermitian: a hermitian such that $\langle u(x) \mid x\rangle \geq 0$ for all x. Positive hermitians are of extreme importance, since the order structure of $\mathbb{R}$ compensates for the deficiencies of topology, for instance in questions of convergence of series. Positive hermitians have a spectrum included in $\mathbb{R}^+$. The typical positive hermitian is a product uu^*; indeed, they are all of that form and one can even suppose u to be in turn a positive hermitian: the capital fact is that a hermitian has a square root.

The standard analogy is as follows: operators are a « non-commutative » version of their spectrum, in other words, hermitians are the « non-commutative reals », the unitaries playing the part of the complex arguments[8] $e^{i\theta}$. By the way, the *polar decomposition* expresses any operator as the product $u = \psi \cdot |u|$ of a module (the positive hermitian $|u| := \sqrt{u^*u}$) and an isometry (partial, however). Indeed, $u = \psi \cdot |u| = |u^*| \cdot \psi$.

[8]While unitaries are closed under product, hermitians under sum, normal operators do not socialise: normal operators form a hybrid class reduced in practice to its two useful subcases, unitaries and hermitians.

Theorem 68. *Let $\mathcal{C}^+$ be the set of positive hermitians of $\mathcal{C}$;*

(i) *$\mathcal{C}^+$ is a closed convex cone in $\mathcal{C}$;*

(ii) *If $u, -u \in \mathcal{C}^+$, then $u = 0$.*

Hermitians can thus be ordered by: $u \leq v$ iff $v - u$ is positive.

Theorem 69. *A hermitian u can be uniquely expressed as the difference $u^+ - u^-$ of two positive hermitians such that $u^+u^- = u^-u^+ = 0$. Moreover $\|u\| = \sup(\|u^+\|, \|u^-\|)$.*

This in no way establishes a lattice structure on hermitians; indeed, in $\mathcal{B}(\mathbf{H})$, two hermitians have a supremum exactly when they are comparable!

Theorem 70. *If $u \in \mathcal{C}$, the following properties are equivalent:*

(i) *$u \in \mathcal{C}^+$;*

(ii) *$u = v^2$ for a v hermitian;*

(iii) *$u = v^*v$ for an arbitrary v.*

Moreover in (ii), *v can in turn be chosen positive, in case the choice is unique.*

When $\mathcal{C}$ is of the form $\mathcal{B}(\mathbf{H})$, one can add the following equivalents:

(iv) $u = v^*v$ for an operator v in some $\mathcal{B}(\mathbf{H}, \mathbf{K})$;

(v) $\langle u(x) \mid x \rangle \geq 0$ for all x in $\mathbf{H}$.

Corollary 70.1. *If u is positive, then v^*uv is positive. If $u, v \in \mathcal{C}^+$ commute, then $uv \in \mathcal{C}^+$.*

Proof. Always think of the square root: $v^*uv = (v^*\sqrt{u})(\sqrt{u}v) = (\sqrt{u}v)^*(\sqrt{u}v)$. □

17.A.5 Commutative case. If u is hermitian, its spectrum is real; moreover, if P is a polynomial, one verifies that $\|P(u)\| = \sup\{|P(x)|; x \in \mathrm{Sp}(u)\}$. Using Stone–Weierstraß, this equality enables one to define $f(u)$ for any continuous function $f \in \mathbb{C}(\mathrm{Sp}(u))$ from $\mathrm{Sp}(u)$ to $\mathbb{C}$; for instance, if u is positive, the function $\sqrt{x}$ yields $\sqrt{u}$. We have just defined an isometry between $\mathbb{C}(\mathrm{Sp}(u))$ and the C^*-algebra generated by u; u corresponds to the canonical inclusion from $\mathrm{Sp}(u)$ into $\mathbb{C}$. This is known as *spectral calculus*.

This extends to the case where u is normal: there is still an isomorphism between the C^*-algebra generated by u and $\mathbb{C}(\mathrm{Sp}(u))$. This can be seen as a sort of diagonalisation. More generally, any commutative C^*-algebra is isomorphic to $\mathbb{C}(X)$ for a certain compact set X. If an operator is normal, it is enough to look at its spectrum to know whether it is hermitian, unitary, etc.

17.A.6 Partial isometries

Definition 90 (Partial isometry). u – not necessarily normal – is a *partial isometry* when uu^* is a projection.

Proposition 47. *If uu^* is a projection, then u^*u is a projection too.*

Proof. Using Proposition 44, one sees that $\mathrm{Sp}(u^*u) \subset \{0, 1\}$; since u^*u is hermitian, one can apply spectral calculus: the canonical inclusion ι of $\mathrm{Sp}(u^*u)$ in $\mathbb{C}$ enjoys $\iota^2 = \iota$, hence $(u^*u)^2 = u^*u$. □

Which can be obtained in another way:

Proposition 48. *u is a partial isometry iff $u = uu^*u$.*

Proof. From $u = uu^*u$ one easily gets $uu^* = (uu^*)^2$ (and $u^*u = (u^*u)^2$). Conversely, if uu^* is a projection, $(u - uu^*u)(u^* - u^*uu^*) = uu^* - (uu^*)^2 - (uu^*)^2 + (uu^*)^3 = 0$. The nullity of the expression aa^* entails that of a, since $\|a\|^2 = \|aa^*\|$. Hence $u - uu^* = 0$. □

Definition 91 (Domain, Image). If u is a partial isometry, the *domain* (resp. *image*) of u is the closed subspace associated with the projection u^*u (resp. uu^*). In fact u establishes an isometric bijection between its domain and its image.

One will similarly toy with and prove that, if u is normal and $\mathrm{Sp}(u) \subset \{-1, 0, 1\}$, then u is a *partial symmetry*, i.e., that u is a hermitian such that u^2 is a projection (equivalently: $u^3 = u$).

17.A.7 Finite dimension. If one extracts from the preceding the information that concerns finite dimension, one gets:

- Normal, e.g., hermitian, unitary, operators are diagonalisable in an orthonormal basis. Which means that a hermitian or unitary matrix is written UDU^* for well-chosen unitary U and diagonal D.

- A hermitian operator corresponds to the case where D has real entries; a positive operator corresponds to positive entries. A unitary operator diagonalises through a D whose diagonal entries lie in the unit circle.

This is obtained through $\mathbb{C}(X)$: in a finite dimension, the compact X is a finite set (with the discrete topology).

In a finite dimension, the trace is defined as the sum of all diagonal coefficients in no matter which matrix representation. It is *cyclic*:

$$\mathrm{tr}(uv) := \mathrm{tr}(vu) \tag{17.44}$$

hence independent from the basis: $\mathrm{tr}(u) = \mathrm{tr}(v^{-1}vu) = \mathrm{tr}(vuv^{-1})$.

Every matrix is triangulable: in an appropriate basis[9],

$$M = \begin{bmatrix} \lambda_{11} & \lambda_{12} & \dots & \lambda_{1n} \\ 0 & \lambda_{22} & \dots & \lambda_{2n} \\ \dots & \dots & \dots & \dots \\ 0 & 0 & \dots & \lambda_{nn} \end{bmatrix};$$

then

$$e^M = \begin{bmatrix} e^{\lambda}_{11} & \mu_{12} & \dots & \mu_{1n} \\ 0 & e^{\lambda}_{22} & \dots & \mu_{2n} \\ \dots & \dots & \dots & \dots \\ 0 & 0 & \dots & e^{\lambda}_{nn} \end{bmatrix},$$

hence:

Proposition 49. $\det(e^M) = e^{\mathrm{tr}(M)}$.

In finite dimension, the quantity $\|u\|_1 := \mathrm{tr}(\sqrt{uu^*})$ defines the *trace-norm*:

$$\begin{aligned} |\mathrm{tr}(u)| &\leq \|u\|_1, \\ \|uvw\|_1 &\leq \|u\| \cdot \|v\|_1 \cdot \|w\|. \end{aligned} \tag{17.45}$$

Indeed the two norms are put in duality by

$$\begin{aligned} \|u\| &= \sup\{|\mathrm{tr}(uv)|;\ \|v\|_1 \leq 1\}, \\ \|v\|_1 &= \sup\{|\mathrm{tr}(uv)|;\ \|u\| \leq 1\}. \end{aligned} \tag{17.46}$$

$\|I\|_1 = n$, the dimension of the space; it tends to infinity when the dimension increases. In infinite dimension, one must therefore restrict oneself to *trace-class operators*, which form a two-sided ideal (Section 20.D.1), thus do not include invertible operators such as the « twist ». Trace-class operators do not seem to be suited for the interpretation of proofs.

17.A.8 The GNS construction. A.k.a. Gel'fand–Neumark–Segal; any C^*-algebra is a subalgebra of some $\mathcal{B}(\mathbf{H})$. This result mainly rests on the possibility of transforming a C^*-algebra into a Hilbert space: this is known as the GNS construction.

Let ρ be a *state*, i.e., a linear functional from the C^*-algebra $\mathcal{C}$ into $\mathbb{C}$ satisfying $\rho(I) = 1$, $\rho(u) \geq 0$ for u positive. A state is hermitian: $\rho(u^*) = \overline{\rho(u)}$, hence, if u is hermitian $\rho(u)$ is real. Above all, a state is monotonic: if $u \leq v$ are hermitian, then $\rho(u) \leq \rho(v)$.

[9] Orthonormal if one wants it.

Given a state on $\mathcal{C}$, the formula

$$\langle u \mid v \rangle := \rho(v^* u) \tag{17.47}$$

defines a pre-Hilbert space, whose separation/completion is denoted by $\mathcal{C}_\rho$. Every $f \in \mathcal{C}$ operates (on the left) on the pre-Hilbert space $\mathcal{C}$ by: $f(u) := fu$ and one sees that $\|f(u)\|^2 = \rho(u^* f^* f u) \leq \|f\|^2 \cdot \rho(u^* u)$ (use $f^* f \leq \|f\|^2 \cdot I$, etc. and monotonicity). The left action of f can thus be extended into an operator of norm at most $\|f\|$ on the separation/completion $\mathcal{C}_\rho$. One thus obtains a morphism from the C^*-algebra $\mathcal{C}$ into the space $\mathcal{B}(\mathcal{C}_\rho)$, which one calls a *representation*.

One must sum up « many » GNS representations so as to eventually get one which is *faithful*, i.e., does not decrease the norm.

By the way, not that the norm of a C^*-algebras is *algebraically* definable:

(i) Since $\|u\|^2 = \|uu^*\|$, one can restrict to positive hermitians.

(ii) In that case, the norm equals the *spectral radius*

$$\|u\| = \inf\{\lambda; u - \lambda I \text{ not invertible}\}.$$

Thus, if a morphism φ of C^*-algebras decreases the norm, it does it on some positive hermitian, of which it shrinks the spectrum: $u - \lambda I$ is not invertible, while $\varphi(u) - \lambda I$ was. With a little bit of spectral calculus, one finds $v \neq 0$ such that $(u - \lambda I)v = 0$, hence $\varphi(v) = 0$.

A $*$-*isomorphism*, i.e., an *injective* morphism of C^*-algebras, is therefore isometric. In particular, the following notions are equivalent:

- Faithful representation.
- Injective representation.

A *simple* C^*-algebras, i.e., without a non-trivial closed bilateral ideal, thus admits only one stellar semi-norm. Which is, for instance, the case of the matrix algebras $\mathcal{M}_n(\mathbb{C})$ and their direct limit, the CAR algebra (Section 20.C.1).

Chapter 18
Nets and duality

We shall revisit proof-nets, especially the *correctness criterion*. What follows (expounded in a paper impossible to find, [39]) essentially dates back to the early times of linear logic and constitutes the *Pons Asinorum* towards Geometry of Interaction (GoI).

18.1 Duality and correctness

18.1.1 Switchings and counterproofs. A *paraproof* (Section 12.5.3) of A is a test for $\sim A$, seen as a sort of proof. We shall be interested in a specific sort of *tests*, the switchings of proof-nets. Consider for instance:

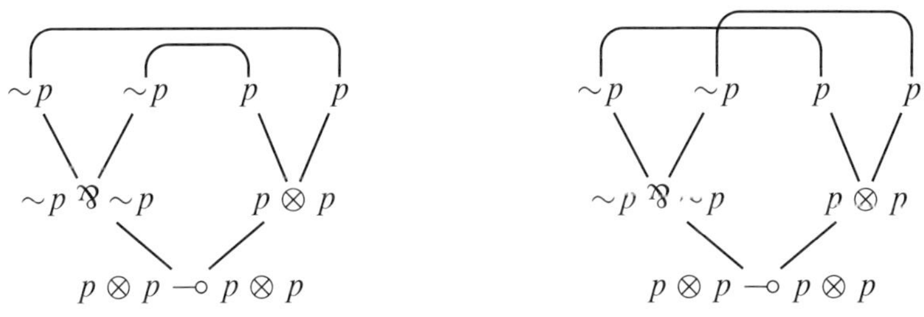

(18.1)

We took the most simplistic hypotheses: etaspanded multiplicative logic without cut. In other terms, we have three links, ⅋, $\otimes$ and an axiom link between atoms. What is thus a proof of $A := p \otimes p \multimap p \otimes p$ by means of a net? One has not much freedom: each formula is a conclusion of a well-defined link, corresponding to its first symbol. When the formula is a literal p or $\sim p$, it is one conclusion of an axiom link of which we may *a priori* ignore the other conclusion. Thus, the formula A, which has four atoms $\sim p$, $\sim p$, p, p has only two proofs, corresponding to the two ways of displaying axiom links between the atoms.

One could rewrite this in a simpler way, by only retaining the four atoms, labeled 1, 2, 3, 4: it is a matter of two graphs, one with the edges 14, 23, the other with 13, 24. To seek a proof of A is to seek a graph between the vertices 1, 2, 3, 4 subject to certain constraints:

(i) The edges must relate matching atoms, p with $\sim p$, q with $\sim q$. We shall ignore this constraint, which only concerns proofs, not paraproofs. Indeed, one will allow arbitrary axioms taking the form of boxes with atomic conclusions: *daimons*, so to speak.

(ii) The correctness criterion must be verified; it is where the formula A intervenes, which links together the atoms by means of connectives: depending on the choice of A, one does not get the same switchings.

Let us come back to the example (18.1): by erasing the axiom links, one finds oneself with the graphs induced by the various switchings of A; each of the four choices yields three connected components. They differ, but, in the four cases, they dispatch the atoms in the same way: $\{1\}, \{2\}, \{3, 4\}$. The solution of my problem consists in all partitions of the set $\{1, 2, 3, 4\}$ enjoying the correctness criterion. There are of course the two we already know, $\{1, 4\}, \{2, 3\}$ and $\{1, 3\}, \{2, 4\}$, but also $\{1, 2, 3\}, \{4\}$ and $\{1, 2, 4\}, \{3\}$, etc., which correspond to paraproofs of A.

Which suggests the following definition:

Definition 92 (Incidence graph). Let I be a non-empty set. Two partitions $\mathcal{P}$ and $\mathcal{Q}$ of I induce a bipartite graph, called an *incidence graph*:

Vertices: on one hand, the elements of $\mathcal{P}$, on the other hand, those of $\mathcal{Q}$.

Edges: the edges between $a \in \mathcal{P}$ and $b \in \mathcal{Q}$ are the elements of $a \cap b$.

$\mathcal{P}$ and $\mathcal{Q}$ are *polar*, notation $\mathcal{P} \perp \mathcal{Q}$, when their incidence graph is a tree, i.e., is connected and acyclic.

There are $\sharp(I)$ edges; if $\mathcal{P} \perp \mathcal{Q}$, $a \in \mathcal{P}$, $b \in \mathcal{Q}$, then $\sharp(a \cap b) \leq 1$.

We will consider, as usual, sets (here, of partitions on a given set I) equal to their own bipolar. I come to the point of multiplicative connectives; suppose that the *carriers* I, J are disjoint and that A, B are sets of partitions equal to their bipolars, whose polars are non-empty:

Theorem 71 (Bipolar). *The set* $A \otimes B := \{\mathcal{P} \cup \mathcal{Q}; \mathcal{P} \in A,\ \mathcal{Q} \in B\}$ *is equal to its bipolar.*

Proof. Let $\mathcal{R}$ be a partition of $I \cup J$ such that $\mathcal{R} \in {\sim}{\sim}(A \otimes B)$. Choose partitions $\mathcal{S}, \mathcal{T}$ in respectively ${\sim}A, {\sim}B$. If $i \in I, j \in J$; one can define $\mathcal{S} \sqcup_{ij} \mathcal{T}$ as made of the elements of $\mathcal{S}$ and $\mathcal{T}$, up to one exception: the component of i in $\mathcal{S}$ and that of j in $\mathcal{T}$ have been glued to form a single element of the partition $\mathcal{S} \sqcup_{ij} \mathcal{T}$. One successively verifies that:

- $\mathcal{S} \sqcup_{ij} \mathcal{T} \in {\sim}(A \otimes B)$. Indeed, the gluing of the component of i with that of j glues two disjoint trees into a tree.
- Hence $\mathcal{R} \perp \mathcal{S} \sqcup_{ij} \mathcal{T}$. In particular, one sees that i, j are not in the same component of $\mathcal{R}$.

- Therefore $\mathcal{R} = \mathcal{P} \cup \mathcal{Q}$, the union of a partition of I and a partition of J. $\mathcal{R} \perp (\mathcal{S} \sqcup_{ij} \mathcal{T})$ implies $\mathcal{P} \perp \mathcal{S}$ and $\mathcal{Q} \perp \mathcal{T}$. □

Note the use of the *locative product* (14.11). This result should be put side by side with the internal completeness of multiplicatives (Section 14.2.4).

18.1.2 The criterion. Theorem 71 is close to the correctness criterion. If one interprets a formula by a set of partitions equal to its bipolar, the *sequentialisation* theorem precisely says that this interpretation is *logically* correct. This is indeed because the switchings of A provide us with a *prepolar* for A. This works as follows:

- Note that an appropriate switching can always link the conclusion A – the « ground » – with an atom given in advance.
- The switch ⅋ puts together two partitions of the atoms without mixing them, on one hand that which is above A, on the other hand that which is above B. It therefore behaves like a tensor product, which is natural, since it exists in the « dual » world. This being said, this switch puts into contact with the « ground », as we please, either a class of A, or a class of B.
- The tensor is without switching. This being said, it glues through the *ground* $A \otimes B$ the class linked with A and the class linked with B. It thus exactly reproduces the construction $\mathcal{S} \sqcup_{ij} \mathcal{T}$ of Theorem 71.

This prepolar is not everything: thus, if one forms a ⅋ between A (located in $\{1, \ldots, N\}$) and $\sim A$ (delocated in $\{N+1, \ldots, 2N\}$), one finds the partition $\{\{1, N+1\}, \{2, N+2\}, \ldots, \{N, 2N\}\}$, not in the prepolar of $\sim A \otimes A$.

A switching literally induces a « proof » of the negation. Beginning with sequent calculus, one writes the conclusion, namely $\vdash \sim A$. One recursively proceeds from the conclusion: if one has written a sequent $\vdash \sim \Gamma$, one tries to make it appear as the conclusion of a rule:

⊗: if $\Gamma = C \otimes D, \Delta$, one writes a rule « ⅋ »:

$$\frac{\vdash \sim C, \sim D, \Delta}{\vdash \sim C \parr \sim D, \Delta}$$

⅋: if $\Gamma = C \parr D, \Delta$, one writes a rule « ⊗ »:

$$\frac{\vdash \sim C, \Delta \quad \vdash \sim D}{\vdash \sim C \otimes \sim D, \Delta} \quad \text{or} \quad \frac{\vdash \sim C \quad \vdash \sim D, \Delta}{\vdash \sim C \otimes \sim D, \Delta}$$

depending on the switch. One sees that the switch « gives » the full context to the subformula it selects.

Atoms: if Γ is made of atoms, one admits it as an axiom, a sort of *daimon*.

All of this can eventually be rewritten by forgetting sequents, as a multiplicative nets with boxes (*daimons*) of atomic conclusions. One could, by the way, treat what precedes, i.e., the duality between partitions associated to switchings, by means of cuts between the associated nets, which reduce to cuts between *daimons* solved by gluing of boxes. Due to the connectedness/acyclicity of the *incidence graph*, the cut-net eventually reduces to the empty graph, i.e., it remains nothing.

One could also investigate subtyping. For instance, $p \otimes (q \parr r)$ admits as prepolar the set formed of the partitions $\{\{p,q\},\{r\}\}$ and $\{\{p,r\},\{q\}\}$; it is therefore reduced to $\{\{p\},\{q,r\}\}$. But $(p \otimes q) \parr r$ has a smaller prepolar, reduced to $\{\{p,q\},\{r\}\}$ and is therefore formed of $\{\{p\},\{q,r\}\}$ and $\{\{q\},\{p,r\}\}$. This is why the implication $(p \otimes q) \parr r \vdash p \otimes (q \parr r)$ is incorrect, see (11.8). The incorrectness of the net written there corresponds to the incidence graph between $\{\{q\},\{p,r\}\}$ and $\{\{q\},\{p,r\}\}$.

18.2 The original criterion

The original criterion ([37], analysed in [39]) was not stated in terms of partitions, but in terms of *trips*, i.e., permutations. This criterion was relegated to the middle ground for pedagogical reasons. This being said, it offers a larger latitude and above all, by replacing partitions – which run into no mathematical concept – with permutations, it opens the door to operator algebras, since a permutation is a *unitary operator* on a finite-dimensional Hilbert space.

18.2.1 Trips. The idea is easily understood from the version « graph ». One will travel through the net, by following the edges; since it is a tree, one will be forced to make round trips. Hence each formula of the net will be visited twice, once « upwards », notation[1] $\uparrow A$, a second time « downwards », notation $\downarrow A$. The travel instructions are as follows[2]:

Conclusion: from $\downarrow A$ to $\uparrow A$.

Cut link: from $\downarrow A$ to $\uparrow {\sim} A$ and from $\downarrow {\sim} A$ to $\uparrow A$.

Axiom link: from $\uparrow A$ to $\downarrow {\sim} A$ and from $\uparrow {\sim} A$ to $\downarrow A$.

$\parr$ link: one must set a switch left/right:

- **Left:** from $\uparrow (A \parr B)$ to $\uparrow A$, from $\downarrow A$ to $\downarrow (A \parr B)$ and from $\downarrow B$ to $\uparrow B$.
- **Right:** from $\uparrow (A \parr B)$ to $\uparrow B$, from $\downarrow B$ to $\downarrow (A \parr B)$ and from $\downarrow A$ to $\uparrow A$.

$\otimes$ link: one must also set a switch left/right:

[1] Without any relation to the ludic shift of Section 13.9.

[2] We mean by « left » the switch going up from the conclusion to the left premise.

Left: from $\uparrow(A \otimes B)$ to $\uparrow A$, from $\downarrow A$ to $\uparrow B$ and from $\downarrow B$ to $\downarrow(A \otimes B)$.

Right: from $\uparrow(A \otimes B)$ to $\uparrow B$, from $\downarrow B$ to $\uparrow A$ and from $\downarrow A$ to $\downarrow(A \otimes B)$.

Let us proceed quickly, resting upon the simplified version (Section 11.3):

- The correctness criterion is the absence of a *short trip*, i.e., the fact that this travel is done in a single « long trip ».
- It is enough to bring back the criterion to its « graph » version. For this, we consider a « $\otimes$ » link; if we have dug it well, the switch of the link does not speak of the link, but precisely of what occurs *outside* the link. We try to determine, in the cyclic course constituted by a long trip, the order of passage out of the link: we see that, after $\uparrow A$, we cannot directly come back through $\downarrow B$ (right switching), nor through $\uparrow(A \otimes B)$ (left switching); we thus come back through $\downarrow A$. Similarly, after $\uparrow B$ we come back through $\downarrow B$. Finally, after $\downarrow(A \otimes B)$, we come back through $\uparrow(A \otimes B)$. In other words, we always come back to where we exited.
- It is thus a matter of travel in a connected and acyclic graph. The two switchings of the $\otimes$ exclude configurations $\updownarrow A \ldots \updownarrow B \ldots \updownarrow A \ldots \updownarrow B \ldots$, which are not « tree courses ». *A contrario*, the restriction to only one of the two switchings will enable considerations inaccessible to the criterion « graph ». This is what underlies *non-commutative* logic (Section 18.B).

18.2.2 Duality

Definition 93 (Polar permutations). Let I be a non-empty set. Two permutations σ, τ of I are *polar*, notation $\sigma \perp \tau$, when the product $\sigma\tau$ is cyclic.

If $I = \{1, \ldots, N\}$, $\sigma \perp \tau$ means that $\sigma\tau(1), (\sigma\tau)^2(1), \ldots, (\sigma\tau)^N(1)$ are pairwise distinct. It is immediate that $\sigma \perp \tau \Leftrightarrow \tau \perp \sigma$.

This condition, which is a translation of the correctness criterion, leads to developments of which we already saw a simplified version in Section 18.1. The passage from the original to the simplified version is easy to understand: if σ is a permutation of I, then it splits as a sum of cyclic permutations. In other terms, one can write $I = I_1 \cup \cdots \cup I_k$, with the I_k disjoint and non-empty and such that the $\sigma \restriction I_i$ are cyclic; one replaces σ with the partition $\mathcal{P} = \{I_1, \ldots, I_k\}$. To prove the analogue of Theorem 71, one must glue two permutations σ, τ: one defines $\rho = \sigma \sqcup_{ij} \tau$ as the disjoint union of both permutations, with a small modification, $\rho(i) = \tau(j)$ and $\rho(j) = \sigma(i)$, which glues the cycles of i and j.

A permutation can easily be *self-polar*, for instance a circular permutation of three elements, whose square remains circular, hence cyclic. One avoids this sort of thing with partitions, since one implicitly considers *all* permutations corresponding to a given partition.

18.2.3 Execution. Let us now consider the elimination of a cut between σ of carrier $|\Gamma| \cup |A|$ (corresponding to a proof of $\vdash \Gamma, A$) and τ of carrier $|A| \cup |\Delta|$ (corresponding to a proof of $\vdash \sim A, \Delta$). The sets $|\Gamma|, |A|, |\Delta|$ are pairwise disjoint, moreover we assume that $|\Gamma| \cup |\Delta| \neq \emptyset$. The final output of cut-elimination will be a permutation ρ of $|\Gamma| \cup |\Delta|$. Coming back to the sources, i.e., to cut-elimination in proof-nets (Section 11.2.5), we see that the reduction of the cuts $\otimes / \invamp$ does nothing but propagate them upwards. Once this propagation is completed, we are left with cuts between axiom links, which can be eliminated by « shortening » the links, i.e., by composition of the permutations.

The computation of ρ is done as follows: we take an argument $i \in |\Gamma| \cup |\Delta|$ and either σ, or τ applies, this in an exclusive way. Say that $i \in |\Gamma|$, hence $\sigma(i) \in |\Gamma| \cup |A|$. If $\sigma(i) \in |\Gamma|$, we define $\rho(i) := \sigma(i)$. Otherwise we have entered the « combat zone » $|A|$ and we thus form $\tau\sigma(i) \in |A| \cup |\Delta|$; if $\tau\sigma(i) \in |\Delta|$, we define $\rho(i) := \tau\sigma(i)$. Otherwise we resume with $\sigma\tau\sigma(i) \in |\Gamma| \cup |A|$, etc. We alternate σ and τ until we exit. We necessarily exit, without any hypothesis[3]: indeed, if the iterations stay forever in $|A|$, it is because of a loop $(\tau\sigma)^n(a) = a, a \in |A|$; but since the functions are injective, one cannot access the loop from outside, in particular from i. The first value obtained outside $|A|$ is by definition $\rho(i)$.

Moreover, observe that the logical hypotheses forbid *de facto* the existence of an internal loop. Indeed, if σ', τ' are respectively polar to $\vdash \Gamma$ and $\vdash \Delta$, then $\sigma \cup \tau' \perp \sigma' \cup \tau$, which excludes any cycle internal to $|A|$ – which would logically constitute a *vicious circle*, see (11.1).

The Tortoise Principle (Section 3.A.2), enables us to slightly simplify the structure: we can suppose that $|\Gamma| = \emptyset$, which by the way corresponds to the original cut, the *Modus Ponens*. We can write

$$\rho := |\Delta|(\tau \cup \tau\sigma\tau \cup \tau\sigma\tau\sigma\tau \cup \cdots)|\Delta|, \tag{18.2}$$

where $|\Delta|$ stands for the (partial) identity permutation of $|\Delta|$. Partial permutations, union of functions, this does not look very nice: let us change horses.

18.2.4 The execution formula. Let us consider the Hilbert space $\mathbb{C}^{|A| \cup |\Delta|}$. Total or partial, any function f from $|A| \cup |\Delta|$ in itself induces an *operator*:

$$\varphi\Big(\sum_{i \in |A| \cup |\Delta|} \lambda_i \mathbf{e}_i\Big) = \sum_i \lambda_i \mathbf{e}_{f(i)}. \tag{18.3}$$

Moreover, when the function is injective, the operator is indeed a *partial isometry* (Definition 90), thus of norm 1, except when f is completely undefined – in which case $\varphi = 0$. In particular, $\sigma, \tau, |\Delta|$ induce operators of $\mathbb{C}^{|A| \cup |\Delta|}$; σ, τ become unitaries, $|\Delta|$ becomes a projection. Our formula (18.2) is rewritten almost identically

[3] See Section 19.3.4 for an operator-theoretic explanation.

as

$$\rho := |\Delta|(\tau + \tau\sigma\tau + \tau\sigma\tau\sigma\tau + \cdots)|\Delta|. \tag{18.4}$$

At least in the logical case where there is no cycle internal to $|A|$: the central sum is finite, one can write it, as we please, $\tau.(1-\sigma\tau)^{-1}$ or $(1-\tau\sigma)^{-1}\tau$: this is because $\sigma\tau$ is *nilpotent*. The inversion suggests the solution of a linear equation; this is the *feedback equation*, which is written, w.r.t. the direct sum decomposition $\mathbb{C}^{|A|} \oplus \mathbb{C}^{|\Delta|}$,

$$\begin{aligned} \tau(y \oplus x) &= y' \oplus x', \\ \sigma(y') &= y. \end{aligned} \tag{18.5}$$

Finally, $\rho(x) = x'$, i.e., only retains the emerging part of the *iceberg*. The « introspective » part y (or y') corresponds to the computation process.

This equation underlies the *Geometry of Interaction* which will occupy the last part of this book.

18.A Trips and coherent spaces

Trips in proof nets necessitate two visits of each formula, $\uparrow A$ « question » and $\downarrow A$ « answer ». This machinery will be used to yield the coherent interpretation of a net, without transiting through sequentialisation. This is a matter of trying to *understand* the correctness criterion by *using* it, but in no way of establishing a new result.

Thus, let $\mathfrak{R}$ be a multiplicative etaspanded cut-free proof-net of conclusion A and $p, q, r, \ldots$, be the variables of A. One substitutes for these variables coherent spaces $X, Y, Z, \ldots$, of non-empty carriers. The interpretation of $\mathfrak{R}$ is made of sequences, $(x_1, \ldots, x_n)$, indexed by the literals of A and such that:

- If the literal of index i corresponds to a variable – say p – or its negation, then x_i is in the associated coherent space – say $x_i \in |X|$.
- If an axiom link relates the literals of indices i and j, then $x_i = x_j$.

Theorem 72. *The set of sequences associated to $\mathfrak{R}$ is a clique.*

One supposes that $(x_1, \ldots, x_n)$ and $(x'_1, \ldots, x'_n)$ are two such sequences; one will show their coherence. For this, one will remark that any formula B of the net induces two subsequences $(x_i, \ldots, x_j)$ and $(x'_i, \ldots, x'_j)$ of the coherent space associated with B: these two sequences are simply denoted B and B'.

Supposing that $(x_1, \ldots, x_n) \asymp (x'_1, \ldots, x'_n)$, I will eventually conclude that $(x_1, \ldots, x_n) \frown (x'_1, \ldots, x'_n)$, by means of a trip: one starts with $\uparrow A$ with the hypothesis $A \asymp A'$ and one comes back to $\downarrow A$, with the conclusion $A \frown A'$. The trip is induced by switchings, chosen in such a way that, when in $\uparrow B$, then $B \asymp B'$, when in $\downarrow B$, then $B \frown B'$.

Atom: if one is in $\uparrow a$ and if $a \asymp a'$ and if one then goes to b through an axiom link, hence in $\downarrow b$, one get $b \frown b'$, because of linear negation.

Link ⅋, upwards: if one enters for the first time in $B ⅋ C$ from below, then $B ⅋ C \asymp B' ⅋ C'$. Hence $B \asymp B'$, $C \asymp C'$. One sets the switch on « left ». Then one moves to $\uparrow B$, where $B \asymp B'$; later, one will arrive in $\downarrow C$, so $C \frown C'$, hence $C = C'$, then one proceeds to come down through $\downarrow B$, which yields $B \frown B'$; hence $B = B'$ and, finally, since $B ⅋ C = B' ⅋ C'$, one can complete the third passage by going to $\downarrow B ⅋ C$.

Link ⅋, downwards: if one enters for the first time in $B ⅋ C$ from above, say, through B, hence $B \frown B'$. If $B ⅋ C \frown B' ⅋ C'$, one switches to « left » so as to go down: when coming back, one has coherence, thus equality. Otherwise, $B = B'$ and $C \smile C'$, exactly what is needed to follow a « right ». Later, reaching $\downarrow C$, one gets $C \frown C'$, a contradiction.

Link ⊗, upwards: one has $B \otimes C \asymp B' \otimes C'$ and one can assume *strict* incoherence, in case one side is *strictly* incoherent, say $C \smile C'$. One switches to « right ». When back through $\downarrow C$, one gets $C \frown C'$, a contradiction.

Link ⊗, downwards: if one enters for the first time in $B \otimes C$ from above, say, through B, hence $B \frown B'$. If $B \otimes C \frown B' \otimes C'$, one can switch to « right » and go down: when going up one gets equality, ... Otherwise, this is because we have $C \smile C'$ and one switches to « left »; after $\uparrow C$, one comes back to $\downarrow C$, hence $C \frown C'$, a contradiction.

18.B Non-commutative logic

The non-commutative logic of Ruet and Abrusci [4] is a very interesting system, albeit slightly experimental and isolated. As to terminology: the expression « non-commutative logic » is far better than « non-commutative linear logic ». Indeed, layer -1 considerations tell us that contraction expresses idempotency which implies – in practice – commutativity. In general, the accumulation of adjectives conveys an impression of marginality; this must therefore be avoided when it is – like here – a pleonasm.

Non-commutative logic unfortunately bears little relation with the non-commutative geometry of Connes [15]; one can however remark that its correctness criterion is stated in terms of trips and not in terms of graphs: it therefore stands slightly closer to functional analysis.

18.B.1 Order varieties

Contexts. What makes the superiority of non-commutative logic over the cyclic logic of Section 11.1.3, is that it does not exclude commutativity: besides the commutative tensor $A \otimes B$, there is the non-commutative one $A \olessthan B$, both combining harmoniously, in particular

$$A \otimes B = A \olessthan B \ \cap \ A \ogreaterthan B. \tag{18.6}$$

One must therefore allow a certain amount of commutativity and consider that contexts are partially ordered. In the purely non-commutative case, we acknowledged the interest of *cyclicity*, i.e., of orders, up to a cyclic permutation. The notion of an *order variety* will be to partial orders what the oriented circle is to the segment. Taking as reference the relation circle/segment, one eventually reaches the following covenant:

Presentation: every partial order defines (*presents*) a unique order variety.

Existence: every order variety can be presented by a partial order.

Unicity: under certain hypotheses, this presentation is unique.

Main results. One can recover the definition of order varieties from the synthesis that follows. Γ, Δ stand for finite partial orders and γ, δ for their carriers. $\Gamma < \Delta$, $\Gamma \parallel \Delta$ stand for their series or parallel sum (assuming the carriers disjoint) and $\Gamma + \Delta$ for any of the three sums $\Gamma < \Delta$, $\Gamma > \Delta$, $\Gamma \parallel \Delta$.

Presentation: to each partial order Γ is associated a variety $v(\Gamma)$ with the same carrier. One has $v(\Gamma < \Delta) = v(\Gamma \parallel \Delta)$ – which corresponds to cyclicity – hence non-unicity of the presentation.

Unicity: if one splits the carrier of the variety V into non-empty γ, δ, then the Γ, Δ such that $V = v(\Gamma + \Delta)$ are unique. By what precedes, the sum « $+$ » is irrelevant, $<, >, \parallel$.

Existence: they do exist when $\sharp(\delta) = 1$: if I choose a point A in the carrier, I can write $V = v(\Gamma + A)$ for a certain partial order Γ (necessarily unique) on the complementary of A. In other terms, an order variety is a structure that can be seen as a partial order every time I remove a point (i.e., I choose a viewpoint).

One can thus always *focalise* on a point A in an order variety, which induces a *unique* ordered context Γ. Observe that A is in no position (series, parallel) w.r.t. its context, hence the single cut rule (and the single negation). In the rules of multiplicative disjunctions, we must place two formulas A, B w.r.t. a context; one defines a partial order Δ between A and B depending on whether the disjunction is commutative or not. Then one must be able to present the sequent by $\Gamma + \Delta$; this is not always possible, but, when it works, Γ remains unique.

The rules. Order varieties constitute an abstract, synthetic, notion. In practice we are more at ease with analytical notions: let us thus present a sequent by means of a partial series/parallel order. The most general structural rule enables us to replace $\vdash \Gamma$ with $\vdash \Gamma'$ when $v(\Gamma') \subset v(\Gamma)$. Which reduces two replacements:

(i) In the series/parallel decomposition of Γ, a « series » becomes a « parallel ». Thus $(A < B) \parallel C$ can be replaced with $A \parallel B \parallel C$.

(ii) An *external* parallel becomes a series: $A \parallel B \parallel C$ becomes $A < (B \parallel C)$.

In particular the inclusion of presentations is neither necessary nor sufficient: thus $A < (B \parallel C)$, which is finer than $(A < B) \parallel C$ defines a strictly smaller variety.

The logical rules are easy to recover:

Pars: from $\vdash \Gamma + \Delta$, with $\Delta = \{A, B\}$, depending whether Δ is ordered $(A < B)$ or not $(A \parallel B)$, we conclude $\vdash \Gamma + A \ltimes B$ or $\vdash \Gamma + A ⅋ B$.

Tensors: from $\vdash \Gamma + A$ and $\vdash B + \Delta$ one concludes $\vdash (\Gamma \parallel \Delta) + A \otimes B$ or $\vdash (\Gamma > \Delta) + A \oslash B$.

Identity: axiom and cut are indifferent to all order nuances: $\vdash A + \sim A$ for the axiom; as to the cut, from $\vdash \Gamma + A$ and $\vdash \sim A + \Delta$, it yields $\vdash \Gamma + \Delta$.

The solution. The solution can be recovered from the constraint

$$v(a < b < c) = v(c < a < b) = v((a < b) \parallel c).$$

An *order variety* is a finite carrier equipped with a ternary relation $a < b < c$ (« b between a and c ») enjoying:

Irreflexivity: $a < b < c$ implies $a \neq c$.

Cyclicity: $a < b < c$ implies $b < c < a$.

Transitivity: $a < b < c$ and $a < c < d$ imply $a < b < d$.

Repartition: if $a < b < c$ and $d \neq a, b, c$, then $d < b < c$ or $a < d < c$ or $a < b < d$.

$v(\Gamma)$ is defined as the set of $a < b < c$ with $a < b$ and either $b < c$ *modulo* Γ or c incomparable to a, b *modulo* Γ; of course $v(\Gamma)$ must be closed under circular permutations. We can verify the conditions without difficulty, but this is sometimes tedious (repartition condition). Our covenant is respected:

Presentation: one immediately sees that $v(\Gamma < \Delta) = v(\Gamma \parallel \Delta)$.

Unicity: the unicity of the writing $V = v(\Gamma < \Delta) = v(\Gamma \parallel \Delta)$ is obtained as follows: if I select $d \in \delta$, then $a < b$ (*modulo* Γ) iff $d < a < b \in V$.

Existence: the existence of the writing $V = v(\Gamma + d)$ comes from the remark that $d < a < b \in V$ always defines an order relation $a < b$ on γ, namely Γ. But one must still show the equality $V = v(\Gamma + d)$, which reduces to:

- The inclusion $v(\Gamma + d) \subset V$: for instance if $a < b$ and $c \parallel a, b$ in Γ, one has $d < a < b \in V$; repartition yields $d < c < b \in V$ or $d < a < c \in V$ (impossible cases) or $a < b < c \in V$.
- $v(\Gamma + d) \supset V$ is also proved by repartition: if $a < b < c \in V$, $a, b, c \neq d$, then say that $d < b < c \in V$; if $b < a$ *modulo* Γ, then $d < b < a \in V$ and by transitivity $d < c < a \in V$ hence $b < c < a$ *modulo* Γ, which yields $a < b < c \in v(\Gamma + d)$. One reaches the same conclusion from $a < c$ *modulo* Γ. Finally, the only remaining possibility is that a is not comparable to b, c *modulo* Γ, but this again yields $a < b < c \in v(\Gamma + d)$.

Various. The topsy-turvying of orders corresponds to topsy-turvying of varieties, i.e., to the replacement of cycles $a < b < c$ with $c < b < a$.

The repartition condition says that one can put d in place of a or b or c in the cycle $a < b < c$. Of course, this does not forbid sharper repartitions, e.g. $a < d < b$ (which implies $a < d < c$ and $d < b < c$).

18.B.2 The criterion. Keeping in mind that $A \otimes B$ is a « subtype » of $A \oslash B$, that switchings are the proofs of the negation, it appears that:

$\oslash$**:** the non-commutative Tensor retains only one switching, i.e., « right ».

$\ltimes$**:** the non-commutative Par has one more, slightly weird: from $\downarrow A$, go to $\downarrow A \ltimes B$; from $\uparrow A \ltimes B$ go to $\uparrow B$.

The additional switching is *partial*: indeed, once in $\downarrow B$, the cycle breaks down; similarly, there is no way of reaching $\uparrow A$.

In any case, a proof-structure whose conclusion is a single formula A is correct when there is a sole uninterrupted cycle, the one passing through the conclusion. The relation with order varieties is as follows: in the presence of several conclusions, partially ordered, if the variety contains $B < C < D$, then any cycle will follow the order $\ldots \uparrow B, \ldots \uparrow C, \ldots \uparrow D \ldots$. In other words, it is a matter of constraint on the course.

An example: say that the unique conclusion A is indeed $(B \ltimes C) \parr D$. Let us look at the « subnet » of conclusions B, C, D: switching $\ltimes$ « ordinarily », it is easily seen that the unique long trip actually passes through B, C, D; but in which order? Let us imagine a course BDC. I switch $\parr$ on « right », $\ltimes$ on the « third position ». Then $\downarrow B$ goes to $\uparrow C$, $\downarrow D$ goes to $\downarrow A$, $\uparrow A$ goes to $\uparrow D$; $\uparrow B$ and $\downarrow C$ are « lost ».

By hypothesis, the order of passage is BDC, in other terms, after $\uparrow B$ comes $\downarrow D$, after $\uparrow D$ comes $\downarrow C$, after $\uparrow C$ comes $\downarrow B$. Let us look at what happens with our switching: $\uparrow A \uparrow D \ldots \downarrow C$, lost! At the other end: $\uparrow B \ldots \downarrow D \downarrow A$, lost too. There is still a long trip, but internal, $\uparrow C \ldots \downarrow B \uparrow C$.

The criterion is sequentialisable in the sole cut-free case. Since it is preserved by normalisation, this does not affect the theory: as long as nets do normalise, one is truly interested in the sequentialisation of the sole cut-free ones.

Even if non-commutative logic – which is more a poetical idea than an urgent need – were bound to disappear, one would at least remember the order varieties and, as to our methodological reflection, the idea of speaking of a cycle passing through the conclusion, the « ground ». This means that the conclusion plays a distinguished role, which was not the case for the interpretation initiated in Section 18.2.2.

Part VI

Geometry of interaction

Chapter 19
The feedback equation

The approach of Chapter 18 (proofs as permutations) is limited to multiplicatives; in order to take care of other connectives, especially exponentials, one must go beyond mere permutations: a permutation of $|A|$ can be seen as a unitary operator of the Hilbert space $\mathbb{C}^{|A|}$ and the execution as the solution of a linear equation on this space (Section 18.2.4). *Geometry of interaction* (GoI) generalises this approach by interpreting proofs by bounded operators and execution, i.e., cut-elimination, by the solution of an equation, the *feedback equation*.

19.1 Basic examples

19.1.1 From syllogistics to feedback. Let us forget for a while the preparatory work done in Chapter 18 and let us try to interpret logic inside Hilbert spaces. The syllogism *Barbara*, (Section 11.B.2), that one can rewrite as

$$\frac{A \multimap B \quad B \multimap C}{A \multimap C} \tag{19.1}$$

seems to deserve a category-theoretic interpretation. Indeed, let us associate Hilbert spaces $\mathbf{E}, \mathbf{F}, \mathbf{G}$ to the formulas A, B, C and assume that the proofs of $A \multimap B$ and $B \multimap C$ do correspond to operators $u \in \mathcal{B}(\mathbf{E}, \mathbf{F}), v \in \mathcal{B}(\mathbf{F}, \mathbf{G})$. The syllogism is then interpreted as composition: the conclusion corresponds to $w := v \circ u \in \mathcal{B}(\mathbf{E}, \mathbf{G})$. Moreover, the axiom $A \multimap A$ is rightfully interpreted as the identity of $\mathcal{B}(\mathbf{E})$. Which enables one to indulge in syllogistics – and nothing more, by the way!

Let

$$U := \begin{pmatrix} 0 & u^* \\ u & 0 \end{pmatrix}, \quad V := \begin{pmatrix} 0 & v^* \\ v & 0 \end{pmatrix}, \quad W := \begin{pmatrix} 0 & w^* \\ w & 0 \end{pmatrix}$$

be operators in $\mathcal{B}(\mathbf{E} \oplus \mathbf{F})$, $\mathcal{B}(\mathbf{F} \oplus \mathbf{G})$, $\mathcal{B}(\mathbf{E} \oplus \mathbf{G})$, written as block matrices (Section 19.A.2). These operators are called *chiasmi*: they are of the form $\sigma(a \oplus a^*)$, with a crossing of inputs/outputs operated by $\sigma(x \oplus y) := y \oplus x$. σ is the analogue of the *twist* (Example 4, Section 17.5.2).

W is indeed the solution of a linear equation. Given $x \in \mathbf{F}, z \in \mathbf{H}$, one defines W by $W(x \oplus z) := x' \oplus z'$, where x', z' are given by

$$\begin{aligned} U(x \oplus y) &= x' \oplus y', \\ V(y' \oplus z) &= y \oplus z' \end{aligned} \tag{19.2}$$

with the unique solution: $y' = u(x), z' = v(u(x)), y = v^*(z), x' = u^*(v^*(z))$. This is a *feedback* equation: the outputs y', y are reinjected – here, one more chiasmus – as inputs y, y'.

Incidentally, observe that category-theoretic compositionality – i.e., associativity – consists in equating the various solutions of the feedback equation between three chiasmi (here, $w \in \mathcal{B}(\mathbf{G}, \mathbf{K})$):

$$\begin{aligned} U(x \oplus y) &= x' \oplus y', \\ V(y' \oplus z) &= y \oplus z', \\ W(z' \oplus t) &= z \oplus t'. \end{aligned} \tag{19.3}$$

One can either solve the equation between U, V, then solve the equation between this solution and W, or solve the equation in a single step: due to the existence and unicity of the solution, the result is the same.

Among the siblings of the syllogism, the *Modus Ponens*

$$\frac{B \quad B \multimap C}{C} \tag{19.4}$$

appears as the particular case of (19.1) with $\mathbf{E} = \mathbf{0}$. This rule is universal – this is why Hilbert-style systems basically rely on it –, hence not limited to chiasmi ($\mathbf{E} = \mathbf{0}$ would lead to $U = 0$!). Which suggests that we seek the generic proof of $A \multimap B$ among arbitrary 2×2 matrices. In that setting, equations of the type (19.2) correspond to the generalisation of category-theoretic composition, but *by no way* reduce to it: while (19.2) is trivial in the case of chiasmi[1], its general « solution », which involves unbounded operators (Section 19.4), is far from being trivial.

19.1.2 Invertibility and nilpotency. Let us experiment with a naïve approach to the feedback equation between $U = \begin{pmatrix} U_{11} & U_{12} \\ U_{21} & U_{22} \end{pmatrix}$ and $V = \begin{pmatrix} V_{22} & V_{23} \\ V_{32} & V_{33} \end{pmatrix}$ respectively acting on $\mathbf{E} \oplus \mathbf{F}$ and $\mathbf{F} \oplus \mathbf{G}$:

$$\begin{aligned} x' &= U_{11}(x) + U_{12}(y), \\ y' &= U_{21}(x) + U_{22}(y), \\ y &= V_{22}(y') + V_{23}(z), \\ z' &= V_{32}(y') + V_{33}(z). \end{aligned} \tag{19.5}$$

The high school method makes use of recursive substitutions, so as to eliminate y, y', i.e., the index 2:

$$x' = U_{11}(x) + U_{12}(V_{23}(z) + V_{22}(U_{21}(x) + U_{22}(V_{23}(z) + V_{22}(U_{21}(x) + \cdots))));$$

[1] A typical example of nilpotency, see next section.

similarly for z', which suggests that the solution is given by

$$\begin{aligned} x' &= W_{11}(x) + W_{13}(z), \\ z' &= V_{31}(x) + V_{33}(z) \end{aligned} \tag{19.6}$$

with (φ stands for the identity map of $\mathbf{F}$)

$$\begin{aligned} W_{11} &= U_{11} + U_{12}V_{22}(\varphi - U_{22}V_{22})^{-1}U_{21}, \\ W_{13} &= U_{12}(\varphi - V_{22}U_{22})^{-1}V_{23}, \\ W_{31} &= V_{32}(\varphi - U_{22}V_{22})^{-1}U_{21}, \\ W_{33} &= V_{33} + V_{32}(\varphi - U_{22}V_{22})^{-1}U_{22}V_{23} \end{aligned} \tag{19.7}$$

which is quite correct assuming the invertibility (in $\mathbf{F}$) of $\varphi - U_{22}V_{22}$; remember that $\varphi - U_{22}V_{22}$ and $\varphi - V_{22}U_{22}$ are simultaneously invertible and that $V_{22}(\varphi - U_{22}V_{22})^{-1} = (\varphi - V_{22}U_{22})^{-1}V_{22}$, $(\varphi - U_{22}V_{22})^{-1}U_{22} = U_{22}(\varphi - V_{22}U_{22})^{-1}$.

The solution by iterated substitutions is correct too, provided these substitutions eventually eliminate y, y', i.e., that the expansion (19.6) is finite, which corresponds to the absence of arbitrary long monomials $U_{12}(V_{22}U_{22})^n$. This is clearly the case when $U_{22}V_{22}$ (equivalently, $V_{22}U_{22}$) is nilpotent. Nilpotency ensures the invertibility of $\varphi - U_{22}V_{22}$, thus the solvability of the equation. In Section 19.5 *infra*, we shall see that syntactical normalisation translates as a symbolic solution of the feedback equation by recursive substitutions, so that nilpotency expresses the convergence of this process.

In terms of cut-systems, (*infra*)), nilpotency translates as « σh nilpotent ». But, notwithstanding its logical import, nilpotency is not that well-behaved in GoI: for instance, it does not socialise with associativity, contrarily to invertibility. Furthermore, the feedback equation between operators of norm at most 1 admits a fully general *solution* (Section 19.4).

19.2 Cut-systems

19.2.1 The Tortoise strikes back. We shall reduce to the case where U, V are hermitian and $\mathbf{E} = \mathbf{0}$; U thus appears as the *feedback* of a *cut-system*[2]

Hermiticity. Starting with an equation (19.2) with U, V arbitrary, let us add duplicates $\mathbf{E}_1, \mathbf{F}_1, \mathbf{G}_1$ of the spaces $\mathbf{E}, \mathbf{F}, \mathbf{G}$; the equations

$$\begin{aligned} U(x \oplus y) &= x_1' \oplus y_1', \\ U^*(x_1 \oplus y_1) &= x' \oplus y', \\ V(y_1' \oplus z_1) &= y \oplus z', \\ V^*(y' \oplus z) &= y_1 \oplus z_1' \end{aligned} \tag{19.8}$$

[2]Another instance of the Tortoise Principle: a simpler problem on a bigger space.

relate the hermitians $F := \begin{bmatrix} 0 & U^* \\ U & 0 \end{bmatrix}$ and $G := \begin{bmatrix} 0 & V^* \\ V & 0 \end{bmatrix}$. Its solution immediately induces a solution of the original equation: $W(x \oplus z_1) := x_1' \oplus z' \ldots$ and of the equation involving U^*, V^* adjoint to (19.2).

The feedback. If $U \in \mathcal{B}(\mathbf{E} \oplus \mathbf{F})$, $V \in \mathcal{B}(\mathbf{F} \oplus \mathbf{G})$, introduce $\mathbf{H} := \mathbf{E} \oplus \mathbf{F} \oplus \mathbf{F}_1 \oplus \mathbf{G}$, where $\mathbf{F}_1$ is a duplicate of $\mathbf{F}$ as well as $h := U \oplus V_1$, which abusively denotes the direct sum of U and the « delocated » version of V. (19.2) is obviously equivalent to

$$h(x \oplus y \oplus y' \oplus z) = x' \oplus y' \oplus y \oplus z', \tag{19.9}$$

indeed a *Modus Ponens* equation:

$$\frac{A \multimap A \quad (A \multimap A) \multimap (B \multimap C)}{B \multimap C} \tag{19.10}$$

between h and the chiasmus $\chi := \begin{pmatrix} 0 & U \\ U^* & 0 \end{pmatrix}$ of $\mathbf{F} \oplus \mathbf{F}_1$, with $U(x) := x_1$. χ, extended into a *partial symmetry* σ of the whole space, is a *feedback*.

19.2.2 Cut-systems

Definition 94 (Partial symmetries). A *partial symmetry* of $\mathbf{H}$ is a hermitian operator σ such that $\sigma^3 = \sigma$.

The spectral calculus (Section 17.A.5) puts σ in correspondence with the inclusion ι between $\mathrm{Sp}(\sigma)$ and $\mathbb{C}$ which thus verifies $\iota^3 = \iota$. Which shows that the spectrum is discrete: $\mathrm{Sp}(\sigma) \subset \{-1, 0, +1\}$. The eigenspaces $\sigma_{-1}, \sigma_0, \sigma_{+1}$ of the values -1, 0, $+1$ correspond to the projections: $1/2(\sigma^2 + \sigma)$, $I - \sigma^2$, $1/2(\sigma^2 - \sigma)$. One will note as $\mathbf{S}$ the sum of the subspaces σ_{-1} and σ_{+1}, i.e., the space corresponding to the projection σ^2; $\mathbf{R}$ will stand for σ_0. Hence $\mathbf{H} = \mathbf{R} + \mathbf{S}$, the direct sum of the *kernel* and the *carrier* of σ.

Definition 95 (Cut-systems). A *cut-system* is a three-tuple $(\mathbf{H}, h, \sigma)$ such that:

- $\mathbf{H}$ is a complex Hilbert space.
- h is a hermitian of $\mathbf{H}$, of norm at most 1.
- σ (the *cut*, the *loop*, the *feedback*) is a partial symmetry: $\sigma = \sigma^* = \sigma^3$.

Hence the *feedback equation*: given $x \in \mathbf{R}$, find $x' \in \mathbf{R}$, $y \in \mathbf{S}$, such that

$$h(x + y) = x' + \sigma(y). \tag{19.11}$$

It is the « visible » part of the equation, the *result* x', which interests us; the *introspective* part y will correspond to the computation *process*. The presence of this hidden component ensures that this belongs in layer -3. This equation is indeed a variant of (19.2) reformulated by means of (19.9).

19.2.3 Discussion. The restriction on the norm, $\|h\| \leq 1$ is the only known way to get *general* results on the feedback equation, in particular to get a *normal form*. Whatever reasons (e.g., non-determinism) one may find against this restriction, its relinquishment would reduce GoI to a catalogue of cooking recipes. This restriction is compatible with the Tortoise transformations of last the section. In other words, we can first replace a feedback equation between U, V s.t. $\|U\|, \|V\| \leq 1$ with another one between U', V' hermitian, $\|U'\|, \|V'\| \leq 1$. This hermitian equation is stronger (since it is the gluing of two independent equations); since all feedback equations are « solvable », this is indeed not a genuine restriction. The equation between U', V' can in turn be replaced with an equivalent feedback equation between h, σ; observe that h is hermitian and that $\|h\| \leq 1$.

Since positive operators enjoy exceptional properties linked to the square root, one could think of a further restriction of the feedback equation: h positive. I thus acknowledge having been tempted by the replacement of identity chiasmi with their positive parts; syntactical normalisation thus translates as an inequality $h' \leq \sigma[\![h]\!]$, which is far from being unpleasant. But I found myself too far from the base camp and I turned back. Anyway, there is something to dig for here: for instance, if, for $n \geq 0$, $P_n := \begin{bmatrix} n/n+1 & 1/n+1 \\ 1/n+1 & n/n+1 \end{bmatrix}$, the equation (19.2) between $U := P_m$ and $V := P_n$ admits the solution $W := P_{m+n}$. P_0, being the identity chiasmus, is not positive; on the other hand, the P_n are positive for $n \geq 1$, thus P_1 is the positive part of P_0. N equations involving P_1 yield the result P_N, thus keeping track of the computation process.

19.3 Solving the equation

19.3.1 Normal forms. The normal form corresponds to the visible, *extraspective*, part of the equation:

Definition 96 (Normal form). Under the proviso of existence and unicity of the solution of equation (19.11), the *normal form* of the system is the operator $\sigma[\![h]\!](x) := x'$.

Observe that $\|\sigma[\![h]\!]\| \leq 1$; indeed, $\|x'\|^2 + \|\sigma(y)\|^2 \leq \|x\|^2 + \|y\|^2$ and, since $\|\sigma(y)\| = \|y\|$, $\|x'\|^2 \leq \|x\|^2$. In the same way, one can see that $\sigma[\![h]\!]$ is hermitian. Therefore, the normal form can be seen as the trivial cut-system: $(\mathbf{R}, \sigma[\![h]\!], 0)$, a « cut-free » system, whose feedback is null.

Which prompts the principle of *iterability* of the normal form: if τ is a feedback of $\mathbf{R}$, then one can form – under the proviso of existence and unicity – the normal form $\tau[\![\sigma[\![h]\!]]\!]$. At the cost of a small abuse of notation, one can assume that τ is defined on the full space $\mathbf{H}$, in which case $\sigma + \tau$ is a feedback too, i.e., a partial symmetry.

Definition 97 (Independence). σ, τ are *independent* when $\sigma\tau = 0$ $(= \tau\sigma)$, in other words when their carriers $\mathbf{S}$, $\mathbf{T}$ are orthogonal.

The solution of equation (19.3) in two steps is written as $\tau[\![\sigma[\![h]\!]]\!]$ or $\sigma[\![\tau[\![h]\!]]\!]$ and its solution in one step is written as $(\sigma + \tau)[\![h]\!]$. Under reasonable hypotheses, these three expressions will be equal: this is *associativity*.

Associativity is all that remains of layer -2; one can use it in a regressive way and build categories based upon GoI, « traced » if one insists upon wasting paper. But the associativity of GoI includes a case inaccessible to ruminants: the decomposition $\sigma = \sigma^+ + \sigma^-$, as the sum of two *lopsided* feedbacks, *positive* – a projection – or *negative* – the opposite of a projection. This decomposition makes sense neither in set-theory (graphs, etc.) nor in category theory (morphisms, etc.), not to speak of logic, since it cuts out objects in a « forbidden » way; it is however the most important technicality of this chapter. Thus, in dimension 3, the exchange of indices $1, 2$ splits as the sum of two independent feedbacks:

$$\begin{bmatrix} 0 & 1 & 0 \\ 1 & 0 & 0 \\ 0 & 0 & 0 \end{bmatrix} = \begin{bmatrix} 1/2 & 1/2 & 0 \\ 1/2 & 1/2 & 0 \\ 0 & 0 & 0 \end{bmatrix} + \begin{bmatrix} -1/2 & 1/2 & 0 \\ 1/2 & -1/2 & 0 \\ 0 & 0 & 0 \end{bmatrix}. \tag{19.12}$$

19.3.2 Invertibility. Let $(\mathbf{H}, h, \sigma)$ be a cut-system; we decompose the space as the direct sum $\mathbf{H} = \mathbf{R} + \mathbf{S}$ between the kernel and the carrier of σ; this induces the block expressions: $h = \left(\begin{smallmatrix} a & b^* \\ b & c \end{smallmatrix}\right)$, $\sigma = \left(\begin{smallmatrix} 0 & 0 \\ 0 & s \end{smallmatrix}\right)$; s is a symmetry of $\mathbf{S}$.

Definition 98 (Invertibility). $(\mathbf{H}, u, \sigma)$ is *invertible* when $s - c$ is invertible as an endomorphism of $\mathbf{S}$.

Theorem 73 (Normal form). *If $(\mathbf{H}, h, \sigma)$ is invertible, then the feedback equation admits the normal form $\sigma[\![h]\!] := a + b^*(s - c)^{-1}b$ as its unique solution.*

Proof. One wants to solve the equation $\left(\begin{smallmatrix} a & b^* \\ b & c-s \end{smallmatrix}\right)\left(\begin{smallmatrix} x \\ y \end{smallmatrix}\right) = \left(\begin{smallmatrix} x' \\ 0 \end{smallmatrix}\right)$. Since $s - c$ is invertible, it is immediate that $x' = (a + b^*(s - c)^{-1}b)(x)$, $y = ((s - c)^{-1}b)(x)$ is a solution of the equation. It is indeed *the* solution: if $\left(\begin{smallmatrix} a & b^* \\ b & c-s \end{smallmatrix}\right)\left(\begin{smallmatrix} 0 \\ y \end{smallmatrix}\right) = \left(\begin{smallmatrix} x' \\ 0 \end{smallmatrix}\right)$, then $(c - s)(y) = 0$, hence $y = 0$ and $x' = 0$. Hence the visible part of the solution: $\sigma[\![h]\!] = a + b^*(s - c)^{-1}b$. □

Remark 2. The shape of the solution makes explicit the hermitian character of the solution. The dependency $h \rightsquigarrow \sigma[\![h]\!]$ is norm-continuous; we shall soon see that it is also *order-continuous* (Corollary 52.3, *infra*).

The logical interpretation contents itself with invertibility; indeed the *termination* of computations translates as the *nilpotency* of sc, see Section 19.5, which implies that $(s - c)^{-1} = s + scs + scscs + \cdots + s(cs)^n$, a finite sum.

19.3.3 Associativity. Given independent feedbacks σ, τ, there are several ways of obtaining a normal form for $(\mathbf{H}, h, \sigma + \tau)$. Either form at once $(\sigma + \tau)[\![h]\!]$, or first normalise $(\mathbf{H}, h, \sigma)$, which yields $\sigma[\![h]\!]$, then normalise $(\mathbf{R}, \sigma[\![h]\!], \tau)$, which yields $\tau[\![\sigma[\![h]\!]]\!]$. One can work in the opposite order, which leads to $\sigma[\![\tau[\![h]\!]]\!]$. The associativity theorem relieves us, all of these lead to the same result:

Theorem 74 (Associativity). *Assume σ, τ independent and write $\mathbf{H} = \mathbf{R} + \mathbf{S} + \mathbf{T}$. Then $(\mathbf{H}, h, \sigma + \tau)$ is invertible iff $(\mathbf{H}, h, \tau)$ and $(\mathbf{R} + \mathbf{S}, \tau[\![h]\!], \sigma)$ are invertible. Moreover*

$$(\sigma + \tau)[\![h]\!] = \sigma[\![\tau[\![h]\!]]\!]. \tag{19.13}$$

The *proof* of the theorem essentially reduces to the following lemma:

Lemma 74.1. *Let $f = \begin{pmatrix} c & d^* \\ d & e \end{pmatrix}$ (w.r.t. a decomposition $\mathbf{S} + \mathbf{T}$) be a hermitian of norm at most 1 and s, t be symmetries of $\mathbf{S}$, $\mathbf{T}$. Then $g := \begin{pmatrix} s-c & -d^* \\ -d & t-e \end{pmatrix}$ is invertible iff $t - e$ and $s - (c + d^*(t-e)^{-1}d)$ are invertible in $\mathbf{T}$ and $\mathbf{S}$ respectively.*

Proof. Hermitian (more generally, normal) operators admit *approximate* eigenvectors, see [63], p. 183. Thus, if $t - e$ is not invertible, there is a sequence $(y_n) \in \mathbf{T}$, with $\|y_n\| = 1$, such that $(t-e)(y_n) \to 0$; since $\|t(y_n)\| = 1$, one sees that $\|e(y_n)\| \to 1$, hence, from $\|f\| \le 1$, $d(y_n) \to 0$. One deduces that $g(y_n) \to 0$, hence g is not invertible.

Defining $c' := c + d^*(t-e)^{-1}d$ and assuming now $s - c'$ not invertible, there is a sequence $(x_n) \in \mathbf{S}$ of norm 1 vectors such that $(s - c')(x_n) \to 0$. But c' is a normal form, which means that one can find $y_n \in \mathbf{T}$ (unique) such that $g(x_n + y_n) = (s - c')(x_n)$; then $g(x_n + y_n) \to 0$.

We just have proved that the condition is necessary; the explicit formula

$$g^{-1} = \begin{pmatrix} (s-c')^{-1} & (s-c')^{-1}d^*(t-e)^{-1} \\ (t-e)^{-1}d(s-c')^{-1} & (t-e)^{-1} + (t-e)^{-1}d(s-c')^{-1}d^*(t-e)^{-1} \end{pmatrix}$$

establishes sufficiency. □

19.3.4 Injectivity. With the same notations as before, $\mathbf{H} = \mathbf{R} + \mathbf{S}$, $h = \begin{pmatrix} a & b^* \\ b & c \end{pmatrix}$, $\sigma = \begin{pmatrix} 0 & 0 \\ 0 & s \end{pmatrix}$.

Definition 99 (Injectivity). The system $(\mathbf{H}, h, \sigma)$ is *injective* if $s - c$ is injective as an endomorphism of $\mathbf{S}$.

Invertible systems are injective, but the converse is not true: in infinite dimension, injectivity does not entail surjectivity. Which is by the way what differentiates *spectrum* and *eigenvalues*: an operator that is injective but not surjective has 0 in its spectrum, while there is no eigenvector for 0[3].

[3] It however admits an *unbounded* inverse (Section 19.A.3).

Definition 100 (Deadlock). The *shortcut* of $(\mathbf{H}, h, \sigma)$ is the subspace $\mathbf{Z} \subset \mathbf{S}$ defined $\mathbf{Z} := \ker(s - c)$; its (non-zero) elements are called *deadlocks*.

Proposition 50. *$I_{\mathbf{Z}}$ commutes with σ, s, c, h.*

Proof. If $z \in \mathbf{Z}$, then we have $c(z) = s(z)$, hence $\|c(z)\| = \|z\|$; we deduce that $\langle c^2(z) \mid z\rangle = \langle c(z) \mid c(z)\rangle = \langle z \mid z\rangle$; since $\|c\| \leq 1$, we have the equality case of Cauchy–Schwarz (Theorem 64), which forces $c^2(z) = \lambda z$ and since c^2 is positive, $c^2(z) = z$. We immediately deduce that $s(c(z)) = c(c(z))$, i.e., that $c(z) \in \mathbf{Z}$. Hence c, s send $\mathbf{Z}$ into $\mathbf{Z}$, which means that they have block matrices of the form $\left(\begin{smallmatrix} u & 0 \\ 0 & w \end{smallmatrix}\right)$ w.r.t. the decomposition $\mathbf{H} = \mathbf{Z} + \mathbf{Z}^{\perp}$; since these operators are hermitian, $v = 0$. I_{Z} thus commutes with s, c and therefore with σ, h. □

Proposition 50 says that the deadlock $\mathbf{Z}$ is innocuous: one can always « remove » it by replacing h with $(I_{\mathbf{Z}^{\perp}})h$ without changing anything to the feedback equation – but for the fact that the introspective component y becomes unique.

The deadlocks exactly correspond to the short trips of the proof-net criterion (Section 18.2.1), which come in turn from configurations of the type (11.1). We have already observed their « autistic » character, their non-communication with the outside: this is the meaning of Proposition 50.

In a finite dimension, injectivity equals invertibility and one can thus solve the equation by « removing the shortcut ».

19.3.5 Computational size. Let $(\mathbf{H}, h, \sigma)$ be a cut-system and let $\mathbf{Z}$ be its shortcut. Two solutions of the feedback equation (19.11) $x' + y$, $x'' + y'$ for the same datum x are such that $x' = x''$: from

$$h(0 + (y - y')) = (x' - x'') + \sigma(y - y') \tag{19.14}$$

and $\|h\| \leq 1$, $\|\sigma(y - y')\| = \|y - y'\|$, one gets $\|x' - x''\| = 0$, thus $x' = x''$. But the two answers $x' + y$, x', y' can still differ by their introspective components y, y', in case $y - y'$ is a deadlock.

Proposition 50 yields a sort of unicity:

(i) If $x' + y$ is a solution for the input x, $x' + \mathbf{Z}(y)$ is also a solution.

(ii) $x' + \mathbf{Z}(y)$ is – amongst all solutions $x' + y'$ for the input x – that of smallest norm.

(iii) This smallest solution consists in « removing the shortcut », i.e., in replacing $(\mathbf{H}, h, \sigma)$ with $(\mathbf{H}, I_{\mathbf{Z}^{\perp}}h, \sigma)$.

Definition 101 (Termination). The cut-system $(\mathbf{H}, h, \sigma)$ *terminates* when equation (19.11) has a solution $x' + y$ for any input $x \in \mathbf{R}$. In case one defines the *execution* operator $\text{ex}(h, \sigma)$ from $\mathbf{R}$ into $\mathbf{H}$: it associates to $x \in \mathbf{R}$ the vector[4] $x + \mathbf{Z}(y)$.

The typical case for termination is invertibility; in this case one has $\text{ex}(h, \sigma) = I_{\mathbf{R}} + (\sigma - c)^{-1}b$. But there are many other cases, e.g., $b = 0$.

Definition 102 (Size). When $(\mathbf{H}, h, \sigma)$ terminates, its *computational size* is defined by

$$\text{size}(h, \sigma) := \|\text{ex}(h, \sigma)\|. \tag{19.15}$$

Termination is a rather accidental feature; this is why the next theorem is less powerful than its model, Theorem 74. In what follows, $\tau[\![h]\!]$ refers to the literal solution and does not necessarily agree with the general « solution » to be defined *infra* (Section 19.4).

Theorem 75 (Associativity of size). *If σ, τ are independent and if $(\mathbf{H}, h, \tau)$ and $(\mathbf{R} + \mathbf{S}, \tau[\![h]\!], \sigma)$ terminate, then $(\mathbf{H}, h, \sigma + \tau)$ terminates too; moreover*

$$\text{size}(h, \sigma + \tau) \leq \text{size}(\tau[\![h]\!], \sigma) \cdot \text{size}(h, \tau). \tag{19.16}$$

Our definition of *size* can be grasped from the naïve solution of (19.11) by means of the series expansion

$$\sigma[\![h]\!] = I_{\mathbf{R}}(h + h\sigma h + h\sigma h\sigma h + \cdots)I_{\mathbf{R}}, \tag{19.17}$$

a correct formula when σh, i.e., sc, is nilpotent:

$$\text{ex}(h, \sigma) = I_{\mathbf{R}} + b^*sb + b^*scsb + b^*scscsb + \cdots + b^*s(cs)^{n-2}b \tag{19.18}$$

where n is the smallest integer such that $(\sigma h)^n = 0$ (the « order of nilpotency » of sc). Then $\text{size}(h, \sigma) \leq n$; in practice, e.g., for a converging normalisation in – say – system $\mathbf{F}$, $\text{size}(h, \sigma) \sim \sqrt{n}$.

One should perhaps replace the size with its logarithm: Danos observed in his PhD (see, e.g., [20]) that, in λ-calculus, the « order of nilpotency » of the cut-system associated with t is exponential in the normalisation length of t. In this case, Theorem 75 involves a sum in place of a product.

19.4 The normal form

Although the feedback equation has no general solution, one can nevertheless extend the notion of *normal form* to arbitrary cut-systems so as to preserve a certain number of properties such as associativity. W.l.o.g. one can restrict to *injective* systems: the normal form of a general system $(\mathbf{H}, h, \sigma)$ can be defined as that of its « deadlock-free » version $(\mathbf{Z}^{\perp}, hI_{\mathbf{Z}^{\perp}}, \sigma I_{\mathbf{Z}^{\perp}})$ (Section 19.3.4).

[4] This is not a typo, I didn't mean $x' + \mathbf{Z}(y)$.

19.4.1 Positive feedbacks. We first extend the normal form to the case of a positive feedback. In what follows, we consider an injective system $(\mathbf{H}, h, \pi)$, with $\pi^2 = \pi$, $\mathbf{H} = \mathbf{R} + \mathbf{P}$, $h = \left(\begin{smallmatrix} a & b^* \\ b & c \end{smallmatrix}\right)$, $I = \left(\begin{smallmatrix} \rho & 0 \\ 0 & \pi \end{smallmatrix}\right)$. Since h is injective, $\pi - c > 0$ admits an injective square root whose range is dense in $\mathbf{P}$. $\sqrt{\pi - c}$ thus admits an inverse γ, usually partial, defined on a dense subset (the range of $\sqrt{\pi - c}$) and with a closed graph, in other words an *unbounded* operator (Section 19.A.3).

Proposition 51. *γb is a bounded operator such that $\|\gamma b\| \leq \|\rho - a\|^{1/2}$; dually, $b^*\gamma$ is* closable *and its closure is $(\gamma b)^*$.*

Proof. From $h \leq I$, one gets $2\Re(\langle x \mid b^*(y)\rangle) \leq \langle(\rho - a)(x) \mid x\rangle + \langle(\pi - c)(y) \mid y\rangle$, which forces the discriminant $|\langle x \mid b^*(y)\rangle|^2 - \langle(\rho - a)(x) \mid x\rangle\langle(\pi - c)(y) \mid y\rangle$ to be ≤ 0, hence $\|b^*(y)\| \leq \|\rho - a\|^{1/2} \cdot \|(\sqrt{\pi - c})(y)\|$ by making $x := b^*(y)$.

If $y := (\sqrt{\pi - c})(z)$, one gets $\|b^*\gamma(z)\| \leq \|\rho - a\|^{1/2}\|z\|$. Since b^* is total and γ has a dense domain, $b^*\gamma$ has a dense domain too; it thus extends to an operator φ of norm bounded by $\|\rho - a\|^{1/2}$, its *closure*. By continuity, $\varphi\sqrt{\pi - c} = b^*$, hence $\sqrt{\pi - c} \cdot \varphi^* = b$: γb is indeed the bounded operator φ^*. □

Definition 103 (Normal form). The *normal form* of $(\mathbf{H}, h, \pi)$ is defined by $\pi[\![h]\!] := a + (\gamma b)^*\gamma b$.

Remark 3. If δ is an unbounded operator such that $\delta^*\delta = (\pi - c)^{-1}$, $\delta\sqrt{\pi - c}$ is easily shown to be an isometry φ of $\mathbf{P}$. Hence $a + (\delta b)^*\delta b = a + (\gamma b)^*\varphi^*\varphi\gamma b = \pi[\![h]\!]$. In other words, the normal form is equal to $a + (\gamma b)^*\gamma b$, where γ is the inverse of any « square root » of $\pi - c$, positive or not, i.e., is such that $\gamma^*\gamma = (\pi - c)^{-1}$.

Proposition 52. *Definitions* 103 *and* 96 *are consistent. Moreover, the normal form of Definition* 103 *is monotonic and commutes with directed suprema.*

Proof. In the invertible case, the surjective $\sqrt{\pi - c}$ admits a bounded inverse: $\pi[\![h]\!] = a + b^*(\pi - c)^{-1}b = a + (b^*(\pi - c)^{-1/2})((\pi - c)^{-1/2}b)$. In other terms,

$$\langle\pi[\![h]\!](x) \mid x\rangle = \langle a(x) \mid x\rangle + \|(\pi - c)^{-1/2}b(x)\|^2 \qquad (19.19)$$

which agrees with Definition 103. The statements concerning order will be proven in the restricted case of systems with the same components a, b, i.e., only c varies: this is a tremendous simplification; see the proof of Corollary 52.3 below for the full case. First of all, in the invertible case, observe that $c \leq d$ implies $\pi - d \leq \pi - c$, hence $(\pi - c)^{-1} \leq (\pi - d)^{-1}$; see [63], Proposition 4.2.8., for the behaviour of the hermitian order w.r.t. inversion and square roots. From this we get monotonicity w.r.t. the coefficient c.

The normal form is indeed order-continuous w.r.t. c in the invertible case: if c_i is a directed increasing net and $c := \sup_i c_i$, then the limit c is indeed a *strong* one (Proposition 58), hence, due to the strong continuity of the products on balls,

$\pi[\![h]\!] = a + b^*(\pi - c)^{-1}b = a + b^*(\pi - \lim_i c_i)^{-1}b = \lim_i \pi[\![h_i]\!]$, provided the systems h_i and h are invertible. The same holds for directed infima.

c being arbitrary, define, for $0 < \lambda < 1$, $c_\lambda := \lambda c + (\lambda - 1)\pi$, so that c is the supremum of the c_λ with the $\pi - c_\lambda$ invertible.

Lemma 52.1. *If y is in the range of $\sqrt{\pi - c}$, then*

$$\|(\pi - c)^{-1/2}(y)\| = \sup \|(\pi - c_\lambda)^{-1/2}(y)\|.$$

Proof. $u_\lambda := \sqrt{\pi - c}\,(\pi - c_\lambda)^{-1}\sqrt{\pi - c}$ is an increasing family of hermitians bounded by π. It therefore admits a supremum $u \leq \pi$ which is (Proposition 58) a *strong* limit. In the commutative von Neumann algebra generated by c: $(\pi - c_\lambda)\sqrt{\pi - c}(\pi - c_\lambda)^{-1}\sqrt{\pi - c} = \sqrt{\pi - c}(\pi - c_\lambda)(\pi - c_\lambda)^{-1}\sqrt{\pi - c} = \pi - c$ admits as a limit, by the strong continuity of the bounded product, $(\pi - c)u$, which proves $u = \pi$. If $y = \sqrt{\pi - c}(x)$, then

$$\begin{aligned}
&\sup \|(\pi - c_\lambda)^{-1/2}(y)\| \\
&\quad = \sup\langle(\pi - c_\lambda)^{-1/2}\sqrt{\pi - c}(x) \mid (\pi - c_\lambda)^{-1/2}\sqrt{\pi - c}(x)\rangle \\
&\quad = \sup\langle u_\lambda(x) \mid x\rangle = \|x\|^2.
\end{aligned}$$

□

The normal form commuting with directed suprema in the invertible case admits a unique extension to general systems commuting with directed suprema. But the normal form of Definition 103 commutes with the particular supremum $h = \sup_\lambda h_\lambda$, thus must coincide with *the* extension by direct suprema.

We are still in want of a complete proof of the result in the positive case. Assume that $\mathbf{K} = \mathbf{R}_1 \oplus \mathbf{H}$ and let φ be an isometry between $\mathbf{R}$ and $\mathbf{R}_1$. Then, if $(\mathbf{H}, h, \pi)$ is an invertible system, so is $(\mathbf{K}, k, I)$, where I stands for the projection on $\mathbf{H}$ and

$$k := \begin{pmatrix} 0 & \epsilon \cdot \varphi & 0 \\ \epsilon \cdot \varphi^* & \epsilon^2 \cdot a & \epsilon \cdot b^* \\ 0 & \epsilon \cdot b & c \end{pmatrix}$$

with $\epsilon > 0$ small enough to keep the norm ≤ 1, i.e., $\epsilon \leq \sqrt{2}/2$. Then anything relative to the system $(\mathbf{H}, h, \pi)$ can be transferred to $(\mathbf{K}, k, I)$, especially the normal form: $I[\![k]\!] = \epsilon^4\sigma[\![h]\!]$. Order features translate as well, i.e., $h \leq h'$ iff $k \leq k'$, etc. Indeed, a generic order relation $h \leq h'$ becomes one of the restricted sort considered in the proof of Proposition 52.

Thank you so much, Honorable Tortoise! □

Corollary 52.1. $\|\pi[\![h]\!]\| \leq 1$.

Proof. True in the invertible case and preserved by directed suprema. □

Corollary 52.2. *The normal form is associative.*

Proof. Yet another argument by order continuity. □

Corollary 52.3. *The normal form $\sigma[\![h]\!]$ (σ arbitrary) is monotonic and order-continuous in the invertible case.*

Proof. First observe that the exact analogue of the proposition holds for negative feedbacks. Hence, writing $\sigma = \sigma^+ + \sigma^-$, associativity yields full monotonicity and order continuity in the invertible case. □

Theorem 76. *The normal form of Definition* 103 *is the unique extension of the invertible case commuting with directed suprema.*

Proof. The non-trivial point is not unicity, but the existence of such an extension, the endeavour of Proposition 52. □

19.4.2 The resolvant. We now consider a general feedback σ, written as the sum of two independent *lopsided* feedbacks, i.e., positive or negative: $\sigma = \sigma^+ + \sigma^-$, in which we note that $\sigma = \pi - \nu$; π, ν are thus projections; and we can pose that $\rho := I - \sigma^2$, hence $I = \rho + \pi + \nu$, which corresponds to a decomposition $\mathbf{H} = \mathbf{R} + \mathbf{P} + \mathbf{N}$ w.r.t. which

$$h = \begin{pmatrix} a & b^* & d^* \\ b & c & e^* \\ d & e & f \end{pmatrix}.$$

$\pi - c$ and $\nu + f$ being positive thus admit positive square roots, which are *injective* by hypothesis: these square roots admit inverses γ, φ, which are *partial* operators, defined on dense subspaces (the ranges of $\sqrt{\pi - c}$, $\sqrt{\nu + f}$) and with closed graphs (the inverse has the « same » graph).

Proposition 53. *The coefficients of the block matrix*

$$\begin{pmatrix} a & b^*\gamma & d^*\varphi \\ \gamma b & 0 & \gamma e^*\varphi \\ \varphi d & \varphi e \gamma & 0 \end{pmatrix}$$

are densely defined and ***closable****.*

Proof. Following Proposition 51 (and a symmetrical use of the inequality $-I \leq h$) we easily conclude that γb, γe^*, φd, φe are plain bounded operators while $b^*\gamma$, $e\gamma$, $d^*\varphi$, $e^*\varphi$ are densely defined and closable into the respective adjoints of the former. Consider $\gamma e^*\varphi$; writing it $(\gamma e^*)\varphi$, one sees that it is densely defined. On the other hand, it is included in $\gamma\overline{(e^*\varphi)}$, which is of closed graph, indeed its closure. *Idem* for the coefficient $\varphi e \gamma$. □

Definition 104 (Resolvant). The *resolvant* of the system $(\mathbf{H}, h, \sigma)$ is defined as the block matrix

$$\operatorname{res}(h,\sigma) := \begin{pmatrix} a & \overline{b^*\gamma} & \overline{d^*\varphi} \\ \gamma b & 0 & \overline{\gamma e^*\varphi} \\ \varphi d & \overline{\varphi e \gamma} & 0 \end{pmatrix} = \begin{pmatrix} a & \beta^* & \delta^* \\ \beta & 0 & \varepsilon^* \\ \delta & \varepsilon & 0 \end{pmatrix}.$$

All coefficients are bounded operators, but perhaps $\varepsilon, \varepsilon^*$. Remember that $\pi + \varepsilon^*\varepsilon$ and $\nu + \varepsilon\varepsilon^*$ are (right) invertible, with positive inverses of norms bounded by 1; also, $\varepsilon(\pi + \varepsilon^*\varepsilon)^{-1}$ and $\varepsilon^*(\nu + \varepsilon\varepsilon^*)^{-1}$ are bounded operators. The « lax system » $(\mathbf{H}, \operatorname{res}(h,\sigma), \sigma)$ is invertible: indeed, the matrix $\begin{pmatrix} \pi & -\varepsilon^* \\ -\varepsilon & -\nu \end{pmatrix}$ admits the (right) inverse $\begin{pmatrix} (\pi+\varepsilon^*\varepsilon)^{-1} & -\varepsilon^*(\nu+\varepsilon\varepsilon^*)^{-1} \\ -\varepsilon(\pi+\varepsilon^*\varepsilon)^{-1} & -(\nu+\varepsilon\varepsilon^*)^{-1} \end{pmatrix}$.

Definition 105 (Normal form). The *normal form* $\sigma[\![h]\!]$ of $(\mathbf{H}, h, \sigma)$ is defined as $\sigma[\![\operatorname{res}(h,\sigma)]\!]$, i.e.,

$$a + \beta^*(\pi + \varepsilon^*\varepsilon)^{-1}\beta - \beta^*\varepsilon^*(\nu + \varepsilon\varepsilon^*)^{-1}\delta - \delta^*\varepsilon(\pi + \varepsilon^*\varepsilon)^{-1}\beta - \delta^*(\nu + \varepsilon\varepsilon^*)^{-1}\delta.$$

Theorem 77 (Associativity).

$$\sigma^-[\![\sigma^+[\![h]\!]]\!] = \sigma^+[\![\sigma^-[\![h]\!]]\!]. \tag{19.20}$$

Proof. We shall prove that $\pi[\![(-\nu)[\![h]\!]]\!] = \sigma[\![h]\!]$ by using the (obvious) associativity of the lax system $(\mathbf{H}, \operatorname{res}(h,\sigma), \sigma)$. If $(-\nu)[\![\operatorname{res}(h,\sigma)]\!] = \begin{pmatrix} f & g^* \\ g & k \end{pmatrix}$ and $F := \begin{pmatrix} \rho & 0 \\ 0 & \sqrt{\pi - c} \end{pmatrix}$, it is immediate that $\pi - (-\nu)[\![h]\!] = F(\pi - (-\nu)[\![\operatorname{res}(h,\sigma)]\!])F$. Now, $\pi[\![(-\nu)[\![h]\!]]\!] = f + ((\pi - H)^{-1/2}\sqrt{\pi - c}\cdot g)^*(\pi - H)^{-1/2}\sqrt{\pi - c}\cdot g$, with $\pi - H := \sqrt{\pi - c}\cdot(\pi - k)\sqrt{\pi - c}$. Indeed, $(\pi - H)^{-1/2}$ can be replaced with any « square root » of $(\pi - H)^{-1}$ (Remark 3), typically with $(\pi - k)^{-1/2}(\pi - c)^{-1/2}$:

$$\pi[\![(-\nu)[\![h]\!]]\!] = f + ((\pi - k)^{-1/2}g)^*(\pi - k)^{-1/2}g = \pi[\![(-\nu)[\![\operatorname{res}(h,\sigma)]\!]]\!] = \sigma[\![h]\!].$$

□

Corollary 77.1. *The normal form is associative.*

Proof. Write $\sigma + \tau = \sigma^+ + \sigma^- + \tau^+ + \tau^-$ and combine the theorem with the associativity of the lopsided cases (Corollary 52.2). □

19.4.3 Digression: the Lebesgue integral. The feedback equation (19.11) is an *impossible* problem. Our normal form bears some analogy to the Lebesgue integral, the noted solution of another impossible problem: the integration of discontinuous functions.

(i) The Riemann integral is a continuous linear form on the Banach space $\mathbb{R}([0,1])$ of real-valued continuous functions on $[0,1]$. Our normal form is slightly of the same kind: the function $\sigma[\![h]\!]$ is defined on certain operators h (invertible systems) of norm ≤ 1. One could even suppose $\mathbf{R}$ of dimension 1, which would make our output belong to $\mathbb{R}$. An essential difference, however: the dependency is not linear; on the other hand, the result remains bounded: $\|\sigma[\![h]\!]\| \leq 1$.

(ii) The next step consists in remarking that the Riemann integral is monotonic, thus to extend it, by passage to suprema, to lower semi-continuous (l.s.c.) functions, which are the suprema of continuous functions. In Section 19.4.1 we actually extended the normal form to directed suprema of invertible systems (l.s.i. systems, *infra*).

(iii) Of course, only a monotonic function can thus be extended: both the Riemann integral and the normal form are increasing. Moreover the extension must be compatible with what has already been defined. In the case of the integral, one uses Dini's theorem: if f_n is a monotonic sequence of continuous functions with a *continuous* supremum f, then the convergence $f_n \to f$ is uniform, i.e., normwise. The same is true of the normal form which commutes to invertible suprema of invertible systems. The idea is the same as in the previous case: it is a matter of topological *bonification* due to the order structure. In the case of Dini, from simple to uniform, here from weak to strong (Proposition 58, *infra*).

(iv) A last step must be performed: after the supremum, the infimum (of suprema). That's all: in the Lebesgue case, any measurable function is equivalent to an infimum of l.s.c. functions. For us too, since any system is a directed infimum of l.s.i. systems.

(v) But the answers are different: the Lebesgue integral can proceed with the second extension because the extension to l.s.c. already commutes with infima. On the other hand, the extension of the normal form to l.s.i. does not commute to infima: this is because we are in a « non-commutative » world. This is why we had to introduce the resolvant.

19.4.4 Semi-invertibility. Let us mention, without proof, a few basics about semi-invertibility.

Definition 106 (Semi-invertibility). The system $(\mathbf{H}, h, \sigma)$ is *lower semi-invertible* (l.s.i.) iff $(\mathbf{H}, h, \sigma^-)$ is invertible.

Proposition 54. (i) *If* $(\mathbf{H}, h, \sigma)$ *is l.s.i. and* $h \leq k$, *then* $(\mathbf{H}, k, \sigma)$ *is l.s.i.*

(ii) *If* $(\mathbf{H}, k, \sigma)$ *is l.s.i.,* $h \leq k$, $0 < \lambda < 1$, *then* $(\mathbf{H}, (1-\lambda)h + \lambda k, \sigma)$ *is l.s.i.*

We symmetrically define *upper semi-invertible* (u.s.i.) systems.

Proposition 55. *The system* $(\mathbf{H}, h, \sigma)$ *is invertible iff it is both u.s.i. and l.s.i.*

Corollary 55.1. *If* $h \le k$ *are such that* $(\mathbf{H}, h, \sigma)$ *is u.s.i. and* $(\mathbf{H}, k, \sigma)$ *is l.s.i., then* $(\mathbf{H}, 1/2(h+k), \sigma)$ *is invertible.*

Since it is easy to mistake l.s.i. and u.s.i., let us say that the l.s.i. are rather « large », that they socialise with directed suprema and positive feedbacks: indeed, if π is positive, $\pi^- = 0$, hence any system $(\mathbf{H}, h, \pi)$ is l.s.i.

The affinity between l.s.i. and directed suprema comes from:

Proposition 56. *Any l.s.i. system can be written as a directed supremum of an invertible system.*

Proof. $(\mathbf{H}, -I, \sigma)$ is u.s.i., hence, for $0 < \lambda < 1$, $(\mathbf{H}, (\lambda-1)I + \lambda h, \sigma)$ is invertible by Corollary 55.1. We define $h_\lambda := (\lambda - 1)I + \lambda h$: it is an increasing family with supremum h (see also Lemma 52.1). □

Corollary 56.1. *Any system can be written as a directed supremum of u.s.i. systems.*

Theorem 78 (Semi-invertible case). *The normal form extends to l.s.i. systems so as to preserve directed suprema. This extension is monotonic and associative.*

Remark 4. The extension to l.s.i. systems does not commute to directed infima. In [53], section 6.6., I gave the example of a decreasing family $(\mathbf{H}, h_n, \pi)$, with π positive – thus a l.s.i. family – such that $\pi[\![\inf_n h_n]\!] < \inf_n \pi[\![h_n]\!]$.

A symmetric extension to u.s.i. systems by passage to infima is possible. These two extensions are compatible; indeed, if $h \le k$ are such that $(\mathbf{H}, h, \sigma)$ is u.s.i. and $(\mathbf{H}, k, \sigma)$ is l.s.i., then $(\mathbf{H}, 1/2(h+k), \sigma)$ is invertible, so $\sigma[\![h]\!] \le \sigma[\![1/2(h+k)]\!] \le \sigma[\![k]\!]$. This being said, this extension is not associative *a priori*: the normal form is well defined for all *lopsided* feedbacks, either positive or negative, since in that case any system is semi-invertible. We can easily get the inequality

$$\sigma^-[\![\sigma^+[\![h]\!]]\!] \le \sigma^+[\![\sigma^-[\![h]\!]]\!]. \tag{19.21}$$

Theorem 77 indeed establishes equality by introducing a non-trivial idea, the resolvant.

19.4.5 The normal form theorem

Theorem 79 (Normal form). *The normal form of Definition* 105 *is characterised by the following properties:*

(i) *It solves the equation in the invertible case.*

(ii) *It commutes to directed suprema of l.s.i. and directed infima of u.s.i.*

(iii) *It is associative, i.e.,* $(\sigma + \tau)[\![h]\!] = \sigma[\![\tau[\![h]\!]]\!]$, *for all* σ, τ *independent.*

Proof. Obvious. □

Theorem 79 is quite amenable; for instance, we can easily prove:

Proposition 57.

$$(-\sigma)[\![-h]\!] = -\sigma[\![h]\!]. \tag{19.22}$$

19.5 The first GoI

19.5.1 The programme. Geometry of Interaction originates in the paper [39] (Section 18.2), conceptualised in a programme [43], that found its first realisation in the paper [41]: the GoI of system **F**, indeed of second-order linear propositional logic minus the additives. GoI was extended to pure λ-calculus in [44]; later on, additives were accommodated in [47].

One can contend that this first GoI was taking place in $\mathcal{B}(\mathbf{H})$, i.e., in a trivial von Neumann algebra (of type $\mathbf{I}_\infty$), thus with little chance of running into an « iconoclast » rereading of logic, i.e., of exponentials (Chapter 16). Still relying on the feedback equation, a second GoI, dwelling in the *hyperfinite factor* (of type $\mathbf{II}_1$), has thus been developed, see Chapter 21.

19.5.2 Generalities. To each proof is associated a cut-system: let us introduce the notation $\vdash \Gamma, [\Xi]$ to speak of the sequent $\vdash \Gamma$ proven by means of cuts on Ξ. The idea of this first GoI is to dispose of one Hilbert space for each formula of Γ, Ξ, always the same, $\mathbf{H}$, of infinite dimension and separable; $\mathbf{H}$ is thus isomorphic to ℓ^2, the isomorphism depending upon an orthonormal basis $(\mathbf{e}_n)$ (Section 17.A.1).

A proof of $\vdash \Gamma, [\Xi]$, i.e., of $\vdash \Gamma$ with cuts on the Ξ, will be interpreted by a cut-system $(\mathbf{H}^{\Gamma,\Xi}, h, \sigma_\Xi)$, where:

- $\mathbf{H}^{\Gamma,\Xi}$ is the sum of several copies of $\mathbf{H}$, one for each element of Γ, Ξ.
- h is a bounded operator on the previous space; which can be written as a matrix of type $(\Gamma, \Xi) \times (\Gamma, \Xi)$, with coefficients in $\mathcal{B}(\mathbf{H})$.
- Ξ is made of an even number of formulas $A, {\sim}A, B, {\sim}B, \ldots$ paired by the cut rule. σ_Ξ is thus the matrix whose coefficients are null but those corresponding to two indices $A, {\sim}A$ paired by a cut: $m_{A\,{\sim}A} = I$.

In [41] the following is (*grosso modo*) proven:

Theorem 80 (Normal form). *If t is a closed term of system* **F** *of normal form u, if* $(\mathbf{H}, h, \sigma)$ *and* $(\mathbf{H}, k, 0)$ *are the respective interpretations of t and u, then:*

(i) *σh is nilpotent.*

(ii) $\sigma[\![h]\!] = k$.

Proof. The sequence of reductions leading to the normal form constitutes a solution of the feedback equation by recursive substitutions, which translates as the nilpotency of σh, a phenomenon already observed in Sections 18.2.4 and 19.1.2. In terms of proof-nets, the reduction of cuts translates as follows:

- The logical cuts, e.g., $\otimes/\parr$, are reduced by replacing the system with an isomorphic one. Beware, we are in infinite dimension and this replacement is everything but innocent: thus it augments the size of matrices, hence that of Ξ, i.e., the number of cuts.

- The cuts between identity axioms are chains of chiasmi of which (19.2), (19.3) are examples for the lengths $N = 2, 3$. If we translate (19.2), (19.3) as cut-systems, we see that σh is nilpotent: $\sigma h = 0$ or $(\sigma h)^2 = 0$. More generally, for a chain of N chiasmi, we will get $(\sigma h)^{N-1} = 0$. □

Remark 5. The statement is approximative, since $\sigma[\![h]\!]$ is not quite k (see Section 19.5.6).

19.5.3 Identity group. The identity axiom $\vdash A, {\sim}A$ is interpreted by the matrix $\left[\begin{smallmatrix} 0 & I \\ I & 0 \end{smallmatrix}\right]$.

The cut between cut-free proofs M of $\vdash \Gamma, A$ and N of $\vdash {\sim}A, \Delta$ consists in the juxtaposition of both matrices, so as to get a matrix $\left(\begin{smallmatrix} M & 0 \\ 0 & N \end{smallmatrix}\right)$ of type $(\Gamma, A, {\sim}A, \Delta) \times (\Gamma, A, {\sim}A, \Delta)$, together with the feedback $\sigma_{A\,{\sim}A}$:

$$\begin{bmatrix} 0 & \dots & 0 & 0 & \dots & 0 \\ \dots & \dots & \dots & \dots & \dots & \dots \\ 0 & \dots & 0 & I & \dots & 0 \\ 0 & \dots & I & 0 & \dots & 0 \\ \dots & \dots & \dots & \dots & \dots & \dots \\ 0 & \dots & 0 & 0 & \dots & 0 \end{bmatrix}.$$

corresponding to the swapping of A and ${\sim}A$. If M, N already have cuts $\sigma'_{\Xi}, \sigma_{\Xi''}$, we define $\sigma_{\Xi} := \sigma'_{\Xi} + \sigma_{\Xi''} + \sigma_{A\,{\sim}A}$.

In what follows, we restrict, for reasons of legibility, to the cut-free case.

19.5.4 Multiplicatives. If we were following, instead of a matricial technique, a block technique, there would not be the slightest problem.

Take for instance the case of a « ⅋ », written blockwise: from a proof of – say – $\vdash A, B, C$ written as

$$\begin{pmatrix} u_{AA} & u_{AB} & u_{AC} \\ u_{BA} & u_{BB} & u_{BC} \\ u_{CA} & u_{CB} & u_{CC} \end{pmatrix}, \tag{19.23}$$

we would pass to $\vdash A ⅋ B, C$ by means of

$$\begin{pmatrix} \begin{pmatrix} u_{AA} & u_{AB} \\ u_{BA} & u_{BB} \end{pmatrix} & \begin{pmatrix} u_{AC} \\ u_{BC} \end{pmatrix} \\ \begin{pmatrix} u_{CA} & u_{CB} \end{pmatrix} & \begin{pmatrix} u_{CC} \end{pmatrix} \end{pmatrix}. \tag{19.24}$$

Similarly for the « ⊗ »: from proofs of $\vdash D, {\sim}A$ and $\vdash {\sim}B, E$ written as

$$v = \begin{pmatrix} v_{DD} & v_{D\sim A} \\ v_{\sim AD} & v_{\sim A\sim A} \end{pmatrix} \quad \text{and} \quad w = \begin{pmatrix} w_{\sim B\sim B} & w_{\sim BE} \\ w_{E\sim B} & w_{EE} \end{pmatrix}, \tag{19.25}$$

we would pass to $\vdash D, {\sim}A \otimes {\sim}B, E$:

$$\begin{pmatrix} \begin{pmatrix} v_{DD} \end{pmatrix} & \begin{pmatrix} v_{D\sim A} & 0 \end{pmatrix} & \begin{pmatrix} 0 \end{pmatrix} \\ \begin{pmatrix} v_{\sim AD} \\ 0 \end{pmatrix} & \begin{pmatrix} v_{\sim A\sim A} & 0 \\ 0 & w_{\sim B\sim B} \end{pmatrix} & \begin{pmatrix} 0 \\ w_{\sim BE} \end{pmatrix} \\ \begin{pmatrix} 0 \end{pmatrix} & \begin{pmatrix} 0 & w_{E\sim B} \end{pmatrix} & \begin{pmatrix} w_{EE} \end{pmatrix} \end{pmatrix}. \tag{19.26}$$

In both cases, we group the A and the B in the same block.

But these blocks have but a subjective value, in other words, a cut between (19.24) and (19.26) is the same thing as a cut between (19.23) and

$$\begin{pmatrix} v_{DD} & v_{D\sim A} & 0 & 0 \\ v_{\sim AD} & v_{\sim A\sim A} & 0 & 0 \\ 0 & 0 & w_{\sim B\sim B} & w_{\sim BE} \\ 0 & 0 & w_{E\sim B} & w_{EE} \end{pmatrix}, \tag{19.27}$$

i.e., two cuts between (19.23) and the two matrices of (19.25). It is indeed what we already did with *permutations* (Section 18.2.3).

This is not tenable for system **F**: *polymorphism*, i.e., subtyping, asks for a space independent from logic. One would like, for instance, to replace $A ⅋ B$ with a variable X which has « forgotten » that it stands for $A ⅋ B$. Which is only possible if all coefficients become homogeneous and this is why one replaces block matrices with plain ones.

The « Hilbert hotel » yields a solution, under the form of an isometry of $\mathbf{H} \oplus \mathbf{H}$ into $\mathbf{H}$. Taking inspiration from Section 14.1.2 and the delocations φ, ψ of (14.8), we can define the partial isometries p, q of the space ℓ^2:

$$\begin{aligned} p\Big(\sum \lambda_n \mathbf{e}_n\Big) &:= \sum \lambda_n \mathbf{e}_{3n}, \\ q\Big(\sum \lambda_n \mathbf{e}_n\Big) &:= \sum \lambda_n \mathbf{e}_{3n+1} \end{aligned} \tag{19.28}$$

of respective adjoints

$$\begin{aligned} p^*\Big(\sum \lambda_n \mathbf{e}_n\Big) &:= \sum \lambda_{3n} \mathbf{e}_n, \\ q^*\Big(\sum \lambda_n \mathbf{e}_n\Big) &:= \sum \lambda_{3n+1} \mathbf{e}_n. \end{aligned} \tag{19.29}$$

We see that the following equations are satisfied:

$$p^* p = q^* q = I, \tag{19.30}$$

$$p^* q = q^* p = 0. \tag{19.31}$$

These equations are those of the isometry of $\mathbf{H} \oplus \mathbf{H}$ *into* $\mathbf{H}$: $x \oplus y \rightsquigarrow p(x) + q(x)$. An isometry *onto* $\mathbf{H}$ – e.g., $2n, 2n+1$ – would correspond to

$$pp^* + qq^* = I \tag{19.32}$$

of which (19.31) is a weak version.

Let us conclude: p, q enable us to embed – say – matrices of type $(A, B, C) \times (A, B, C)$ in the matrices of type $(A \parr B, C) \times (A \parr B, C)$, by contracting two indices, thus

$$\begin{pmatrix} u_{AA} & u_{AB} & u_{AC} \\ u_{BA} & u_{BB} & u_{BC} \\ u_{CA} & u_{CB} & u_{CC} \end{pmatrix}$$

becomes

$$\begin{pmatrix} pu_{AA}p^* + pu_{AB}q^* + u_{BA}p^* + u_{BB}q^* & pu_{AC} + qu_{BC} \\ u_{CA}p^* + u_{CB}q^* & u_{CC} \end{pmatrix}.$$

The previous transformation – which works as well on the « Tensor » side – is an isomorphism of C^*-algebras, which preserves everything, but the identity (equation 19.32 would be required). Concretely, this means that two cuts between $\vdash \Gamma, A$ and $\vdash B, \Delta$ and $\vdash {\sim}A, {\sim}B, \Pi$, represented by a matrix of type $(\Gamma, A, B, \Delta, {\sim}A, {\sim}B, \Pi) \times (\Gamma, A, B, \Delta, {\sim}A, {\sim}B, \Pi)$ and a feedback swapping $A, {\sim}A$ and $B, {\sim}B$ are replaced, using the isometry, with the following: a matrix of type $(\Gamma, A \parr B, \Delta, {\sim}A \otimes {\sim}B, \Pi) \times (\Gamma, A \parr B, \Delta, {\sim}A \otimes {\sim}B, \Pi)$ and a feedback exchanging $A \parr B$ and ${\sim}A \otimes {\sim}B$. Isometricity ensures that the normal form of both systems is the same, which corresponds to the reduction of a cut $\otimes/\parr$. Two remarks:

- The refusal of (19.32) is a refutation of etaspansion, but not really honest: nobody forbids me from taking a surjective isometry, of the kind $(2n, 2n+1)$. On the other hand, the definition of Section 17.5.4 is the only possible one; moreover, without involving shady hotels with infinitely many dimensions: by staying in dimension 2.

- One can see p, q as the maintenance of an operational stack. p, q respectively *push* 0, 1, while p^*, q^* *pull* them. This means that, if the top of the stack is a « 0 », then p^* removes it, otherwise, it is not defined. Which justifies the equations (19.30) and (19.31), but in no way (19.32) which would for instance require a stack of infinite depth.

19.5.5 Exponentials. A chunk of [41]: its discovery took more than one year, the problem being with contraction, i.e., duplication. One tensorises with a « message space » isomorphic to $\mathbf{H}$: u is « promoted » into $!u := I \otimes u$. We will take care of duplications by means of the $p \otimes I, q \otimes I, \ldots$, which *commute* to the $I \otimes u$.

But we are not in $\mathbf{H} \otimes \mathbf{H}$, we are in $\mathbf{H}$. The previous operations are therefore « brought back in $\mathbf{H}$ » by means of an isomorphism Φ between $\mathbf{H} \otimes \mathbf{H}$ and $\mathbf{H}$. In particular, there is an internal « tensor product »: $u \odot v := \Phi(u \otimes v)\Phi^*$.

Which enables us to interpret rules of « constant depth » (Section 16.4.1). To get dereliction, we must be able to recover, from $!u$, a copy of u, such as done by the operator d, which satisfies $d^*(!u)d = u$. One must also interpret the principle $!A \vdash !!A$, which consists in exchanging $I \odot (I \odot u)$ with $(I \odot I) \odot u = I \odot u$. Which is ensured by the unitary operator t: $t(u \odot (v \odot w))t^* = (u \odot v) \odot w$. Note that t is a sibling of the associativity functors of monoidal categories (Section 9.A.2).

Let us fix an orthonormal basis $(\mathbf{e}_n)$ of $\mathbf{H}$; any partial injective function from $\mathbb{N}$ into $\mathbb{N}$ induces a partial isometry of $\mathbf{H}$, $u(\sum \lambda_n \mathbf{e}_n) := \sum \lambda_n \mathbf{e}_{\varphi(n)}$, which we already used for p, q, p^*, q^*.

- The isomorphism Φ between $\mathbf{H} \otimes \mathbf{H}$ and $\mathbf{H}$ is induced by a bijection $(m, n) \rightsquigarrow \langle m, n \rangle$ between $\mathbb{N} \times \mathbb{N}$ and $\mathbb{N}$.

- If $u \in \mathcal{B}(\mathbf{H})$, then $I \otimes u \in \mathcal{B}(\mathbf{H} \otimes \mathbf{H})$, which can be brought back into $!u := \Phi(I \otimes u)\Phi^* \in \mathcal{B}(\mathbf{H})$. If u is induced by a partial injection φ from $\mathbb{N}$ into itself, then $!u$ is induced by $!\varphi(\langle m, n \rangle) := \langle m, \varphi(n) \rangle$.

- A possible choice for d is the operator induced by the function $n \rightsquigarrow \langle 1, n \rangle$.

- t corresponds to the bijection $\langle m, \langle n, p \rangle \rangle \rightsquigarrow \langle \langle m, n \rangle, p \rangle$.

The rules are interpreted as follows (the matrix u corresponds to the premise, the matrix v corresponds to the conclusion):

Weakening: from $\vdash \Gamma$ to $\vdash \Gamma, ?A$: $v_{ij} := u_{ij}$ if $i, j \neq ?A$, $v_{ij} := 0$ else.

Contraction: from $\vdash \Gamma, ?A, ?A$ to $\vdash \Gamma, ?A$: similar to the conclusion $\vdash \Gamma, ?A \parr ?A$; but, instead of $p, q, \ldots$, use $\Phi(p \otimes I)\Phi^*$, $\Phi(q \otimes I)\Phi^*$, etc.

Dereliction: from $\vdash \Gamma, A$ to $\vdash \Gamma, ?A$: let D be the diagonal matrix $D_{CC} := I$ for $c \in \Gamma$, $D_{AA} := d$. Then $v := DuD^*$.

Promotion: from $\vdash \Gamma, A$ to $\vdash ?\Gamma, !A$ is easily interpreted: $v_{ij} := \Phi(I \otimes u_{ij})\Phi^*$. But this is not enough, one also needs:

Burying: from $\vdash \Gamma, ??A$ to $\vdash \Gamma, ?A$: let T be the diagonal matrix with coefficients $T_{CC} := I$ when $c \in \Gamma$, $T_{AA} := t$. Then $v := TuT^*$.

We easily check that normalisation rules are « correctly » interpreted, provided one restricts to exponential cuts whose premise $\vdash ?\Gamma, !A$ is without context, i.e., with $\Gamma = \emptyset$. In the general case, there is a slight difference between GoI and syntactic normalisation:

19.5.6 « Mistakes » of GoI. W.r.t. exponentials, GoI makes mistakes. Thus, it « forgets » erasing. Take for instance a proof of $\vdash ?B, A$, obtained by weakening, thus of the form $\begin{bmatrix} 0 & 0 \\ 0 & u \end{bmatrix}$, that we cut with a proof $\begin{bmatrix} a & b \\ c & d \end{bmatrix}$ of $\vdash A, !\sim B$. The solution of the feedback equation yields the result: $\begin{bmatrix} a & 0 \\ 0 & u \end{bmatrix}$, while, if we were following syntax, the coefficient a should disappear too.

Thus, GoI makes small mistakes w.r.t. syntax. But this is not that bad, indeed:

- If the conclusion is a Π^1 formula which does not use « ? », there is no mistake, even in the presence of exponential cuts. Hence, by associativity, GoI does not contradict syntactical normalisation.
- Moreover, GoI corresponds to *optimal reduction*, see *infra*. In other terms, who makes a mistake? Not the one you may think...

19.5.7 Strong nilpotency. The extension of GoI to pure λ-calculus is a rather interesting endeavour. W.r.t. the work on system **F**, it only requires a few typing mistakes! This extension had an unexpected consequence: Lamping had devised a rather hermetic algorithm of *optimal reduction* for λ-calculus. Gonthier [1] was then able to translate the algorithm in GoI, thus proving the mathematical correctness of Lamping's ideas.

Typing mistakes, typically $A = (A \Rightarrow A)$, posed a problem to this first GoI, which was relying too much on logical correctness: one cannot release operators in the wilderness and say « please normalise ». In the absence of any *general* result on the feedback equation, the hypothesis of *strong* nilpotency: $(\sigma h)^n$ strongly converges to 0, $\langle (\sigma h)^n(x) \mid x \rangle \to 0$ provided a fall-back solution. λ-calculus was interpreted by strongly nilpotent cut-systems: this was enough to get a normal form enjoying the right properties.

Twenty years later, I consider this strong nilpotency as an incompetent theoretisation, a cheap way to ground the constructions of [44]. Nowadays, I would ground them on Section 21.B.1, e.g., Proposition 90, see Remark 13.

19.5.8 Digression: the dictionary. We saw that normalisation translates in terms of nilpotency, a very natural idea in logic: let us use the metaphor of a *dictionary*.

This dictionary is perfect in its kind: from a set of basic knowledge, it enables one to define a whole vocabulary, say, technical. Which means that, taking a word μ, the dictionary defines it clearly, with perhaps references to other words; we are sent back to these other words and, recursively, to other references. A good dictionary is the one that does not endlessly walk us from reference to reference.

Which we can express, as we please, in two ways:

- The reflexive/transitive closure of the relation « μ refers to the word ν » is an order relation.
- Define the matrix $(a_{\mu\nu})$: $a_{\mu\nu} = 1$ when the word μ refers to ν, $a_{\mu\nu} = 0$ else; then $(a_{\mu\nu})$ is *nilpotent*.

19.5.9 Additives. Additives are difficult to cope with in GoI; this is why they were absent from the first paper. The reason is technically obvious: GoI interprets the tensor product as a direct sum of spaces, thus producing a « logarithmic simplification ». But where to find a *logarithm* for the sum? It took a certain time to work out an extension [47] based upon the idea of an *idiom*, i.e., a *private* message space. Being private, the messages are not really shared: this is *communication without comprehension*. This idea turns out to be central, see Chapter 20 *infra*.

19.5.10 Limitations of the first GoI. Fundamentally, the first GoI makes use of the language of operator algebras without really respecting their spirit. Thus, one never steps out of partial isometries, moreover those arising as partial permutations of a distinguished basis. We hardly get more than a preposterous dressing of a theory of the partial permutations of $\mathbb{N}$[5]. This being said, this criticism affects the implementation, not the concept of GoI, which only needs a more exciting *milieu* than the one induced by logical proofs. Our brief incursion into the quantum world freed us from this limited standpoint, as we freed from *Kindergarten* topology with coherent spaces.

At a deeper level, I do believe that there was a mistake of a reductionist nature: we have a propensity to think that galaxies are made of stars, thus that an operator

[5]Practical jokers had even the bad taste to rewrite GoI inside Scott domains, which is vexing, albeit not totally undeserved.

is less primitive than a Hilbert space, etc.[6] Which is pure illusion: one can perfectly contend that a von Neumann algebra is a more primitive structure than the space on which it operates – which is perhaps but a convenient reification. In other terms, rather than interpreting logic by operators, one will try to interpret it by von Neumann algebras.

19.A Complements on operators

19.A.1 Topologies. On $\mathcal{B}(\mathbf{H})$, there are several topologies, in particular:

Norm: the norm $\|u\| = \sup\{\|u\|(x); \|x\| \leq 1\}$ makes product and adjunction continuous.

Strong: (u_i) converges *strongly* to u when for all $x \in \mathbf{H}$, $\|u_i(x) - u(x)\|$ converges to 0. The strong topology is slightly weird: indeed, adjunction is not strongly continuous. On the other hand $u, v \rightsquigarrow uv$ is strongly continuous, provided the argument u remains bounded in norm.

Weak: (u_i) converges *weakly* to u when for all $x \in \mathbf{H}$, $\langle u_i(x) \mid x\rangle$ converges to $\langle u(x) \mid x\rangle$. Adjunction is weakly continuous, but the product is only *separately* continuous, which is of limited interest. As a compensation, the unit ball $B_1(\mathbf{H}) := \{u; \|u\| \leq 1\}$ is weakly compact.

So to speak, strong convergence is « columnwise » (which explains why the adjoint is not continuous), while weak convergence is « coefficientwise ». The inequality: $|\langle u(x) \mid x\rangle| \leq \|u(x)\| \cdot \|x\| \leq \|u\| \cdot \|x\|^2$ explains the relative strengths of these topologies. Remember that a stronger topology has more open sets, thus makes convergence more difficult.

A directed and bounded family of hermitians admits a (weak) supremum:

$$\langle h(x) \mid x\rangle := \sup_{i \in I}\langle h_i(x) \mid x\rangle. \tag{19.33}$$

This gives us a sort of « Dini's theorem », see [63], p. 307:

Proposition 58. *If $h = \sup_{i \in I} h_i$, then $h_i \to h$ in the strong topology.*

In mathematics, function spaces always bear several topologies (four, five on $\mathcal{B}(\mathbf{H})$ for instance) and vehicles to make them communicate. For instance a one-parameter group of unitaries, $(u_t)_{t \in \mathbb{R}}$, is usually assumed *strongly* continuous,

---

[6] Jurassic foundations look like a scientistic version of Genesis: the first day sets, then natural numbers, next reals, then complex, eventually function spaces… nay the rest of the seventh day provided by the fulfilment of Hilbert's Program.

which is more liberal than norm continuity, while still compatible with composition. This is indeed equivalent to *weak* continuity:

$$\begin{aligned}\|(u_t - 1)(x)\|^2 &= \langle (u_t)(x) \mid (u_t)(x)\rangle + \langle x \mid x\rangle - 2\Re(\langle u_t(x) \mid x\rangle)\\ &= -2\Re(\langle (u_t - 1)(x) \mid x\rangle).\end{aligned}$$

Similarly, a strongly closed *convex* set is weakly closed.

This contrasts with the arrogant naïveté of certain logicians who contend that they can reduce everything to a single topology, not even Hausdorff! Such a single topology is an all-terrain vehicle that is good for everything – thus for nothing. The topology à la Scott on $\mathbb{R}$ is the one making continuous the l.s.c. functions: its open sets are the $]a, +\infty[$, $a \in [-\infty, +\infty]$; since it is not Hausdorff, it offers little interest. Mathematicians (Dini, Proposition 58) are perfectly aware of the interest of order-continuity, but prefer to use it as a *bonification* between *good* topologies rather than introducing a mediocre topology devoted to the sole l.s.c. (and not even adapted to u.s.c. functions).

But all of this is a matter of good taste; and I can hear the sophism: « How do you define good taste ? »

19.A.2 Matrices and blocks.

Two styles of matrices are useful:

Block matrices: in the case of a direct sum decomposition $\mathbf{H} = \sum_1^n \mathbf{H}_i$, if α_i is the orthogonal projection of $\mathbf{H}_i$, one can write any operator f of $\mathbf{H}$ as the sum $\sum_{ij} \alpha_i f \alpha_j$. For $n = 2$: $f = \begin{pmatrix}\alpha_1 f \alpha_1 & \alpha_1 f \alpha_2\\ \alpha_2 f \alpha_1 & \alpha_2 f \alpha_2\end{pmatrix}$. These block matrices compose in the usual way, i.e., $h_{ik} = \sum_j f_{ij} g_{jk}$. But they are not « true » matrices.

Plain matrices: they correspond to an isomorphism $\mathcal{B}(\mathbf{H}) \simeq \mathcal{B}(\mathbf{K}) \otimes M_n(\mathbb{C})$. In other terms, in $\begin{bmatrix} f_{11} & f_{12}\\ f_{21} & f_{22}\end{bmatrix}$ (remark the different graphical style), the coefficients belong to $\mathcal{B}(\mathbf{K})$ and not to $\mathcal{B}(\mathbf{H})$.

One can relate these notions in a particular case: if one is given partial isometries α_{ij}, such that $\alpha_{ii} = \alpha_i, \alpha_{ji} = \alpha_{ij}^*, \alpha_{ik} = \alpha_{jk}\alpha_{ij}$. Then f can be written as the « true » matrix (with coefficients in $\mathcal{B}(\mathbf{H}_1)$), $f_{ij} = \alpha_{1i} f \alpha_{j1}$, e.g., $f = \begin{bmatrix}\alpha_{11} f \alpha_{11} & \alpha_{11} f \alpha_{21}\\ \alpha_{12} f \alpha_{11} & \alpha_{12} f \alpha_{21}\end{bmatrix}$.

19.A.3 Unbounded operators.

If $u \in \mathcal{B}(\mathbf{H}, \mathbf{K})$, the graphs $\Gamma(-u) \subset \mathbf{H} \oplus \mathbf{K}$ and $\Gamma(u^*) \subset \mathbf{K} \oplus \mathbf{H}$ are linked by the relation

$$\Gamma(u^*)^{\mathrm{op}} = \Gamma(-u)^{\perp}. \tag{19.34}$$

If u is a *partial* operator from (a subspace of) $\mathbf{H}$ into $\mathbf{K}$, the adjoint (defined by (19.34) makes appear two dual hypotheses:

Density: the domain of u is dense. In this case, u^* is a partial operator of a closed graph.

Closure: the graph of u is closed: in this case, u^* is densely defined, but need not be a graph; which obviously subsists when u is *closable*, i.e., when the closure $\overline{\Gamma(u)}$ is still a graph. u has a closed graph when $x_n \to x$ and $u(x_n) \to y$ imply $u(x)$ is defined and $u(x) = y$. u is closable when $x_n \to 0$ and $u(x_n) \to y$ imply $y = 0$.

An *unbounded operator* is a partial operator with dense domain and closed graph. From what precedes, an unbounded operator has an unbounded adjoint. The typical unbounded operator $f \rightsquigarrow df/dx$ is the inverse of an injective operator of $\mathcal{B}(\mathbf{H})$, the *primitive*. By the closed graph theorem, a total unbounded operator is bounded.

Unbounded operators compose poorly, however:

- ub is closed when u is closed and b is bounded. If u is densely defined, ub need not be densely defined.
- bu is densely defined when u is densely defined and b is bounded. If u is closed, bu need not even be *closable*.

As a closed subset of $\mathbf{H} \oplus \mathbf{K}$, Γ_u is a Hilbert space; the bounded map (of norm 1) $\varphi(x \oplus y) := x$ from Γ_u to $\mathbf{H}$ admits an adjoint φ^* of the same norm: if $x \in \mathrm{dom}(u)$, $y \in \mathbf{H}$, $\langle x \mid y \rangle = \langle x \oplus u(x) \mid \varphi^*(y) \rangle$. With $\varphi^*(y) = \theta(y) \oplus u(\theta(y))$, we get: $\langle x \mid y \rangle = \langle x \mid \theta(y) \rangle + \langle u(x) \mid u(\theta(y)) \rangle = \langle x \mid (I + u^*u)(\theta(y)) \rangle$. Thus θ is the inverse of $I + u^*u$ (indeed, $(I + u^*u)\theta = I$, while $\theta(I + u^*u)$ admits I as closure). Hence the non-trivial proposition that u^*u is densely defined and closed ([63], Section 2.7.). From $(I + u^*u)\theta = I$, we get the fact that $u(I + u^*u)^{-1}$ is a bounded operator. Those facts were used in Definition 104 of the *resolvant*.

19.A.4 An analogy. The departure bounded/unbounded is somewhat reminiscent of another one, recursive/partial recursive. The closed graph theorem ensures that a *total* unbounded operator is bounded; in the same way, a total semi-recursive function is computable. This analogy exposes the inanity of non-monotonic « logics »: rendering total a semi-computable algorithm, nay « completing » an incomplete theory, as allegedly done in AI, is an operation of the same nature as the completion of an unbounded operator into a total one. Indeed, the axiom of choice enables one to construct an *algebraic* supplement for the carrier thus rendering the operator total, while losing all its topological properties: this supplement bears the same hideous look as the (non-computable) complement of the domain $\{n; f(n) \downarrow\}$ of a partial recursive function. The idea of an algebraic supplement is, by the way, so dumb that textbooks of mathematics do not waste paper refuting it. While, if no good logician indulges in non-monotonicity, this nonsense is nevertheless tolerated in international conferences... Which means that logic has still progress to make to keep up with scientific standards.

There is something fascinating in unbounded operators, which is due to their total absence of categorical status: they are by no means *morphisms*. Which hints that their limitations could be exploited (in a way more elaborated than my rather simple-minded analogy) to explain incompleteness, complexity etc., i.e., *why the world is not transparent*.

Chapter 20
Babel Tower vs. Great Wall

This chapter introduces, in a rather informal way, the notion of *idiom*.

20.1 Idioms

20.1.1 Superposition. From proofs of $\vdash C, A$ and $\vdash C, B$ written blockwise as

$$u = \begin{pmatrix} u_{CC} & u_{CA} \\ u_{AC} & u_{AA} \end{pmatrix} \quad \text{and} \quad v = \begin{pmatrix} v_{CC} & v_{CB} \\ v_{BC} & v_{BB} \end{pmatrix} \tag{20.1}$$

the only reasonable candidate for a proof of $\vdash C, A \,\&\, B$ is

$$w := \begin{pmatrix} u_{CC} + v_{CC} & u_{CA} & v_{CB} \\ u_{AC} & u_{AA} & 0 \\ v_{BC} & 0 & v_{BB} \end{pmatrix}. \tag{20.2}$$

which badly fails[1]: the entries u_{CC} and v_{CC} are irreversibly mingled.

In the context of proof-nets, this problem of *superposition* (Section 11.C.4) was solved by the introduction of boolean *eigenvariables* and *slices*: the adaptation of this idea in GoI leads to *idioms*.

The idea is that a GoI operator u no longer dwells in a vN algebra $\mathcal{A}$ (in old-style GoI, the only one so far introduced, $\mathcal{A} = \mathcal{B}(\mathbf{H})$), but in a tensor product $\mathcal{A} \otimes \mathcal{D}$, where $\mathcal{D}$ is another vN algebra, the *idiom* of u. Idiom-free GoI thus corresponds to $\mathcal{D} = \mathbb{C}$. The pattern of interaction is not quite the same and expresses the *privacy* of the idiom.

In the additive case just considered, u, v now carry their own idioms $\mathcal{D}, \mathcal{E}$; the idiom of w will be the direct sum $\mathcal{D} \oplus \mathcal{E}$. The algebra $\mathcal{A} \otimes (\mathcal{D} \oplus \mathcal{E})$ can be written $(\mathcal{A} \otimes \mathcal{D}) \oplus (\mathcal{A} \oplus \mathcal{E})$, thus

$$w := \begin{pmatrix} u_{CC} \oplus v_{CC} & u_{CA} \oplus 0 & 0 \oplus v_{CB} \\ u_{AC} \oplus 0 & u_{AA} \oplus 0 & 0 \oplus 0 \\ 0 \oplus v_{BC} & 0 \oplus 0 & 0 \oplus v_{BB} \end{pmatrix}. \tag{20.3}$$

This is the basic idea at work in [47].

---

[1] Believe it or not, this « solution » has been published; the authors just restricted the analogue of Theorem 80 to the fragment *without* additives. Think of those new portable phones that can be used as mustard pots… as long as one puts no mustard in them!

20.1.2 Multiplicatives. Idioms, originally introduced to handle additive operations, must now be maintained in all cases, including the basic one, multiplicatives. For instance *quid* of the construction (19.26) of Section 19.5.4? v, w now have respective idioms $\mathcal{D}, \mathcal{E}$: $v \in \mathcal{A} \otimes \mathcal{D}, w \in \mathcal{A} \otimes \mathcal{E}$. We consider the following $*$-isomorphisms from respectively $\mathcal{A} \otimes \mathcal{D}$, $\mathcal{A} \otimes \mathcal{E}$ to $\mathcal{A} \otimes (\mathcal{D} \otimes \mathcal{E})$:

$$(a \otimes b)^{\ddagger} := a \otimes (b \otimes I_{\mathcal{E}}), \tag{20.4}$$

$$(a \otimes c)^{\dagger} := a \otimes (I_{\mathcal{D}} \otimes c). \tag{20.5}$$

These isomorphisms are merely a tensorisation with $I_{\mathcal{D}}$ (or $I_{\mathcal{E}}$) together with a few tensorial isomorphisms (commutativity, associativity). (19.27) becomes

$$\begin{pmatrix} v^{\ddagger}_{DD} & v^{\ddagger}_{D\sim A} & 0 & 0 \\ v^{\ddagger}_{\sim AD} & v^{\ddagger}_{\sim A\sim A} & 0 & 0 \\ 0 & 0 & w^{\dagger}_{\sim B\sim B} & w^{\dagger}_{\sim BE} \\ 0 & 0 & w^{\dagger}_{E\sim B} & u^{\dagger}_{EE} \end{pmatrix} \tag{20.6}$$

of idiom $\mathcal{D} \otimes \mathcal{E}$.

20.1.3 Identity group. The identity axiom is treated as before, idiom-free (i.e., $\mathcal{D} = \mathbb{C}$). With the notations of Section 19.5.3, the cut between $M \in \mathcal{A} \otimes \mathcal{D}$ and $N \in \mathcal{A} \otimes \mathcal{E}$, treated in the spirit of the tensor rule, becomes $\begin{pmatrix} M^{\ddagger} & 0 \\ 0 & N^{\dagger} \end{pmatrix}$. The feedback still exchanges indices $A, \sim A$ (but the algebra is now $\mathcal{A} \otimes \mathcal{D} \otimes \mathcal{E}$).

20.1.4 Communication without comprehension. The idiom must be seen as a personal system of reference, by nature private. This is why the two protocols of communication between idioms are *without comprehension*, in the basic sense of « understanding »:

Additive, $\mathcal{D} \oplus \mathcal{E}$: the idioms ignore each other: they violently refuse to cooperate. With obvious notations, $(a \otimes (d \oplus 0)) \cdot (b \otimes (0 \oplus e)) = 0$.

Multiplicative, $\mathcal{D} \otimes \mathcal{E}$: the idioms have no aggressiveness against each other: they interfere in a mode of polite indifference. This is the mode of communication proper (i.e., the cut rule) and this deserves a long discussion. With obvious notations, $(a \otimes (d \otimes 1)) \cdot (b \otimes (1 \otimes e)) = ab \otimes (d \otimes e))$.

Such a behaviour can be observed in real life; for instance – assuming I know strictly nothing about Japanese –, given a text in that language, I can duplicate, burn, mail it, but perform strictly no operation related to its contents. *Idem* with psychology: my internal idiom is a private system of references for colours, sensations etc. What I perceive as « blue » is definitely personal and cannot be shared with anybody. The miracle is that I can nevertheless communicate about colours, for instance by creating a word that I can share; which, by the way, implies that not everything in communication is idiomatic.

Logically speaking, the typical idiomatic artefact is a *variable*; indeed *bound*, since free variables are bound to be… bound. When I write $\int f(x)dx$, the meaning of x is limited to the scope of the symbol $\int$. Bound variables socialise by renaming; we all learned how to rename variables, so as to avoid interferences, i.e., comprehension: $\sum_i a_i \times \sum_i b_i = \sum_{ij} a_i b_j$, the replacement of i with j being the exact syntactical counterpart of the tensorisation of idioms. This example shows the deep necessity of idioms; also, due to the efficiency of the idiomatic communication at work in mathematics, it shows that one can create an *illusion* of comprehension.

Coming back to psychology, think of famous couples (writers, composers, singers, scientists, etc.). Very often, one of them does 90% of the job, hence the question « what is my partner good for ? » and the catastrophic conclusion « I will continue alone »! The director Jean Renoir once said that he was using other people as walls « to throw the ball back »; that this was not a matter of getting ideas from others, only of speeding up his mind: after all, it was his *own* ball he was getting back.

This prompts an interpretation of communication, inspired from the idiomatic form of GoI. I can fancy myself as dwelling in the space $\mathcal{A} \otimes \mathcal{D}$, which decomposes as the tensor product of the algebra $\mathcal{A}$ which represents common channels, typically the five senses, and the idiom $\mathcal{D}$, forever personal. When I communicate with a « wall » (whose dwelling space is $\mathcal{A} \otimes \mathcal{E}$), I manage to pass information through the common channels, but this information (belonging to $\mathcal{D}$) has no sense for the wall, which treats it generically and throws it back to me, but *through an unexpected channel*; I thus get back my own information (which I can understand) in a *location* not chosen by me. Taking into account that the process goes on and, moreover, that the « wall » is not that passive, that its own idiom takes a symmetrical part in the communication, one can guess why this process might be so efficient. In particular, the minor partner of a collaboration might be the one that throws back the ball at the most unexpected place; whatever the potentialities of the major partner, he may never be able to realise them without the right « wall ». This applies to the benefit we take in various interactions with others – after all, it is but one's sole benefit. And also to this puzzle of my childhood: why, although so ticklish, was I unable to tickle myself?

In the decomposition $\mathcal{A} \otimes \mathcal{D}$, $\mathcal{A}$ is locative; since shared, it is rigid: without a locative substrate, communication could not be established. While $\mathcal{D}$ is spiritual: if I replace the idiom with an isomorphic copy – this includes automorphisms of $\mathcal{D}$ –, the interaction with the outer world is not affected.

20.2 The Babel Tower

20.2.1 The contraction rule. If $u \in \mathcal{A} \otimes \mathcal{D}$ represents a proof π of A, then the proof $\pi \otimes \pi$ of $A \otimes A$ is represented by $\begin{bmatrix} u^{\ddagger} & 0 \\ 0 & u^{\dagger} \end{bmatrix}$, of idiom $\mathcal{D} \otimes \mathcal{D}$. This makes duplication impossible: one needs a matrix F of type $(\sim A, A, A) \times (\sim A, A, A)$, which, « cut » with u, yields $\pi \otimes \pi$ (or some idiomatic variant); if F is of idiom $\mathcal{E}$, the output $[F]u$ of the cut is of idiom $\mathcal{E} \otimes \mathcal{D}$, in which we have no hope of identifying – independently of $\mathcal{D}$ – two commuting subalgebras isomorphic to $\mathcal{D}$!

20.2.2 Perennialisation. If u is idiom-free, i.e., if $\mathcal{D} = \mathbb{C}$, then, $\pi \otimes \pi$ becomes $\begin{bmatrix} u & 0 \\ 0 & u \end{bmatrix}$ and contraction becomes possible. Indeed, using the idiom $\mathcal{E} := \mathcal{M}_2(\mathbf{C})$, the following matrix F implements duplication:

$$F := \begin{bmatrix} \begin{bmatrix} 0 & 0 \\ 0 & 0 \end{bmatrix} & \begin{bmatrix} 1 & 0 \\ 0 & 0 \end{bmatrix} & \begin{bmatrix} 0 & 0 \\ 1 & 0 \end{bmatrix} \\ \begin{bmatrix} 1 & 0 \\ 0 & 0 \end{bmatrix} & \begin{bmatrix} 0 & 0 \\ 0 & 0 \end{bmatrix} & \begin{bmatrix} 0 & 0 \\ 0 & 0 \end{bmatrix} \\ \begin{bmatrix} 0 & 1 \\ 0 & 0 \end{bmatrix} & \begin{bmatrix} 0 & 0 \\ 0 & 0 \end{bmatrix} & \begin{bmatrix} 0 & 0 \\ 0 & 0 \end{bmatrix} \end{bmatrix}. \tag{20.7}$$

Indeed, a cut between F and $\begin{bmatrix} u \end{bmatrix}$ yields, after normalisation, the result

$$[F]u = \begin{bmatrix} \begin{bmatrix} u & 0 \\ 0 & 0 \end{bmatrix} & \begin{bmatrix} 0 & 0 \\ 0 & 0 \end{bmatrix} \\ \begin{bmatrix} 0 & 0 \\ 0 & 0 \end{bmatrix} & \begin{bmatrix} u & 0 \\ 0 & 0 \end{bmatrix} \end{bmatrix}, \tag{20.8}$$

an idiomatic variant of $\pi \otimes \pi$, indeed a *rescaling*, see Section 21.4.1.

Hence the interpretation of exponentials (Chapter 21): use an isomorphism $\Phi : \mathcal{A} \otimes \mathcal{D} \hookrightarrow \mathcal{A}$ and replace $u \in A$ with the idiom-free $!u := \Phi(u)$.

20.2.3 The two infinities. $!A$ expresses infinity through the iterated contraction $!A \multimap A \otimes \cdots \otimes A$. In GoI, this involves a double infinity: *quantitative*, « Great Wall » vs. *qualitative*, « Babel Tower »; while the former deals with $\mathcal{A}$, the latter deals with $\mathcal{D}$.

The Great Wall. Sticking to the matricial representation, the n-ary logical tensor $A \otimes \cdots \otimes A$ involves $n \times n$ matrices. We can see this growth of matrices as purely spatial. They, by the way, fit our naïve image of the infinity « one brick after another », thus the image of the Great Wall. Mathematically speaking, this infinity is quantified by *dimension*. In linear algebra, a space is finite-dimensional when not isomorphic with a subspace: this definition leads to *finite* von Neumann algebras (Section 20.B).

A finite vN algebra, typically the *hyperfinite factor*, Section 20.C.2, is not finite in the standard sense: its cardinality is the power of the continuum and, as a plain algebra, its dimension is infinite. However, *seen from the inside*, it is finite in a very reasonable sense. What is interesting in this approach to finitism is that, unlike the natural numbers, the totality is itself finite. Finitism is, so to speak, a limitation on the possibilities of creating new dimensions in an internal way, e.g., in GoI.

A finite vN algebra is an algebra with a *trace*. Based upon the trace (indeed the determinant), one can develop a duality that enables one to reconstruct the logical aspects missing in the first GoI.

The Babel Tower. $A \otimes \cdots \otimes A$ also involves n-ary tensor powers of the idiom space. Here lies another infinity, much more cryptic, that of the idioms, hence the image of the Babel Tower. Keeping in mind the blind spot of infinity, i.e., the necessity of an *indirect* approach to it, let us observe that the idiom is what opposes duplication. In other words:

$$\textit{resource} = \textit{idiom}$$

By the way, when two operators interact through a cut or a tensor, their idioms are tensorised, in analogy to the addition of resources.

The idiomatic infinity should therefore be seen as the unlimited possibility of creating new idioms, i.e., « fresh » messages. Due to its intrinsic unicity, the hyperfinite factor has a lot of automorphisms, most of them *external*: this external character limits the inner creation of new idioms.

Of the two forms of infinity, the qualitative form is presumably the most important,... and the most delicate to cope with!

20.2.4 Quantitative vs. qualitative. The idea is thus that the world is doubly infinite, like a notebook with a quantitative infinity of pages that one can « describe » with a qualitative infinity of nuances. The encoding of language, images, sounds – from Gödel to modern computers – translates (i.e., betrays: *traduttore, traditore*) this second infinity to natural numbers, i.e., reduces it to quantitative infinity.

This *reduction* of qualitative to quantitative infinity is problematic. Without adhering to any sort of spiritualism, here lies something of an opposition matter/spirit: the quantitative reduction of the language is furiously reminiscent of the reduction of thought to its material support, the brain.

Computer scientists seem to be aware of the irreducibility of one to another. Thus, Milner's π-calculus[2] [83] distinguishes channels through which messages

[2]Based upon a synchretism between various ideas, some already (and better) treated by linear logic, others still in want of a logical conceptualisation. The analogy with pure λ-calculus, of which it claims

transit. One of the basic operations consists in the creation of « fresh » names: the creation *ad libitum* of new names is typical of qualitative infinity.

20.3 A new finitism?

20.3.1 Jurassic finitism. The ancient, hilbertien, finitism relies on an impossible hypothesis. Indeed, by acknowledging only the natural numbers, it puts in front the *infinite* set $\mathbb{N}$ of integers. And there is no way to avoid this basic misunderstanding; this is by the way the deep meaning of the incompleteness theorem: « finitism is not finite ». There is no great divide finite/infinite that would separate « finitistic », « predicativistic », etc., methods from the others; there remain only more or less religious, sectarian postures: one pretends. Witness the discussion list « fom » (for foundations of mathematics), which currently discusses the hottest matters, typically the latest developments of Hilbert's Program. After all, in 1452 in Byzantium, the question at stake was that of the sex of angels, not that of the Turks who were camping beyond the ramparts.

How to produce a new finitism without self-blinding, like the aforementioned Doctors of the Law, or ramble like the unfortunate Essenin-Volpin? The task seems impossible, almost a contradiction in terms. The answer – at least, the hope of an answer – lies in the *hyperfinite factor* (Section 20.C.2). Functional analysts style it as doubly finite (finite and hyperfinite), while the logical tradition – dogmatism –, relying on its sole cardinality $2^{\aleph_0}$, would rather depict it as an infinite monster[3]. This is true if handled from outside; but, internally speaking, it is from many aspects a finite object.

I shall try to present the two ideas of *finiteness* at work in von Neumann algebras. The novelty is that we consider structures which are infinite from the outside, but finite from the inside. This has nothing to do with some sort of non-standard doohickeys, of the sort cherished by logicians: nothing is more standard, « mainstream », than the hyperfinite factor. And, by the way, since « non-standard » refers to a preexisting standard, what could be the value of non-standard foundations?

20.3.2 Internal vs. external. One of the most stubborn foundational prejudices is linked to the distinction internal/external, that many mathematicians have tried to negate, then to corner. From Cantor to Gödel, everything is internal – if not technically, at least ideologically: this is the meaning of the various encodings which reduce everything to natural numbers. The various paradoxes (Cantor, Burali-Forti, Russell, Richard, Gödel, Turing, etc.) show that the idea of a submarine that

to be the « parallel » version is abusive: the π-calculus is still in want of a structuration, of a dorsal spine.

[3]A statement which is, by the way, foundationally suspect: cardinality makes sense only if we freeze the concepts into their set-theoretic reifications.

embarks everything, including the submarine[4] itself, is nonsense: not everything can be internal. These paradoxes come from the same matrix, *diagonalisation* (Section 2.1.2) and produce counterexamples, efficient albeit not convincing, since hard to understand[5]. Things could hardly have been different, since *everything* reasonable was embarked; when the submarine is loaded more modestly, one gets more interesting refutations. Thus, $\sqrt{2}$ provides a magnificent counterexample to the pythagorician dogma « everything is rational ».

Instead of opening his paw and freeing the nut (Section 15.1.2), the monkey introduced the *meta*, a concession to externality, limited to the limbs of the significant. By adding to a formal system a consistency formula – which hardly means anything – to make a *meta-system* of it, one relegates the external to the department of anomalies: the formula that means nothing, the non-standard integer, the class which is not a set, etc. There is something bizarre in the very form of those bizarreries, which result from the basic postulate « everything is internal »: a sort of tumor created by the medication.

Thus, rather than internalisation at any price, which runs – backwards – into a limited and untractable form of externality, one must admit, from the start, the coexistence of an internal viewpoint with an external viewpoint; in particular, that a natural object (not a non-standard doohickey) such as the hyperfinite factor might be finite or infinite, depending upon the viewpoint, internal or external, one adopts. By the way, vN algebras have a specific technique of (limited) internalisation, *crossed products* (Section 20.B.3, *infra*).

20.3.3 The finiteness of language. In the foundational vulgate[6], the language is infinite; indeed, one must be able to create an infinity of copies, of « occurrences », of the same symbol. But is this reasonable? Should we try to do it concretely, physically, one would need a *delocating machine*, which would deposit copies of the symbol at distances of 1 km, 2 km,... We see that, as long as thought is concerned, we are ready to admit that the machine thus launched will never come back to its starting point. However, in the physical world, it is what happens! And I remember my shock, when, still a child, I heard about the impossibility of getting farther and farther from Earth. Finiteness, hyperfiniteness might be the analogues of the finiteness of space transposed in the realm of thought. And no sophist will convince me that there should be – internally speaking – infinitely many distinct symbols: let him show them to me, and, might as well, the infinity of galaxies supposedly created by God!

Between the hypothesis of an actually finite language – with a limited stock of possible expressions –, something mathematically intractable and the unbridled

[4]Audiberti, *Le retour du divin*: « La prison appelée la vie enferme toutes les prisons. »

[5]A good pretext for certain « scientists » to call them into question.

[6]Apollinaire: « La terre plate à l'infini/ Comme avant Galilée »; here, the language is flat at infinity.

infinity at work in logic, the two notions of finiteness of vN algebras provide, if not a definite answer, at least a hint at what could be a mature finitistic approach to language. *Finitism* allows – depending on their *size* – as many quantitatively independent artifacts as desired. *Hyperfinitism* is more delicate to grasp; if the meaningful logical *actions* were part of a denumerable group, this group should then be *amenable*, i.e., of « tame growth » in a certain sense, for instance *locally finite* (Section 20.C.3 *infra*).

One can contend that the achievements of GoI are still far from proving my point. But it is better to see a dim light than nothing at all... especially when the problem is to shine a light on a blind spot.

20.A Von Neumann algebras

The general solution of the feedback equation (Section 19.4) involves inversions, norm limits, square roots, which all are C^*-algebras operations; and also directed suprema which are not C^*-algebraic. A C^*-algebra with directed suprema is called a *von Neumann algebra*.

20.A.1 Introduction: commutative case

Theorem 81 (Extremely disconnected spaces). *Let X be a compact space; the following conditions are equivalent:*

(i) *Any bounded family of $\mathbb{C}(X)$ has a supremum in $\mathbb{C}(X)$.*

(ii) *The closure of an open set of X is still open.*

In this case we say that X is ***extremely disconnected*** *(e.d.).*

Beware: the supremum in $\mathbb{C}(X)$ need not be *pointwise*. For instance, in $\mathbb{C}([0,1])$ the supremum of the $f_n(x) := x^{1/n}$ equals $f(x) := 1$, while the pointwise sup is discontinuous.

Just like Scott domains, e.d. spaces are eccentric topologies: while the former are too coarse, the latter are too discrete. While the former were lattices in disguise, the latter are indeed measure spaces. Thus, the clopen sets of an e.d. topology form a complete boolean algebra: it is enough to define $\bigsqcup_i O_i := \overline{\bigcup_i O_i}$.

If (M,μ) is a measured space, then $\mathcal{L}^\infty(M,\mu)$ is a C^*-algebra admitting bounded suprema; by the way, those are pointwise suprema. As a C^*-algebra, $\mathcal{L}^\infty(M,\mu)$ is thus written as $\mathbb{C}(X)$ for an e.d. space X. X is unique, while two *equivalent* measures (M,μ), (M,μ') – i.e., with the same negligible sets – will define the same algebra. Since e.d. spaces are primarily complete boolean algebras, one can pull back any space $\mathbb{C}(X)$ to the form $\mathcal{L}^\infty(M,\mu)$. Although this expression is not unique, it has the immense advantage of involving *natural* notions: compare ℓ^∞ with $\mathbb{C}(\beta\mathbb{N})$), the space of continuous functions over a bizarre space, the Stone–Čech compactification of $\mathbb{N}$. Thus:

- The « von Neumann » spirit differs from the C^* spirit: continuity is not measurability. To disguise a measurable function as a continuous function on a warped space is possible, but dishonest.
- The « completion to the suprema » of a C^*-algebra has no intrinsic sense. Indeed, to pass from $\mathbb{C}(X)$ to $\mathcal{L}^\infty(X,\mu)$ depends upon a measure μ; but, on the sole space $[0,1]$, there are many non-equivalent diffuse measures.
- In the non-commutative case, the completion depends upon the choice of a *faithful* representation, such a thing being seldom unique. A faithful representation indeed enables us to imbed our stellar algebra in a space $\mathcal{B}(\mathbf{H})$, naturally provided with bounded suprema. The GNS construction (Section 17.A.8) constructs a representation from a state ρ. This representation is faithful when the state is itself faithful, i.e., enjoys $\rho(uu^*) = 0 \Rightarrow u = 0$. Amongst faithful states, the *traces*, such that $\rho(uv) = \rho(vu)$, are sorts of « non-commutative measures »: This establishes a link with the commutative case.
- We are actually interested in C^*-algebras with bounded directed suprema. But we are merely able to work on $\mathcal{B}(\mathbf{H})$; this is why we restrict to those algebras (the W^*-algebras) which are isomorphic – as C^*-algebras – to a subalgebra of some $\mathcal{B}(\mathbf{H})$, *the isomorphism respecting directed suprema.* Since W^*-algebras have no intrinsic definition, one falls back onto the *concrete* subalgebras of some $\mathcal{B}(\mathbf{H})$; even if they are taken up to a $*$-isomorphism. By the way, when dealing with $*$-isomorphisms of vN algebras, make sure that they are *normal*, i.e., ultraweakly continuous (*infra*); this means that they preserve directed suprema, a condition automatically fulfilled by surjective $*$-isomorphisms.

20.A.2 The predual

Definition 107 (Von Neumann algebras). A *von Neumann algebra* (or vN algebra) is a sub-C^*-algebra $\mathcal{A} \subset \mathcal{B}(\mathbf{H})$ closed under bounded directed suprema.

There are many characterisations of von Neumann algebras, e.g.:

(i) The closure in the strong topology.

(ii) The closure in the weak topology.

(iii) The equality to the *bicommutant*.

This last characterisation is the most popular. If $\mathcal{A} \subset \mathcal{B}(\mathbf{H})$ is closed under adjunction, the *commutant* $\mathcal{A}^c := \{u \in \mathcal{B}(\mathbf{H}); \forall a \in A\ au = ua\}$ is a von Neumann algebra; indeed $\mathcal{A}^c = \mathcal{A}^{ccc}$.

Example 7. $\mathcal{L}^\infty(M, \mu)$ operates by multiplication over $\mathcal{L}^2(M, \mu)$; one easily checks that it is equal to its commutant, hence to its bicommutant: this is a maximal commutative sub-algebra.

As W^*-algebras, von Neumann algebras are exactly the *dual* C^*-algebras, i.e., those isomorphic to the dual of some Banach space. The *predual* of the vN algebra $\mathcal{A}$, unique up to isomorphism, consists of the *ultraweakly* continuous forms, often styled *normal*, i.e., weakly continuous on the unit ball of $\mathcal{A}$. Indeed, a vN algebra is the norm dual of the space of weakly continuous forms, which is not complete; its completion is the space of ultraweakly continuous forms. For instance, the predual of ℓ^∞ is ℓ^1, while the weakly continuous forms on ℓ^∞ correspond to the dense subspace of almost null sequences.

Unless finite-dimensional, a vN algebra, e.g., ℓ^∞, is not separable; however, its predual, e.g., ℓ^1, may be separable. There are indeed three equivalent « separability » conditions for a vN algebra $\mathcal{A}$:

- The predual of $\mathcal{A}$ is separable.
- There is a denumerable (weakly or strongly) dense subset in $\mathcal{A}$.
- $\mathcal{A}$ is $*$-isomorphic with $\mathcal{B} \subset \mathcal{B}(\mathbf{H})$, with $\mathbf{H}$ separable.

20.A.3 Factors. If a von Neumann algebra splits as a sum $\mathcal{B} + \mathcal{C}$, the respective neutral elements of $\mathcal{B}$ and $\mathcal{C}$ are *central* projections $\mathcal{A}$. The center of $\mathcal{A}$ is trivial (i.e., equal to $\mathbb{C}I$) when its only projections are $0, I$, which means that the algebra $\mathcal{A}$ is « connected », i.e., cannot be decomposed as a sum.

Definition 108 (Factors). A *factor* is a von Neumann algebra whose center is trivial, i.e., consists in the scalar multiples of the identity.

The theory of von Neumann algebras reduces to the study of factors. Indeed, the center of a von Neumann algebra $\mathcal{A}$ is a von Neumann algebra (as the commutant of $\mathcal{A} \cup \mathcal{A}^c$), thus of the form $\mathcal{L}^\infty(M, \mu)$. $\mathcal{A}$ can be written as a sum of factors – a sort of integral – indexed by its center. When $\mathcal{A}$ is commutative, these factors are all isomorphic to $\mathbb{C}$.

Definition 109 (Comparison of projections). Between the projections of a von Neumann algebra $\mathcal{A}$, one defines the preorder relation $\preccurlyeq$, with associated equivalence $\sim$:

$$\pi \sim \pi' \iff \exists u\ (u^*u = \pi \text{ and } uu^* = \pi'), \tag{20.9}$$
$$\pi \preccurlyeq \pi' \iff \exists \pi''\ (\pi = \pi\pi'' \text{ and } \pi'' \sim \pi'). \tag{20.10}$$

Theorem 82 (Type). *If $\mathcal{A}$ is a factor, the preorder $\preccurlyeq$ is total.*

This induces a classification of factors over a separable Hilbert space:

$\mathbf{I}_n$: order type $\{0, \ldots, n\}$.

$\mathbf{I}_\infty$: order type $\mathbb{N} \cup \{+\infty\}$.

$\mathbf{II}_1$: order type $[0, 1]$.

$\mathbf{II}_\infty$: order type $[0, +\infty]$.

III: order type $\{0, +\infty\}$.

The symbol « $+\infty$ » has a special meaning; it denotes the class of the identity, when the identity is *infinite*, i.e., not alone in its equivalence class.

Type **I** matches the algebras which are $*$-isomorphic to $\mathcal{B}(\mathbf{H})$, with $\mathbf{H}$ of hilbertian dimension n or $\aleph_0$. Those algebras are of little interest, in the sense that the theory was not intended for them. The preorder $\preccurlyeq$ corresponds to the comparison of hilbertian dimensions, anyway bounded by $\aleph_0 = \infty$.

The type **III** can be further refined into $\mathbf{III}_\lambda$ $(0 \leq \lambda \leq 1)$ a subclassification obtained by Connes [15], but which is out of the scope of our present interests.

20.A.4 Tensor products. If $\mathcal{A} \subset \mathcal{B}(\mathbf{H}), \mathcal{B} \subset \mathcal{B}(\mathbf{K})$, then $\mathcal{A} \otimes \mathcal{B} \subset \mathcal{B}(\mathbf{H} \otimes \mathbf{K})$ is defined as the bicommutant of the set of simple tensors $u \otimes v, u \in \mathcal{A}, v \in \mathcal{B}$.

A tensor product of factors is still a factor, of type:

- $\mathbf{I}_m \otimes \mathbf{I}_n = \mathbf{I}_{mn}$, for $m, n \leq \infty$.
- $\mathbf{I} \otimes \mathbf{II} = \mathbf{II} \otimes \mathbf{II} = \mathbf{II}$.
- $\mathbf{I} \otimes \mathbf{III} = \mathbf{II} \otimes \mathbf{III} = \mathbf{III} \otimes \mathbf{III} = \mathbf{III}$.

The type **II** is so far the most interesting for us. Since $\mathbf{II}_\infty$ resolves as a tensor product $\mathbf{II}_1 \otimes \mathbf{I}_\infty$, $\mathbf{II}_1$ is the real novelty.

By the way, the commutant of a factor is of the same gross type (**I**, **II**, **III**).

20.B Finite algebras

20.B.1 The trace

Definition 110 (Finiteness). $\mathcal{A}$ is *finite* when I stands alone in its equivalence class: $uu^* = I \Rightarrow u^*u = I$.

The finite factors are those of type $\mathbf{I}_n (n < \infty)$ and $\mathbf{II}_1$. A tensor product of finite factors is still finite, thus:

- $\mathbf{I}_n \otimes \mathbf{II}_1 = \mathbf{II}_1 \otimes \mathbf{II}_1 = \mathbf{II}_1$ $(n < \infty)$.

In a finite factor, given projections $\pi, \pi' \neq 0$, define a « euclidian division »: $\pi = \pi'_1 + \cdots + \pi'_n + \pi"$ with $\pi'_1 \sim \cdots \sim \pi'_n \sim \pi', \pi" \preccurlyeq \pi', \pi" \not\sim \pi'$, which one writes $\pi \sim n.\pi' + \pi"$; n and the *remainder* $\pi"$ (up to $\sim$) are unique. This enables us to define the dimension of a projection by a continued fraction. Typically, π is of dimension $1/2$ when $\pi \sim I - \pi$. We must see the type $\mathbf{II}_1$ as « another » cardinality, in a sort of « pointless » setting: since every projection can be halved, we never reach the atoms! Dimension extends by linearity to linear combinations of projections, then to the full algebra by ultraweak continuity. This is the *trace*: finite algebras are algebras with a trace.

Definition 111 (Trace). In the vN algebra $\mathcal{A}$, a *trace* is an ultraweakly continuous state τ such that

$$\tau(uv) = \tau(vu).$$

The trace is thus an element of the *predual*.

Proposition 59. *A factor is finite iff it admits a trace (necessarily unique).*

For factors of type $\mathbf{I}_n$, the trace (in the sense of Definition 111) is obtained by renormalising the usual algebraic trace: $\tau(u) = 1/n \cdot \mathrm{Tr}(u)$.

20.B.2 Algebra of a discrete group. If $\mathfrak{G}$ is a discrete (i.e., finite or denumerable) group, the space of complex linear combinations of elements of $\mathfrak{G}$ is the *convolution* ring $\mathcal{A}(\mathfrak{G})$ of $\mathfrak{G}$:

$$\Big(\sum_g x_g \cdot g\Big) * \Big(\sum_h y_h \cdot h\Big) := \sum_{gh=k} x_g y_h \cdot k. \tag{20.11}$$

Here, the coefficients x_g, etc. are almost all null. The convolution product can be extended to infinite sums in two remarkable cases:

$\ell^1(\mathfrak{G})$: the convolution product sends $\ell^1(\mathfrak{G}) \times \ell^1(\mathfrak{G})$ into $\ell^1(\mathfrak{G})$.

$\ell^2(\mathfrak{G})$: the convolution product sends $\ell^2(\mathfrak{G}) \times \ell^2(\mathfrak{G})$ into $\ell^\infty(\mathfrak{G})$.

Definition 112 (Algebra of a group). The group algebra of $\mathfrak{G}$ is defined as

$$\mathcal{A}[\mathfrak{G}] := \{x \in \ell^2(\mathfrak{G})\ ;\ \forall y \in \ell^2(\mathfrak{G})\ \ x * y \in \ell^2(\mathfrak{G})\}. \tag{20.12}$$

$x \in \mathcal{A}[\mathfrak{G}]$ induces an operator on the space $\ell^2(\mathfrak{G})$; $\mathcal{A}[\mathfrak{G}]$ is thus identified with a subalgebra of $\mathcal{B}(\ell^2(\mathfrak{G}))$, indeed a vN algebra, since the commutant of the right convolutions $r_g(y) := y * g$.

The neutral element is the unit 1 of $\mathfrak{G}$, the adjoint being given by

$$\Big(\sum_g x_g \cdot g\Big)^* = \sum_g \overline{x_{g^{-1}}} \cdot g). \qquad (20.13)$$

$\mathcal{A}[\mathfrak{G}]$ admits the *trace*:

$$\mathrm{tr}\Big(\sum_g x_g \cdot g\Big) := x_1. \qquad (20.14)$$

Proposition 60. *$\mathcal{A}[\mathfrak{G}]$ is a finite algebra.*

Proof. If $uu^* = I$, then the projection $\sum_g x_g \cdot g := u^*u$ is of trace 1; but $\mathrm{tr}(\sum_g x_g \cdot g) = x_1 = \sum |x_g|^2$, hence $x_1 = 1, x_g = 0$ for $g \neq 1$. □

Definition 113 (i.c.c. groups). $\mathfrak{G}$ is with *infinite conjugacy classes* (i.c.c.) iff, for all $g \in \mathfrak{G}, g \neq 1$, the set $\{h^{-1}gh; h \in \mathfrak{G}\}$ of conjugates of g is infinite.

Proposition 61. *The algebra $\mathcal{A}[\mathfrak{G}]$ of an i.c.c. group is a type $\mathbf{II}_1$ factor.*

Proof. If $\sum_g x_g \cdot g$ is in the center of $\mathcal{A}[\mathfrak{G}]$ and $g \neq 1$, then $x_g = x_{h^{-1}gh}$ is constant on its conjugacy class; since $\sum_g x_g \cdot g \in \ell^2(\mathfrak{G})$, $x_g = 0$. □

Observe that $\ell^2(\mathfrak{G} \times \mathfrak{H}) \simeq \ell^2(\mathfrak{G}) \otimes \ell^2(\mathfrak{H})$, hence

$$\mathcal{A}[\mathfrak{G} \times \mathfrak{H}] \simeq \mathcal{A}[\mathfrak{G}] \otimes \mathcal{A}[\mathfrak{H}]. \qquad (20.15)$$

20.B.3 Crossed products. Let $\mathfrak{G}$ be a discrete group and α be an automorphic representation of $\mathfrak{G}$ on $\mathcal{A}$, i.e., a homomorphism associating to any $g \in \mathfrak{G}$ an automorphism α_g of the vN algebra $\mathcal{A} \subset \mathcal{B}(\mathbf{H})$. On the Hilbert space $\mathbf{H} \otimes \ell^2(\mathfrak{G})$, we can consider:

- For $u \in \mathcal{A}$, the operators $\tilde{\alpha}(u)(x \otimes g) := \alpha_{g^{-1}}(u)(x) \otimes g$.
- For $g \in \mathfrak{G}$ the operators $\ell_g(x \otimes h) := x \otimes gh$.

Definition 114 (Crossed product). The crossed product $\mathcal{A} \rtimes_\alpha \mathfrak{G}$ is the vN subalgebra of $\mathcal{A} \otimes \mathcal{A}[\mathfrak{G}]$ generated by (i.e., the bicommutant of) the $\tilde{\alpha}(u)$ and the ℓ_g.

We can easily check that

$$\ell_g \tilde{\alpha}(u) \ell_g^* = \tilde{\alpha}(\alpha_g(u)). \qquad (20.16)$$

The $\tilde{\alpha}(u)$ thus generate a vN algebra isomorphic with $\mathcal{A}$, and the conjugations $u \rightsquigarrow \ell_g u \ell_g^*$ act as the original α_g. In other terms, $\mathcal{A} \rtimes_\alpha \mathfrak{G}$ is the vN algebra obtained from $\mathcal{A}$ by « internalising » the α_g. A typical example is a tensor product $\mathcal{A} \otimes \mathcal{A}$, with an action of the group $\mathbb{Z}_2$ given by $\alpha(1) = \sigma$, where σ is the *twist*, $\sigma(u \otimes v) := v \otimes u$.

Proposition 62. *If $\mathcal{A}$ is a factor and the α_g are outer automorphisms for $g \neq 1$, then $\mathcal{A} \rtimes_\alpha \mathfrak{G}$ is a factor.*

If $\mathfrak{G}, \mathfrak{H}$ are denumerable groups and α is an automorphic representation of $\mathfrak{G}$ on $\mathfrak{H}$, then one defines:

Definition 115 (Semi-direct product). The semi-direct product $\mathfrak{H} \rtimes_\alpha \mathfrak{G}$ is the cartesian product $\mathfrak{H} \times \mathfrak{G}$ equipped with the group law: $(h, g)(h', g') := (h \cdot \alpha_g(h'), g \cdot g')$.

In such a situation, α induces an automorphic representation of $\mathfrak{G}$ in $\mathcal{A}[\mathfrak{H}]$, still noted α and:

Proposition 63.

$$\mathcal{A}[\mathfrak{H}] \rtimes_\alpha \mathfrak{G} \sim \mathcal{A}[\mathfrak{H} \rtimes_\alpha \mathfrak{G}]. \tag{20.17}$$

Consistently with (20.17) and Proposition 62: if $\mathfrak{H}$ is i.c.c. and the α_g are outer for $g \neq 1$, then $\mathfrak{H} \rtimes_\alpha \mathfrak{G}$ is i.c.c.

20.C Hyperfinite algebras

20.C.1 The CAR algebra

Definition 116 (Hyperfiniteness). The vN algebra $\mathcal{A}$ is *hyperfinite* if there is an increasing sequence $\mathcal{A}_n$ of finite-dimensional subalgebras such that a is the closure (weak, strong, or the bicommutant) of the union $\bigcup_n \mathcal{A}_n$.

If $n = mk$, one can embed $\mathcal{M}_m(\mathbb{C})$ into $\mathcal{M}_n(\mathbb{C})$ by replacing each coefficient of an $m \times m$ matrix with a $k \times k$ diagonal matrix; which can be pedantically expressed by means of the isomorphism $\mathcal{M}_n(\mathbb{C}) \sim \mathcal{M}_m(\mathbb{C}) \otimes \mathcal{M}_k(\mathbb{C})$. A sequence of integers $n_0 \mid n_1 \mid \ldots \mid n_i \mid \ldots$, each of them dividing the next one, thus defines an inductive system of C^*-algebras; its direct limit is characterised, up to isomorphism by the « exponent », finite or infinite, of each prime number in the family (n_i), see [64]. In particular, if $n_i = 2^i$, one gets the *CAR algebra*.

Definition 117 (CAR algebra). The CAR algebra is defined as the direct limit of the matrix algebras $\mathcal{M}_{2^n}(\mathbb{C})$.

Since matrix algebras are simple, the embeddings $\mathcal{M}_m(\mathbb{C}) \hookrightarrow \mathcal{M}_{mk}(\mathbb{C})$ are isometric (Section 17.A.8), hence the algebraic direct limit is naturally equipped with a C^* norm; its completion is the CAR algebra.

Alternative presentation: given a Hilbert space $\mathbf{H}$, consider the C^* algebra $\mathbf{CAR}(\mathbf{H})$ generated by the *creators* $\kappa(a)(a \in \mathbf{H})$ and their adjoints, the *annihilators* $\zeta(a)(a \in \mathbf{H})$, subject to the *canonical anticommutation relations*:

$$\kappa(a)\zeta(b) + \kappa(b)\zeta(a) = \langle a \mid b \rangle I, \tag{20.18}$$

$$\kappa(a)\kappa(b) + \kappa(b)\kappa(a) = 0, \tag{20.19}$$

$$\zeta(a)\zeta(b) + \zeta(b)\zeta(a) = 0. \tag{20.20}$$

$\mathbf{CAR}(\mathbb{C}^n)$ is of dimension 2^{2n}, the same as $\mathcal{M}_{2^n}(\mathbb{C})$, to which it is indeed isomorphic, see [64], exercise 10.5.88. The CAR algebra of Definition 117 thus appears as $\mathbf{CAR}(\ell^2)$.

Given a state ρ on the CAR algebra, the GNS construction (Section 17.A.8) provides a representation, hence a completion of the algebra into a vN algebra which is thus hyperfinite. Depending on ρ, one gets various non-isomorphic factors (one in each type $\mathbf{I}_n$, $\mathbf{I}_\infty$, $\mathbf{II}_1$, $\mathbf{II}_\infty$, $\mathbf{III}_\lambda$ $(0 < \lambda \leq 1)$ and infinitely many of type $\mathbf{III}_0$). The $\mathcal{M}_{2^n}(\mathbb{C})$ thus form a direct system of vN algebras without a direct limit.

20.C.2 The hyperfinite factor. Among all states, one is more natural than the others: the renormalised traces $\tau_{2^n}(a) := 2^{-n}\mathrm{Tr}(a)$ admit as direct limit a state τ of the CAR algebra. The vN algebra corresponding to this state τ is thus finite-while infinite-dimensional, i.e., of type $\mathbf{II}_1$. A celebrated result is the following ([64], theorem 12.2.1):

Theorem 83 (Murray–von Neumann). *Up to isomorphism, there is only one hyperfinite factor of type* $\mathbf{II}_1$.

Definition 118 (Hyperfinite factor). The *hyperfinite factor* $\mathcal{R}$ is the unique hyperfinite factor of type $\mathbf{II}_1$.

This means that the default type for a hyperfinite factor (which can indeed be of any type) is $\mathbf{II}_1$.

20.C.3 Amenable groups. Among[7] the many characterisations of hyperfiniteness, the most important is due to Connes:

Theorem 84 (Injectivity). *A vN algebra* $\mathcal{A}$ *is hyperfinite iff it is* injective, *i.e., if there is a linear projection* Π *of norm* 1 *of* $\mathcal{B}(\mathbf{H})$ *onto* $\mathcal{A}$.

Proposition 64 (Tomiyama, 1957). *If* Π *is a linear projection of* $\mathcal{B}(\mathbf{H})$ *onto* $\mathcal{A}$ *such that* $\|\Pi(u)\| \leq \|u\| (u \in \mathcal{B}(\mathbf{H}))$, *then* Π *is a* conditional expectation, *i.e.:*

(i) Π is positive: $\Pi(u) \geq 0$ when $u \geq 0$.

(ii) $\Pi(I) = I$.

(iii) If $a, b \in \mathcal{A}$, $u \in \mathcal{B}(\mathbf{H})$, then $\Pi(aub) = a\Pi(u)b$.

Coming back to group algebras, $\mathcal{A}[\mathfrak{G}]$ is injective iff $\mathfrak{G}$ is *amenable*:

[7] Many thanks to Georges Skandalis for his invaluable help!

Definition 119 (Amenability). An *invariant mean* on $\mathfrak{G}$ is a state on $\ell^\infty(\mathfrak{G})$ which is left invariant:

$$\mu\Big(\sum_g x_g \cdot g\Big) = \mu\Big(\sum_g x_g \cdot hg\Big). \tag{20.21}$$

$\mathfrak{G}$ is *amenable* iff it admits an invariant mean.

Amenability is remarkably stable; amenable groups do include:

Finite groups: take the mean proper, $\mu(s) := 1/\sharp(\mathfrak{G}) \cdot \sum s(g)$.

Commutative groups: typically $\mathbb{Z}$.

Directed unions: if $\mathfrak{G} = \bigcup \mathfrak{G}_n$ is the union of an increasing family of amenable groups, then $\mathfrak{G}$ is amenable.

Subgroups and quotients: let $\mathfrak{H} \subset \mathfrak{G}$ be two groups; if $\mathfrak{G}$ is amenable, so is $\mathfrak{H}$, as well as $\mathfrak{G}/\mathfrak{H}$ if $\mathfrak{H}$ is distinguished. Conversely, if $\mathfrak{H}$ is distinguished, if $\mathfrak{H}$ and $\mathfrak{G}/\mathfrak{H}$ are amenable, so is $\mathfrak{G}$. In particular:

Direct sum: $\mathfrak{G} \oplus \mathfrak{H}$ is amenable iff $\mathfrak{G}$ and $\mathfrak{H}$ are amenable.

Semi-direct product: $\mathfrak{H} \times \{1\}$ is a distinguished subgroup of $\mathfrak{H} \rtimes_\varphi \mathfrak{G}$, moreover $\mathfrak{H} \rtimes_\varphi \mathfrak{G}/\mathfrak{H} \times \{1\}$ is isomorphic to $\mathfrak{G}$. Hence $\mathfrak{H} \rtimes_\varphi \mathfrak{G}$ is amenable when $\mathfrak{G}$ and $\mathfrak{H}$ are amenable.

Remark 6. The typical example of a non-amenable group is that of a free group $\mathcal{F}_2$ with two generators. Let[8] μ be an invariant mean, let a, b be the two generators and S be the set of reduced words beginning with a non-zero power of b; then $\mathcal{F}_2 = S \cup bS$, hence $\mu(S) \geq 1/2$, while S, aS, a^2S are disjoint, hence $\mu(S) \leq 1/3$. However:

Proposition 65. *There exists an i.c.c. amenable group containing a copy of the free* monoid *with two generators.*

Proof. Let $\mathfrak{H} := \mathbb{Z}^{\mathbb{Z}}$, the sum of denumerably many copies of $\mathbb{Z}$, which is amenable as a commutative group. Let $\mathfrak{G} := \mathbb{Z}$ with the automorphic representation $\alpha_n((x_m)) := (x_{m+n})$. Then $\mathtt{1} := ((\delta_{m0}), 0)$ (with $\delta_{00} = 1$, $\delta_{m0} = 0$ $(m \neq 0)$), $\mathtt{r} := ((0_m), 1)$ generate a free monoid. Indeed, $\mathtt{r}^{\mathtt{x}_0}\mathtt{1}\dots\mathtt{1}\mathtt{r}^{\mathtt{x}_k} = ((x_i), k)$, with $x_i = 0$ when $i \notin \{0, \dots, n\}$. □

Amenability is indeed equivalent to the existence of a Følner sequence, i.e., of an increasing family X_n of finite subsets of $\mathfrak{G}$ such that, for any $g \in \mathfrak{G}$,

$$\lim_{n\to\infty} \sharp(gX_n \setminus X_n)/\sharp(X_n) = 0. \tag{20.22}$$

The most natural example of a Følner sequence is that of an increasing sequence of finite subgroups:

[8] Exercise taken from [64], 8.7.30.

Definition 120 (Local finiteness). $\mathfrak{G}$ is *locally finite* when any finite subset of $\mathfrak{G}$ generates a finite subgroup.

The typical example of a locally finite group consists of the permutations of $\mathbb{N}$ which leave all but a finite number of points unchanged.

Given a Følner sequence and an ultrafilter $\mathfrak{U}$, one can define an invariant mean by $\mathfrak{U}$, $\mu(f) := \lim_{\mathfrak{U}} \sum_{g \in X_n} f(g)/\sharp(X_n)$. This shows the status of invariant means (and conditional expectations): just a convenient way to speak of Følner sequences (or approximation by finite-dimensional algebras). The price for this convenience is an unbridled use of the axiom of choice, but this use is so little involved that it reduces to a pure *façon de parler*.

20.D The determinant

20.D.1 Determinant vs. trace. We have already observed (Section 17.6.1) that *quantum coherent spaces* do not step out of type $\mathbf{I}_n$: the type $\mathbf{I}_\infty$ admits no trace and, as to type $\mathbf{II}_1$, it does not even allow the interpretation of the identity axiom! However, it is possible to work in type $\mathbf{II}_1$, provided one replaces the trace with the *determinant*: instead of $\mathrm{tr}(uv)$, one uses $\det(I - uv)$.

In general, $\det((a_{ij})) := \sum_\sigma (-1)^\sigma a_{1\sigma(1)} \dots a_{n\sigma(n)}$ sums up all « exhaustive travels » in the « graph » represented by a; and $\det(I + a)$ sums up all travels, exhaustive or not.

Given a Hilbert space $\mathbf{H}$, the (antisymmetric) *Fock space* $\Lambda\mathbf{H}$ is defined as the direct sum $\bigoplus_n \Lambda_n\mathbf{H}$, with $\Lambda_0\mathbf{H} := \mathbb{C}$, $\Lambda_1\mathbf{H} := \mathbf{H}$; for $n \geq 2$, $\Lambda_n\mathbf{H}$ is the separation/completion of the n-fold algebraic tensor power of $\mathbf{H}$ (the symbol « $\otimes$ » being replaced with « $\wedge$ ») w.r.t. the unique sequilinear form such that

$$\langle x_1 \wedge \dots \wedge x_n \mid y_1 \wedge \dots \wedge y_n \rangle := \det(\langle x_i \mid y_j \rangle). \tag{20.23}$$

If $\mathbf{H}$ admits the basis $\{\mathbf{e}_i ; 1 \leq i \leq n\}$, $\Lambda\mathbf{H}$ is of dimension 2^n and admits the basis $\{\bigwedge_{i \in I} \mathbf{e}_i \; ; \; I \in \wp(\{1, n\})\}$. In particular, assuming a triangular w.r.t. an orthonormal basis $\mathbf{e}_1, \dots, \mathbf{e}_n$ of $\mathbf{H}$, with diagonal entries $\lambda_1, \dots, \lambda_n$, one sees that $\det(I + a) = (1 + \lambda_1) \dots (1 + \lambda_n)$, hence

$$\mathrm{tr}(\Lambda(a)) = 1 + \sum_{1 \leq i \leq n} \lambda_i + \sum_{1 \leq i < j \leq n} \lambda_i \lambda_j + \dots + \lambda_1 \dots \lambda_n = \det(I + a). \tag{20.24}$$

Thus, $\det(I - uv) = \mathrm{tr}(\Lambda(-uv)) = \mathrm{tr}(\Lambda(iu)\Lambda(iv))$. The Fock space is indeed the space of all travels, exhaustive or not, that one can perform in $\mathbf{H}$. *Modulo* the replacement $u \rightsquigarrow \Lambda iu$, the determinant reduces to the trace, which thus appears as its *explicitation*.

This is yet another occurrence of the reduction of the potential to a catalogue, a *list of possibilities*, in the (old-)fashion of those debilitating Kripke models. But immanent justice strikes back: this replacement *diverges*, for deep reasons linked to the passage at the limit in von Neumann algebras. The equation $\mathrm{tr}(\Lambda(a)) = \det(I + a)$ can only subsist in a type $\mathbf{I}_\infty$ algebra, where precisely the trace no longer exists; indeed, the factor $\mathcal{B}(\ell^\infty)$ of type $\mathbf{I}_\infty$ is *semi-finite*: one can define a trace for the sole *trace-class* operators. But the typical operators of GoI, being partial isometries, are not of trace class.

Reminder: if $\mathbf{H}$ is separable and $u \in \mathcal{B}(\mathbf{H})$, $u \geq 0$, the quantity

$$\mathrm{Tr}(u) := \sum_{n\in\mathbb{N}} \langle u(\mathbf{e}_n) \mid \mathbf{e}_n \rangle \tag{20.25}$$

is an element of $[0, +\infty]$ not depending upon the orthonormal basis $\{\mathbf{e}_n \ ; \ n \in \mathbb{N}\}$.

Definition 121 (Trace-class operators). $u \in \mathcal{B}(\mathbf{H})$ is of *trace class* if $\mathrm{Tr}(\sqrt{uu^*}) < +\infty$, in case one defines $\mathrm{Tr}(u) \in \mathbb{C}$ by (20.25), the choice of the orthogonal base $\{\mathbf{e}_n \ ; \ n \in \mathbb{N}\}$ still being irrelevant.

20.D.2 The Fuglede–Kadison determinant. In what follows, $\mathcal{A}$ is a finite factor, thus admitting a unique trace $\mathrm{tr}(\cdot)$.

Theorem 85 (Fuglede & Kadison, [29]). *If $u \in \mathcal{A}$ is invertible, define*

$$\det(u) := e^{\mathrm{tr}(\log(|u|))}. \tag{20.26}$$

The determinant thus defined is multiplicative, monotonic and commutes to directed infima. The determinant can then be extended to the full $\mathcal{A}$ and is still multiplicative, monotonic and commuting to directed infima.

Proof. One easily restricts to the unit ball. Then $|u| := \sqrt{u^*u}$ and $|u|^2 = I - a$, with $a \geq 0$ and $\|a\| < 1$. Then $\det(I - a) = e^{-\mathrm{colog}(I-a)}$, with

$$\mathrm{colog}(I - a) := a + a^2/2 + a^3/3 + \cdots . \tag{20.27}$$

If $v \geq 0$, $\|v\| < 1$ and $v^*v = I - b$, then $\det(v^*u^*uv) = \det(I - (b + v^*av))$. Then

$$\det(v^*u^*uv) = \det(I - a)\det(I - b) \tag{20.28}$$

can be established using the power series expansion for $\|a\|$ small enough, relying on the sole property $\mathrm{tr}(xy) = \mathrm{tr}(yx)$; one informally justifies it by the well-known formula, valid in finite dimension n:

$$\det(u) = |\mathrm{Det}(u)|^{1/n}. \tag{20.29}$$

From the equation $\det(v^*(I - \lambda a)v) = \det(I - \lambda a)\det(I - b)$ which holds for λ small enough, one gets (20.28) by analytic continuation. (20.28) yields the multiplicativity of the alternative definition $\det^2(u) := \det(u^*u)$ of the determinant. Replacing u, v with $(u^*u)^{1/4}$, we get $\det^2(u) = \det(u)^2$, hence the multiplicativity of det.

If $0 \leq a \leq I(1-\epsilon)$, then $\operatorname{colog}(I-a) \geq 0$ and $\det(I-a) \leq 1$. If $0 \leq u \leq v$, then $u = w^*vw$ for some w, $\|w\| \leq 1$; $\det(u) = \det(v)\det(w^*w) \leq \det(v)$, hence the monotonicity of det. If $u = \inf_i u_i$ is the supremum of a directed system of positive hermitians, then u is the strong limit of the u_i (Proposition 58) and similarly, due to the strong continuity of the product on balls, u^n is the strong limit of the u_i^n; hence $\operatorname{colog}(I-u)$ is the strong limit of the $\operatorname{colog}(I-u_i)$; since the trace is normal, $\det(I-u) = \lim_i \det(I-u_i) = \inf_i \det(I-u_i)$.

The unique extension commuting to directed infima is such that, for $0 \leq a \leq I$:

$$\operatorname{colog}(\det(I-a)) = \operatorname{tr}(a) + \operatorname{tr}(a^2)/2 + \operatorname{tr}(a^3)/3 + \cdots \in [0, +\infty] \tag{20.30}$$

Hence $\det(u) = \det(u^*)$, $\det(uu^*) = \det(u^*u)$. If $a, b \geq 0$ and b is invertible,

$$\begin{aligned}\det(\sqrt{b}a\sqrt{b}) &= \inf_{\epsilon \to 0} \det(\sqrt{b}(\epsilon I + a)a\sqrt{b}) \\ &= \det(b) \inf_{\epsilon \to 0} \det(\epsilon I + a) \\ &= \det(a)\det(b).\end{aligned}$$

From $\det(uu^*) = \det(u^*u)$, we get $\det(\sqrt{a}b\sqrt{a}) = \det(a)\det(b)$. The same argument, now applied to $a, b \geq 0$, yields $\det(\sqrt{a}b\sqrt{a}) = \det(a)\det(b)$ in full generality. $\det^2(u) := \det(u^*u)$ is therefore multiplicative; in particular, $\det^2(u) = \det(u)^2$, as before, etc. □

Chapter 21
Finite GoI

The essential reference for this chapter is [55].

21.1 Projects

21.1.1 Associativity.

Definition 122 (Closed cut-system). A cut-system $(\mathbf{H}, h, \sigma)$ is *closed* when $\sigma^2 = I$.

The normal form $(\mathbf{0}, 0, 0)$ of a closed system hardly makes sense; however, if $\mathcal{A} \subset \mathcal{B}(\mathbf{H})$ is a finite factor, one can contend that the actual normal form of the cut-system $(\mathbf{H}, h, \sigma)$ is $\det(\sigma - h)$. Because the determinant associates with the normal form:

Theorem 86 (Associativity). *The normal form* (*Section* 19.4) *associates with the determinant: if* σ, τ *are independent and* $(\mathbf{H}, h, \sigma + \tau)$ *is closed, then, with* $\mathbf{S} := \sigma^2$, $\mathbf{T} := \tau^2$,

$$\det(\sigma + \tau - u) = \det(\mathbf{S} + \tau - \sigma[\![u]\!]) \cdot \det(\sigma + \mathbf{T} - \mathbf{S}u\mathbf{S}). \tag{21.1}$$

Proof. We first establish a few facts about determinants:

Lemma 86.1. (i) $\det(u) \in \{0, 1\}$ *when* u *is a partial isometry;* $\det(u) = 1$ *iff* u *is unitary.*

(ii) $\det(u^*) = \det(u)$.

(iii) $\det(I - uv) = \det(I - vu)$.

(iv) $\det(I - u) = 1$ *when* u *is nilpotent.*

Proof. (i) Use (20.30).

(ii) Used in the proof of Theorem 85.

(iii) If $v = \sigma$ is invertible, then one has $\det(I - u\sigma) = \det(\sigma(I - u\sigma)\sigma^{-1}) = \det(I - \sigma u)$. If $v = h$ is hermitian, choose $\alpha \in \mathbb{C} \setminus \mathbb{R}$ such that $I + \alpha u$ is invertible; since $-\alpha \notin \mathrm{Sp}(h)$, $\alpha I + h$ is invertible too. Then $I - uh = (I + \alpha u) - u(\alpha I + h)$ and $\det(I - uh) = \det(I + \alpha u)\det(I - (I + \alpha u)^{-1}u(\alpha I + h))$; using the previous case (with $\sigma := \alpha I + h$) and the commutation between u and $(I + \alpha u)^{-1}$, we get $\det(I - uh) = \det(I + \alpha u)\det(I - (\alpha I + h)u(I + \alpha u)^{-1})$. Hence $\det(I - uh) = \det((I - (\alpha I + h)u(I + \alpha u)^{-1})(I + \alpha u)) = \det(I - hu)$. In general, write v as the product hs of a hermitian and a partial isometry s; if t is any partial isometry from $I - s^*s$ to $I - ss^*$, then $\sigma := s + t$ is unitary and $v = h\sigma$. Then $\det(I - uh\sigma) = \det(I - \sigma uh) = \det(I - h\sigma u)$.

(iv) Let π be the projection of the closure of the range of u; then $\det(I - u) = \det(I - \pi u) = \det(I - u\pi)$. If $u^2 = 0$, then $u\pi = 0$ and we are done; otherwise, redo the same thing with $u\pi$, etc. □

Observe that $(\sigma + \tau - u)(I - \mathbf{Z}) = \sigma + \tau - u$, $(\sigma + \mathbf{T} - \mathbf{S}u\mathbf{S})(I - \mathbf{Z}) = \sigma + \mathbf{T} - \mathbf{S}u\mathbf{S}$, where $\mathbf{Z}$ is the *deadlock* (Section 19.3.4); if $\sigma - \mathbf{S}u\mathbf{S}$ is not injective, then, by (i) of the lemma, $\det(I - \mathbf{Z}) = 0$, hence both sides of (21.1) are null.

If, w.r.t. the block decomposition $I = \mathbf{S} + \mathbf{T}$, $u = \left(\begin{smallmatrix} a & b^* \\ b & c \end{smallmatrix}\right)$, then

$$\sigma + \tau - u = \begin{pmatrix} \sigma - a & -b^* \\ -b & \tau - c \end{pmatrix}, \quad \mathbf{S} + \tau - \sigma[\![u]\!] = \begin{pmatrix} \mathbf{S} & 0 \\ 0 & \tau - \sigma[\![u]\!] \end{pmatrix},$$
$$\sigma + \mathbf{T} - \mathbf{S}u\mathbf{S} = \begin{pmatrix} \sigma - a & 0 \\ 0 & \mathbf{T} \end{pmatrix}.$$

Assuming $\sigma \geq 0$, i.e., $\sigma = \pi$ and $\pi - a$ injective, then

$$\begin{pmatrix} \pi - a & -b^* \\ -b & \tau - c \end{pmatrix} = \begin{pmatrix} \sqrt{\pi - a} & 0 \\ 0 & \mathbf{T} \end{pmatrix} \begin{pmatrix} \pi & 0 \\ \overline{-b(\pi - a)^{-1/2}} & \mathbf{T} \end{pmatrix} \begin{pmatrix} \pi & 0 \\ 0 & \tau - \sigma[\![u]\!] \end{pmatrix}$$
$$\begin{pmatrix} \pi & -(\pi - a)^{-1/2} b^* \\ 0 & \mathbf{T} \end{pmatrix} \begin{pmatrix} \sqrt{\pi - a} & 0 \\ 0 & \mathbf{T} \end{pmatrix}.$$

From the multiplicativity of the determinant and the fact that triangular matrices are of the form $I - u$ with u nilpotent, we get (21.1); $\overline{-b(\pi - a)^{-1/2}}$, the closure of $-b(\pi - a)^{-1/2}$ is the adjoint of the bounded operator $-(\pi - a)^{-1/2} b^*$.

The same holds if $\sigma = -\nu \leq 0$. The full case follows from the associativity of the normal form (Theorem 79):

$$\begin{aligned} \det(\sigma + \tau - u) &= \det(\pi - \nu + \tau - \pi[\![u]\!]) \cdot \det(I - \pi u \pi) \\ &= \det(\mathbf{S} + \tau - \sigma[\![u]\!]) \cdot \det(\sigma + \mathbf{T} - \nu\pi[\![u]\!]\nu) \cdot \det(I - \pi u \pi) \\ &= \det(\mathbf{S} + \tau - \sigma[\![u]\!]) \cdot \det(\sigma + \mathbf{T} - \mathbf{S}u\mathbf{S}) \end{aligned}$$

together with the lopsided case already treated. □

21.1.2 The adjunction. We replace $\det(\cdot)$ with its cologarithm $\mathrm{ldet}(\cdot)$; when $\|u\| < 1$ and $u = u^*$:

$$\mathrm{ldet}(I - u) = \mathrm{tr}(u) + \mathrm{tr}(u^2)/2 + \mathrm{tr}(u^3)/3 + \cdots. \tag{21.2}$$

In general, if $\|u\| < 1$, then $\mathrm{ldet}(I - u)$ is not real, thus fails to be the cologarithm of $\det(I - u)$. However, if $\|u\|, \|v\| < 1$ and u, v are hermitian, then $\mathrm{ldet}(I - uv) =$

$\mathrm{ldet}(I - vu) = \mathrm{ldet}(I - uv)^*$, hence $\mathrm{ldet}(I - uv) \in \mathbb{R}$; furthermore, if $u, v \geq 0$, then $\mathrm{ldet}(I - uv) = \mathrm{ldet}(I - \sqrt{v}u\sqrt{v}) \geq 0$. In particular, when $(\mathbf{H}, \sigma, u)$ is closed and invertible, then

$$\mathrm{ldet}(\sigma - u) = \mathrm{tr}(u\sigma) + \mathrm{tr}((u\sigma)^2)/2 + \mathrm{tr}((u\sigma)^3)/3 + \cdots . \tag{21.3}$$

Let us reformulate associativity (Theorem 86) in the context of the application of a « function » to an argument. We want to relate $[F]A \cdot B$ with $F(A + B)$ in a way analogous to the fundamental adjunction of QCS (Section 17.5.1):

$$\mathrm{tr}(((\Phi)f) \circ g) = \mathrm{tr}(\Phi \circ (f \otimes g)). \tag{17.26}$$

Assume that, *modulo* a block decomposition $I = \boldsymbol{a} + \boldsymbol{b}$,

$$F := \begin{pmatrix} F_{11} & F_{12} \\ F_{21} & F_{22} \end{pmatrix}, \quad G := \begin{pmatrix} A & 0 \\ 0 & B \end{pmatrix} (= A + B).$$

Consider the normal form $[F]A$ of Section 19.4); when $I - F_{11}A$ is invertible,

$$[F]A := F_{22} + F_{21}(I - F_{11}A)^{-1} \cdot F_{12}\,. \tag{21.4}$$

The associativity of the cut-system $(\mathbf{H} \oplus \mathbf{H}, F \oplus G, \left[\begin{smallmatrix} 0 & \boldsymbol{a} \\ \boldsymbol{a} & 0 \end{smallmatrix}\right])$ is rephrased as

$$\mathrm{ldet}\begin{pmatrix} \boldsymbol{a} - F_{11}A & -F_{12}B \\ -F_{21}A & \boldsymbol{b} - F_{22}B \end{pmatrix} = \mathrm{ldet}(I - [F]A \cdot B) + \mathrm{ldet}(I - FA) \tag{21.5}$$

(observe that $\mathrm{ldet}(I - FA) = \mathrm{ldet}(I - F_{11}A)$.)

Compared to (17.26), notice the additional term $\mathrm{ldet}(I - FA)$ which makes the equation non-homogeneous; in practice, FA is often nilpotent, which may explain why this term has no analogue in (17.26). This additional term – or rather its absence in (17.26) – may also explain why the QCS paradigm does not generalise to infinite dimension.

In order to obtain a satisfactory adjunction, one must homogenise: instead of an operator, one introduces the pair of a *wager* $w \in]-\infty, +\infty]$, the set of possible values for the cologarithm of a positive real, and an operator, notation $w + U$. Define $[f + F](a + A) := f + a + \mathrm{ldet}(I - F) + [F]A$, then

$$\begin{aligned}(a + b) + f + \mathrm{ldet}(I - F(A + B)) \\ = (f + a + \mathrm{ldet}(I - FA)) + b + \mathrm{ldet}(I - [F]A \cdot B);\end{aligned} \tag{21.6}$$

thus, defining $\ll c + C \mid d + D \gg := c + d + \mathrm{ldet}(I - CD)$:

Theorem 87 (Adjunction). *The application $[f + F](a + A)$ is characterised by*

$$\ll f + F \mid (a + b) + (A + B) \gg \; = \; \ll [f + F](a + A) \mid b + B \gg . \tag{21.7}$$

Proof. (21.7) is Theorem 86. It remains to show that $d + D \rightsquigarrow \ll c + C \mid d + D \gg$ determines $d + D$. First, $\ll c + C \mid 0 + 0 \gg = c$ determines c; then, $D \rightsquigarrow \mathrm{ldet}(I - CD)$ determines C: indeed, since $\mathrm{ldet}(I - (\lambda C)D) = \lambda(\mathrm{tr}(CD) + o(\lambda))$, $D \rightsquigarrow \mathrm{ldet}(I - CD)$ determines $D \rightsquigarrow \mathrm{tr}(CD)$. The latter dependency is linear; if $\mathrm{tr}(CD) = 0$ for all D, then $\mathrm{tr}(D^2) = 0$, hence $D = 0$ by the faithfulness of the trace. □

The adjunction (21.6) can thus be used as an abstract definition of the normal form in finite factors.

21.1.3 Traces and determinants. Trace and determinant make sense in any *finite* vN algebra. Three properties have been used:

Cyclicity: $\mathrm{tr}(uv) = \mathrm{tr}(vu)$ yields the mutiplicativity of the trace.

Positivity: $\mathrm{tr}(uu^*) \geq 0$, subsumed by $\mathrm{tr}(I) = 1$, yields the monotonicity of the trace.

Normality: ultraweak continuity yields the extension to directed infima.

In a finite algebra, there are non-zero elements of the predual which are positive and cyclic. Indeed, the most general notion of trace for a finite algebra is that of a *central trace*, [64] chapter 8: a cyclic and normal *conditional expectation* (Section 20.C.3) from $\mathcal{A}$ onto its center. In the particular case of an algebra with a finite-dimensional center, which we can write $\mathcal{A} = \bigoplus_{i \in I} \mathcal{A}_i$, I being the set of minimal projections of $\mathcal{A}$, so that $\mathcal{A}_i := i\mathcal{A}i$, the central trace associates to $u \in \mathcal{A}$ the element $\sum \mathrm{tr}_{a_i}(iui) \cdot i$ of the center.

Any normal and cyclic form on $\mathcal{A}$ is written as $\varphi(u) = f(\mathrm{tr}(u))$, where f is a linear form on the center of $\mathcal{A}$: if $\mathcal{A} = \bigoplus_{i \in I} \mathcal{A}_i$, $\varphi(u) = \sum f_i \mathrm{tr}_{a_i}(iui)$.

For reasons that find their origin in ludics, especially the half-baked *bihaviours*, Section 14.B.3, it is important to consider non-positive traces, i.e., to replace *positivity* with the weaker:

Hermiticity: $\mathrm{tr}(u) = \mathrm{tr}(u^*)$.

Definition 123 (Pseudo-trace). If $\mathcal{A}$ is a finite vN algebra, a *pseudo-trace* is an element α of the predual of $\mathcal{A}$, which is hermitian, cyclic, faithful (see *infra*)) and such that $\alpha(I) \neq 0$.

Faithfulness generalises the notion of a faithful state: if α is hermitian, then $\mathcal{A}$ splits into a direct sum $\mathcal{A} = \mathcal{A}_- \oplus \mathcal{A}_0 \oplus \mathcal{A}_+$, with $\alpha(uu^*) < 0, = 0, > 0$ for $u \neq 0, u \in \mathcal{A}_-, \mathcal{A}_0, \mathcal{A}_+$; α is *faithful* when $\mathcal{A}_0 = 0$. There is a small problem with the determinant: if u splits as $u_- \oplus u_+$, then $\det(u) := \det(u_-)\det(u_+)$ can take the undetermined value $(+\infty)0$. Indeed the sign of $\alpha(I)$ determines this ambiguous case while staying multiplicative. In terms of cologarithms, $\mathrm{ldet}(u) := \mathrm{ldet}(u_-) + \mathrm{ldet}(u_+)$, with

$$-\infty + (+\infty) = +\infty \quad (\text{if } \alpha(I) > 0),$$
$$-\infty + (+\infty) = -\infty \quad (\text{if } \alpha(I) < 0).$$

Proposition 66. *If the finite vN algebras $\mathcal{A}, \mathcal{B}$ are equipped with pseudo-traces α, β, then*

$$\mathrm{ldet}_{\lambda\alpha}(I - u) = \lambda \mathrm{ldet}_\alpha(I - u), \tag{21.8}$$
$$\mathrm{ldet}_{\alpha\oplus\beta}(I - u \oplus v) = \mathrm{ldet}_\alpha(I - u) + \mathrm{ldet}_\beta(I - v), \tag{21.9}$$
$$\mathrm{ldet}_\beta(I - \varphi(u)) = \mathrm{ldet}_\alpha(I - u), \tag{21.10}$$
$$\mathrm{ldet}_{\alpha\otimes\beta}(I - (u \otimes \pi)) = \mathrm{ldet}_\alpha(I - u) \cdot \beta(\pi). \tag{21.11}$$

With $u \in \mathcal{A}, v \in \mathcal{B}$; in (21.8) *$\lambda \in \mathbb{R}$, in* (21.11) *$\pi$ is a projection of $\mathcal{B}$, in* (21.10) *φ is a* normal *$*$-isomorphism from $\mathcal{A}$ to $\mathcal{B}$ such that $\alpha = \beta \circ \varphi$.*

Proof. Obvious. In (21.10), φ need not be unital, i.e., enjoy $\varphi(I_{\mathcal{A}}) = I_{\mathcal{B}}$: we only use $\alpha(I_{\mathcal{A}})\beta(I_{\mathcal{B}}) > 0$. Modulo this remark, (21.11) follows from (21.8) and (21.10), using $\varphi(u) := (\beta(\pi))^{-1} \cdot u \otimes \pi$. □

21.1.4 Idioms. GoI is now *idiomatic*; operators dwell in tensor products of the form $\mathcal{R} \otimes \mathcal{D}$, where $\mathcal{D}$ is a finite-dimensional algebra and $\mathcal{R}$ is the hyperfinite factor. When relating two operators through a tensor or a cut, the idioms must be tensorised: from $A \in \mathcal{R} \otimes \mathcal{A}$ and $B \in \mathcal{R} \otimes \mathcal{B}$, we form $A^\ddagger, B^\dagger \in \mathcal{R} \otimes (\mathcal{A} \otimes \mathcal{B})$ in the obvious way, see (20.4, 20.5). Moreover, the idioms come with *pseudo-traces* (Section 21.1.3): if A, B are given with pseudo-traces α, β, then $\mathcal{A} \otimes \mathcal{B}$ is equipped with the pseudo-trace $\alpha \otimes \beta$.

Moreover, operators are given together with wagers: for reasons of homogeneity, when changing the idioms, $a + A, b + B$ must be replaced with $a \cdot \beta(I_{\mathcal{B}}) + A^\ddagger$, $b \cdot \alpha(I_{\mathcal{A}}) + B^\dagger$. Which explains the restriction $\alpha(I) \neq 0$ on pseudo-traces. By the way, the final restriction on wagers is that $a \in \mathbb{R} \cup \{\alpha(I_{\mathcal{A}}) \cdot \infty\}$.

Of course, we could have followed the alternative way, and formed the idiom $\mathcal{B} \otimes \mathcal{A}$, with a strictly isomorphic result; indeed, the canonical $*$-isomorphism $\varphi\colon \mathcal{R} \otimes (\mathcal{A} \otimes \mathcal{B}) \mapsto \mathcal{R} \otimes (\mathcal{B} \otimes \mathcal{A})$, combined with Proposition 66 (21.10) shows that $\mathrm{ldet}(I - A^\ddagger B^\dagger) = \mathrm{ldet}(I - B^\ddagger A^\dagger)$. This last formula is very hard to read – not to speak of writing it! Although its contents is rather trivial: a common idiom has been created by tensorisation, period.

Remember that idioms are, so to speak, the *bound variables* (Section 20.1.4) of GoI. In logic, a bureaucratic discipline called *α-conversion* and specially devoted to the handling of bound variables, has been introduced. α-conversion is so boring, so devoid of interest, that I didn't pay any attention to it within this book, thus writing

$(\lambda xx)\lambda xx$ instead of the correct $(\lambda xx)\lambda yy$. I therefore propose to do the same with idioms, thus ignoring the superscripts $A^{\ddagger}$, $B^{\dagger}$. But we need first to indulge in some α-conversion, GoI-style!

Definition 124 (Projects). Let $\mathcal{R}$ be the hyperfinite factor and let $\mathcal{A}$ be a finite dimensional vN algebra. A *project* $\mathfrak{a} = a \cdot\!+\!\cdot \alpha + A$ of *idiom* $\mathcal{A}$ consists in the following data:

- A pseudo-trace α on $\mathcal{A}$.
- A « real » number $a \in \mathbb{R} \cup \{\alpha(I_{\mathcal{A}}) \cdot \infty\}$, the *wager*.
- A hermitian operator $A \in \mathcal{R} \otimes \mathcal{A}$ of norm ≤ 1, the *plot*.

The notation $a \cdot\!+\!\cdot \alpha + A$ is incorrect: it mentions neither $\mathcal{A}$, which is however determined as the source space of α, nor the carrier $\boldsymbol{a}$, to be introduced below.

Definition 125 (α-conversion). If $\mathfrak{a} = a \cdot\!+\!\cdot \alpha + A$ is a project of idiom $\mathcal{A}$, if φ is a $*$-isomorphism of $\mathcal{A}$ into another idiom $\mathcal{B}$ such that $\beta \circ \varphi = \lambda\alpha$ $(\lambda \in \mathbb{R})$, then $\varphi(\mathfrak{a}) := \lambda a \cdot\!+\!\cdot \beta + \varphi(A)$ is a *variant* of $\mathfrak{a}$, an *isovariant* if $\lambda = 1$. More generally, two projects are *variants* when they have a common variant in the previous sense.

If $\mathcal{B} \subset \mathcal{A}$ is a (unital) subalgebra of $\mathcal{A}$ such that $A \in \mathcal{R} \otimes \mathcal{B}$, then $\mathfrak{a}$ is a variant of $a \cdot\!+\!\cdot \alpha \restriction \mathcal{B} + A \restriction \mathcal{R} \otimes \mathcal{B}$.

Proposition 67. *Among all unital subalgebras $\mathcal{B} \subset \mathcal{A}$ such that $A \in \mathcal{R} \otimes \mathcal{B}$, there is a smallest one, the* minimal idiom *of* $\mathfrak{a}$.

Proof. Let $\mathcal{A}_0$ be the subalgebra generated by the $\theta(A)$, where $\Theta \colon \mathcal{R} \otimes \mathcal{A} \mapsto \mathcal{A}$ is induced by an element θ of the predual of $\mathcal{R}$, i.e., $\Theta(u \otimes v) := \theta(u) \otimes v$. □

Two variants have therefore isomorphic minimal idioms.

GoI is built so as to be variant-independent; this is why A, B are replaced with their variants $A^{\ddagger}$, $B^{\dagger}$; one might as well have chosen the variants $A^{\dagger}$, $B^{\ddagger}$, or variants involving a bigger idiom, e.g., some $\mathcal{A} \otimes \mathcal{B} \otimes \mathcal{C}$.

Definition 126 (Extraneousness). Two projects $\mathfrak{a}, \mathfrak{b}$ with the same idiom and pseudo trace $\mathcal{A}, \alpha$ are *alien* when their respective minimal idioms $\mathcal{A}_0, \mathcal{B}_0 \subset \mathcal{A}$ commute to each other and are such that, for all $u \in \mathcal{A}_0, v \in \mathcal{B}_0$,

$$\alpha(u) \cdot \alpha(v) = \alpha(uv) \cdot \alpha(I_{\mathcal{A}}). \tag{21.12}$$

The two extraneousness conditions are independent: if $\mathcal{A}_0 = \mathcal{B}_0$ is commutative, then they commute, but (21.12) is most likely to fail; conversely, in dimension 2, the projections $\pi := \left[\begin{smallmatrix} 1 & 0 \\ 0 & 0 \end{smallmatrix}\right]$ and $\nu := \left[\begin{smallmatrix} 1/4 & \sqrt{5}/4 \\ \sqrt{5}/4 & 3/4 \end{smallmatrix}\right]$ generate commutative algebras $\mathcal{A}_0, \mathcal{B}_0$ enjoying (21.12), but which do not commute.

The typical example is that of $\mathcal{A} \otimes I_{\mathcal{B}}, I_{\mathcal{A}} \otimes \mathcal{B} \subset \mathcal{A} \otimes \mathcal{B}$:

Proposition 68. *With the hypotheses and notations of Definition* 126, $\mathcal{A}_0 \otimes \mathcal{B}_0$ *is isomorphic to the algebra generated by* $\mathcal{A}_0 \cup \mathcal{B}_0$, *the isomorphism* φ *being such that* $\varphi(u \otimes I_{\mathcal{B}_0}) = u, \varphi(I_{\mathcal{A}_0} \otimes v) = v$ *for* $u \in \mathcal{A}_0, v \in \mathcal{B}_0$.

The construction of $A^{\ddagger}$, $B^{\dagger}$ is thus a way to build alien variants of A, B. Extraneousness is a sophisticated version of α-conversion, whose technical contents is the absence of interference, of « comprehension », between the idioms.

We shall therefore work, not quite with projects, but with equivalent classes (w.r.t. variance). When combining projects in a multiplicative way (which includes cut), we shall select alien elements in the respective classes. The resulting object will be well-defined up to variance.

21.1.5 Morphology. Morphological questions are central in logic. However, there is a tendency to take morphology as a *ready-made*. Typically, in the naïve opposition between the object (the semantics) and the syntax. Of course, we need such a sort of opposition, but not that dumb! The cognitive duality, whatever it is, does not go without saying, it must be *constructed.* This was, by the way, the primal sense of *constructivism*, before it became the name of a Chapel. And remember that *essentialism* is basically a morphological *simplism* (Section 1.A.2): for instance this XIXth century idiocy of the four races, ordered as White, Yellow, Black and Red (!!!) which presented as a biological absolute what was hardly more than the contingent excuse for colonialism, slavery and extermination.

Layers **−1**, **−2**, **−3** (Section 7.1) are based on several dualities:

−1: models vs. proofs, truly lame!

−2: objects vs. morphisms, very interesting, however still unsatisfactory.

−3: operators vs. operators: this is GoI.

These dualities rest upon: mutual exclusion (layer **−1**, leading to consistency); the trace (layer **−2**) leading to QCS, Chapter 17. Layer **−3** rests upon the determinant. However, this is only gross morphology: studying the relation of the determinant with the feedback equation, we discovered *wagers*; additives and exponentials prompt the use of idioms. This is the morphology which appears first and leads to *conducts*, i.e., sets of projects.

More refined morphologies, i.e., restrictions on projects, must be considered. They do not change the nature of interaction: this was already the case with idioms, which restrict interaction to *alien* projects.

For instance, additive principles are not satisfactorily handled by conducts. A *polarisation* negative/positive must be introduced: a *negative* conduct is a conduct made of wager-free projects. Polarised conducts yield an interesting account of logic, including the first natural explanation of *iconoclasm*, i.e., of **ELL**-like exponentials (Chapter 16).

Strangely enough, the GoI polarisation differs radically from the one introduced in Chapter 12: thus, the tensor product becomes negative. This alternative polarisation is very robust in the sense that it hardly demands any change of polarity: most logical operations, including implication (except the two disjunctions) are perfectly happy within negative polarity. In order to change polarities, one must introduce another morphological hypothesis, *lateralisation* left/right, which instills as an analogue the first action of *ludics* (Section 13.2.2) in *right* conducts. These lateralised conducts are called *behaviours*.

The coexistence of several morphologies could be a way to explain the aporia of proof-nets: multiplicative proof-nets belong to the conduct morphology, while additive proof-nets need a more refined morphology (e.g., polarised conduct) to be properly handled.

21.2 Conducts

From now on, $\mathcal{R}$ is the hyperfinite factor of type $\mathbf{II}_\infty$. The reason for this minor modification is explained in the next section.

21.2.1 Carriers

Definition 127 (Carriers). A *carrier* $\boldsymbol{a} \in \mathcal{R}$ is a *finite* projection. If $\boldsymbol{a}$ is a carrier, let $\mathcal{R}_{\boldsymbol{a}} : \boldsymbol{a}\mathcal{R}\boldsymbol{a} = \{u \in \mathcal{R} \,;\, \boldsymbol{a}u = u\boldsymbol{a} = u\}$. A project $\mathfrak{a} = a \cdot\!+\!\cdot \alpha + A$ of idiom $\mathcal{A}$ is of carrier $\boldsymbol{a}$ when $A \in \mathcal{R}_{\boldsymbol{a}} \otimes \mathcal{A} = (\boldsymbol{a} \otimes I_{\mathcal{A}})(\mathcal{R} \otimes \mathcal{A})(\boldsymbol{a} \otimes I_{\mathcal{A}})$. Two carriers $\boldsymbol{a}, \boldsymbol{b}$ are *disjoint* when $\boldsymbol{ab} = 0$ $(= \boldsymbol{ba})$.

Carriers take into account the *locative* aspects of GoI. The replacement of type $\mathbf{II}_1$ with $\mathbf{II}_\infty$ ensures that we have no worry about the existence of « enough » disjoint carriers. But, at the price of some inconvenience, e.g., assuming carriers to be « small enough », we could stay within type $\mathbf{II}_1$.

The hyperfinite factor of type $\mathbf{II}_\infty$ is unique up to isomorphism (yet another result of Connes [15]). It admits a *semi-finite* trace, unique up to renormalisation: $\mathrm{tr}' = \lambda \mathrm{tr}$ for some $\lambda > 0$; one chooses such a trace once and for all. When $\boldsymbol{a} \neq 0$ is a carrier, then $\mathcal{R}_{\boldsymbol{a}}$ is of type $\mathbf{II}_1$, thus isomorphic to the hyperfinite factor of that type; the only minor detail is that $\mathrm{tr} \upharpoonright \mathcal{R}_{\boldsymbol{a}}$ is not normalised, since $\mathrm{tr}(\boldsymbol{a}) > 0$ has no reason to be equal to 1, but this hardly matters!

Although we should write expressions of the form $\mathrm{ldet}(\boldsymbol{a} \otimes \alpha(I_{\mathcal{A}}) - AB)$, etc., we shall content ourselves with $\mathrm{ldet}(I - AB)$, which is less pedantic and, anyway, perfectly correct if we think twice.

21.2.2 Duality

Definition 128 (Duality). Let $\mathfrak{a} := a \cdot\!+\!\cdot \alpha + A$, $\mathfrak{b} := b \cdot\!+\!\cdot \alpha + B$ be *alien* projects of carrier $\boldsymbol{a}$; we define

$$\ll \mathfrak{a} \,|\, \mathfrak{b} \gg := a + b + \operatorname{ldet}(I - AB). \tag{21.13}$$

$\mathfrak{a}$ and $\mathfrak{b}$ are *polar*, notation $\mathfrak{a} \perp \mathfrak{b}$ iff $\ll \mathfrak{a} \,|\, \mathfrak{b} \gg \neq 0, \pm\infty$.

The determinant is relative to the pseudo-trace $(\mathrm{tr} \restriction \mathcal{R}_{\boldsymbol{a}}) \otimes \alpha$.

An explicit formulation of (21.13), when $\mathfrak{a} := a \cdot\!+\!\cdot \alpha + A$, $\mathfrak{b} := b \cdot\!+\!\cdot \beta + B$, still of the same carrier $\boldsymbol{a}$, are not supposed to be alien:

$$\ll \mathfrak{a} \,|\, \mathfrak{b} \gg := a\beta(I_{\mathcal{B}}) + b\alpha(I_{\mathcal{A}}) + \operatorname{ldet}(I - A^{\ddagger} B^{\dagger}). \tag{21.14}$$

The equivalence between the two notions follows from the obvious:

Proposition 69. *If* $\mathfrak{a} \perp \mathfrak{b}$ *(in the sense of* (21.14)) *and* $\mathfrak{a}'$, $\mathfrak{b}'$ *are variants of* $\mathfrak{a}$, $\mathfrak{b}$, *then* $\mathfrak{a}' \perp \mathfrak{b}'$.

As a corollary:

Proposition 70.

$$\mathfrak{a} \perp \mathfrak{b} \quad \Leftrightarrow \quad \mathfrak{b} \perp \mathfrak{a}$$

Proof. By Lemma 86.1 (iii) and Proposition 69. □

In (21.13) $\ll \mathfrak{a} \,|\, \mathfrak{b} \gg \in \mathbb{R} \cup \{\alpha(I_{\mathcal{A}}) \cdot \infty\}$; polarity thus excludes the two values $0, \infty$. One should see this exclusion as an analogue of the correctness property of proof-nets (Section 11.3.3): connectedness and acyclicity respectively corresponding to the exclusion of the outputs 0 and ∞.

Definition 129 (Conducts). A *conduct* **A** of carrier $\boldsymbol{a}$ is a « set » of projects of carrier $\boldsymbol{a}$ equal to its bipolar.

Of course, due to the use of arbitrary idioms, a conduct *cannot* be a set, but this remark is pure nonsense. Up to variance, conducts do form a set.

21.2.3 Partial projects. Besides the standard duality, there is a coarser one, based upon $\ll \mathfrak{a} \,|\, \mathfrak{b} \gg \neq \infty$, and whose antagonists are styled *partial*. Indeed, making full use of non-positive pseudo-traces, a conduct generates a vector space and the map $\ll \cdot \,|\, \cdot \gg$ extends into a bilinear form.

In what follows, **A** is a conduct of carrier $\boldsymbol{a}$.

Definition 130 (Partial projects). If we relax faithfulness and the requirement that $\alpha(I_{\mathcal{A}}) \neq 0$ (we can thus even afford to have $\mathcal{A} = 0$), we obtain *partial projects*. If $\mathfrak{a}_i := a_i \cdot\!+\!\cdot \alpha_i + A_i$ are partial projects of idioms $\mathcal{A}_i$, if $\lambda_i \in \mathbb{R}$ $(i = 1, \dots, n)$, we define

$$\sum_1^n \lambda_i \cdot \mathfrak{a}_i := \sum_1^n \lambda_i a_i \cdot\!+\!\cdot \bigoplus_1^n \lambda_i \alpha_i + \bigoplus_1^n A_i \tag{21.15}$$

of idiom $\bigoplus_1^n \mathcal{A}_i$. The set $\wp\mathbf{A}$ of *partial projects of* $\mathbf{A}$ is the closure of $\mathbf{A}$ under finite linear combinations[1].

The binary function $\ll\cdot|\cdot\gg$ naturally extends into a function from $\wp\mathbf{A} \times \wp\sim\mathbf{A}$ into $\mathbb{R}$, for instance by means of the formula (21.14). We define the equivalence relation $\equiv_{\mathbf{A}}$ on $\wp\mathbf{A}$:

$$\mathfrak{a} \equiv_{\mathbf{A}} \mathfrak{b} \;:\Longleftrightarrow\; \forall \mathfrak{c} \in \sim\mathbf{A} \quad \ll\mathfrak{a}\,|\,\mathfrak{c}\gg = \ll\mathfrak{b}\,|\,\mathfrak{c}\gg . \tag{21.16}$$

The typical case is that of an isovariant (Definition 125): $\mathfrak{a} \equiv_{\mathbf{A}} \varphi(\mathfrak{a})$.

Theorem 88 (Linearisation). *The quotient* $\ell\mathbf{A} := \wp\mathbf{A}/{\equiv_{\mathbf{A}}}$ *is a real vector space. The application* $\ll\cdot|\cdot\gg$ *from* $\ell\mathbf{A} \times \ell\sim\mathbf{A}$ *to* $\mathbb{R}$ *is bilinear.*

Proof. In (21.16), one can replace « $\forall \mathfrak{d} \in \sim\mathbf{A}$ » with « $\forall \mathfrak{d} \in \wp\sim\mathbf{A}$ ». □

Definition 131 (Internal completeness). An *ethics* of carrier $\boldsymbol{a}$ is any « set » $\mathbf{E}$ of projects of carrier $\boldsymbol{a}$; $\mathbf{E}$ generates a conduct, namely the bipolar $\mathbf{A} := \sim\sim\mathbf{E}$. The ethics $\mathbf{E}$ is said to be *complete* when any equivalence class of projects in $\mathbf{A}$ has a witness in $\mathbf{E}$:

$$\forall \mathfrak{a} \in \mathbf{A}\ \exists \mathfrak{e} \in \mathbf{E} \quad \mathfrak{a} \equiv_{\mathbf{A}} \mathfrak{e}. \tag{21.17}$$

Theorem 89 (Ethic lemma). *If* $\alpha(I_{\mathcal{A}}) = \beta(I_{\mathcal{B}})$, *one can replace in* (21.16) « $\sim\mathbf{A}$ » *with* « $\mathbf{E}$ » , *where* $\mathbf{E}$ *is any ethics for* $\sim\mathbf{A}$.

Proof. Let $\mathfrak{d} \in \mathbf{A}$; then $\mathfrak{a} \equiv_{\mathbf{A}} \mathfrak{b}$ iff for all $\lambda \in \mathbb{R}$, $\lambda\mathfrak{a} - \lambda\mathfrak{b} + \mathfrak{d} \in \mathbf{A} = \sim\mathbf{E}$, thus, iff for all $\mathfrak{c} \in \mathbf{E}$, $\ll\mathfrak{a}\,|\,\mathfrak{c}\gg = \ll\mathfrak{b}\,|\,\mathfrak{c}\gg$. □

The condition « $\alpha(I_{\mathcal{A}}) = \beta(I_{\mathcal{B}})$ » makes sure that $\lambda\mathfrak{a} - \lambda\mathfrak{b} + \mathfrak{d}$ is a project; in the polarised case, *rescaling* (Definition 139) renders this restriction pointless.

21.2.4 Images and projections. The inclusion $\mathbf{A} \subset \mathbf{A} \oplus \mathbf{B}$ cannot make sense *stricto sensu* for questions of carrier. However, a project of carrier $\boldsymbol{a}$ can be seen as a project of carrier $\boldsymbol{a} + \boldsymbol{b}$. Hence the notion of *injection*, which is not problematic at the level of projects or ethics. But, injection does not commute at all with negation; its converse, *projection*, is better behaved.

Let $\boldsymbol{a}$, $\boldsymbol{b}$ be carriers, then:

[1] $\mathbf{A}$ is anyway closed under non-zero homotheties.

Definition 132 (Images). If $\Phi \in \mathcal{R}$, $\|\Phi\| \leq 1$, is such that $\Phi = \boldsymbol{b}\Phi$, the *image* under Φ of a project $\mathfrak{a} = a \cdot\!+\!\cdot \alpha + A$ of carrier $\boldsymbol{a}$ is the project $\Phi(\mathfrak{a}) := a \cdot\!+\!\cdot \alpha + (\Phi \otimes I_{\mathcal{A}})A(\Phi^* \otimes I_{\mathcal{A}})$ of carrier $\boldsymbol{b}$. If $\mathbf{E}$ is an ethics of carrier $\boldsymbol{a}$, its *image* under Φ is the ethics $\Phi(\mathbf{E}) := \{\Phi(\mathfrak{a}); \mathfrak{a} \in \mathbf{E}\}$ of carrier $\boldsymbol{b}$.

Example 8. The natural example is that of a projection $\Phi := \boldsymbol{b}$; two subcases are of interest:

Projection: if $\boldsymbol{b} \subset \boldsymbol{a}$, then $\boldsymbol{b}(\mathfrak{a})$, denoted by $\mathfrak{a}_{\boldsymbol{b}}$, is the *projection* of $\mathfrak{a}$ on $\boldsymbol{b}$; if $\mathbf{E}$ is an ethics of carrier $\boldsymbol{a}$, its *projection* on $\boldsymbol{b}$ is $\mathbf{E}_{\boldsymbol{b}} := \{\mathfrak{a}_{\boldsymbol{b}}; \mathfrak{a} \in \mathbf{E}\}$.

Injection: if $\boldsymbol{a} \subset \boldsymbol{b}$, then $\boldsymbol{b}(\mathfrak{a})$ $(= \mathfrak{a})$, denoted by $\mathfrak{a}_{\boldsymbol{b}}$, is the *injection* of $\mathfrak{a}$ in $\boldsymbol{b}$; if $\mathbf{E}$ is an ethics of carrier $\boldsymbol{a}$, its *injection* in $\boldsymbol{b}$ is $\mathbf{E}_{\boldsymbol{b}} := \{\mathfrak{a}_{\boldsymbol{b}}; \mathfrak{a} \in \mathbf{E}\}$.

Definition 133 (Faithfulness). A subcarrier $\boldsymbol{b} \subset \boldsymbol{a}$ is $\mathbf{A}$-*faithful* (w.r.t. a conduct $\mathbf{A}$ of carrier $\boldsymbol{a}$) when, for all $\mathfrak{a} \in \mathbf{A}$, $\mathfrak{a}_{\boldsymbol{b}} \in \mathbf{A}$ and $\mathfrak{a}_{\boldsymbol{b}} \equiv \mathfrak{a}$.

Proposition 71. *Let* $\boldsymbol{b} \subset \boldsymbol{a}$ *be* $\mathbf{A}$*-faithful; then:*

(i) *For all* $\mathfrak{a} \in \wp\mathbf{A}$, $\mathfrak{a}_{\boldsymbol{b}} \in \wp\mathbf{A}$ *and* $\mathfrak{a}_{\boldsymbol{b}} \equiv \mathfrak{a}$.

(ii) $\boldsymbol{b}$ *is* $\sim\!\mathbf{A}$*-faithful.*

(iii) $\mathbf{A}_{\boldsymbol{b}}$ *is a conduct.*

(iv) $(\sim\!\mathbf{A})_{\boldsymbol{b}} = \sim\!(\mathbf{A}_{\boldsymbol{b}})$.

Proof. (i) Immediate.

(ii) Using (i) and the general property

$$\ll \mathfrak{a}_{\boldsymbol{b}} \,|\, \mathfrak{b} \gg \;=\; \ll \mathfrak{a} \,|\, \mathfrak{b}_{\boldsymbol{b}} \gg \;=\; \ll \mathfrak{a}_{\boldsymbol{b}} \,|\, \mathfrak{b}_{\boldsymbol{b}} \gg . \tag{21.18}$$

(ii) By the $\sim\!\mathbf{A}$-faithfulness of $\boldsymbol{b}$ and (21.18), the polar of $(\sim\!\mathbf{A})_{\boldsymbol{qp}}$, i.e., the set of projects of carrier $\boldsymbol{a}$ polar to the $\mathfrak{b}_{\boldsymbol{b}}$ $(\mathfrak{b} \in \sim\!\mathbf{A})$ is equal to $\mathbf{A}$. Hence the polar of $(\sim\!\mathbf{A})_{\boldsymbol{b}}$ is equal to $\mathbf{A}_{\boldsymbol{b}}$. This also proves (iv). □

Proposition 72.

$$\mathfrak{a}_{\boldsymbol{b}} \perp \mathfrak{b} \iff \mathfrak{a} \perp \mathfrak{b}_{\boldsymbol{b}} \iff \mathfrak{a}_{\boldsymbol{b}} \perp \mathfrak{b}_{\boldsymbol{b}}.$$

Example 9. If $\boldsymbol{b} \subset \boldsymbol{a}$, a conduct $\mathbf{A}$ of carrier $\boldsymbol{b}$ induces the two dual injections of carrier $\boldsymbol{a}$: $\sim\sim\!\mathbf{A}_{\boldsymbol{a}}$ and $\sim\!(\sim\!\mathbf{A})_{\boldsymbol{a}}$. However, $\boldsymbol{b}$ faithfully projects both injections onto $\mathbf{A}$.

21.3 The social life of conducts

21.3.1 Multiplicatives. Let $\boldsymbol{a}$, $\boldsymbol{b}$ be disjoint carriers.

Definition 134 (Application). If $\mathfrak{a} := a \cdot\!+\!\cdot \alpha + A$ and $\mathfrak{f} := f \cdot\!+\!\cdot \alpha + F$ are alien projects of respective carriers $\boldsymbol{a}, \boldsymbol{a} + \boldsymbol{b}$ and idiom $\mathcal{A}$, one defines the project $[\mathfrak{f}]\mathfrak{a}$ of carrier $\boldsymbol{b}$ and idiom $\mathcal{A}$:

$$[\mathfrak{f}]\mathfrak{a} := f + a + \operatorname{ldet}(I - FA) \cdot\!+\!\cdot \alpha + [F]A \tag{21.19}$$

where $[F]A$ has been defined in Section 21.1.2.

An explicit formulation of the same thing, when $\mathfrak{a} := a \cdot\!+\!\cdot \alpha + A$ and $\mathfrak{f} := f \cdot\!+\!\cdot \varphi + F$ are not assumed to be alien, is the project of idiom $\mathcal{F} \otimes \mathcal{A}$:

$$[\mathfrak{f}]\mathfrak{a} := f\alpha(I_{\mathcal{A}}) + a\varphi(I_{\mathcal{F}}) + \operatorname{ldet}(I - F^{\ddagger}A^{\dagger})) \cdot\!+\!\cdot (\varphi \otimes \alpha) + [F^{\ddagger}]A^{\dagger}. \tag{21.20}$$

Definition 135 (Multiplicatives). If $\mathbf{A}$, $\mathbf{B}$ are conducts of carriers $\boldsymbol{a}$, $\boldsymbol{b}$, one defines the conducts of carrier $\boldsymbol{a} + \boldsymbol{b}$:

$$\mathbf{A} \multimap \mathbf{B} := \{\mathfrak{f} \in \mathbf{A}\ ;\ \forall \mathfrak{a} \in \mathbf{A}\ \ [\mathfrak{f}]\mathfrak{a} \in \mathbf{B}\}, \tag{21.21}$$

$$\mathbf{A} \otimes \mathbf{B} := {\sim}(\mathbf{A} \multimap {\sim}\mathbf{B}), \tag{21.22}$$

$$\mathbf{A} \parr \mathbf{B} := {\sim}\mathbf{A} \multimap \mathbf{B}. \tag{21.23}$$

Theorem 90 (Adjunction).

$$\mathbf{A} \otimes \mathbf{B} \ = \ {\sim}{\sim}(\mathbf{A} \odot \mathbf{B}) \ := \ {\sim}{\sim}\{\mathfrak{a} \otimes \mathfrak{b}\ ;\ \mathfrak{a} \in \mathbf{A},\ \mathfrak{b} \in \mathbf{B}\} \tag{21.24}$$

with $\mathfrak{a} \otimes \mathfrak{b} := a + b \cdot\!+\!\cdot \alpha + (A + B)$ *when* $\mathfrak{a} \in \mathbf{A}, \mathfrak{b} \in \mathbf{B}$ *are alien.*

Proof. Not quite a surprise: this is a by-product of Theorem 87. □

Corollary 90.1. *The tensor product is commutative, associative, with neutral element the conduct* $\mathsf{T} := \{0 \cdot\!+\!\cdot \alpha + 0\ ;\ \alpha$ *pseudo-trace on some idiom* $\mathcal{A}\}$ *of carrier* 0.

The neutral element of $\parr$ is $\mathbf{0} := \{a \cdot\!+\!\cdot \alpha + 0\ ;\ a \neq 0,\ \alpha$ pseudo-trace… $\}$.

Remark 7. It is useful to rephrase the previous results in terms of *ethics*: $\mathbf{E} \odot \mathbf{F}$ and $\mathbf{E} \multimap \mathbf{F}$ can still be defined when $\mathbf{E}$, $\mathbf{F}$ are ethics. Observe that

$${\sim}{\sim}(\mathbf{E} \odot \mathbf{F}) = ({\sim}{\sim}\mathbf{E}) \otimes ({\sim}{\sim}\mathbf{F}), \tag{21.25}$$

$$\mathbf{E} \multimap ({\sim}{\sim}\mathbf{F}) = ({\sim}{\sim}\mathbf{E}) \multimap ({\sim}{\sim}\mathbf{F}). \tag{21.26}$$

Proposition 73.

$$\wp(\mathbf{A} \multimap \mathbf{B}) = \wp\mathbf{A} \multimap \wp\mathbf{B}.$$

21.3.2 Quantifiers. Let $\mathbb{I} \neq \emptyset$ be a non-empty index set (usually uncountable).

Definition 136 (Universal quantification). If $\mathbf{A}[i]$ $(i \in \mathbb{I})$ is a family of conducts of carrier $\boldsymbol{a}$, then $\forall i \in \mathbb{I}\ \mathbf{A}[i]$ is the conduct of carrier $\boldsymbol{a}$ defined by

$$\forall i \in \mathbb{I}\ \mathbf{A}[i] := \bigcap_i \mathbf{A}_i . \tag{21.27}$$

The definition makes sense because of:

Proposition 74. *Any intersection of conducts of carrier* $\boldsymbol{a}$ *is a conduct of carrier* $\boldsymbol{a}$.

Proof. Since an intersection of polars is the polar of a union:

$$\bigcap_i \sim\mathbf{E}[i] = \sim\bigcup_i \mathbf{E}[i].$$ □

Proposition 75.

$$\wp\forall i \in \mathbb{I}\ \mathbf{A}[i] \ = \ \bigcap_i \wp\mathbf{A}_i$$

Theorem 91 (Distributivity).

$$\mathbf{A} \multimap \forall i \in \mathbb{I}\ \mathbf{B}[i] = \forall i \in \mathbb{I}\ (\mathbf{A} \multimap \mathbf{B}[i]). \tag{21.28}$$

Remark 8. Existential quantification is defined dually as

$$\exists i \in \mathbb{I}\ \mathbf{A}[i] := \sim\sim\bigcup_{i \in \mathbb{I}} \mathbf{A}[i].$$

In terms of ethics, the following remark is useful:

$$\sim\sim\bigcup_{i \in \mathbb{I}} \mathbf{E}_i \ = \exists i \in \mathbb{I}\ \sim\sim\mathbf{E}[i]. \tag{21.29}$$

Second order quantification is treated in annex 21.A.

21.3.3 Additives. Let $\boldsymbol{a}$, $\boldsymbol{b}$ be disjoint carriers.

Definition 137 (Additives). If $\mathbf{A}$, $\mathbf{B}$ are conducts of respective carriers $\boldsymbol{a}$, $\boldsymbol{b}$, we define:

$$\mathbf{A} \oplus \mathbf{B} := \sim\sim(\mathbf{A}_{\boldsymbol{a}+\boldsymbol{b}} \cup \mathbf{B}_{\boldsymbol{a}+\boldsymbol{b}}), \tag{21.30}$$

$$\mathbf{A} \,\&\, \mathbf{B} := \sim(\sim\mathbf{A}_{\boldsymbol{a}+\boldsymbol{b}}) \cap \sim(\sim\mathbf{B}_{\boldsymbol{a}+\boldsymbol{b}}). \tag{21.31}$$

Proposition 76. *The two definitions are dual, i.e.,*

$$\sim(\mathbf{A} \oplus \mathbf{B}) = \sim\mathbf{A} \,\&\, \sim\mathbf{B}.$$

Additives are commutative, associative, with as respective neutrals, the void conduct $(\oplus)$ *and the full conduct* $(\&)$ *of carrier* 0.

Little more can be said; a good transition towards *polarisation*.

21.4 Polarised conducts

21.4.1 Polarisation

Definition 138 (Daimon). If $a \in \mathbb{R}$, the project $\mathfrak{Dai}_a := a \cdot{+}\cdot 1 + 0$, of idiom $\mathbb{C}$ and pseudo-trace $1(z) := z$ is called a *daimon*; proper if $a \neq 0$.

Definition 139 (Polarised conducts). A conduct $\mathbf{A}$ is *positive* when it enjoys the following:

Daimon: $\mathbf{A}$ contains all proper *daimons* $\mathfrak{Dai}_a$, $a \neq 0$.

Rescaling: if $a, b \neq 0$ and $a \cdot{+}\cdot \alpha + A \in \mathbf{A}$, then $b \cdot{+}\cdot \alpha + A \in \mathbf{A}$.

Negative conducts are defined as the polars of positive conducts; a conduct is *polarised* when it is either positive or negative.

Proposition 77. *A conduct* $\mathbf{A}$ *is* negative *iff it enjoys the following:*

Wager: *all projects of* $\mathbf{A}$ *are* wager-free, *i.e., with a null wager.*

Rescaling: *if* $\mathfrak{a} \in \mathbf{A}$, *then* $\mathfrak{a} + \lambda \mathfrak{Dai}_0 \in \mathbf{A}$.

Proof. $\ll a \cdot{+}\cdot 1 + 0 \,|\, b \cdot{+}\cdot \beta + B \gg = a\beta(I_{\mathcal{B}}) + b$; since $\beta(I_{\mathcal{B}}) \neq 0$, it turns out that $b \cdot{+}\cdot \beta + B$ is polar to all proper *daimons* iff $b = 0$. If $\mathfrak{a} := a \cdot{+}\cdot \alpha + A \in \mathbf{A}$ with $a \neq 0$, then $\ll \mathfrak{a} \,|\, \mathfrak{b} + \lambda \mathfrak{Dai}_0 \gg = \ll \mathfrak{a} \,|\, \mathfrak{b} \gg + \lambda a \neq 0$ for all $\lambda \neq -\beta(I_{\mathcal{B}})$ iff $\ll \mathfrak{a} \,|\, \mathfrak{b} \gg = a\beta(I_{\mathcal{B}})$, i.e., iff the component $\mathrm{ldet}(I - A^{\ddagger} B^{\dagger})$ is null. Hence the equivalence between the two rescaling conditions. □

Remark 9. Negative rescaling can be understood as the closure under non-unital *variants* (Definition 125); by the way, negative rescaling holds for positive conducts too.

Corollary 77.1. *If an ethics* $\mathbf{A}$ *is positive (in the obvious sense), so is its bipolar.*

Proof. The conditions « daimon » and « rescaling » induce by duality conditions « wager » and « rescaling » on $\sim\mathbf{A}$ which, in turn, induce conditions « daimon » and « rescaling » on $\sim\sim\mathbf{A}$. □

Corollary 77.2. *All non-wager-free projects of a positive conduct* $\mathbf{A}$ *are homothetic as elements of the vector space* $\ell\mathbf{A}$.

Proof. If $a \neq 0$, then $\ll \mathfrak{a} \,|\, \mathfrak{b} \gg = a\beta(I_{\mathcal{B}}) = \ll \mathfrak{Dai}_a \,|\, \mathfrak{b} \gg$ (proof of Proposition 77), hence $\mathfrak{a} \equiv_{\mathbf{A}} \mathfrak{Dai}_a$. The $\mathfrak{Dai}_a$ are pairwise homothetic. □

Definition 140 (Proper conducts). A positive conduct $\mathbf{A}$ is *proper* when it does not contain the improper *daimon* $\mathfrak{Dai}_0$. A negative conduct $\mathbf{A}$ is *proper* when non-empty.

Proposition 78. *The polar of a proper polarised conduct is proper.*

Proof. $\ll 0 \cdot{+}\cdot 1 + 0 \,|\, 0 \cdot{+}\cdot \beta + B \gg = 0$, hence a mutual exclusion. □

21.4.2 Polarisation of multiplicatives. Polarisation is reasonably compatible with multiplicative constructions, although their « table of polarities » is quite unexpected.

Proposition 79. *Assume that* **A**, **B** *are polarised conducts with disjoint carriers* $\boldsymbol{a}$, $\boldsymbol{b}$*:*

(i) *If both are negative,* $\mathbf{A} \otimes \mathbf{B}$ *is negative; and proper if both are proper.*

(ii) *If* **A** *is positive, if* **B** *is negative and proper, then* $\mathbf{A} \otimes \mathbf{B}$ *is positive; and proper in case* **A** *is proper.*

(iii) *If both are positive, then* $\mathbf{A} \otimes \mathbf{B}$ *is positive and unproper.*

Proof. $- \otimes - = -$**:** immediate.

$+ \otimes - = +$**:** if **A** is positive, if **B** is proper negative, let $\mathfrak{b} := b \cdot\!+\!\cdot \beta + B \in \mathbf{B}$; then $\mathfrak{Dai}_a \otimes (\mathfrak{b} + \lambda \mathfrak{Dai}_0) \in \mathbf{A} \otimes \mathbf{B}$. If $\mathfrak{c} := c \cdot\!+\!\cdot \gamma + C \in \sim\!(\mathbf{A} \otimes \mathbf{B})$, then $\ll \mathfrak{Dai}_a \otimes (\mathfrak{b} + \lambda \mathfrak{Dai}_0) \,|\, \mathfrak{c} \gg = a\gamma(I_{\mathcal{C}}) + c\beta(I_{\mathcal{B}} + \lambda) + \mathrm{ldet}(I - B^{\ddagger}C^{\dagger})$ can be nullified by an appropriate choice of a and λ, unless $c = 0$: this proves « daimon ». Moreover, since $\mathbf{A} \odot \mathbf{B}$ (Theorem 90) obviously enjoys rescaling, so does its bipolar. If **A** is proper and $\mathfrak{c} \in \sim\!\mathbf{A}$ is total, then it is immediate (this is indeed *weakening* see Section 21.5.1 *infra*) that $\mathfrak{c} \in \sim\!(\mathbf{A} \otimes \mathbf{B})$.

$+ \otimes + = u$**:** $\mathbf{A} \otimes \mathbf{B}$ contains all $(a+b) \cdot\!+\!\cdot 1 + 0$. The tensor product contains all the $(a+b) \cdot\!+\!\cdot 1 + 0$, with $a, b \neq 0$, hence $0 \cdot\!+\!\cdot 1 + 0$. Dually, $[\mathfrak{f}]\cdot$ cannot send all the $a \cdot\!+\!\cdot 1 + 0$ into something wager-free. □

Consistently Proposition 78, the neutral $\top$ of Corollary 90.1 is negative.

Let us restate the polarity table for linear implication in the *proper* case:

(i) If **A** is negative and **B** is positive, then $\mathbf{A} \multimap \mathbf{B}$ is positive.

(ii) If **A**, **B** have the same polarity, then $\mathbf{A} \multimap \mathbf{B}$ is negative.

(iii) If **A** is positive and **B** is negative, then $\mathbf{A} \multimap \mathbf{B}$ is unproper.

In terms of cotensor, the important thing is that an n-ary « par » $\mathbf{A}_1 \parr \ldots \parr \mathbf{A}_n$ of proper polarised conducts is proper iff at most one of them is negative, consistently with the maintenance of *sequents* in ludics (Chapter 13).

21.4.3 Additives. Let $\boldsymbol{a}$, $\boldsymbol{b}$ be disjoint carriers; if $\mathfrak{f}, \mathfrak{g}$ are wager-free projects of carrier $\boldsymbol{a} + \boldsymbol{b}$, define $\mathfrak{f} \,\&\, \mathfrak{g} := \mathfrak{f} + \mathfrak{g}$, provided $\alpha(I_{\mathcal{A}}) + \beta(I_{\mathcal{B}}) \neq 0$.

Proposition 80. *If* **A**, **B** *are positive conducts of respective carriers* $\boldsymbol{a}$, $\boldsymbol{b}$, *then*

$$\sim\!\{\mathfrak{f} \,\&\, \mathfrak{g} \,;\, \mathfrak{f} \in \sim\!\mathbf{A}, \mathfrak{g} \in \sim\!\mathbf{B}\} = \sim\sim\!\mathbf{A}_{\boldsymbol{a}+\boldsymbol{b}} \cup \sim\sim\!\mathbf{B}_{\boldsymbol{a}+\boldsymbol{b}}. \tag{21.32}$$

Proof. If $\mathfrak{a} \perp \mathfrak{f} \,\&\, \mathfrak{g}$, then $\ll \mathfrak{a} \mid x\mathfrak{f} + \lambda\mathfrak{Dai}_0 \gg + \ll \mathfrak{a} \mid y\mathfrak{g} + \mu\mathfrak{Dai}_0 \gg \neq 0$, for any $x, y \neq 0$ and *ad hoc* λ, μ. Then $x \ll \mathfrak{a} \mid \mathfrak{f} \gg + y \ll \mathfrak{a} \mid \mathfrak{g} \gg \neq 0$ for all $x, y \neq 0$, hence one and only one among $\ll \mathfrak{a} \mid \mathfrak{f} \gg$, $\ll \mathfrak{a} \mid \mathfrak{g} \gg$ is non-zero. Since $\mathfrak{f}$, $\mathfrak{g}$ are not related, the choice is always the same, i.e., either $\mathfrak{a} \in {\sim\sim}\mathbf{A}_{\boldsymbol{a}+\boldsymbol{b}}$ or $\mathfrak{a} \in {\sim\sim}\mathbf{B}_{\boldsymbol{a}+\boldsymbol{b}}$. The converse inclusion is almost immediate. □

Theorem 92 (Disjunction property). $\mathbf{A}_{\boldsymbol{a}+\boldsymbol{b}} \cup \mathbf{B}_{\boldsymbol{a}+\boldsymbol{b}}$ *is a complete ethics for* $\mathbf{A} \oplus \mathbf{B}$.

Theorem 93 (Mystery of incarnation). $\{\mathfrak{a} \,\&\, \mathfrak{b}\ ;\ \mathfrak{a} \in \mathbf{A}, \mathfrak{b} \in \mathbf{B}\}$ *is a complete ethics for* $\mathbf{A} \,\&\, \mathbf{B}$.

Proof. Both results are immediate, *modulo* the *ethic lemma* (Theorem 89). □

Corollary 93.1. *If* **A**, **B** *are positive (resp. negative) and proper, so is* $\mathbf{A} \oplus \mathbf{B}$ *(resp.* $\mathbf{A} \,\&\, \mathbf{B}$*).*

Proposition 81. & *is (literally) commutative, associative, with as unit the tensor unit* $\top$.

21.4.4 Distributivity. For questions of carrier, ⅋ cannot *literally* distribute over &. However, If $\boldsymbol{a}, \boldsymbol{b}, \boldsymbol{c}, \boldsymbol{d}, \boldsymbol{e}, \boldsymbol{f}, \boldsymbol{g}$ are disjoint carriers and u (resp. v) is a partial isometry from $\boldsymbol{a} + \boldsymbol{b}$ (resp. $\boldsymbol{a} + \boldsymbol{c}$) onto $\boldsymbol{d} + \boldsymbol{e}$ (resp. $\boldsymbol{f} + \boldsymbol{g}$) s.t. $u\boldsymbol{a} = \boldsymbol{d}u$ (resp. $v\boldsymbol{a} = \boldsymbol{f}v$), consider $\mathfrak{Distr} := 0 \cdot\!+\!\cdot (1 \oplus 1) + ((u + u^*) \oplus (v + v^*))$ of idiom $\mathbb{C} \oplus \mathbb{C}$. Then, if **A**, **B**, **C** are negative conducts of respective carriers $\boldsymbol{a}, \boldsymbol{b}, \boldsymbol{c}$:

(i) $\mathfrak{Distr} \in (\mathbf{A} \multimap (\mathbf{B} \,\&\, \mathbf{C})) \multimap (u(\mathbf{A} \multimap \mathbf{B}) \,\&\, v(\mathbf{A} \multimap \mathbf{C}))$.

(ii) $\mathfrak{Distr} \in (u(\mathbf{A} \multimap \mathbf{B}) \,\&\, v(\mathbf{A} \multimap \mathbf{C})) \multimap (\mathbf{A} \multimap (\mathbf{B} \,\&\, \mathbf{C}))$.

(iii) If $\mathfrak{f} = f \cdot\!+\!\cdot \varphi + F \in \mathbf{A} \multimap (\mathbf{B} \,\&\, \mathbf{C})$, then $[\mathfrak{Distr}]([\mathfrak{Distr}]\mathfrak{f}) = f \cdot\!+\!\cdot (\varphi \oplus \varphi \oplus \varphi \oplus \varphi) + ((\boldsymbol{a} + \boldsymbol{b})F(\boldsymbol{a} + \boldsymbol{b}) + \boldsymbol{a}F\boldsymbol{a} + \boldsymbol{a}F\boldsymbol{a} + (\boldsymbol{a} + \boldsymbol{c})F(\boldsymbol{a} + \boldsymbol{c}))$, which is $\equiv$ to $\mathfrak{f}$ when $f = 0$.

(iv) If $\mathfrak{g} = g \cdot\!+\!\cdot \psi + G \in u(\mathbf{A} \multimap \mathbf{B}) \,\&\, v(\mathbf{A} \multimap \mathbf{C})$, then $[\mathfrak{Distr}]([\mathfrak{Distr}]\mathfrak{g}) = g \cdot\!+\!\cdot (\psi \oplus \psi \oplus \psi \oplus \psi) + ((\boldsymbol{d} + \boldsymbol{e})G(\boldsymbol{d} + \boldsymbol{e}) + \boldsymbol{d}G\boldsymbol{d} + \boldsymbol{f}G\boldsymbol{f} + (\boldsymbol{f} + \boldsymbol{g})G(\boldsymbol{f} + \boldsymbol{g}))$, which is $\equiv$ to $\mathfrak{g}$ when $g = 0$.

$\mathfrak{Distr}$ therefore implements a sort of distributivity, up to $\equiv$. By the way, the possibility of neglecting the parasitic expressions $\boldsymbol{a}F\boldsymbol{a}$, $\boldsymbol{d}G\boldsymbol{d}$, $\boldsymbol{f}G\boldsymbol{f}$ is a pure result of polarisation: for instance, if $f = 0$ and $\mathfrak{a} = 0 \cdot\!+\!\cdot \alpha + A \in \mathbf{A}$, the wager of $[\mathfrak{f}]\mathfrak{a}$ must be 0, hence $\mathrm{ldet}(I - A^{\ddagger}F^{\dagger}) = 0$.

21.5 Exponentials

The polarised exponentials turn out to be **ELL**-style (Chapter 16).

21.5.1 Structural rules. Polarisation enables weakening in the positive case.

Proposition 82. *If* $\mathfrak{c} \in \mathbf{A} \otimes \mathbf{B}$, *where* $\mathbf{B}$ *is of carrier* $\boldsymbol{b}$ *and* $\mathbf{A}$ *is negative, then* $\mathfrak{c}_{\boldsymbol{b}} \in \mathbf{B}$.

Proof. A project $\mathfrak{f} \in \sim\mathbf{B}$ induces a « function » $\mathfrak{f}_{\boldsymbol{a}+\boldsymbol{b}} \in \mathbf{A} \multimap \sim\mathbf{B}$: since $\mathfrak{a} \in \mathbf{A}$ is wager-free, $[\mathfrak{f}_{\boldsymbol{a}+\boldsymbol{b}}]\mathfrak{a} = \mathfrak{f}$. Now, $\ll \mathfrak{f} \mid \mathfrak{c}_{\boldsymbol{b}} \gg = \ll \mathfrak{f}_{\boldsymbol{a}+\boldsymbol{b}} \mid \mathfrak{c} \gg \neq 0, \infty$. □

As observed in Section 20.2.1, contraction fails in the presence of idioms: this would require something like $\mathcal{F} \otimes \mathcal{A} \sim \mathcal{A} \otimes \mathcal{A}$ for all A, hopeless!

21.5.2 Perennial conducts

Definition 141 (Perenniality). A project is *perennial* when of the form $0 \cdot + \cdot 1 + A$. A *perennial* ethics is a negative ethics made of rescalings of perennial projects. A *perennial* conduct is the bipolar of a perennial ethics; it is therefore negative.

Let $\boldsymbol{a}, \boldsymbol{b}, \boldsymbol{c}$ be disjoint carriers and let u, v be partial isometries between $\boldsymbol{a}$ and (respectively) $\boldsymbol{b}, \boldsymbol{c}$. Consider the idiom $\mathcal{M}_2(\mathbb{C})$ and, as pseudo-trace, the normalised trace tr on $\mathcal{M}_2(\mathbb{C})$, so that $\mathcal{R} \otimes \mathcal{M}_2(\mathbb{C}) \simeq \mathcal{M}_2(\mathcal{R})$. We define the project $\mathfrak{Contr} := 0 \cdot + \cdot \operatorname{tr} + \begin{bmatrix} u+u^* & v \\ v^* & 0 \end{bmatrix}$. Then:

Theorem 94 (Contraction). *If* $\mathbf{A}$ *is a perennial conduct of carrier* $\boldsymbol{a}$, *then* $\mathfrak{Contr}$ *implements a rescaling of the map* $\mathfrak{a} \rightsquigarrow u(\mathfrak{a}) \otimes v(\mathfrak{a})$. *In particular* $\mathfrak{Contr} \in \mathbf{A} \multimap (u(\mathbf{A}) \otimes v(\mathbf{A}))$.

Proof. If $\mathfrak{a} = 0 \cdot + \cdot 1 + A$, then $[\mathfrak{Contr}]\mathfrak{a} = 0 \cdot + \cdot \operatorname{tr} + \begin{bmatrix} uAu^* + vAv^* & 0 \\ 0 & 0 \end{bmatrix}$, which is a rescaling of $u(\mathfrak{a}) \otimes v(\mathfrak{a})$ by means of the map $z \rightsquigarrow \begin{bmatrix} z & 0 \\ 0 & 0 \end{bmatrix}$. □

Notice the use of a non-commutative idiom.

21.5.3 Perennialisation. Here $\mathcal{H}$ stands for the hyperfinite factor (of type $\mathbf{II}_1$). The idea of perennialisation is first to standardise idioms, replacing them – when possible – with the hyperfinite factor, then to exploit the isomorphism $\mathcal{R} \otimes \mathcal{H} \simeq \mathcal{R}$.

Definition 142 (Standardisation). A project $\mathfrak{a} := a \cdot + \cdot \alpha + A$ is *connected* when its pseudo-trace is positive and $\alpha(I_{\mathcal{A}}) = 1$. In which case we can introduce « isovariants » of $\mathfrak{a}$, namely the $\varphi(\mathfrak{a})$, where φ is any $*$-isomorphism of $\mathcal{A}$ into $\mathcal{H}$ such that $\operatorname{tr} \circ \varphi = \alpha$. The « projects » obtained in this way from connected projects are styled *standardised*.

Definition 143 (Perennialisation). A *perennialisation* is a normal $*$-isomorphism Ω from $\mathcal{R} \otimes \mathcal{H}$ into $\mathcal{R}$.

Definition 144 (Standard bang). $\mathfrak{G}$ being the amenable group of Proposition 65, let $M \subset |\mathfrak{G}|$ be the monoid generated by $\mathtt{l}, \mathtt{r}$. If $\mathcal{H}^{[X]}$ denotes the X-fold tensor power of $\mathcal{H}$ and the crossed product $\mathcal{H}^{[|\mathfrak{G}|]} \rtimes \mathfrak{G}$ refers to the automorphic action $g(\bigotimes_h u_h) := \bigotimes_h u_{gh}$, define

$$\Omega\colon \mathcal{R} \otimes \mathcal{H} \simeq \mathcal{R} \otimes \mathcal{H}^{[M]} \subset \mathcal{R} \otimes (\mathcal{H}^{[|\mathfrak{G}|]} \rtimes \mathfrak{G}) \simeq \mathcal{R}.$$

Definition 145 (Exponentials). Let Ω be a perennialisation. If $\mathfrak{a} = a \cdot\!+\!\cdot \mathrm{tr} + A$ is a standardised project, one defines the project $!_\Omega \mathfrak{a} := a \cdot\!+\!\cdot 1 + \Omega(A)$ of carrier $\Omega(\boldsymbol{a} \otimes I_{\mathcal{H}})$. If $\mathbf{A}$ is a negative conduct of carrier $\boldsymbol{a}$, one defines the ethics $\sharp_\Omega \mathbf{A} := \{!_\Omega \mathfrak{a}; \mathfrak{a} = a \cdot\!+\!\cdot \mathrm{tr} + A \in \mathbf{A}\}$ and the negative conduct $!_\Omega \mathbf{A} := {\sim\sim}\sharp_\Omega \mathbf{A}$, both of carrier $\Omega(\boldsymbol{a} \otimes I_{\mathcal{H}})$.

When Ω is the standard perennialisation, one uses the notation $!\mathbf{A}$.

Theorem 95 (Exponentiation).

$$!_\Omega(\mathbf{A} \mathbin{\&} \mathbf{B}) = !_\Omega \mathbf{A} \otimes !_\Omega \mathbf{B}. \tag{21.33}$$

Proof. Using Remark 7, equation (21.25), the right-hand side can be replaced with ${\sim\sim}(\sharp_\Omega \mathbf{A} \odot \sharp_\Omega \mathbf{B})$. Since $\mathbb{C} \otimes \mathbb{C} = \mathbb{C}$, $\sharp_\Omega \mathbf{A} \odot \sharp_\Omega \mathbf{B} = \sharp_\Omega \mathbf{A} \mathbin{\&} \sharp_\Omega \mathbf{B} = \sharp_\Omega(\mathbf{A} \mathbin{\&} \mathbf{B})$. □

Remark 10. In order to get (21.33), the carriers of $\mathbf{A}$, $\mathbf{B}$ must be disjoint, hence $\mathbf{A} \mathbin{\&} \mathbf{B}$ cannot be defined when the carriers intersect like in $(\mathbf{A} \parr \mathbf{C}) \mathbin{\&} (\mathbf{B} \parr \mathbf{C})$. Hence the loss of literal distributivity.

21.5.4 Promotion. The standard bang allows *contextual* promotion « from $\Gamma \vdash A$, get $!\Gamma \vdash !A$ ».

Theorem 96 (Promotion). *The principle* $!A, !(A \multimap B) \vdash !B$ *can be implemented in GoI.*

Proof. Let $\boldsymbol{a}$, $\boldsymbol{a}'$, $\boldsymbol{b}$, $\boldsymbol{b}'$ be disjoint carriers, let u, v be partial isometries from $\boldsymbol{a}$ to $\boldsymbol{a}'$ and from $\boldsymbol{b}$ to $\boldsymbol{b}'$. If $\mathbf{A}$, $\mathbf{B}$ are negative conducts of respective carriers $\boldsymbol{a}$, $\boldsymbol{b}$, we are seeking an inhabitant of $(!\mathbf{A} \otimes !(u(\mathbf{A}) \multimap \mathbf{B})) \multimap !v(\mathbf{B})$. Indeed, $\mathfrak{c} := 0 \cdot\!+\!\cdot 1 + (u + u^* + v + v^*)$ inhabits $(\mathbf{A} \otimes (u(\mathbf{A}) \multimap \mathbf{B})) \multimap v(\mathbf{B})$ and sends $0 \cdot\!+\!\cdot \alpha + A, 0 \cdot\!+\!\cdot \varphi + F$ to $0 \cdot\!+\!\cdot \alpha \otimes \varphi + v([F^\dagger] u(A^\ddagger))$.

« Banging » $\mathfrak{c}$ basically means internalising the operations $(\cdot)^\dagger$, $(\cdot)^\ddagger$. For this, observe that the sets $\mathtt{l}M$ and $\mathtt{r}M$ are disjoint, because M is a free monoid. In particular $\mathcal{H}^{[\mathtt{l}M \cup \mathtt{r}M]} \simeq \mathcal{H}^{[\mathtt{l}M]} \otimes \mathcal{H}^{[\mathtt{r}M]}$. It is therefore possible to internalise $(\cdot)^\dagger$, $(\cdot)^\ddagger$ by the conjugations w.r.t. the unitaries $\mathtt{l}, \mathtt{r}$.

We thus define $!\mathfrak{c} := 0 \cdot\!+\!\cdot 1 + \Omega(u \otimes \mathtt{r}^*\mathtt{l} + u^* \otimes \mathtt{l}^*\mathtt{r} + v \otimes \mathtt{r} + v^* \otimes \mathtt{r}^*)$. □

Corollary 82.1. *Contextual promotion works for « ! ».*

Proof. Assume that, say $A, B \vdash C$; then one gets $A \multimap (B \multimap C)$, and, since context-free promotion is free of charge, $!(A \multimap (B \multimap C))$. Now, the theorem yields $!A, !B, !(A \multimap (B \multimap C)) \vdash !C$, hence, by a cut, $!A, !B \vdash !C$. □

21.6 Lateralised logic

21.6.1 The witness theorem

Definition 146 (Witnesses). If $\boldsymbol{p} \in \mathcal{R}$ is a carrier, one defines the sets

$$\mathbf{Z}_{\boldsymbol{p}} := \{a \cdot\!+\!\cdot (\lambda \oplus -\mu) + (\boldsymbol{u} \oplus \boldsymbol{v})\ ;\ \boldsymbol{u}, \boldsymbol{v} \subset \boldsymbol{p}, \lambda \mathrm{tr}(\boldsymbol{u}) = \mu \mathrm{tr}(\boldsymbol{v}), \lambda, \mu > 0, a \neq 0\},$$

$$\mathbf{P}_{\boldsymbol{p}} := \{0 \cdot\!+\!\cdot 1 + \boldsymbol{u}\ ;\ 0 \neq \boldsymbol{u} \subset \boldsymbol{p}\},$$

and the conduct $¥\boldsymbol{p} := \sim\sim(\mathbf{Z}_{\boldsymbol{p}} \cup \mathbf{P}_{\boldsymbol{p}})$ of carrier $\boldsymbol{p}$.

Theorem 97 (Witness). *The conducts $¥\boldsymbol{p}$ are positive and proper; moreover:*

(i) $¥0$ *is the positive neutral* $\mathbf{0} := \sim\top$.

(ii) *If $\boldsymbol{p} \neq 0$, then $\mathbf{Z}_{\boldsymbol{p}} \cup \{0 \cdot\!+\!\cdot x + \boldsymbol{p}\ ;\ x \neq 0\}$ is a complete ethics for $¥\boldsymbol{p}$.*

(iii) *If $0 \cdot\!+\!\cdot \alpha + A \in ¥\boldsymbol{p}$ is connected, then A is a (non-zero) projection.*

(iv) *If $\boldsymbol{p}, \boldsymbol{q}$ are disjoint, then $¥\boldsymbol{p} \oplus ¥\boldsymbol{q} = ¥\boldsymbol{p} \parr ¥\boldsymbol{q}$.*

Proof. For $b \neq 0$, $\mathfrak{b} := b \cdot\!+\!\cdot (2 \oplus -1) + (0 \oplus 0) \in \mathbf{Z}_{\boldsymbol{p}}$; if $\mathfrak{a} = a \cdot\!+\!\cdot \alpha + A \in \sim\mathbf{Z}_{\boldsymbol{p}}$, $\ll \mathfrak{a} \mid \mathfrak{b} \gg = a + b\alpha(I_{\mathcal{A}}) \neq 0$ for all $b \neq 0$: this forces $\mathfrak{a}$ to be wager-free. Moreover, since $\mathbf{Z}_{\boldsymbol{p}} \cup \mathbf{P}_{\boldsymbol{p}}$ enjoys positive rescaling, so does its bipolar: we conclude that $¥\boldsymbol{p}$ is positive. Moreover, observe that $\sim\mathbf{Z}_{\boldsymbol{p}} \cap \sim\mathbf{P}_{\boldsymbol{p}}$ is proper: $\ll 0 \cdot\!+\!\cdot 1 + \boldsymbol{p}/2 \mid 0 \cdot\!+\!\cdot \beta + \boldsymbol{q} \gg = \lambda \mathrm{tr}(\boldsymbol{q})\beta(I_{\mathcal{B}}) \log 2$; from this, it follows that $0 \cdot\!+\!\cdot 1 + \boldsymbol{p}/2 \in \sim\mathbf{Z}_{\boldsymbol{p}} \cap \sim\mathbf{P}_{\boldsymbol{p}}$. Thus, $¥\boldsymbol{p} = \sim(\sim\mathbf{Z}_{\boldsymbol{p}} \cap \sim\mathbf{P}_{\boldsymbol{p}})$ is proper as well.

(i) There are not that many conducts of an empty carrier: $¥0$ being positive and proper, it must be equal to $\mathbf{0}$. The remaining items being either vacuous or trivial in the case of null carriers, we assume that $\boldsymbol{p}, \boldsymbol{q} \neq 0$.

(ii) Since $¥\boldsymbol{p} \subset \sim\mathbf{Z}_{\boldsymbol{p}}, 0 \cdot\!+\!\cdot \mathrm{tr}(\boldsymbol{v}) + \boldsymbol{u} \equiv_{¥\boldsymbol{p}} 0 \cdot\!+\!\cdot \mathrm{tr}(\boldsymbol{u}) + \boldsymbol{v}$ for all non-zero $\boldsymbol{u}, \boldsymbol{v} \subset \boldsymbol{p}$, so $¥\boldsymbol{p} = \sim\sim(\mathbf{Z}_{\boldsymbol{p}} \cup \{0 \cdot\!+\!\cdot x + \boldsymbol{p}\ ;\ x \neq 0\})$. If $\mathfrak{b} \in ¥\boldsymbol{p}$ is wager-free, if $\mathfrak{a}, \mathfrak{a}' \in \sim¥\boldsymbol{p}, \lambda \in \mathbb{R}$, then $\mathfrak{a} + \lambda\mathfrak{a}' + \lambda c \mathfrak{Dai}_0 \in \sim\mathbf{Z}_{\boldsymbol{p}}$, with $c := -\alpha(I_{\mathcal{A}})$. If $\ll \mathfrak{b} \mid \mathfrak{a} \gg + \lambda \ll \mathfrak{b} \mid \mathfrak{a}' \gg = 0$, then $\ll \mathfrak{b} \mid \mathfrak{a} + \lambda\mathfrak{a}' + \lambda c \mathfrak{Dai}_0 \gg = 0$ and $\ll 0 \cdot\!+\!\cdot 1 + \boldsymbol{p} \mid \mathfrak{a} + \lambda\mathfrak{a}' + \lambda c \mathfrak{Dai}_0 \gg = \ll 0 \cdot\!+\!\cdot 1 + \boldsymbol{p} \mid \mathfrak{a} \gg + \lambda \ll 0 \cdot\!+\!\cdot 1 + \boldsymbol{p} \mid \mathfrak{a}' \gg = 0$ hence $\mathfrak{b} \equiv_{¥\boldsymbol{p}} 0 \cdot\!+\!\cdot x + \boldsymbol{p}$ for some $x \neq 0$.

(iii) If $\mathfrak{a} = 0 \cdot\!+\!\cdot \alpha + A \in ¥\boldsymbol{p}$ ($\alpha > 0$), then $\ll \mathfrak{a} \mid 0 \cdot\!+\!\cdot 1 + \lambda \gg = c\, \mathrm{colog}(1 - \lambda)$, hence $(\mathrm{tr} \otimes \alpha)(A^n) = c$. Since A is hermitian, $0 \leq A^2 \leq I$, hence $0 \leq A^4 \leq A^2$; since $\mathrm{tr} \otimes \alpha$ is faithful and positive, $(\mathrm{tr} \otimes \alpha)(A^4 - A^2) = 0$ yields $A^4 = A^2$: A^2 is a projection and the partial symmetry A is the difference $A = A^+ - A^-$ of two projections s.t. $A^+A^- = 0$; then $A^2 = A^+ + A^-$ and $(\mathrm{tr} \otimes \alpha)(A - A^2) = 0$ yields $A^- = 0$, i.e., $A^2 = A$.

(iv) Let $\mathfrak{f} = f \cdot\!+\!\cdot \varphi + F \in {¥}\boldsymbol{p} \parr {¥}\boldsymbol{q}$, $\mathfrak{a} = 0 \cdot\!+\!\cdot \alpha + A \in \sim{¥}\boldsymbol{p}$, $\mathfrak{b} = 0 \cdot\!+\!\cdot \beta + B \in \sim{¥}\boldsymbol{q}$, with $\varphi(I_{\mathcal{F}}) = \alpha(I_{\mathcal{A}}) = \beta(I_{\mathcal{B}}) = 1$; write $\ll \mathfrak{f} \mid \mathfrak{a} \otimes \mathfrak{b} \gg = f + g_{\mathfrak{a}} + h_{\mathfrak{b}} + k_{\mathfrak{a}\mathfrak{b}}$, with $g_{\mathfrak{a}} := \mathrm{ldet}(I - FA)$, $h_{\mathfrak{b}} := \mathrm{ldet}(I - FB)$. If $x, y \neq 0$, then $\ll \mathfrak{f} \mid \mathfrak{a}_\lambda + x - 1\mathfrak{Dai}_0 \otimes \mathfrak{b}_\mu + y - 1\mathfrak{Dai}_0 \gg = fxy + g_{\mathfrak{a}} y + h_{\mathfrak{b}} x + k_{\mathfrak{a}\mathfrak{b}} \neq 0$, hence one and only one of the four reals $f, g_{\mathfrak{a}}, h_{\mathfrak{b}}, k_{\mathfrak{a}\mathfrak{b}}$ is non-zero. If $\mathfrak{a}' \in \sim{¥}\boldsymbol{p}$ with $\alpha'(I_{\mathcal{A}'}) = 1$, define $\mathfrak{a}'' := \mathfrak{a} + \mathfrak{a}' - \mathfrak{Dai}_0$ or $\mathfrak{a}'' := \mathfrak{a} - \mathfrak{a}' + \mathfrak{Dai}_0$ so that $\mathfrak{a}'' \in \sim{¥}\boldsymbol{p}$; for the three cases $\mathfrak{a}/\mathfrak{b}, \mathfrak{a}'/\mathfrak{b}, \mathfrak{a}''/\mathfrak{b}$, select one among f, g, h, k, therefore always the same. Same remark for the argument $\mathfrak{b}$, hence we conclude that the departure $f/g/h/k$ is independent of the arguments $\mathfrak{a}, \mathfrak{b}$. Assume $k \neq 0$; let $\mathfrak{x} := 0 \cdot\!+\!\cdot 1 + x\boldsymbol{p} \in \sim{¥}\boldsymbol{p}$, $\mathfrak{y} = 0 \cdot\!+\!\cdot 1 + y\boldsymbol{q} \in \sim{¥}\boldsymbol{q}$ $(x, y \in]0, 1[)$. Then $k_{xy} = \mathrm{ldet}(I - xyF_{12}(I - yF_{22})^{-1}F_{21}(I - xF_{11})^{-1})$. The convergence radius of the analytical function $k_x : y \rightsquigarrow k_{xy}$ tends to infinity when $x \to 0$; but $k_x(y) = \ll [\mathfrak{f}]\mathfrak{x} \mid \mathfrak{y} \gg = \mathrm{colog}(I - c_x y)$ has the convergence radius 1, contradiction. Three cases remain:

f: then $\mathfrak{f} \equiv \mathfrak{Dai}_f$.

g: then $\mathfrak{f} \in \sim\sim({¥}\boldsymbol{p})_{\boldsymbol{p}+\boldsymbol{q}}$.

h: then $\mathfrak{f} \in \sim\sim({¥}\boldsymbol{q})_{\boldsymbol{p}+\boldsymbol{q}}$.

Hence, using weakening, ${¥}\boldsymbol{p} \parr {¥}\boldsymbol{q} = {¥}\boldsymbol{p} \oplus {¥}\boldsymbol{q}$. □

21.6.2 The first action. In ludics (Chapter 13), an essential role is devoted to *actions*: thus, in a behaviour $\mathbf{A} \oplus \mathbf{B}$, the first action of a proper design chooses between $\mathbf{A}$ and $\mathbf{B}$. In GoI, the role of first actions is played by connected projects of the form $0 \cdot\!+\!\cdot \alpha + A$, with A a non-zero projection. If $\mathbf{B} \subset {¥}\boldsymbol{p}$ is a conduct of carrier $\boldsymbol{p}$, its connected projects are of the required form by Theorem 97 (iii): such a conduct may be seen as a « space of first actions ». It must be noticed that, whereas ${¥}\boldsymbol{p}$ admits, up to equivalence, at most one first action, it is no longer the case with $\mathbf{B} \subset {¥}\boldsymbol{p}$ whose equivalence is usually coarser than the one induced by $\equiv_{{¥}\boldsymbol{p}}$. A few examples may help:

(i) When $\boldsymbol{p} \neq 0$, ${¥}\boldsymbol{p}$ admits (up to equivalence) exactly one first action.

(ii) If $\boldsymbol{p} \cdot \boldsymbol{q} = 0$, the first actions of $\sim\sim({¥}\boldsymbol{p})_{\boldsymbol{p}+\boldsymbol{q}} \subset {¥}(\boldsymbol{p}+\boldsymbol{q})$ are those of ${¥}(\boldsymbol{p})$. Indeed, if $\mathfrak{a} = 0 \cdot\!+\!\cdot \alpha + A \in \sim\sim({¥}\boldsymbol{p})_{\boldsymbol{p}+\boldsymbol{q}}$ with $A \cdot (\boldsymbol{q} \otimes I_{\mathcal{A}}) \neq 0$, it is easy to find a project in $\sim({¥}\boldsymbol{p})_{\boldsymbol{p}+\boldsymbol{q}}$ not polar to $\mathfrak{a}$.

(iii) If $\boldsymbol{p} \cdot \boldsymbol{q} = 0$, $\boldsymbol{p}, \boldsymbol{q} \neq 0$, then ${¥}\boldsymbol{p} \oplus {¥}\boldsymbol{q} = \sim\sim({¥}\boldsymbol{p})_{\boldsymbol{p}+\boldsymbol{q}} \cup \sim\sim({¥}\boldsymbol{q})_{\boldsymbol{p}+\boldsymbol{q}}$ by Theorem 92. Thus the first actions $0 \cdot\!+\!\cdot \alpha + A$ of ${¥}\boldsymbol{p} \oplus {¥}\boldsymbol{q}$ split into two equivalence classes: either $A \subset \boldsymbol{p} \otimes I_{\mathcal{A}}$ or $A \subset \boldsymbol{q} \otimes I_{\mathcal{A}}$.

(iv) The case of ${¥}\boldsymbol{p} \parr {¥}\boldsymbol{q}$ is reduced to the previous one by Theorem 97 (iv).

21.6.3 Lateralisation. The remarkable stability of positive subconducts of witnesses is the missing link between GoI and ludics; it is now possible to define *behaviours*, which are sorts of « conducts with a first action », thus allowing *changes of polarity*.

Definition 147 (Behaviours). If $\boldsymbol{p} \subset \boldsymbol{a} \in \mathcal{R}$ are carriers, a *right behaviour of base* $\boldsymbol{p}$ and carrier $\boldsymbol{a}$ is a positive conduct $\mathbf{A}$ of carrier $\boldsymbol{a}$ such that $\mathbf{A}_{\boldsymbol{p}} \subset \text{¥}\boldsymbol{p}$.

Polars of right behaviours are called *left behaviours*; indeed a left behaviour of base $\boldsymbol{p}$ is a negative conduct containing $(\sim\text{¥}\boldsymbol{p})_{\boldsymbol{a}}$.

Consistently with the change of expression (left/right $\rightsquigarrow$ negative/positive), this refined form of polarisation is styled *lateralisation*.

Example 10. The simplest example of a right behaviour of base $\boldsymbol{p}$ and carrier $\boldsymbol{p}$ is $\text{¥}\boldsymbol{p}$, in particular $\mathbf{0} := \text{¥}0$, the disjunctive neutral.

Proposition 83. *Let* $\mathbf{E}$, $\mathbf{F}$ *be ethics of respective carriers* $\boldsymbol{a} \supset \boldsymbol{p}$*; then*

$$\mathbf{E}_{\boldsymbol{p}} \subset \mathbf{F} \iff (\sim\mathbf{F})_{\boldsymbol{a}} \subset \sim\mathbf{E}. \tag{21.34}$$

Proof. Both sides are equivalent to $\forall \mathfrak{a} \in \mathbf{E}\ \forall \mathfrak{b} \in \sim\mathbf{F}\ \ \mathfrak{a} \perp \mathfrak{b}$. □

Corollary 83.1. *Let* $\mathbf{E}$ *be an ethics for the positive conduct* $\mathbf{A}$ *of carrier* $\boldsymbol{a}$ *and assume that* $\boldsymbol{p} \subset \boldsymbol{a}$ *is such that* $\mathbf{E}_{\boldsymbol{p}} \subset \text{¥}\boldsymbol{p}$. *Then* $\mathbf{A}$ *is a right behaviour of base* $\boldsymbol{p}$.

Proof. $\mathbf{E}_{\boldsymbol{p}} \subset \text{¥}\boldsymbol{p} \iff \sim(\text{¥}\boldsymbol{p})_{\boldsymbol{a}} \subset \sim\mathbf{E} = \sim\mathbf{A} \iff \mathbf{A}_{\boldsymbol{p}} \subset \text{¥}\boldsymbol{p}$. □

21.7 The social life of behaviours

21.7.1 Multiplicatives

Right case

Definition 148 (Right tensor product). If $\mathbf{A}, \mathbf{B}$ are right and left behaviours of bases $\boldsymbol{p}, \boldsymbol{q}$ and disjoint carriers $\boldsymbol{a}, \boldsymbol{b}$, then $\mathbf{A} \otimes \mathbf{B}$ is the positive conduct of carrier $\boldsymbol{a} + \boldsymbol{b}$, indeed a right behaviour of base $\boldsymbol{p}$, of Definition 135.

One defines, *mutatis mutandis*, the « par » of two behaviours of opposite lateralities, which is a negative conduct, indeed a left behaviour of base $\boldsymbol{p}$.

Proposition 84. $\mathbf{A} \otimes \mathbf{B}$ *is a right behaviour of base* $\boldsymbol{p}$.

Proof. By weakening, $\sim\mathbf{A} \subset \sim\mathbf{A} \parr \sim\mathbf{B}$; hence $(\mathbf{A} \otimes \mathbf{B})_{\boldsymbol{a}}$ is included in (thus, equal to) $\mathbf{A}$. Then $(\mathbf{A} \otimes \mathbf{B})_{\boldsymbol{p}} = (\mathbf{A}_{\boldsymbol{a}})_{\boldsymbol{p}} = \mathbf{A}_{\boldsymbol{p}} \subset \text{¥}\boldsymbol{p}$. □

Remark 11. The result persists when $\mathbf{B}$ is a plain negative conduct.

Left case

Definition 149 (Left tensor product). If **A**, **B** are left behaviours of bases $\boldsymbol{p}, \boldsymbol{q}$ and disjoint carriers $\boldsymbol{a}$, $\boldsymbol{b}$, then $\mathbf{A} \otimes \mathbf{B}$ is the negative conduct of carrier $\boldsymbol{a} + \boldsymbol{b}$, indeed a left behaviour of base $\boldsymbol{p} + \boldsymbol{q}$, of Definition 135. One defines, *mutatis mutandis*, the « par » of two right behaviours, which is indeed a right behaviour of base $\boldsymbol{p} + \boldsymbol{q}$.

Lemma 85.1. *If* **E**, **F** *are ethics of disjoint carriers* $\boldsymbol{a}$, $\boldsymbol{b}$ *and* $\boldsymbol{p} \subset \boldsymbol{b}$, *then*

$$(\mathbf{E} \multimap \mathbf{F})_{\boldsymbol{a}+\boldsymbol{p}} \subset \mathbf{E} \multimap \mathbf{F}_{\boldsymbol{p}}.$$

Proof. Immediate; see Remark 7 for the definition of $\mathbf{E} \multimap \mathbf{F}$. □

Proposition 85. $\mathbf{A} \otimes \mathbf{B}$ *is a left behaviour of base* $\boldsymbol{p}$.

Proof. Dually, assume that **A**, **B** are right behaviours; the lemma yields

$$(\mathbf{A} \parr \mathbf{B})_{\boldsymbol{p}+\mathbf{c}} \subset \mathbf{A}_{\boldsymbol{p}} \parr \mathbf{B}_{\boldsymbol{q}} \subset \yen \boldsymbol{p} \parr \yen \mathbf{c} \subset \yen(\boldsymbol{p} + \boldsymbol{q}).$$ □

21.7.2 Delateralisation

Definition 150 (Shift). If **A** is a left behaviour of carrier $\boldsymbol{a}$ and base $\boldsymbol{p}$, if $\boldsymbol{s}$ is a non-zero carrier disjoint from $\boldsymbol{a}$, one defines the *right shift* $\downarrow_{\boldsymbol{s}} \mathbf{A} := \yen\boldsymbol{s} \otimes \mathbf{A}$, a right behaviour of base $\boldsymbol{s}$ and carrier $\boldsymbol{a} + \boldsymbol{s}$. One defines, *mutatis mutandis*, the left shift $\uparrow_{\boldsymbol{s}} \mathbf{a} := \sim\yen\boldsymbol{s} \parr \mathbf{A}$ of a right behaviour.

Theorem 98 (Delateralisation). *The usual logical principles of the shift can be implemented in behaviours.*

Proof. We treat the case of a context $\text{Ł} = \mathbf{B}, \mathbf{C}$, where **B**, **C** are right behaviours of bases $\boldsymbol{q}, \boldsymbol{r}$. We assume that **A** is a behaviour of base $\boldsymbol{p}$. We assume that the three carriers and $\boldsymbol{s}$ are pairwise disjoint.

Right case: if $\mathfrak{a} = 0 \cdot + \cdot \alpha + A \in \mathbf{A} \parr \mathbf{B} \parr \mathbf{C}$, where **A** is a left behaviour, then $\downarrow_{\boldsymbol{s}} \mathfrak{a} := 0 \cdot + \cdot \alpha + (\boldsymbol{s} \otimes I_{\mathcal{A}} + A) \in \downarrow_{\boldsymbol{s}} \mathbf{A} \parr \mathbf{B} \parr \mathbf{C}$.

Left case: if $\mathfrak{a} = 0 \cdot + \cdot \alpha + A \in \mathbf{A} \parr \mathbf{B} \parr \mathbf{C}$ is connected, then

$$(\mathbf{A} \parr \mathbf{B} \parr \mathbf{C})_{\boldsymbol{p}+\boldsymbol{q}+\boldsymbol{r}} = \mathbf{A}_{\boldsymbol{p}} \parr \mathbf{B}_{\boldsymbol{q}} \parr \mathbf{C}_{\boldsymbol{r}} \subset \yen\boldsymbol{p} \parr \yen\boldsymbol{q} \parr \yen\boldsymbol{r} = \yen\boldsymbol{p} \oplus \yen\boldsymbol{q} \oplus \yen\boldsymbol{r}.$$

By Theorem 97, $\mathfrak{a}_{\boldsymbol{p}+\boldsymbol{q}+\boldsymbol{r}} = 0 \cdot + \cdot \alpha + c$, with c a non-zero projection included in one of $\boldsymbol{p} \otimes I_{\mathcal{A}}$, $\boldsymbol{q} \otimes I_{\mathcal{A}}$ or $\boldsymbol{r} \otimes I_{\mathcal{A}}$. Let n be such that $n \cdot \mathrm{tr}(\boldsymbol{s}) \geq \mathrm{tr}(\boldsymbol{p}), \mathrm{tr}(\boldsymbol{q}), \mathrm{tr}(\boldsymbol{r})$, and let φ be the (non-unital) $*$-isomorphism from $\mathcal{A}$ to $\mathcal{A}' := \mathcal{M}_n(\mathbb{C}) \otimes \mathcal{A} = \mathcal{M}_n(\mathcal{A})$: $\varphi(u) := \begin{bmatrix} u & 0 & \dots \\ 0 & 0 & \dots \\ \dots & \dots & \dots \\ \dots & \dots & \dots \end{bmatrix}$, the pseudo-trace on $\mathcal{A}'$ being $\mathrm{Tr} \otimes \mathcal{A}$.

Then, by Remark 9, $\varphi(\mathbf{a}) := 0 \cdot + \cdot \alpha' + (I \otimes \varphi) A (I \otimes \varphi^*) \in \mathbf{A} \parr \mathbf{B} \parr \mathbf{C}$. Let $u \in \mathcal{R} \otimes \mathcal{A}'$ be a partial isometry of domain c and image $\boldsymbol{s}' \otimes I_{\mathcal{A}'}$, with $\boldsymbol{s}' \subset \boldsymbol{s}$. Then $\uparrow_{\boldsymbol{s}} \mathfrak{a} := 0 \cdot + \cdot \alpha' + ((I \otimes \varphi) A (I \otimes \varphi^*) + u + u^*) \in \uparrow_{\boldsymbol{r}} \mathbf{A} \parr \mathbf{B} \parr \mathbf{C}$. Indeed, if $\mathfrak{b} \in \sim\mathbf{A}$, then $[\uparrow_{\boldsymbol{r}} \mathfrak{a}](\downarrow_{\boldsymbol{r}} \mathfrak{b}) = [\mathfrak{a}]\mathfrak{b}$. □

21.7.3 Quantifiers

Definition 151 (Quantifiers). If $\mathbf{A}[i] (i \in \mathbb{I})$ is a non-empty family of behaviours of the same carrier $\boldsymbol{a}$, the same base $\boldsymbol{p}$ and the same laterality, one defines the quantifier $\forall i \in \mathbb{I}\ \mathbf{A}[i] := \bigcap_{i \in \mathbb{I}} \mathbf{A}[i]$, which turns out to be another behaviour of the same carrier, base and lateralisation.

One defines dually $\exists i \in \mathbb{I}\ \mathbf{A}[i] := \sim\sim\bigcup_{i \in \mathbb{I}} \mathbf{A}[i]$.

Proposition 86. $\forall i \in \mathbb{I}\ \mathbf{A}[i]$ *is a behaviour.*

Proof. **Right:** if $i_0 \in \mathbb{I}$, then $(\forall i \in \mathbb{I}\ \mathbf{A}[i])_{\boldsymbol{p}} \subset \mathbf{A}[i_0]_{\boldsymbol{p}} \subset ¥\boldsymbol{p}$.

Left (dually): if the $\mathbf{A}[i]$ are right behaviours, then $\bigcap_i \mathbf{A}[i] \subset ¥\boldsymbol{p}$. By the corollary to Proposition 83, $\exists i \in \mathbb{I} \mathbf{A}[i]$ is a right behaviour. □

Remark 12. « $\uparrow_s$ » being an instance of « par », it distributes over universal quantification.

21.7.4 Additives

Definition 152 (Additives). If **A**, **B** are behaviours of the same lateralisation, of bases $\boldsymbol{p}, \boldsymbol{q}$ and disjoint carriers $\boldsymbol{a}, \boldsymbol{b}$, Definition 137 yields conducts of carrier $\boldsymbol{a}+\boldsymbol{b}$, indeed a behaviour of base $\boldsymbol{p}+\boldsymbol{q}$ with the same lateralisation as **A**, **B**.

Proposition 87. *If* **A**, **B** *are right behaviours, so is* $\mathbf{A} \oplus \mathbf{B}$.

Proof. If $\mathbf{E} := \mathbf{A}_{\boldsymbol{a}+\boldsymbol{b}} \cup \mathbf{B}_{\boldsymbol{a}} + \boldsymbol{b}$, then $\mathbf{E}_{\boldsymbol{p}+\boldsymbol{q}} \subset ¥\boldsymbol{p} \oplus ¥\boldsymbol{q} \subset ¥(\boldsymbol{p}+\boldsymbol{q})$. We conclude using the corollary to Proposition 83. □

21.7.5 Du côté de chez Gustave. The Gustave function (Section 12.1.6) admits the ternary structure $(A \oplus B) ⅋ (A' \oplus B') ⅋ (A'' \oplus B'')$ that coherent spaces cannot disentangle into something simpler, e.g., $(A \oplus B) ⅋ B' ⅋ (A'' \oplus B'')$.

Assuming $(\mathbf{A} \oplus \mathbf{B}) ⅋ \mathbf{B}' ⅋ (\mathbf{A}'' \oplus \mathbf{B}'')$ of base $\boldsymbol{r} := \boldsymbol{a}+\boldsymbol{b}+\boldsymbol{a}'+\boldsymbol{b}'+\boldsymbol{a}''+\boldsymbol{b}''$,

$$
\begin{aligned}
&((\mathbf{A} \oplus \mathbf{B}) ⅋ (\mathbf{A}' \oplus \mathbf{B}') ⅋ (\mathbf{A}'' \oplus \mathbf{B}''))_{\boldsymbol{r}} \\
&\quad \subset (¥\boldsymbol{a} \oplus ¥\boldsymbol{b}) ⅋ (¥\boldsymbol{a}' \oplus ¥\boldsymbol{b}') ⅋ (¥\boldsymbol{a}'' \oplus ¥\boldsymbol{b}'') \\
&\quad \subset ¥\boldsymbol{a} \oplus ¥\boldsymbol{b} \oplus ¥\boldsymbol{a}' \oplus ¥\boldsymbol{b}' \oplus ¥\boldsymbol{a}'' \oplus ¥\boldsymbol{b}''
\end{aligned}
$$

from which we get the existence of the « first action », see Section 21.6.2.

21.7.6 Secularisation. If $\mathbf{A}$ is a left behaviour of carrier $\boldsymbol{a}$ and base $\boldsymbol{p}$ and Ω is a perennialisation, then $!_\Omega \mathbf{A}$ (Definition 145) is a negative conduct, but not a left behaviour. This problem is perhaps the explanation for the other iconoclastic logic, indeed the original one, **LLL** (Chapter 16). In that case, this would definitely show the soundness of the present approach, which manages to explain both light logics out of natural geometric constraints, and not in the usual *Deus ex machina*, i.e., essentialist, way. I just put together a few facts:

(i) Conducts may socialise with behaviours: when $\mathbf{A}$ is a right behaviour and $\mathbf{B}$ is a negative conduct, $\mathbf{A} \otimes \mathbf{B}$ is a right behaviour (Remark 11). It is therefore possible to use ! on the left of an implication: if $\mathbf{A}$, $\mathbf{B}$ are left behaviours, so is $!\mathbf{A} \multimap \mathbf{B}$.

(ii) In terms of sequent calculus, this requires a special maintenance for « ! », e.g., through the familiar *underlining* technique (Section 15.4.2).

(iii) However, due to the want of *dereliction*, it is not reasonable to represent implication by $!\mathbf{A} \multimap \mathbf{B}$, and $!\mathbf{A} \multimap !\mathbf{B}$ is still not a behaviour. The idea is to use a lateralised *subrogate* for « ! », the *secularisation* §.

(iv) Promotion subsists under the weaker form « from $\Gamma \vdash A$, get $!\Gamma \vdash \S A$ » (Γ, A left lateralised).

Definition 153 (Standardisation). In the spirit of Definition 142, a non-connected project can be « standardised » into $\mathfrak{a} := a \cdot\!+\!\cdot (\lambda \mathrm{tr} \oplus \mu \mathrm{tr}) + A$, the « idiom » being now $\mathcal{H} \oplus \mathcal{H}$ and the « pseudo-trace » being $\lambda \mathrm{tr} \oplus \mu \mathrm{tr}$, with $\lambda > 0$, $\mu < 0$.

Definition 154 (Secularisation). If $\mathfrak{a} := a \cdot\!+\!\cdot (\lambda \mathrm{tr} \oplus \mu \mathrm{tr}) + (A \oplus B)$ is a standardised project of carrier $\boldsymbol{a}$, one defines $\bar{\S}_\Omega \mathfrak{a} := a \cdot\!+\!\cdot (\lambda \ominus \mu) + (\Omega(A) \oplus \Omega(B))$.

If $\mathbf{A}$ is a right behaviour of carrier $\boldsymbol{a}$ and base $\boldsymbol{p}$, one defines the right behaviour $\bar{\S}_\Omega \mathbf{A} := \sim\sim\{\bar{\S}_\Omega \mathfrak{a} : \mathfrak{a} \in \mathbb{A}\}$ of carrier $\Omega(\boldsymbol{a} \otimes I_{\mathcal{H}})$ and base $\Omega(\boldsymbol{p} \otimes I_{\mathcal{H}})$.

One symmetrically defines $\S_\Omega \mathbf{A} := \sim\bar{\S}_\Omega \sim\mathbf{A}$; obviously $!_\Omega \mathbf{A} \subset \S_\Omega \mathbf{A}$.

Proposition 88. *If $\mathbf{A}$ is a left behaviour, then $\S_\Omega \mathbf{A}$ is a left behaviour.*

Proof. The result follows follows from

Lemma 88.1. $\bar{\S}\yen \boldsymbol{p} \subset \yen\Omega(\boldsymbol{p} \otimes I_{\mathcal{H}})$.

In this Kamchatka of the book, I feel like skipping one proof, easy anyway. □

21.A Second-order quantification

A purely locative approach would consist in defining, for $r > 0$, $\forall_r \mathbf{X}\, \mathbf{A}[\mathbf{X}]$, where $\mathbf{X}$ varies over all conducts of carrier $\boldsymbol{r}$, where $\boldsymbol{r}$ is a given carrier such that $\mathrm{tr}(\boldsymbol{r}) = r$.

The problem is with the change of « size »: the replacement of $\forall_r \mathbf{X}$ with $\forall_s \mathbf{X}$ is a cinch – using projections – when $0 < s \le r$, but is problematic when $s > r$. Defining[2] $\mathbf{nat} := \forall_1 \mathbf{X}(!(\mathbf{X} \multimap \mathbf{X}) \multimap !(\mathbf{X} \multimap \mathbf{X}))$, we see that **nat** has size $4 > 1$ and cannot be substituted for **X**, thus barring any decent form of recurrence.

The correct definition is semi-locative; in a spirit loosely inspired from the coherent interpretation of second-order quantification (Section 8.3.4), we shall « approximate » conducts by means of conducts of smaller size. One should thus define *variable* conducts and projects.

21.A.1 The negative universe. Instead of a general (and illegible) definition of variability, I will content myself with the case of those negative conducts arising from variables $\mathbf{X}, \mathbf{Y}, \ldots$, the constant $\top$ (conjunctive unit), $\otimes$, &, ! and an *ad hoc* redefinition of implication (to be used throughout this section):

$$\mathbf{A} \multimap \mathbf{B} := \{\mathfrak{f} \in \mathbf{A} \multimap \mathbf{B} \; ; \; \mathfrak{f} \text{ wager-free}\} \tag{21.35}$$

which is such that $\mathbf{A} \multimap \mathbf{B}$ is negative when both $\mathbf{A}, \mathbf{B}$ are negative. It will turn out that second-order universal quantification – still to be defined – is also part of those operations internal to negative conducts.

A sort of *negative universe* has thus been introduced, where no change of polarity is actually needed: an alternative to lateralisation and behaviours. Should we need disjunction, the second-order definition

$$A \oplus B := \forall X \; (A \multimap X) \multimap ((B \multimap X) \multimap X) \tag{21.36}$$

would provide a sort of *ersatz* (see Sections 6.1.3, 12.B.2).

The negative universe is most likely **ELL**-like: usual data translate as

$$\begin{aligned}
\mathbf{bool} &:= \forall X((X \otimes X) \multimap X), \\
\mathbf{bin} &:= \forall X((!(X \multimap X) \otimes !(X \multimap X)) \multimap !(X \multimap X)), \\
\mathbf{nat} &:= \forall X(!(X \multimap X) \multimap !(X \multimap X)).
\end{aligned}$$

21.A.2 Variability. I restrict myself to those conducts obtained by means of $\mathbf{X}$, $\top$, $\otimes$, &, !, $\multimap$, $\forall X$. I try, as much as possible, to minimise the use of isomorphisms; typically, the carriers $\boldsymbol{a}$ under consideration are such that $\Omega(\boldsymbol{a} \otimes I) = \boldsymbol{a}$, where Ω is the *standard bang* of Section 21.5.3.

Supports: we fix, once and for all, a carrier $\boldsymbol{v}$ such that $\mathrm{tr}(\boldsymbol{v}) = 1$. Second-order variables $\mathbf{X}, \mathbf{Y}, \ldots$ will stand for negative conducts of carrier $\boldsymbol{v}$. *Literals* $\varphi(\mathbf{X}), {\sim}\psi(\mathbf{Y}), \ldots$ are obtained by means of *delocations* $\varphi, \psi, \ldots$, i.e., partial isometries of domain I (the full space) and *pairwise disjoint* images

[2]The formula « forgets » the four delocations of **X**.

$\varphi\varphi^*, \psi\psi^*, \dots$, the *supports* of $\varphi(\mathbf{X}), \sim\psi(\mathbf{Y}), \dots$ which are infinite projections containing the carriers $\varphi(\boldsymbol{v}), \sim\psi(\boldsymbol{v}), \dots$: some « extra space » is needed to handle second-order substitution. It is indeed the case that the carrier (resp. the support) of a compound negative conduct **A** is the sum of the carriers (resp. supports) of its literals. In particular, the carrier $\boldsymbol{a}$ of a conduct $\mathbf{A}[\mathbf{X}]$ depending on **X** can symbolically be written $(m+n)\cdot\boldsymbol{v}$, which means « m occurrences (= delocations) of **X** and n occurrences of other literals, free or bound ».

Substitution: the substitution of **B** for **X** in **A** cannot keep the carrier constant, for the simple reason that the carrier $\boldsymbol{b}$ of **B** is a priori distinct from $\boldsymbol{v}$. However, the carrier $\boldsymbol{c}$ of $\mathbf{A}[\mathbf{B}/\mathbf{X}]$ is included in the support of **A**. The carrier of $\mathbf{A}[\mathbf{B}/\mathbf{X}]$ can symbolically be written $m\cdot\boldsymbol{b}+n\cdot\boldsymbol{v}$; if $\boldsymbol{b}$ is symbolically written $p\cdot\boldsymbol{v}$, we get the expression $(mp+n)\cdot\boldsymbol{v}$: a perfectly incorrect – but legible – way to speak of the various isomorphisms at stake. Since $mp+n \leq (m+n)(p+1)$, there is a (non-unital) $*$-isomorphism (noted $\cdot[\mathbf{B}/\mathbf{X}]$) from $\boldsymbol{c}\mathcal{R}\boldsymbol{c}$ into $\boldsymbol{a}\mathcal{R}\boldsymbol{a}\otimes\mathcal{M}_{p+1}(\mathbb{C})$.

Quantification: if $\mathfrak{a} \in \sim\mathbf{A}[\mathbf{B}/\mathbf{X}]$, then $\mathfrak{a}[\mathbf{B}/\mathbf{X}]$ is a project of carrier $\boldsymbol{a}$, provided we consider the component $\mathcal{M}_{p+1}(\mathbb{C})$ of the image of the isomorphism $\cdot[\mathbf{B}/\mathbf{X}]$ as *idiomatic*. $\forall\mathbf{X}\,\mathbf{A}$ is defined as the polar of all $\mathfrak{a}[\mathbf{B}/\mathbf{X}]$, when $\mathfrak{a} \in \sim\mathbf{A}[\mathbf{B}/\mathbf{X}]$ for some negative conduct **B**.

21.A.3 An example: natural numbers.

Proofs of

$$(X \multimap X), \dots, (X \multimap X) \vdash X \multimap X$$

yield matrices M_n; those matrices are plain, i.e., embody the delocations: this explains the coefficients $\boldsymbol{v}$. Typically:

$$M_0 := \begin{bmatrix} 0 & \boldsymbol{v} \\ \boldsymbol{v} & 0 \end{bmatrix}, \quad M_3 := \begin{bmatrix} 0 & 0 & 0 & 0 & 0 & 0 & 0 & \boldsymbol{v} \\ 0 & 0 & \boldsymbol{v} & 0 & 0 & 0 & 0 & 0 \\ 0 & \boldsymbol{v} & 0 & 0 & 0 & 0 & 0 & 0 \\ 0 & 0 & 0 & 0 & \boldsymbol{v} & 0 & 0 & 0 \\ 0 & 0 & 0 & \boldsymbol{v} & 0 & 0 & 0 & 0 \\ 0 & 0 & 0 & 0 & 0 & 0 & \boldsymbol{v} & 0 \\ 0 & 0 & 0 & 0 & 0 & \boldsymbol{v} & 0 & 0 \\ \boldsymbol{v} & 0 & 0 & 0 & 0 & 0 & 0 & 0 \end{bmatrix}.$$

Next, the M_n are perennialised, yielding matrices N_n. The case $n = 3$ involves elements $\mathtt{d}_1, \mathtt{d}_2, \mathtt{d}_3$ of the free monoid, e.g., $\mathtt{r}, \mathtt{rl}, \mathtt{rl}^2$ which are incompatible as suffixes, i.e., such that $\mathtt{ad_i} = \mathtt{bd_j}$ implies $\mathtt{a} = \mathtt{b}$ and $i = j$. For legibility, let us introduce the notation $a := \Omega(\boldsymbol{v} \otimes \mathtt{d}_1)$, $b := \Omega(\boldsymbol{v} \otimes \mathtt{d}_2)$, $c := \Omega(\boldsymbol{v} \otimes \mathtt{d}_3)$:

$$N_0 := \begin{bmatrix} 0 & \boldsymbol{v} \\ \boldsymbol{v} & 0 \end{bmatrix}, \quad N_3 := \begin{bmatrix} 0 & 0 & 0 & 0 & 0 & 0 & 0 & a^* \\ 0 & 0 & b^*a & 0 & 0 & 0 & 0 & 0 \\ 0 & a^*b & 0 & 0 & 0 & 0 & 0 & 0 \\ 0 & 0 & 0 & 0 & c^*b & 0 & 0 & 0 \\ 0 & 0 & 0 & b^*c & 0 & 0 & 0 & 0 \\ 0 & 0 & 0 & 0 & 0 & 0 & c & 0 \\ 0 & 0 & 0 & 0 & 0 & c^* & 0 & 0 \\ a & 0 & 0 & 0 & 0 & 0 & 0 & 0 \end{bmatrix}.$$

Finally comes the contraction/weakening, yielding 4×4 matrices P_n:

$$P_0 := \begin{bmatrix} 0 & 0 & 0 & 0 \\ 0 & 0 & 0 & 0 \\ 0 & 0 & 0 & \boldsymbol{v} \\ 0 & 0 & \boldsymbol{v} & 0 \end{bmatrix}, \quad P_3 := \begin{bmatrix} 0 & A & 0 & B \\ A^* & 0 & C & 0 \\ 0 & C^* & 0 & 0 \\ B^* & 0 & 0 & 0 \end{bmatrix}$$

where $A, B, C, \ldots$ are the 4×4 matrices

$$A := \begin{bmatrix} 0 & b^*a & 0 & 0 \\ 0 & 0 & c^*b & 0 \\ 0 & 0 & 0 & 0 \\ 0 & 0 & 0 & 0 \end{bmatrix}, \quad B := \begin{bmatrix} 0 & 0 & 0 & a^* \\ 0 & 0 & 0 & 0 \\ 0 & 0 & 0 & 0 \\ 0 & 0 & 0 & 0 \end{bmatrix}, \quad C := \begin{bmatrix} 0 & 0 & 0 & 0 \\ 0 & 0 & 0 & 0 \\ 0 & 0 & 0 & 0 \\ 0 & 0 & c & 0 \end{bmatrix}.$$

The idiom of P_n is $\mathcal{M}_{n+1}(\mathbb{C})$, hence the 4×4 matrices A, B, C when $n = 3$.

What we just constructed can be denoted by $P_n[\boldsymbol{v}]$ to emphasise the dependency upon the carrier $\boldsymbol{v}$. Should we perform an extraction on $\mathbf{B}$ of carrier $\boldsymbol{b}$, then P_n should become $P_n[\boldsymbol{b}]$, an element of $!(\mathbf{B} \multimap \mathbf{B}) \multimap !(\mathbf{B} \multimap \mathbf{B})$. The important point is that this extraction can be implemented, using delocations, by a sort of contraction. For instance, if $\operatorname{tr}(\boldsymbol{b}) = 2$, an appropriate variant of the project $\mathfrak{Contr}$ of Section 21.5.2 will do the job: if φ, ψ are partial isometries between $\boldsymbol{v}$ and $\boldsymbol{b}', \boldsymbol{b}''$ such that $\boldsymbol{b} = \boldsymbol{b}' + \boldsymbol{b}''$, define $u := \Omega(\varphi \otimes I)$, $v := \Omega(\psi \otimes I)$.

21.B Truth

21.B.1 Viewpoints. It $\mathbb{R}$ is equipped with the Lebesgue measure μ, if T is a partial measure-preserving bijection from $X \subset \mathbb{R}$ to $Y \subset \mathbb{R}$, then $\widetilde{T}(f) := f \circ T$ defines a bounded operator on $\mathcal{L}^2(\mathbb{R})$. Indeed, $\widetilde{TU} = \widetilde{T}\widetilde{U}$, $\widetilde{T}^* = \widetilde{T^{-1}}$, hence $\widetilde{T}$ is a partial isometry, of domain and image $\widetilde{X}, \widetilde{Y}$ $(= \mathcal{L}^2(X), \mathcal{L}^2(Y))$ where X, Y denote the identity maps of X and Y.

Definition 155 (Viewpoints). A *viewpoint* is a normal representation of $\mathcal{R}$ in $\mathcal{L}^2(\mathbb{R})$, what we (abusively) write $\mathcal{R} \subset \mathcal{B}(\mathcal{L}^2(\mathbb{R}))$ such that, for any $X \subset \mathbb{R}$, with $\mu(X) < \infty$, $\widetilde{X} \in \mathcal{R}$ and $\operatorname{tr}(\widetilde{X}) = \mu(X)$.

Lemma 89.1. *Let* $a \in \boldsymbol{p}\mathcal{R}\boldsymbol{p}$ *with* $\boldsymbol{p}$ *finite and* $\|a\| \leq 1$ *be such that* $\operatorname{tr}(a^n) = 0$ *for all* $n > 0$*; then* $\operatorname{ldet}(I - a) = 0$ *in the two following cases:*

(i) *a is hermitian.*

(ii) *a is a partial isometry.*

Proof. $2\text{ldet}(I - a) = \text{ldet}((I - a)(I - a^*)) = -log4 + \text{ldet}(I - b)$, with $b := 1/4(3I + a + a^* - aa^*)$. Since $\|aa^* - a - a^*\| \leq 3$, $0 \leq b \leq I$ and $\text{ldet}(b) = \sum_{n>0} \text{tr}(b^n)/n$. If a is hermitian, the b^n are polynomials in a and $\text{tr}(b^n) = (3/4)^n$, hence $\text{ldet}(b) = \text{colog}(1 - 3/4) = \log 4$ and $\text{ldet}(I - a) = 0$. If a is a partial isometry, then $b^n = x_n I + y_n(a + a^*) + z_n aa^* + w_n a^* a$ and $\text{tr}(b^n) = x_n + (z_n + w_n)\text{tr}(aa^*)$. The coefficients x_n, z_n, w_n do not depend upon a; in particular, if $a^2 = 0$, then $\text{ldet}(I - a) = 0$ and $\text{ldet}(I - b) = \log 4$, hence the two series $\sum_{n>0} x_n = \log 4$ and $\sum_{n>0}(z_n + w_n)\text{tr}(aa^*) = 0$ are absolutely convergent. The same holds for any partial isometry a. □

Lemma 89.2. *If $\widetilde{T} \in \mathcal{R}$, where T is a partial measure-preserving bijection from $X \subset \mathbb{R}$ ($\mu(X) < \infty$) to $Y \subset \mathbb{R}$, then* $\text{tr}(\widetilde{T}) = \mu(\{x \,;\, T(x) = x\})$.

Proof. If $Z \subset X$ is measurable, let $T_Z : Z \mapsto T(Z)$ be the restriction of T to Z. If $A := \{x \in X \,;\, T(x) \neq x\}$, then $\text{tr}(T) = \text{tr}(T_A) + \mu(X \setminus A)$: it remains to prove that $\text{tr}(T_A) = 0$; in other terms that $\text{tr}(T) = 0$ when $T(x) \neq x$ for all $x \in X$. By the strong continuity of the trace, there is a maximal (up to a negligibility) $Z \subset X$ such that $\text{tr}(T_Z) = 0$. If, up to negligibility, $Z \neq X$, there is a non-negligible $W \subset X \setminus Z$ such that $T(W) \cap W = \emptyset$; it is immediate that $\text{tr}(T_{Z \cup W}) = 0$, contradicting the choice of Z. Hence $\text{tr}(T) = \text{tr}(T_Z) = 0$. □

Proposition 89. *If $\widetilde{T} \in \mathcal{R}$, where T is a partial measure-preserving bijection from $X \subset \mathbb{R}$ ($\mu(X) < \infty$) to $Y \subset \mathbb{R}$; then* $\text{ldet}(I - \widetilde{T}) = 0$ *or* $\text{ldet}(I - \widetilde{T}) = \infty$.

Proof. If the set $\{z \,;\, \exists n > 0\ T^n(z) = z\}$ is of measure 0, Lemma 89.2 yields $\text{tr}(\widetilde{T^n}) = 0$ for all $n > 0$, hence, by Lemma 89.1 (ii), $\text{ldet}(I - \widetilde{T}) = 0$. Otherwise, let $N > 0$ be such that $Z := \{z \,;\, T^N(z) \neq 0\}$ is not negligible. Then, writing $T = T_Z \cup (T \restriction Z)$ and $\widetilde{T} = \widetilde{T_Z} + \widetilde{T \restriction Z}$, with $T_Z \cdot (T \restriction Z)$, we get $\text{ldet}(I - \widetilde{T}) = \text{ldet}(I - \widetilde{T_Z}) + \text{ldet}(I - \widetilde{T \restriction Z})$. By Lemma 89.2, the terms $\text{tr}(\widetilde{T \restriction Z^{kN}})/kN$ are equal to $\mu(Z)/kN$. Hence $\text{ldet}(I - \widetilde{T \restriction Z}) = +\infty$ and $\text{ldet}(I - \widetilde{T}) = +\infty$. □

Let us come back to the feedback equation: if w.r.t. a decomposition $I = \boldsymbol{a} \oplus \boldsymbol{b}$, $F := \begin{pmatrix} F_{11} & F_{12} \\ F_{21} & F_{22} \end{pmatrix}$ and $A = \boldsymbol{a}A\boldsymbol{a}$, then $[F]A = F_{22} + F_{21}A(I - F_{11}A)^{-1}F_{12}$, provided $\boldsymbol{a} - F_{11}A$ is invertible. My contention is that this formula is still valid when $\boldsymbol{a} - F_{11}$ is injective and $(\boldsymbol{a} - F_{11}A)^{-1}F_{12}$ is total, hence (since of closed graph) bounded; I prove it in a particular case:

Lemma 90.1. *If $\boldsymbol{a} - F_{11}A$ is injective,* $\text{ldet}(I - F_{11}A) < +\infty$ *and* $(\boldsymbol{a} - F_{11}A)^{-1}F_{12}$ *is total, then* $[F]A = F_{22} + F_{21}A(\boldsymbol{a} - F_{11}A)^{-1}F_{12}$.

Proof. Let B be such that $\mathbf{b}B\boldsymbol{b} = B$; a standard computation shows that

$$\text{ldet}(I-F\cdot(A+B)) = \text{ldet}(I-F_{11}A)+\text{ldet}(I-(F_{22}+F_{21}A(\boldsymbol{a}-F_{11}A)^{-1}F_{12}B)).$$

Hence, using Theorem 87, $[F]A = F_{22} + F_{21}A(\boldsymbol{a} - F_{11}A)^{-1}F_{12}$. □

Proposition 90. *If $\boldsymbol{a} = \widetilde{X}, \boldsymbol{b} = \widetilde{Y}$ are disjoint carriers and if the partial measure-preserving bijections A, F induce partial isometries $\widetilde{A}, \widetilde{F} \in \mathcal{R}$ of respective carriers $\boldsymbol{a}$ and $\boldsymbol{a}+\boldsymbol{b}$, then $[F]A = \widetilde{U}$ for some partial measure-preserving bijection U.*

Proof. Let $Z := \{x \in \mathbf{R}\ ;\ \exists n > 0\ (F_{11}A)^n(x) = x\}$; one easily reduces the problem to the case where $\mu(Z) = 0$. Consider the partial bijection

$$U := F_{22} \cup F_{21}(A \cup AF_{11}A \cup AF_{11}AF_{11}A \cup \cdots)F_{12};$$

$\boldsymbol{a} + \widetilde{F_{11}A} + \widetilde{F_{11}AF_{11}A} + \cdots$ is a left inverse of $\boldsymbol{a} - F_{11}A$ and $(\boldsymbol{a} + \widetilde{F_{11}A} + \widetilde{F_{11}AF_{11}A} + \cdots)\widetilde{F_{12}}$ comes from a partial bijection and is thus bounded. The result follows from the lemma. □

Remark 13. If $\mu(Z) = 0$, the $(\widetilde{F_{11}A})^N$ tend to 0, strongly: a case of *strong nilpotency*.

21.B.2 Subjective truth. If $\mathcal{R} \subset \mathcal{B}(\mathcal{L}^2(\mathbb{R}))$ is a viewpoint, then a base $\mathbf{e_1}, \ldots, \mathbf{e_n}$ of the idiom $\mathcal{A}$ induces a viewpoint $\mathcal{R}\otimes\mathcal{A} \subset \mathcal{B}(\mathcal{L}^2(\mathbb{R}\times\{1,\ldots,n\}))\ (\simeq \mathcal{B}(\mathcal{L}^2(\mathbb{R})))$.

Definition 156 (Success). $\mathfrak{a} := a\cdot\!+\!\cdot\alpha + A$ is *successful* w.r.t. a viewpoint $\mathcal{R} \subset \mathcal{B}(\mathcal{L}^2(\mathbb{R}))$ when:

(i) The carrier $\boldsymbol{a}$ of $\mathfrak{a}$ is of the form $\widetilde{X}$.

(ii) $\mathfrak{a}$ is wager-free ($a = 0$) and connected ($\alpha > 0$).

(iii) W.r.t. a base $\mathbf{e_1}, \ldots, \mathbf{e_n}$ of the idiom $\mathcal{A}$, $A = \widetilde{T}$ for a certain partial measure-preserving map from a subset of $\mathbb{R} \times \{1,\ldots,n\}$ to a subset of $\mathbb{R} \times \{1,\ldots,n\}$.

Definition 157 (Truth). A conduct $\mathbf{A}$ of carrier $\boldsymbol{a}$ is *true* w.r.t. a viewpoint $\mathcal{R} \subset \mathcal{B}(\mathcal{L}^2(\mathbb{R}))$ when $\boldsymbol{a}$ is of the form $\widetilde{X}$ and $\mathbf{A}$ contains a project $\mathfrak{a}$ successful w.r.t. the viewpoint. $\mathbf{A}$ is *false* when $\sim\!\mathbf{A}$ is true.

Theorem 99 (Compositionality of truth). *If $\boldsymbol{a} = \widetilde{X}$, $\boldsymbol{b} = \widetilde{Y}$, if the conducts $\mathbf{A}$ and $\mathbf{A} \multimap \mathbf{B}$ of respective carriers $\boldsymbol{a}$, $\boldsymbol{a}+\boldsymbol{b}$ are true, then $\mathbf{B}$ is true.*

Proof. Let $\mathfrak{a} := 0\cdot\!+\!\cdot\alpha + \widetilde{T} \in \mathbf{A}$, and let $\mathfrak{f} := 0\cdot\!+\!\cdot\varphi + \widetilde{U} \in \mathbf{A} \multimap \mathbf{B}$. Then $[\mathfrak{f}]\mathfrak{a} := \text{ldet}(I - \widetilde{U}^{\ddagger}_{11}\widetilde{T^{\dagger}})\cdot\!+\!\cdot[\widetilde{U^{\ddagger}}]\widetilde{T^{\dagger}}$. If $\sim\!\mathbf{B}$ is true, then $0\cdot\!+\!\cdot 1 + 0 \in \mathbf{B}$ is successful; if $\mathfrak{b} \in \sim\!\mathbf{B}$, then, using Theorem 87, $\ll [\mathfrak{f}]\mathfrak{a} \mid \mathfrak{b} \gg \neq \infty$ implies $\text{ldet}(I - \widetilde{U}^{\ddagger}_{11}\widetilde{T^{\dagger}}) \neq \infty$. Now, if T, U are measure-preserving partial bijections of

$\mathbb{R} \times \{1, \ldots, n\}$, $\mathbb{R} \times \{1, \ldots, m\}$, then $\widetilde{T}^{\dagger}, \widetilde{U}^{\ddagger}$ come from partial bijections $T^{\dagger}, U^{\ddagger}$ of $\mathbb{R} \times \{1, \ldots, n\} \times \{1, \ldots, m\}$. By Proposition 89, $\operatorname{ldet}(I - \widetilde{U}^{\ddagger}_{11}\widetilde{T^{\dagger}}) = 0$, since the value ∞ has just been excluded. By Proposition 90, $[\widetilde{U^{\ddagger}}]\widetilde{T}^{\dagger}$ is of the form $\widetilde{V}$, hence $[\mathfrak{f}]\mathfrak{a}$ is successful. □

Corollary 99.1 (Subjective consistency). *A conduct cannot be both true and false w.r.t. a given viewpoint.*

Proof. $\sim\mathbf{A} = \mathbf{A} \multimap \mathbf{0}$, where $\mathbf{0} := \{a \cdot\!+\!\cdot \alpha + 0\ ;\ a \neq 0\}$ of carrier 0 is the neutral element of « par », which contains no wager-free project. □

21.B.3 The subjective paradox. A conduct can be true or false depending on the viewpoint:

Proposition 91. *There exists a conduct* $\mathbf{C}$ *and viewpoints* $\mathcal{P}_1, \mathcal{P}_2$ *such that* $\mathbf{C}$ *is true w.r.t.* $\mathcal{P}_1$ *and* $\sim\mathbf{C}$ *is true w.r.t.* $\mathcal{P}_2$.

Proof. Should we define truth in the finite-dimensional case, then a viewpoint would become a plain base. Let $u, v \in \mathcal{M}_3(\mathbb{C})$ be the partial symmetries

$$u := \begin{bmatrix} -1 & 0 & 0 \\ 0 & 1 & 0 \\ 0 & 0 & 0 \end{bmatrix}, \quad v := \begin{bmatrix} 0 & 0 & 0 \\ 0 & \sqrt{2}/2 & \sqrt{2}/2 \\ 0 & \sqrt{2}/2 & -\sqrt{2}/2 \end{bmatrix}$$

hence

$$I - uv = \begin{bmatrix} 1 & 0 & 0 \\ 0 & 1 - \sqrt{2}/2 & \sqrt{2}/2 \\ 0 & 0 & 1 \end{bmatrix},$$

$\det(I - uv) = 1 - \sqrt{2}/2 \neq 0, 1$. $\mathfrak{u} := 0 \cdot\!+\!\cdot 1 + u$ succeeds w.r.t. the base $\{(\sqrt{2}/2, \sqrt{2}/2, 0), (\sqrt{2}/2, -\sqrt{2}/2, 0), (0, 0, 1)\}$, while $\mathfrak{v} := 0 \cdot\!+\!\cdot 1 + v$ succeeds w.r.t. the canonical base. It suffices to define $\mathbf{C} := \sim\{\mathfrak{v}\}$. The argument is made rigourous by replacing $\mathcal{M}_3(\mathbb{C})$ with $\mathcal{M}_3(p\mathcal{R}p) = \mathcal{M}_3(\mathbb{C}) \otimes p\mathcal{R}p$. □

21.C Truth and intersubjectivity

We just defined truth; let us now explain why and how we came to this definition.

21.C.1 Truth definitions. According to Tarski–La Palice:

Truth is the quality of what is true.

which is reminiscent of Molière, making fun of physicians: *L'opium fait dormir à cause de sa vertu dormitive*. This « definition » summarises, with its essentialist arrogance, the blind spot of logic.

It is natural to ask for more imaginative approaches. This was attempted in *ludics*, see Definition 71; when freed from technical contingencies, this definition takes the following shape:

- A formula (*behaviour*) refers to certain argumentations (*designs*), which are opposed to refutations, i.e., counter-argumentations (*counter-designs*).
- In the debate (*dispute*) between argumentation and counter-argumentation, one of the contenders *wins*.
- The truth of a formula is the presence of a convincing argumentation, i.e., which always has the last word[3].

« The last word » must be understood literally: the *daimon* is an abortion.

I was progressively led to doubt as to this perfect (a bit too perfect, indeed) construction. The decisive input came from *quantum coherent spaces* and their handling of booleans, i.e., of the *spin*: booleans are part of logic, even if they are hardly more than an epiphenomenon. One observes:

- The relativisation of the oppositions left/right, up/down, true/false, red/green,... w.r.t. an axis of $\mathbb{R}^3$; together with a real connective, the *antipode*, corresponding to the exchange of the value up/down and, to some extent, to negation.
- This door, just open, slams in our face: there is no satisfactory binary connective, i.e., compatible with the newly discovered freedom (Section 17.4.2). Concretely, a connective « $\rightarrow$ » taking care of implication should take the value $x \rightarrow x := \texttt{true}$, a value which becomes `false` when we reverse the reading axis.

This being said, the idea of reconstructing logic at the level of truth values, i.e., at layer -1 is infantile: witness the failure without appeal of multi-valued « logics ». It seems more reasonable to import the relativising « quantum » aspects from layer -1 to layer -3. The most important modification of our paradigm concerns the notion of *last word*: in the ludic approach, there is a certain manicheism, since gain is defined a priori by the absence of *daimon*; thus, most designs are bastards besmirched with a risk of loss. In spite of the American mythology of the *loser*, I doubt the existence of strategies whose only purpose is to better its opponent. This is why, in subjective truth, there is no a priori winner (Proposition 91).

[3]Which does not mean that *any* argumentation for the formula is convincing.

21.C.2 Subjective truth. What can we ask from a truth definition, besides not coming from the sky? The following criteria, formulated in terms of conducts, seem appropriate:

Compositionality: truth is stable under logical consequence, i.e., by cut or *Modus Ponens*.

Consistency: the conducts **A** and $\sim$**A** cannot be simultaneously true; which, *modulo* compositionality, reduces to « absurd conduct is not true ».

Objectivity: two isomorphic conducts are simultaneously true.

The condition of *objectivity* – whether the isomorphism is internal or external – seem no longer pertinent to me; this is why truth is *subjective*. Once this bone is swallowed, the solution is rather natural:

- It is enough to define a notion of *success* closed under logical consequence and consistent.
- Traditionally, GoI interprets logical proofs by means of *partial symmetries*; it by the way ignores the *wager*, which is a sort of « heating » of the circuit, without an equivalent in logical deduction proper.
- To sum up, a *successful* project is of the form $0 \cdot + \cdot \alpha + A$, where $\alpha > 0$ (non-positive pseudo-traces do not occur when dealing with « real » proofs) and u is a partial symmetry. This is satisfactory from the viewpoint of consistency: the absurdity $\mathbf{0} = \{a \cdot + \cdot 1 + 0\ ;\ a \neq 0\}$ is not true.
- This stumbles on compositionality; indeed, *success* as we just defined it, is not preserved by logical consequence. Typically (Proposition 91) **C** and $\sim$**C** can both contain successful projects.

21.C.3 Relation to earlier ideas. The solution is expounded in Section 21.B.2; by writing $\mathcal{R} \subset \mathcal{B}(\mathcal{L}^2(\mathbb{R}))$ we individualise a sort of « continuous base ». Indeed, these ideas were already present in the year 1987; but handled as *objective* ideas, they had little value. While the awareness of *subjectivity* gives them back a central status.

The first occurrence can be found in Section 18.2.2; proofs are interpreted by the *partial permutations* of a finite set I, which induce partial symmetries of the Hilbert space $\mathbb{C}^I$. The canonical base defines a *viewpoint* – rather the analogue of a viewpoint in a factor of type $\mathbf{I}_n$ – and partial permutations do induce successful projects w.r.t. the base. But this idea soon etiolates, for want of a satisfactory duality: once generalised to partial permutations of $\mathbb{N}$, one gets stuck. One does not know how to deal with operators « badly » positioned w.r.t. the preferred base; while the « quantum » requires their inclusion in the general pattern.

This first reminder enables one to understand the condition of success: admitting the metaphor « viewpoint = base = vertices of a graph », it means that the *plot* u is a graph pairing the vertices.

A second occurrence is to be found in Section 19.5.10: the first GoI works w.r.t. a preferred base, i.e., a viewpoint – here in the factor of type $\mathbf{I}_\infty$. The bleak theoretical standing of this first version lies in the fact that all general results are established under a *success* hypothesis and are thus subjective. Since this subjectivity is endured, they are worse than subjective: *subjectivistic*[4]. These results were indeed not that stupid; but, due to the absence of awareness of subjectivity, I was cornered into an absurd choice between two options:

Subjectivism: formulate all results w.r.t. a « preferred base »: the mathematical robustness of such results is very low.

Objectivism: formulate the results without any reference to a base: this is quite impossible.

This dilemma à la Corneille[5]; I lived with it for 20-odd years. In particular, I always hesitated between two approaches, one rather subjective and discrete, e.g., « $\sigma\tau$ cyclic », the other continuous and objective, e.g., « $0 \leq \mathrm{tr}(uv) \leq 1$ ».

The awareness of subjectivity severed this gordian knot: certain properties (for instance, duality, associativity, the scalar output) belong in the *objective*; others (typically, truth) belong in the *subjective*. This conciliates two approaches, roughly speaking the opposition wave/particle:

Particle: the subjective approach, handled by the discrete, graphs.

Wave: the objective approach, handled by the continuous, real numbers.

In GoI, the basic condition which interests *us* (the subject) is of discrete, particle style, nature: « uv nilpotent », i.e., $(uv)^N = 0$ for a certain N. The first GoI formulated its preservation by logical consequence under *success* conditions, see [41], with the limitations of any base-dependent approach. I now use the « wave » version $\det(I - uv) \neq 0, 1$; this definition is farther from « us », but it is better behaved, since objectively compositional. The two approaches almost merge w.r.t. a *viewpoint*: assuming success, one easily gets nilpotency, at least *strong nilpotency*, see Remark 13.

---

[4]Just as the geocentric system of Ptolemy which, by ignorance of its own subjectivity went astray into the pure subjectivistic delirium of *epicycles*.

[5]Similar dilemmas can be found in the various problematics linked to « artificial intelligence »: the adamant negation of subjectivity (in cognitive problems!) produces a monstrous pseudo-objectivity. Thus, the procedurally correct idea of *negation as failure* (Section 4.D.4), once « objectivisised », becomes the grotesque *closed world assumption*, which claims to define failure *sub specie æternitatis*, independently of any computational process.

21.C.4 Microcosms. It is most important to emphasise the fact that this relativisation of truth takes place outside the set-theoretic, « commutative », framework. A viewpoint is a sort of portable set-theory, a *microcosm*; the *traditional* setting of the logical explanation is combinatorics, i.e., a fixed set-theoretic setting; thus nothing of the previous discussion applies, *stricto sensu*, to this setting, surely valuable, but limited, since non-aware of its own subjectivity.

We can see a microcosm as a place of mutual recognition, which expresses itself through commutations. If – in a slightly reductionistic fashion – one identifies the subject with a collection of apparatuses of world recognition, of *measurement*, represented by hermitian operators, these various operators must *commute* in order to be compatible, i.e., belong in the same « set-theory ». A subject thus induces a commutative von Neumann algebra; for instance a subject focusing on impulsion would harbour all operators « impulsion », which do commute: $p_x p_y - p_y p_x = 0$, but not the associated operators « position », since position and impulsion along the same axis do not commute: $p_x q_x - q_x p_x = -i\hbar I$. Two subjects « recognise each other » when their measurement apparatuses, i.e., their algebras, commute; intersubjectivity leads, by juxtaposition of commutative algebras, to a sort of intersubjective microcosm. This can be translated as some $\mathcal{L}^\infty(\mathbb{R})$, thus leading to *viewpoints*.

21.C.5 Intersubjective gendarmerie. Classical intersubjectivity corresponds, more or less, to the creation of a common lexicon. If one introduces a doubt as to the carrier of this lexicon (its decomposition into significant units), one gets *subjective truth*. But one can also « cement » and introduce the idea of the *ineluctability* – compulsory, infallible and predetermined – of knowledge, at work in epistemic « logic » : as usual, *paralogic* exposed its deep obscurantism.

When the Baghdad cuckolds stone their unworthy spouses, this is done in the name of a « common knowledge »: impossible to be more intersubjective than that! But this gregarism has little to do with the constitution of a collective subject: here, the truth is objective, it suffices to « force into talking ». Subjects become sorts of gendarmes writing reports and the *common knowledge* the trading of « terrorists » between the CIA and the FSB; this conception refers, more generally, to informatic heavy policing and – thanks to George Orwell – to *Big Brother*.

21.C.6 Mathematical significance. Let us go back to logic proper. What is the difference between my notion of truth and *models*? More bluntly, am I calling my cat a dog just to boast about his meowing?

- $\mathcal{L}^\infty(\mathbb{R})$ is a σ-algebra, i.e., a complete boolean algebra. But, besides verbal analogies, the notion of viewpoint does not rely on the aspect true/false of boolean algebras.

- While models belong in layer -1, GoI dwells in layer -3; the duality proof/model (Section 7.1.1) is bleak since based upon mutual exclusion. For us, a theorem will be true from the « right » viewpoint (see *infra*), but can however admit refutations – i.e., be false from some « wrong » viewpoint.

Which is, once the change of paradigm has been swallowed, rather soothing. Think of those reasoning by contraposition which begin with « let x be such that $A[x]$ is false » and that may use extremely involved arguments. The final contradiction destroys this elaboration, which, all of a sudden, loses all material basis. It is obvious that such a refined construction should have its « model ». Subjective truth thus offers us a space of refutation of all formulas, *including established truths*.

Now, what could be the meaning of the refutation of a theorem? I perhaps pushed relativism too far in Proposition 91 and maybe truth in my sense no longer means anything. The subjective paradox looks more natural if one takes into account the existence of another subjective protocol, namely the one associating to a conduct a signification in the *subjective* sense of the term. This signification is induced by an *analysis* of the conduct, i.e., of its recursive decomposition in significative elements. If the subjective choices involved in our analysis are consistent with the viewpoint chosen to define truth, there will be an agreement: my theorem will be true.

Take the example of galilean relativity which says that rest cannot be distinguished from uniform motion. A superficial use of this principle leads to sophisms « no point in hurrying », etc. This baloney collapses when we take into account *two* bodies: while individually at rest w.r.t. their *ad hoc* referential, they can badly collide as we all have experimented with cars!

Speaking of logical consequence, a mathematical result is not this *Aboli bibelot d'inanité sonore*[6] that one shelves to look at it for its beauty. It has usually the faculty of *reproduction*, i.e., of being used and reused as a lemma in other proofs. What says Theorem 87 about compositionality of truth? That lemma **A** will have consequence **B** *provided the spatial decomposition is respected*, i.e., provided one chooses as viewpoint for $\mathbf{A} \multimap \mathbf{B}$ the one induced by the viewpoints already available for **A**, **B**: in other words, the implication $\mathbf{A} \multimap \mathbf{B}$ will make « sense », will transmit truth, only in case it respects the structuration of $\mathbf{A} \multimap \mathbf{B}$ into its signifying (for us) elements **A** and **B**. This agreement between several subjectivities (those accompanying **A**, **B**, $\mathbf{A} \multimap \mathbf{B}$) is the precise meaning of *intersubjectivity*.

[6] One of the most beautiful French verses (Mallarmé).

Envoi. The phantom of transparency

An echo to the title of the book: « the blind spot ».

The transparent world

A deep and pregnant, mostly unspoken, belief in the validity of scientific activity is the subliminal idea that, beyond immediate perception, could exist a world, a layer of reading, completely intelligible, i.e., explicit and immediate. What I will call the fantasy (or phantom) of *transparency*.

Transparency has little to do with poetical ideas (the key of dreams, etc.). It is indeed a unidimensional *underside* of the universe, not always monstrous, but anyway grotesque. Think of the Axis of Evil that is supposedly responsible for all the misery of the world, or of those unbelievable *minority studies* which expose carefully concealed truths: according to *Feminine studies*, Shakespeare was a woman, while *African studies* fantasy him as an Arab, Cheikh Zubayr!

The starting premise, to go beyond the surface, beyond mere appearances, is correct; but, to do so, one imagines an « other side of the mirror » whose delimitations are neat, precise, without the slightest ambiguity: the world is seen as a rebus of which it suffices to find the key. In the transparent world, everything is so immediate, legible, that one no longer needs to ask questions, i.e., to think. This putting in question of the very idea of question leads to the worst idiocies: if answers are so easy to access, is it because God amuses Himself by presenting us with an encoded world for the sole purpose of testing us? Unless men are to blame, whose industry is devoted to dissimulation for non-avowable reasons; such behaviour thus justifies the *question*, a middle-age French word for « torture ».

One must anyway admit that a question need not have answers, that it is not even bound to have any, since a great part of scientific activity consists, precisely, in seeking the *good* questions. Thus, the correspondence between planets and regular polyhedra, of which Kepler was so proud, is not even a wrong hypothesis, it is an absurd connection, which only deserves a shrug of the shoulders, a question that did not deserve to be posed, to be compared to speculations linking the length of a ship with the age of her captain. Transparency stumbles on the questioning as to the interest of questions, next on the difficulty to find the answers to the supposedly good problems. Indeed, answers are, mostly, partial: a half-answer accompanied with a new question. The relation question/answer thus becomes an endless dialogue, an *explicitation* process; it is in this process, which yields no definitive and totalising key, where the afterworld of appearances, i.e., knowledge, is to be found.

Logics of transparency

In logic, the phantom of transparency can be summarised in a word: *semantics*. Before discussing semantics, it is salutary to indulge in bad logic, the logic of those who do not have the words, i.e., the technical know-how: there remains only the music, i.e., the affirmation of this transparent world.

Abduction. « If $A \Rightarrow B$, B needs A, hence $B \Rightarrow A$ ». The figure is currently used by demagogues, typically in this example, heard in 2008: « the people of Neuilly are wealthy *because* they elected a right-wing mayor ». This sort of statement, close to racism, is not a good advertisement for abduction; Sherlock Holmes, with his warped, undoubtedly amusing, deductions, is a more friendly choice: indeed, to analyse the ashes of a cigar and conclude that the criminal is 47, back from India and limps with the left foot is, at least, unexpected! What Sherlock Holmes actually supposes is a world transparent at the level of police and criminal activities, the key to this world being the science of ashes, a sort of necromancy, positive but just as absurd[1]. Such a pseudo-science refers to this afterworld in which all questions are supposed to get their answer. There are however questions which have no room in this too polished (and policed) world, typically those of the form « is this problem well-posed ? ».

The search for possible causes is, however, an ancient and legitimate activity, albeit not a mode of reasoning: this would put appearances in command. Mathematics created a special category for those possible causes, in want of legitimation and, for that reason, in the limbs of reasoning: *conjectures*, interesting hypotheses, upon which one attracts attention. The process of integration of a conjecture in the corpus is complex and by no means requires an inversion of the sense of reasoning.

It should be observed that mathematical *induction* is close to abduction. Etymologically, induction is reasoning by generalisation which, not to be abusive, must be restricted to the emission of conjectures. What one calls mathematical induction is an abduction which moves from possible causes to possible methods of constructions, typically a *universal problem*. Mathematical induction is not, contrary to abduction, a grotesque mistake of reasoning; it is nevertheless a form of transparency.

Non-monotonic logics. « What is not provable is false » (Section 2.3.4): one seeks a completion by adding unprovable statements. We know that this completion (that would yield transparency) is fundamentally impossible, because of undecidability and incompleteness, which has been rightly named: it denotes, not a want with respect to a preexisiting totality, but the fundamentally incomplete nature of the cognitive process[2].

[1]On the other hand, the same Sherlock Holmes is proud to ignore the rotation of Earth around the Sun: this is not part of « positive science ».

[2]To be put in relation with the unbounded operators of functional analysis, intrinsically and desper-

Epistemic logic. « He who stays silent must have something to hide » (see Section 2.3.3): such a slogan – natural from a torturer filling the bathtub – is ludicrous in a logical context. Epistemic logic thus appears as the derisive logical counterpart of totalitarianism.

Explicit mathematics. This tentative bureaucratisation of science is just as exciting as a fiction by Leonid Brezhnev. But, rather than the mediocrity of the approach, we shall question the *oxymoron* « mathematics + explicit ».

Is mathematics, can it be, explicit? Since it is an extreme of thought, there would therefore be an explicit thought. Coming back to the words: in « implicit », there is *imply*, thus implication, logic or not; what is implicit is what we can indirectly access, i.e., through thought. On the other hand, « explicit » refers to explication, *explicitation*: it means direct access.

The major part of human activity belongs in the implicit. Besides thought, one can mention the superb abstraction constituted by money, which evoluted through centuries from gold to paper. An explicit economy would be barter, W. giving his wife to V. in exchange for a cow. In the same way, an explicit mathematics would be a verification of the style $2 + 0 = 2$, of which any mathematician knows that it is not a real theorem, not because of its triviality, but because of the absence of any implicit contents; *a contrario* $x + 0 = x$ has an implicit contents (explicitable by providing a value for the variable, e.g., $x = 2$).

There is the same difference between a verification and a theorem as between a table of logarithms and a pocket calculator: the table proposes a long, but frozen, list of values, whereas the calculator possesses, at least in advance, no answer to the query. By the way, computer scientists, who are people of common sense, never dreamed of an « explicit computer », a sort of monstrous telephone directory.

Semantics

This *newspeak* expression originally referred to a theory of signs, thus of meaning. Semantics turns out to be a fantastic *machine à décerveler*[3] by obfuscation of the sense. This is because of its pretension at materialising this transparent world; the failure of the project runs into intellectual skulduggery. Semantics rests upon the fantasy of a reduction to boolean truth values: obvious, since *one can answer any query*! Observe that the other major dogma of current life « one can compare everything » is to found again in fuzzy logics (which brings us back to the aforementioned indignities) and also in tarskism.

From Frege to Tarski. The distinction sense/denotation, due to Frege, is to some extent, a noble version of the myth of transparency: sense refers to a denotation,

ately partial, see Section 19.A.4.

[3]To remove the brain, after Alfred Jarry.

ideal and definitive. This dichotomy *a priori* excludes any link other than fantasmatic between the two aspects, the sense and its *underside*, the denotation: $A \Rightarrow A$ refers to a denotation which is, by definition, completely alien to us. It is thus impossible to understand how the slightest reasoning is possible: just as Zeno's arrow lingers on, we cannot determine how or why the slightest cognitive act can legitimately be performed.

Even less inspired, Tarski defines the answer to the question as… the answer to the question: thus, the transparent universe would be but a pleonasm of the immediate universe, typically « $A \wedge B$ is true when A is true and B is true ». Therefore the denotation of $A \Rightarrow A$ reduces to the implication between the denotation of A and the denotation of A, which means strictly nothing. The failure of this sort of explanation induces a forward flight: real transparency should be sought, beyond immediate transparency, in a « meta » – this fuel for frozen brains – defined through a super-pleonasm in turn iterable in meta-meta, etc. and, eventually, transfinitely! This theology of transparency is but one more obscurantism.

Kripke models. Compared with the previous idiocies, Kripke models almost look like a conceptual breakthrough. But, if the first encounter, with its perfume of parallel worlds, causes a certain jubilation, this enthousiasm is soon soothed by the absolute barrenness of the object.

The idea underlying this approach is that the potential (which is another name for the implicit) is the sum, the totality, of the possibilities. From the philosophical standpoint, have we ever heard anything more ludicrous? For instance, can one say that a 200 Euro bill is the catalogue of everything we can buy with it? Even neglecting the variability of price, one should make room for discontinued merchandise, or those not yet produced! Indeed, a 200 Euro bill is a *question* whose answer belongs in its protocol of circulation: one can exchange it against a merchandise of nominal value 200 Euro, but also against two 100 Euro bills. The merchandise can, in turn, be partially implicit, witness this DVD reader which requires a disk to proceed. We discover on the way that the explicitation need not be total: it can be purely formal, or even partial; in other terms, the implicit may refer, totally or partially, to other implicits.

Kripke models do crystallise this vision of the potential as a sum of possibilities, hence their paradoxical importance: although faulty, this idea is indeed difficult to refute, since of quasi-universal implementation. Thus, (thanks, Brouwer!), a function will never be a graph, but an implicit structure, a construction, given, for instance, by a program: « give me an input, an argument n and I return you $F(n)$ ». It turns out that one can, nevertheless, « define » F through the associated graph $\{(n, F(n)); n \in \mathbb{N}\}$, which is, *stricto sensu*, a monstrous reduction, but which is also incredibly efficient. Hence the success of set-theory and the concomitant washing up of Brouwer's ideas which became subjectivistic, « intensional » (after « meta », yet another swear word).

Categories. The categorical interpretation of logic (especially, intuitionistic logic) make questions appear as objects, answers as *morphisms*. Typically, the disjunction $A \vee B$ asks the question « A or B ? », whereas the morphisms inhabiting it[4] are proofs of A or proofs of B, hence answers to the question. Categories finally appear as the transparent world of morphisms; answers are combined through composition, i.e., by categorical diagrams: once entered in the realm of answers, everything is free of charge; something else than equality should be introduced to say that composition has a cost: as observed in Section 7.2.1, one side is more commutative than the other. Composition is implemented by an algorithm: this is not transparency – which is but a fantasy –, but a construction, a search, necessarily partial and faulty, of transparency.

The weak point of the categorical approach lies in its *essentialism*: it presupposes the form (to which the expression *morphism* refers), hence cannot analyse it. This being said, contrary to tarskian transparency, the sort of transparency at work in categories is not trivial; the analysis of its limitations yields precious information.

Universal problems. Mathematical induction, the civilised form – since technically impeccable – of abduction, is expressed as the solution of a *universal problem*: in a category, a set of constructors *induces* a destructor whose action amounts to inventorying of all *possible* construction means. The most familiar case is that of natural numbers, whose constructors are zero and the successor and whose destructor is the principle of recurrence. This idea is expedient, much nobler than Kripke models, but summary. For instance, defining integers as the solution of a universal problem makes them *ipso facto* unique: the infinite, etymologically « unfinished », is thus reduced to its explicitation, which yields, in the case of natural numbers, this Great Wall, the set $\{0, 1, 2, 3, \ldots\}$. This reduction turns out to be an aporia, exposed by Gödel's paradox (incompleteness).

We find ourselves in a strange situation; the reflexion on the infinite has been sacrificed on the altar of efficiency to the construction of expedient mathematical tools; just as equal temperament sacrificed natural resonances to the exigencies of piano builders (Section 16.1.2). For most utilisations, such compromises are reasonable, but there are cases where they are disastrous. Typically, the theory of algorithmic complexity cannot develop on such bases: indeed, an algorithm is an explicitation procedure; how can we seriously speak of such a thing in a universe where answers (all answers) exist, long before the corresponding questions have been asked?

From semantics to the cognitive onion

Genesis of the categorical interpretation. The progress of logical thought can be identified with a progressive liberation from the essentialist gangue. Essentialism,

[4]For purists, from the terminal object into it.

this morphological simplism, supposes the anteriority of the explicit over the implicit. This Thomism works marvelously well in classical logic, but fails when one does not stick to one's knitting: since everything proceeds from the sky, one runs into arbitrariness, into sectarianism: witness modal logics, by nature disposable and interchangeable.

Originally, logic is interested in unavoidable truths, in the « laws of thought ». A logical formalism, as it can be found in laborious textbooks – and, *dixit* Kreisel apropos « the » Mendelsohn, popular for that very reason – looks like a list, not that far from a cooking recipe, but succeeds anyway in its task, that of codifying those universal truths.

Schönfinkel, as early as the 1920s, and later Curry would eventually individuate the functional (in fact, algorithmic *ante litteram*) meaning of some of those axioms (and rules): this is Curry's isomorphism, recentered in 1969 by Howard around the works of Gentzen, which established a functional reading of (intuitionistic) logic: a proof of $A \Rightarrow B$ is a function from A to B.

Scott domains. Since the principles of logic are of a frightening generality, the search for spaces harbouring such functions turned out to be quite difficult. The only available solution – functions as set-theoretic graphs – being disqualified for questions of size (monstrous *cardinals*) or algorithmics (not computable): a hammer to crush a fly, moreover antagonistic to the approach. One thus sought morphologic criteria in order not to embark « too many » functions, thus constructing *closed cartesian categories* (CCC): those are indeed the exact category-theoretic formulation of intuitionistic logic.

One was quickly led to restrict the search to topological spaces; the discovery by Scott, around 1969, of a topology making *all* logical operations continuous must be considered as a real breakthrough, the mother of all ulterior developments. Yet a deep gap separates *Scott domains* from « real » topology: it suffices to remark that on these uneven spaces, a separately continuous function $f(x, y)$ is continuous!

The continuity of logical operations expresses the same obsession of transparency, here under the form of a perfect control of logical complexity; whereas the incompleteness theorem, which supposes functions of arbitrary complexity, cannot cope with continuity, unless one fiddles with topology. Let us put it bluntly: non-continuity is the native, tangible, manifestation of incompleteness, of non-transparency.

Coherent spaces. Scott domains are but orders in disguise and continuity means preservation of directed suprema. Consistent with the excessive emphasis put by Tarski on order relations, thus implementing another form of transparency: « everything is comparable ».

If the task is to find an abstract version of directed suprema, direct limits are much better behaved than corny topologies. In particular, because of their playmate, the pull-back, whose preservation is *stability* à la Berry. This led to *coherent spaces* (Chapter 8) and linear logic. Linearity should be understood rather literally, since it has been pushed quite far, typically *coherent Banach spaces* (CBS, Section 15.A.1) and *quantum coherent spaces* (QCS, Chapter 17).

These interpretations interiorise the necessary non-continuity of logic. Thus, in CBS, a (non-linear) function from A to B appears as an analytical function from the *open* ball of A to the *closed* ball of B; the behaviour of an analytical function near its border being erratic, the composition of such functions is *a priori* impossible. Similarly, QCS dwell in finite dimension, due to the disappearance of the *trace* in infinite dimension.

Perfection and transparency. Linear logic reveals a *perfect* layer, corresponding to operations that one performs totally, once and for all, see the *perfective* of slavic tongues. This layer is quite continuous (it involves finite-dimensional spaces) and is thus compatible with a limited amount of transparency. At this layer appears *polarisation*, i.e., the dichotomy negative/positive.

It is indeed an old pragmatic distinction[5], reactivated by linear logic through the works of Andreoli. Thus, a program looks like a dialogue combining questions (negative) and answers (positive). Implication, universal quantification, are negative: for instance, $\forall n(A[n] \Rightarrow B[n])$ means « give me $n = N$ as well as $a[N]$ and I will return you $B[N]$ », provided one can « return » $B[N]$, i.e., B is positive, typically when B is an intuitionistic disjunction $C \vee D$: to return $B[N]$ amounts to returning $C[N]$ (left) or returning $D[N]$ (right), say $D[N]$. This being done, depending on the polarity, positive or negative, of $D[N]$, one must either return more, or ask for fresh information, i.e., resume the questioning.

By the way, the formal manipulation leading from $\forall n(A[n] \Rightarrow (C[n] \vee D[n]))$ to $C[N] \vee D[N]$ has no explicative value: I was able to perform it while knowing nothing about A, C, D, N, etc. Which means that this is only the priming of a process of explicitation. This algorithm admits numerous variants, all derived from *cut-elimination*, the celebrated *Hauptsatz* of Gentzen. The necessity of such a process is common sense: a proof of $C[N] \vee D[N]$ is not likely to be of one of the disjuncts (otherwise, what masochism: enunciate $C[N] \vee D[N]$ instead of $D[N]$!). Those of that very form are the *cut-free* proofs; since there are very few of them in nature (they are, if not explicit, the most explicit possible, hence not legible[6]), one is led to content oneself with a partial elimination: one determines the first *bit* left/right of the disjunction before proceeding.

Explicitation thus presents itself through an interactive and dynamical form, in an intrinsically incomplete fashion. It takes the aspect of an alternation of polarities (negative = question, positive = answer). This idea is well captured by the idea of a *game*. This being said, the game is but a metaphor, suffering from a want of

[5]At work in the *negative fragment* ($\Rightarrow, \wedge, \forall$) of intuitionistic logic.

[6]If a lemma is used three times in a proof, its cut-free version will require three independent subproofs.

mathematical depth and also of the supposed instantaneity of the answer, that can perhaps be neglected in the perfect case, but which is a faulty hypothesis: there is a time of latency, corresponding to the algorithmic complexity of cut-elimination, i.e., of the explicitation process, whose metaphor is the slowness of W. in the story of the Houston Cuckolds (Section 2.3.3).

Ludics. Personally, I prefer the image of a « cognitive onion », of which one strips off the successive skins[7], which is realised, at least partially, by *ludics*. The basic object, the *design* (= delogicalised proof) combines *actions* of alternating polarities: *negative* (questions) and *positive* (answers).

Everything resembling logic in its essentialist aspects, typically the *rule of the game*, which supposes a referee, hence new places of transparency, is expelled from ludics. Thus, *behaviours* are games whose rule are established by *consensus* between designs and counter-designs: everything is permitted, provided one reaches a conclusion (when one the players gives up).

This being said, ludics stumbles on a point: its space of questions/answers (the *actions*) is pre-constituted. Which permits the description of the onion through all ways of peeling it. This set of processes is an ultimate possibility of transparency, of semantics, of which we must find the *blind spot*.

Negation

Coherent spaces. *Linear* negation is the exchange question/answer. It originates from a natural operation of *coherent spaces* that I will justify in terms of a logical onion. Let us consider complete cognitive processes (sequences of questions/answers) of a certain type: a proof will be interpreted as the set of all sequences that can be associated with it. If two such sequences differ, they bifurcate *negatively* (i.e., on different questions), which we denote by $x \frown x'$; this yields the coherence relation. One sees that a proof is thus interpreted by a *clique*.

Linear negation corresponds to the exchange question/answer: $x \frown x'$ means that x, x' bifurcate negatively, while $x \smile x'$ (i.e., $x \not\frown x'$) means that x, x' bifurcate positively (i.e., on answers).

The most important output of *stability* (preservation of pull-backs) is that a clique and an anti-clique (i.e., a clique in the negation) have at most one point in common. Intuitively, two distinct sequences must bifurcate either negatively or positively.

Quantum coherent spaces. The notion of a *point* of a coherent space supposes a preconstitution of the space of questions/answers, i.e., the survival of an architecture

[7]As in *Le retour du divin* by Audiberti: the heroin Martine strips her beloved Ambroise of his coat, revealing a second coat, etc.; of the handsome Ambroise, Martine will eventually hug only the successive coats.

subject/object, albeit seriously amended. Further developments were concerned with the dissolution of the notion of a point, which should exist but in relation with the choice of a subject; this accepted *subjectivity* is the only rampart against subjectivism.

QCS are obtained by replacing points with vectors in a Hilbert space and cliques with hermitian operators. Typically, a finite coherent space generates a finite-dimensional vector space whose cliques become projections.

Among the achievements of QCS in our fight against the ideology of transparency is the refutation of « η-expansion » – as part of a universal problem, of inductive nature – which possesses no *convincing* category-theoretic refutation: QCS can naturally tell the difference between the natural identity and the inductive, reconstituted, identity (Section 17.5.4). Which can be related to the notion of measurement in quantum physics, which is basically the reduction to diagonal form w.r.t. a *distinguished* base; distinguished by what? Not by a « what », by a « who »: the subject.

Geometry of interaction. QCS are unfortunately confined to finite dimension; the cardinality has been replaced with the *trace*, compare the two formulations of the negation:

$$\sharp(a \cap b) \leq 1,$$
$$0 \leq \mathrm{tr}(a \cdot b) \leq 1,$$

but the trace no longer exists, in general, in infinite dimension; furthermore, when it exists (factors of type $\mathbf{II}_1$), it is unfit for QCS (Section 17.6.1). However, the Fock space enables one to replace a trace with a determinant:

$$\mathrm{tr}(\Lambda u) = \det(I + u)$$

where Λu can be seen as the space of all possible travels. But this space « diverges », indeed, is a factor of the wrong type, $\mathbf{I}_\infty$(Section 20.D.1).

Here one fingers the mistake made by the advocates of transparency, from the simplistic Kripke models to the most elaborated category-theoretic interpretations: the dialogue between questions and answers has been replaced with the space of their interactions; if one can hardly refuse this replacement in finite dimension, this reduction of the potential to the *list of possibles* diverges in infinite dimension: it is no longer baloney, it is an impossibility.

If Λu « diverges », $\det(I + u)$ makes perfectly good sense: thus, although there is no space of possible travels, one can however quantify it. This leads to GoI, where negation is handled by

$$\det(I - ab) \neq 0, 1.$$

Logical consequence is thus based upon the three successive adjunctions:

$$\begin{aligned} \sharp(F \cap (a \times b)) &= \sharp([F]a \cap b), \\ \mathrm{tr}(F \cdot (a \otimes b)) &= \mathrm{tr}([F]a \cdot b), \\ \det(I - F(a \oplus b)) &= \det(I - [F]a \cdot b) \cdot \det(I - Fa). \end{aligned}$$

$[F]a$ is the solution of a *feedback equation* (Chapter 19). However, the adjunction in the GoI case is not well-balanced, due to the additional coefficient $\det(I - Fa)$, a sort of *heating* caused by the feedback, indeed a *truth value*.

Truth in becoming. A *project*[8] consists of a truth value $a \in \mathbb{R}$ and an operator A, notation $a \cdot A$. The duality $ab \cdot \det(I - AB) \neq 0, 1$ between projects collapses, in the absence of second components, to

$$ab \neq 0, 1,$$

1 corresponding to « true » and 0 to something like a morphological mistake. This should not advocate yet another fuzzy « logic », which belongs in the same waste paper basket as the previous ones.[9]

A project is the sketch a of a truth value together with a process A for developing it, which supposes an interaction with other processes. The complete development demands a *counter-project* $b \cdot B$ and leads to $ab \cdot \det(1 - AB)$; it is then a matter of *total* explicitation, relative to $b \cdot B$. But most explicitations are partial, thus the application of a function to an argument. Indeed, the various adjunctions, especially the most elaborated one, that of GoI, link the partial explicitation to its becoming, the definitive and complete explicitation which never actually occurs in logic for reasons of consistency.

All this to say that the idea of truth value can recover part of the place that it usurped in the old style of foundations. Simply, it splits into a part already computed (a) and a part in becoming (A) which can *in no way* be reduced to a collection of possible futures.

The X-rays of knowledge. Why is the world not transparent? How to quantify this want of transparency?

These questions are difficult and very little has been achieved in the effort to answer them. Incompleteness, undecidability do refute the most brutal and absolute form of transparency; but they hardly say anything relevant as to more subtle questions, typically « Is it more difficult to find than to check ? », whose

[8] W.r.t. to Chapter 21, we use a simplified notation: no dialect, no logarithms. Hence, $a \cdot A$ stands for $\mathrm{colog}(a) \cdot\!+\!\cdot 1 + A$ and $\mathrm{colog}(a) + \mathrm{colog}(b) + \mathrm{ldet}(I - AB) \neq 0, \infty$ becomes $ab \cdot \det(I - AB) \neq 0, 1$.

[9] Speaking of fuzzy « logic », what is wrong is not the idea of going beyond the boolean truth values $\{0, 1\}$, it is the fact of confining one to the static, dead, domain of truth.

precise formulation is the famous problem $P \stackrel{?}{=} NP$. And complexity theory, a theory without *concepts*, has been totally unable to deal with such questions.

Results of the style « it is impossible to obtain this by that method » are very important in science and, in mathematics, central: think of the irrationality of $\sqrt{2}$, of the transcendence of π, of the unsolvability of the fifth-degree equation. These questions are *retrospective*, in the sense that they deal with the possible ways of obtaining something, e.g., constructing a number. The retrospective tools in logic are very limited: classically, one can show that something is not provable by exhibiting a counter-model, but this approach is too coarse in the sort of problems we are interested in. Sequent calculus admits, using cut-elimination, a limited form of retrospection, namely the last rule of the disjunction property; but the technology is strictly inoperant in the only important case, that of an implication, i.e., of a functional algorithm. The replacement of the blunt classical logic with intuitionistic, linear, light versions, produce systems whose retrospection is a priori closer to the subtle algorithmic points we want to clarify. But – and this is the unwanted counterpart of a stronger retrospection – these systems are « weaker », i.e., they prove less. In particular, there is no way to use, say, **LLL**, as a formal system dealing with low complexity. To give a precise example, the existence of the exponential function, which can be expressed by $\forall m \exists n \log n = m$ (the function log does exist) cannot be proved in **LLL**, but – since **LLL** is « weaker » than usual logic – cannot be disproved either, while we would expect from a complexity-sensitive system to take a position as to the exponential function. This suggests that, besides the familiar logical categories (formulas, proofs, models, etc.), a retrospective category should perhaps be added; but all attempts in this direction have fumbled into abductionist gibberish. By moving proof-theory from combinatorics to operator algebras, I have the hope that some synthetic tool from sophisticated mathematics can be imported and shed some light on these basic logical questions. Wishful thinking? Worth a try anyway.

An old activity like logic can find its justification neither in the preservation of a rather obsolete tradition, nor in technical developments, no matter how heroic and brilliant they might be. Its meaning should be sought in questions of a true logical nature, i.e., dealing with the fundamentals of reasoning. As a central task, the building of a non-fregean theory of cognition, the benchmark for such an endeavour being an updated version of incompleteness: to prove, once and for all, that questions are not the same thing as answers, i.e., the inexistence of those unlikely *X-rays of knowledge*.

NON SI NON LA

Bibliography

The numbers at the end of each item refer to the pages on which the respective work is cited.

[1] M. Abadi, G. Gonthier, and J.-J. Levy, The geometry of optimal lambda reduction. In *Proceedings of the Nineteenth Annual ACM Symposium on Principles of Programming Languages* (*POPL*'92), ACM Press, New York, 15–26. 437

[2] S. Abramsky, R. Jagadeesan, and P. Malacaria, Full abstraction for PCF. Extended abstract, in *Theoretical Aspects of Computer Software* (International Symposium, TACS 94, Sendai, Japan), Masami Hagiya and John C. Mitchell, eds., Lecture Notes in Computer Science 789, Springer-Verlag, Berlin, 1994. 259

[3] V. M. Abrusci, Syllogisms and linear logic. Prépublication, Dipartimento di Informatica, Università Roma III, 2000. 239, 357

[4] V. M. Abrusci and P. Ruet, Non-commutative logic I: the multiplicative fragment. *Annals of Pure and Applied Logic* **101** (2000), 29–64. 45, 217, 410

[5] J.-M. Andreoli and R. Pareschi, Linear objects: logical processes with built-in inheritance. *New Generation Computing* **9** (3–4) (1991), 445–473. 184, 210

[6] A. Asperti, A logic for concurrency. Prépublication, Dipartimento di Informatica, Pisa, 1987. 205

[7] A. Asperti, Light affine logic. In *13th Annual IEEE Symposium on Logic in Computer Science* (Indianapolis, IN, 1998), IEEE Computer Society Press, Los Alamitos, CA, 1998, 300–309. 364

[8] S. Bainbridge, P. J. Freyd, A. Scedrov, and P. J. Scott, Functorial polymorphism. *Theoretical Computer Science* **70** (1990), 35–64. 176

[9] H. P. Barendregt, *The lambda calculus: its syntax and semantics*. Revised edition, North-Holland, Amsterdam, 1984. 103, 298

[10] M. Barr, $*$*-Autonomous categories*. Lecture Notes in Mathematics 752, Springer-Verlag, Berlin, 1979. 195

[11] M. Barr, $*$-Autonomous categories and linear logic. *Mathematical Structures in Computer Science* **1** (1991), 159–178. 195

[12] G. Berry, Stable models of typed λ-calculi. In *Automata, languages and programming* (Udine, 1978), Lecture Notes in Computer Science 62, Springer-Verlag, Berlin, 1978, 72–89. 168, 254

[13] R. Blute and P. Scott, Linear Läuchli semantics. *Annals of Pure and Applied Logic* **77** (1996), 101–142. 236

[14] R. Blute and P. Scott, The shuffle Hopf algebra and non-commutative full completeness. *Journal of Symbolic Logic* **63** (1998), 1413–1436. 236

[15] A. Connes, *Non-commutative geometry*. Academic Press, San Diego, CA, 1994. 392, 410, 453, 469

[16] T. Coquand and G. Huet, The calculus of constructions. *Information and Computation* **76** (1988), 95–120. 114, 128

[17] V. Danos, La logique linéaire appliquée à l'étude de divers processus de normalisation et principalement du λ-calcul. PhD thesis, Université Paris VII, 1990. 349

[18] V. Danos, J.-B. Joinet, and H. Schellinx, LKQ and LKT: sequent calculi for second order logic based upon dual linear decompositions of classical implication. In *Advances in Linear Logic*, J.-Y. Girard, Y. Lafont, and L. Regnier (eds.), London Mathematical Society Lecture Note Series 222, Cambridge University Press, Cambridge, 1995. 339, 351

[19] V. Danos and L. Regnier, The structure of multiplicatives. *Archive for Mathematical Logic* 28 (1989), 181–203. 233

[20] V. Danos and L. Regnier, Proof-nets and the Hilbert space. In *Advances in Linear Logic*, J.-Y. Girard, Y. Lafont, and L. Regnier (eds.), London Mathematical Society Lecture Note Series 222, Cambridge University Press, Cambridge, 1995, 307–328. 425

[21] V. De Paiva, The Dialectica categories. In *Categories in Computer Science and Logic*, John W. Gray and Andre Scedrov (eds.), American Mathematical Society, Providence, RI, 1989, 47–62. 258

[22] R. Di Cosmo, D. Kesner, and E. Polonovski, Proof nets and explicit substitutions. *Mathematical Structures in Computer Science* **13** (2003), 409–450. 349

[23] T. Ehrhard, Hypercoherences: a strongly stable model of linear logic. In *Advances in Linear Logic*, J.-Y. Girard, Y. Lafont, and L. Regnier (eds.), London Mathematical Society Lecture Note Series 222, Cambridge University Press, Cambridge, 1995, 83–108. 270

[24] T. Ehrhard and L. Regnier, The differential lambda-calculus. *Theoretical Computer Science* **309** (2003), 1–41. 341

[25] C. Faggian, Interactive observability in ludics. In *Automata, languages and programming* (Turku, 2004), Lecture Notes in Computer Science 3142, Springer-Verlag, Berlin, 2004, 506–518. 328

[26] W. Felscher, Dialogues, strategies and intuitionistic provability. *Annals of Mathematical Logic* **28** (1985), 217–254. 259, 265

[27] A. Fleury and C. Rétoré, The mix rule. *Mathematical Structures in Computer Science* **4** (1994), 273–285. 236

[28] M.-R. Fleury-Donnadieu and M. Quatrini, First-order in ludics. *Mathematical Structures in Computer Science* **14** (2004), 189–312. 328

[29] B. Fuglede and R.V. Kadison, Determinant theory in finite factors. *Annals of Mathematics* (2) **55** (1952), 520–530. 460

[30] G. Gentzen, Investigations into logical deduction. In *The collected works of Gerhard Gentzen*, M. E. Szabo (ed.), North-Holland, Amsterdam, 1969, 68–131. 41

[31] G. Gentzen, New version of the consistency proof for elementary number theory. In *The collected works of Gerhard Gentzen*, M. E. Szabo (ed.), North-Holland, Amsterdam, 1969, 252–286. 14, 131

[32] G. Gentzen, The consistency of elementary number theory. In *The collected works of Gerhard Gentzen*, M. E. Szabo (ed.), North-Holland, Amsterdam, 1969, 132–213. 14, 257

[33] G. Gentzen, The consistency of the simple theory of types. In *The collected works of Gerhard Gentzen*, M. E. Szabo (ed.), North-Holland, Amsterdam, 1969, 214–222. 128

[34] J.-Y. Girard, Une extension de l'interprétation fonctionnelle de Gödel à l'analyse et son application à l'élimination des coupures dans l'analyse et la théorie des types. In *Proceedings of the Second Scandinavian Logic Symposium* (Oslo, 1970), J. E. Fenstad (ed.), North-Holland, Amsterdam, 1971, 63–92. 115

[35] J.-Y. Girard, Interprétation fonctionnelle et élimination des coupures de l'arithmétique d'ordre supérieur. Thèse de Doctorat d'Etat, Université Paris VII, Paris, 1972. 115

[36] J.-Y. Girard, The system **F** of variable types, fifteen years later. *Theoretical Computer Science* **45** (1986), 159–192. 175

[37] J.-Y. Girard, Linear logic. *Theoretical Computer Science* **50** (1987), 1–102. 164, 233, 406

[38] J.-Y. Girard, *Proof-theory and logical complexity I*. Bibliopolis, Napoli, 1987. 40, 51

[39] J.-Y. Girard, Multiplicatives. In *Logic and computer science: new trends and applications* (Torino, 1986), G. Lolli (ed.), Università di Torino, Rendiconti del seminario matematico dell'università e politecnico di Torino, special issue **1987**, 11–34. 403, 406, 432

[40] J.-Y. Girard, Normal functors, power series and λ-calculus. *Annals of Mathematical Logic* **37** (1988), 129–177. 175, 341

[41] J.-Y. Girard, Geometry of interaction I: interpretation of system F. In *Logic Colloquium '88*, R. Ferro, C. Bonotto, S. Valentini and A. Zanardo (eds.), Studies in Logic and the Foundations of Mathematics 127, North-Holland, Amsterdam, 1989, 221–260. 432, 436, 494

[42] J.-Y. Girard, Le champ du signe. In *Le théorème de Gödel*, Le Seuil, Paris, 1989, 141–171; reprint: Il sogno del segno, in *La prova di Gödel*, Bollati Boringhieri, Torino, 1992, 109-136. 40

[43] J.-Y. Girard, Towards a geometry of interaction. In *Categories in Computer Science and Logic*, J. W. Gray and A. Scedrov (eds.), Contemporary Mathematics 92, American Mathematical Society, Providence, R.I., 1989, , 69–108. 432

[44] J.-Y. Girard, Geometry of interaction II: deadlock-free algorithms. In *Proceedings of COLOG 88*, P. Martin-Löf and G. Mints (eds.), Lecture Notes in Computer Science 417, Springer-Verlag, Heidelberg, 1990, 76–93. 432, 438

[45] J.-Y. Girard, A new constructive logic: classical logic. *Mathematical Structures in Computer Science* **1** (1991), 255–296. 156, 268

[46] J.-Y. Girard, On the unity of logic. *Annals of Pure and Applied Logic* **59** (1993), 201–217. 270

[47] J.-Y. Girard, Geometry of interaction III: accommodating the additives. In *Advances in Linear Logic*, J.-Y. Girard, Y. Lafont, and L. Regnier (eds.), London Mathematical Society Lecture Note Series 222, Cambridge University Press, Cambridge, 1995, 329–389. 432, 438, 443

[48] J.-Y. Girard, Proof-nets: the parallel syntax for proof-theory. In *Logic and Algebra*, A. Ursini and P. Aglianò (eds), Lecture Notes in Pure and Applied Mathematics 180, Marcel Dekker, New York, 1996, 97–124. 247

[49] J.-Y. Girard, Light linear logic. *Information and Computation* **143** (1998), 175–204. 360, 363, 364

[50] J.-Y. Girard, Coherent Banach Spaces : a continuous denotational semantics. *Theoretical Computer Science* **227** (1999), 275–297. 341

[51] J.-Y. Girard, Locus Solum: from the rules of logic to the logic of rules. *Mathematical Structures in Computer Science* **11** (2001), 301–506, 14, 102, 263, 289, 293, 326

[52] J.-Y. Girard, Between logic and quantic: a tract. In *Linear Logic in Computer Science*, T. Ehrhard, J.-Y. Girard, P. Ruet and P. Scott (eds.), London Mathematical Society Lecture Note Series 316, Cambridge University Press, Cambridge, 2004, 346–381. 369

[53] J.-Y. Girard, Geometry of interaction IV: the feedback equation. In *Logic Colloquium '03*, V. Stoltenberg-Hausen and J. Väänänen (eds), Lecture Notes in Logic 24, Association for Symbolic Logic, La Jolla, CA, 2006, 76–117. 431

[54] J.-Y. Girard, La logique comme géométrie du cognitif. In *Logique, géométrie et cognition*, J. B. Joinet (ed.), Presses de la Sorbonne, 2007, 13–29. 14

[55] J.-Y. Girard, Geometry of interaction V: logic in the hyperfinite factor. *Theoretical Computer Science* **412** (2011), 1860–1883. 462

[56] K. Gödel, Über eine bisher noch nicht benützte Erweiterung des finiten Standpunktes. *Dialectica* **12** (1958), 280–287. 129, 258

[57] S. Guerrini, Correctness of multiplicative proof-nets is linear. In *14th Annual IEEE Symposium on Logic in Computer Science* (LICS '99), IEEE Computer Society Press, Los Alamitos, CA, 1999, 454–463. 232

[58] M. Hamano, Pontrjagin duality and full completeness for multiplicative linear logic (without mix). *Mathematical Structures in Computer Science* **10** (2000), 213–259. 325

[59] D. Hilbert, Über die Grundlagen der Logik und die Arithmetik. In *Verhandlungen des Dritten Internationalen Mathematiker-Kongresses in Heidelberg*, 1905. 14

[60] D. Hilbert, Über das Unendliche. *Mathematische Annalen* **95** (1926), 161–190; English translation: J. van Heijenoort, *From Frege to Gödel*, Harvard University Press, Cambridge, Mass., 1967. 14

[61] D. J. D. Hughes and R. D. van Glabbeek, Proof nets for unit-free multiplicative-additive linear logic. Extended abstract, In *18th IEEE Symposium on Logic in Computer Science* (LICS 2003), IEEE Computer Society Press, Los Alamitos, CA, 2003, 1–10. 247

[62] M. Hyland and L. Ong, On Full Abstraction for PCF: I, II and III. *Information and Computation* **163** (2000), 285–408. 259

[63] R. V. Kadison and J. R. Ringrose, *Fundamentals of the theory of operator algebras*. Vol. I, Pure and Applied Mathematics, Academic Press, New York, 1983; corrected exercises in *Fundamentals of the theory of operator algebras*, Vol. III, Birkhäuser, Boston, Mass., 1991. 393, 423, 426, 439, 441

[64] R. V. Kadison and J. R. Ringrose, *Fundamentals of the theory of operator algebras*. Vol. II, Pure and Applied Mathematics 100, Academic Press, Orlando, Florida, 1986; corrected exercises in *Fundamentals of the theory of operator algebras*, Vol. IV, Birkhäuser, Boston, Mass., 1992. 456, 457, 458, 465

[65] M. Kanovitch, The multiplicative fragment of linear logic is NP-complete. Technical report X-91-13, Institute for language, logic and information, Amsterdam, 1991. 205

[66] S. C. Kleene, *Introduction to metamathematics*. North-Holland, Amsterdam, 1952. 131

[67] G. Kreisel, Mathematical logic. In *Lectures in modern mathematics*, Vol. III, T. L. Saaty (ed.), Wiley & Sons, New York, 1965, 99–195. 10, 14, 110

[68] G. Kreisel and A. Levy, Reflection principles and their uses for establishing the complexity of axiomatic systems. *Zeitschrift für Mathematische Logik* **14** (1968), 97–142. 10, 64

[69] Y. Lafont, From proof-nets to interaction nets. In *Advances in Linear Logic*, J.-Y. Girard, Y. Lafont, and L. Regnier (eds.), London Mathematical Society Lecture Note Series 222, Cambridge University Press, Cambridge, 1995, 225–247. 237

[70] Y. Lafont, The undecidability of second order linear logic without exponentials. *Journal of Symbolic Logic* **61** (1996), 541–548. 142

[71] Y. Lafont, Interaction combinators. *Information and Computation* **137** (1997), 69–101. 237

[72] J. Lambek, The mathematics of sentence structures. *American Mathematical Monthly* **65** (1958), 154–169. 195, 217

[73] S. Mac Lane and G. M. Kelly, Coherence in closed categories. *Journal of Pure and Applied Algebra* **1** (1971), 97–140. 195

[74] O. Laurent, Étude de la polarisation en logique. Thèse de doctorat, Université Aix-Marseille II, March 2002. 344, 347

[75] O. Laurent, Polarized proof-nets and $\lambda\mu$-calculus. *Theoretical Computer Science* **290** (2003), 161–188. 352

[76] P. Lincoln, Deciding provability of linear logic formulas. In *Advances in Linear Logic*, J.-Y. Girard, Y. Lafont, and L. Regnier (eds.), London Mathematical Society Lecture Note Series 222, Cambridge University Press, Cambridge, 1995, 109–122. 236

[77] P. Lincoln, J. C. Mitchell, J. C. Shankar, and A. Scedrov, Decision problems for propositional linear logic. In *Proceedings of 31st IEEE symposium on foundations of computer science*, Vol. 2, IEEE Computer Society Press, Los Alamitos, CA, 1990, 662–671. 205

[78] K. Lorenz, Dialogspiele als semantische Grundlage von Logikkalkülen. I, II. *Archiv für Mathematische Logik und Grundlagenforschung* **11** (1968), 32–55, 73–100. 259

[79] P. Lorenzen, Logik und Agon. In *Atti Congresso Internazionale di Filosofia*, Vol. 4, Sansoni, Firenze, 1960, 187–194. 259

[80] P. Martin-Löf, *Intuitionistic type theory*. Bibliopolis, Napoli, 1984. 113

[81] M. Masseron, C. Tollu, and J. Vauzeilles, Generating plans in linear logic. In *Foundations of software technology and theoretical computer science*, K. V. Nori and C. E. Veni Madhavan (eds.), Lecture Notes in Computer Science 472, Springer, Berlin, 1990, 63–75. 332

[82] F. Maurel, Un cadre quantitatif pour la Ludique. Thèse de doctorat, Université Paris VII, November 2004. 342, 350

[83] R. Milner, *Communicating and mobile systems: the π-calculus*. Cambridge University Press, New York, 1999. 447

[84] H. Nickau, Hereditarily sequential functionals. In *Proc. Symp. Logical Foundations of Computer Science: Logic at St. Petersburg*, A. Nerode and Y. V. Matiyasevich (eds.), Lecture Notes in Computer Science 813, Springer-Verlag, Berlin, 1994, 253–264. 259

[85] M. Parigot, $\lambda\mu$-calculus: an algorithmic interpretation of classical natural deduction. In *Proceedings of International Conference on Logic Programming and Automated Reasoning*, Lecture Notes in Computer Science 624, Springer-Verlag, Berlin, 1992, 190–201. 352

[86] G. D. Plotkin, LCF considered as a programming language. *Theoretical Computer Science* **5** (1977), 223–256. 166

[87] D. Prawitz, *Natural deduction, a proof-theoretical study*. Almqvist & Wiksell, Stockholm, 1965. 34, 75

[88] L. Regnier, Une équivalence sur les lambda-termes. *Theoretical Computer Science* **126** (1994), 281–292. 349

[89] B. Russell, Mathematical logic as based on the theory of types. *American Journal of Mathematics* **30** (1908), 222–262. 127

[90] K. Schütte, *Beweistheorie*. Die Grundlehren der mathematischen Wissenschaften 103, Springer-Verlag, Heidelberg, 1960. 66

[91] K. Schütte, Syntactical and semantical properties of simple type theory. *Journal of Symbolic Logic* **25** (1960), 305–326. 51, 61, 128

[92] D. Scott, Data types as lattices. *SIAM Journal of Computing* **5** (1976), 522–587. 149, 151

[93] P. Selinger, Control categories and duality: on the categorical semantics of the lambda-mu calculus. *Mathematical Structures in Computer Science* **11** (2001), 207–260. 352

[94] P. Selinger, Towards a quantum computing language. *Mathematical Structures in Computer Science* **14** (2004), 527–586. 372

[95] W. W. Tait, A non-constructive proof of Gentzen's Hauptsatz for second-order logic. *Bulletin of the AMS* **72** (1966), 980–983. 61

[96] W. W. Tait, Intensional interpretation of functionals of finite type I. *Journal of Symbolic Logic* **32** (1967), 198–212. 121

[97] G. Takeuti, On a generalized logical calculus. *Japanese Journal of Mathematics* **23** (1953), 39–96. 61

[98] K. Terui, Which structural rules admit cut elimination? An algebraic criterion. *Journal of Symbolic Logic* **72** (2007), 738–754. 201

[99] A. van Tonder, A lambda-calculus for quantum computing. Technical report, Department of Physics, Brown University, Providence, RI, July 2003. 372

[100] J. B. Wells, Typability and type checking in system **F** are equivalent and undecidable. *Annals of Pure and Applied Logic* **98** (1999), 111–156. 136

[101] A. N. Whitehead and B. Russell, *Principia Mathematica*. Cambridge University Press, Cambridge, 1910. 127

[102] D. N. Yetter, Quantales and non-commutative linear logic. *Journal of Symbolic Logic* **55** (1990), 41–64. 217

Index

- Abduction, 23, **498**
- Abélard, P., 257
- Abramsky, S., **259**
- Abrusci, M., 196, **239**, 357, **410**
- Absorbers
  - Additive, **308**
  - Multiplicative, **317**
- Absurdity, $\perp$, 6
- Action (ξ, I), ✠, **279**
  - Dummy, **304**
  - Focus, **279**
  - Hidden, **289**
  - Opposite $\widetilde{\kappa}$, **289**
  - Polarity, **279**
  - Proper/improper, **279**
  - Ramification, **279**
  - Visible, **289**
- Action vs. Reaction, **204**
- Additives &, $\oplus$, **183**, **309**, 347
  - CBS, **340**
  - Decomposition, **312**
  - Locative $\bigcap$, $\uplus$, **307**
  - PCS, **374**
  - QCS, **385**
- Adjunction, 155, 168, 181, **464**
  - $\otimes / \multimap$, 195
  - $\& / \Rightarrow$, 187
- AI, 21, 22, **27**, **371**, 441, 494
- Akhenaten, 311
- Amnesty (self-), 20, 21
- Analysis vs. Synthesis, 259, **265**, 274, 277, 328
- Analyticity, 292
- Anderssen, A., 263, 276
- Andreoli, J.-M., 184, **210**, 251, 503
- Antinomy, **98**
  - Burali-Forti, 5, 128
  - Girard, 114, 128
  - Russell, 5, 15, **34**, 46, 88, 103, 282, **361**
- Apollinaire, G., 449
- Application, **100**
- Arborescence, 289
- Aristotle, 12, 209, **240**
- Arithmetic
  - **HA**, **33**, 98, 129
  - **PA**, 21, **33**, 61, 64
  - **RR**, **31**, 38, 129
  - $\mathbf{PA_2}$, **61**
  - Bounded, 360
  - Pressburger, 33, 36
- Arity, **60**, 117
- Artificial
  - Instinct, 27
  - Intelligence, *see* AI
- Aspect, A., 391
- Asperti, A., **205**, **364**
- Associativity, 146, 150, 156, 195, 199, 202, 219, 266, 269, 284, 285, 315, 352, 418, 422
- Atomic weapon, 302
- Audiberti, J., 449, 504
- Autism, 226, 424
- *Automath*, **113**
- Axiom of Choice, 58, 441, 459

- Babel Tower, **447**
- Banach space, **395**
  - Coherent, *see* CBS
  - Reflexive, 340, **395**
- Bankruptcy, 143
- *Barbara*, **238**, 417
- *Barbari*, **240**
- Barendregt, H., 103
- Barr, M., **195**

Base, **274**, 279, **280**, **284**
Bathtub, 371, 499
Beard, 329
Behaviour **G**, 266, **300**, 492
  Alien, 309, **319**
  Connected, **309**
  Disjoint, **309**
  Polarity, **300**
  Principal, **300**
  Sequent, **326**
  vs. Game, **301**
    `Even, Odd`, 273, **302**
    `Me, You`, **302**
Behaviour (GoI), 469, **482**
  Additives, **484**
  Delateralisation, **483**
  Multiplicatives, **482**
  Quantifiers, **484**
  Secularisation, **485**
Berardi, S., **125**
Bernays, P., 12
Berry, G., **167**, **254**, 503
  Order, **167**
Bias i, **272**
Bibliometry, 389
Big Brother, 495
Bihaviour, 326, **327**, 342
Bilocation, *see* Occurrence
Blair, A., 26
Blasphemy, 329
Blind spot, xiii, 11, 62, 67, 99, 131, 138, 159, 228, 331, 358, 447, 450, 492
Block matrix, 417, **440**
Blute, R., **236**
Bolex watches, 41
$\mathfrak{Bomb}^+$, **302**, 318
  Negative $\mathfrak{Bomb}^-$, **318**
Bondage, *see* Predicativity, 159
Bonification, 430, 440
Booleans
  Commutative, **382**
  Quantum, 383, 492
    Antipode, 384, 388, 390, 492
Border conflict, **232**
Borges, J. L., 329
Bourbaki, N., 58
Box $\mathfrak{B}$, **240**, 264, 333, 344, 363
  Conclusions, **240**
  Exponential, **242**
  Nested, 333
Brezhnev, L., 499
Broken watch, 92
Brouwer, L. E. J., **7**, 8, **97**, 258, 500
Bureaucracy, 8, 17
Bush, G. W., 28, 143, 332

C^*-algebra, 393, **397**, 451
  Simple, **402**
  $*$-isomorphism, **402**
Canonicity, 148
Cantor, G., 3, 5, 21, 360
CAR
  Algebra, 402, **456**
  Relations, **456**
Cardinal, 163, 253, 320, 332, 360, 448, 454
  Inaccessible, 5
Category, 10, 11, 143, **147**, 315, 372, 417, 422, **501**
  COH, **179**
  $*$-Autonomous, **195**
  Cartesian, **148**
    Closed, *see* CCC
  Concrete, 252
  Control, **352**
  Degenerate, **150**, **155**, 335
  Direct sum, **154**
  Monoidal, 185, **194**, 436
    Braided, **195**
    Closed symmetric, **195**
    Symmetric, **194**
    Traced, 393, 422
  Morphism, 8, **147**, 442

Composition, 143, **147**
Source, **147**
Target, **147**
Object, 143, **147**
Opposite, 256
Product
Cartesian, **148**
Tensor, **194**
Pull-back, **167**
2-, 145
Cavern myth, 12
CBS, **340**, 375
Separation, 342
CCC, **149**, 271
Classical, **155**
Evaluation, **150**
Logic, **152**
Sets, **150**
Chaos vs. Incompleteness, 369
Cheikh Zubayr, 497
Chiasmus, **417**, 433
Chronicle c, **280**
Coherence, **280**
Comparability, **280**
Positivity, **280**
Propagation, **280**
Proper/improper, **280**, 289
Totality, **280**
Church, A., *see* Theorem (Church–Rosser), 104
Circumscription, 332
Clarke, A. C., 3
Classical logic, 11, 138, **154**, 299, **357**, 391
Category, **334**
Involutivity, 352, **353**
LC, 251, 269, **337**, 351, 353
Hereditarily positive, **339**
LK, **41**, 334
LKT, **LKQ**, **339**, 352, 353
Clique $a \sqsubset X$, **165**, 372
CBS, **340**
Comaximal, **252**
Maximal, **252**
Polarised, **255**
Total, **252**
Winning, **253**
Closed world assumption, *see* Negation as failure, 333, 494
Closure Principle, **296**
Coding, xii, 16, 272
Arbitrary, 168
ASCII, 17
Recursive, 67
à la Scott, 166
of Sequences, **36**
Coherent space $(|X|, \frown\!\!\!\!\smile)$, **164**, **181**, 292, 504
Anticlique, **180**
Carrier, **373**
CC, 169
CCC, **169**
Coherence
$x \frown\!\!\!\!\smile_X y$, **165**
Strict $\frown$, **170**
Faithfulness, **251**
Implication
Intuitionistic, **170**
Linear, **179**
Incoherence
$x \asymp_X y$, **180**
Strict $\smile$, **180**
Linear negation
$\sim X$, **180**
Polarised, **255**
Probabilistic, *see* PCS
Quantum, *see* QCS
System **F**, **171**
Embedding, **172**
Extraction, **173**
Schizophrenia, **174**
Variable clique, **174**
Variable space, **173**
Tensor product, **182**

Totalitarian, **252**
vs. Trips, **409**
Web $|X|$, **165**
Combat zone, 408
Combinator, 109, 111
K, S, **110**
Combinatorics, 68, 495
Combinatory logic, **110**
Comma
Left, **43**
Right, **43**
Common knowledge, 495
Commutant, **452**
Commutation, **144**
Comonoid, **336**, 352
Central morphism, **337**
Dereliction, **337**
Promotion, **337**
Completeness, 67, 97, 140, 251
External, 306
Internal, **304**, 306, 311, 318, 405
Complexity, 20, 236, **359**
$\mathrm{P} \stackrel{?}{=} \mathrm{NP}$, 40, 69, 360, 507
Classical, 74
Intuitionistic, 74
Linear fragments, **205**
Neutrals, **236**
Compositionality, 156, 266, 493
Comprehension, **444**
Conclusion, *see* Base
Concomitance, 391
Communicating vessels, **206**, 391
Extension cord, **207**
Condescension, 239
Conditional expectation, **457**, 465
Conduct **A**, **470**, **493**
Additives, **474**, **476**
Exponentials, **477**
False, **490**
Multiplicatives, **473**, **476**
Negative, **475**
Perennial, **478**
Positive, **475**
Quantifiers, **474**, **485**
True, **490**
Confluence, **80**, 106, 107
Conjecture, 498
Connes, A., 11, 392, 410, 453, 469
Consensus vs. Dissensus, 303
Consistency
Computational, 105, **106**
Logical, 5, 7, **19**, 75, 105, 141, 146, 251, 299
Con($\mathcal{T}$), 18, **19**
Intuitionistic, 142
1−, 18, 20, **37**
Second proof, **65**
Constantin V *Copronymos*, 357
Constatation vs. Reasoning, 24, 371
Constructions (calculus), **114**, **128**
Constructivism, 468
Context, **43**, 44, 102
$\Gamma < \Delta$, **411**
$\Gamma \parallel \Delta$, **411**
Mixed $\Gamma, \underline{\Delta}$, **199**
Contraposition, **72**, **496**
Convolution, **454**
Copernicus, N., 371
Coq, **114**
Coquand, T., 114, **128**
Correctness, 405, **406**
Counter-
Base, **292**
Design, **292**, 303, 492
Model, 259, 260, 303
Proof, 259, 403
Covenant, **142**, 216
CR, *see* Reducibility (candidate)
Crossed product $\mathcal{A} \rtimes_\alpha \mathfrak{G}$, **455**
Cuckold
Baghdad, 24, 36, 207, 371, 495
Houston, 24, 504
Curien, P.-L., **342**

Curry, H. B., *see* Isomorphism (Curry–Howard), **109**, 502
Curtain (mobile), 5, 33
Cut, 262, 264, 266, 274, 276, **283**, 286, 288, 333, 352, 363, 420
  Double, 334
Cut-elimination, *see Hauptsatz*, 417
  Partial, 59
Cut-system ($\mathcal{H}, h, \sigma$), **419, 420**
  Closed, **462**
  Cut-free, **421**
  Injective, **423**
  Invertible, 419, **422**
  Lax, 429
  Nilpotent, **419**
  Positive, **421**
  Semi-invertible
    l.s.i, u.s.i., **430**
  Termination, 422, **424**
Cyclicity, 411

Daimon ✠, $\mathfrak{Dai}$, **0**, **255**, 262, 264, 266, **274, 275**, **300**, 312, 325, 327, 404, 406, 492
  Negative $\mathfrak{Dai}^-$, $\mathbf{0}^-$, **287**, **294**, **300**, **307**
Danos, V., **233**, **425**
De Bruijn, N. J., **102**, 108, **113**, 220
De Morgan, A., 48, 54, 157, 182, 188, 269
De Paiva, V., **258**
Deadlock, **423**
Debray, R., 26
Dedekind, R., 3, 21, 30, 109, 116
Default reasoning, 332
Deficiency, 22
Degeneration, 301
Degree, 54, 361, **365**
  Commutative, **86**
  Cut, **80**, 89
Delocation φ, ψ, 44, 103, 108, 224, 276, 288, 296, **310**, 322, 324, 420, 435
  Partial, **310**
  Polarity, 310
Denotational vs. Operational, 97
Density operator, 381
Desessentialisation, **181**, 271, 373
Design $\mathfrak{D}$, 255, 266, **272**, 492
  Adjoint, **285**
  Arborescence, 280
  Main, **284**
  Material, **301**, 310
  Partial, **282**, 327
  Polarity, **280**
  Positivity, **280**
  Proper/improper, **275**
  Stable order $\mathfrak{D} \subset \mathfrak{E}$, **283**
  Totality, **280**
  Winning, **299**, 312, 325, **492**
Dessein, *see* Design, **277**
Dessin, *see* Design, **267**, **272**
  vs. *Dessein*, **281**
Determinant, 447
  Fuglede–Kadison, **460**
Determinism, 369
Deterrence, 303
Diagonalisation, **15**, 19, 20, 132, 449
Diagram (commutative), **147**
Dialectica, **129**, **258**, 261
Dialectics, *see* Lorenzen
Dictionary, **438**
Dijkstra, E. W., 41
Directory $\mathcal{N}$, **272**
  ¶G, **307**
  Additives, **308**
  Multiplicative, **318**
  $\mathfrak{Dir}$, 303, **306**
Disjunction, *see* Property
  Property, 253, 304
Dispute $[\mathfrak{D} \rightleftharpoons \mathfrak{E}]$, **290**, 299, 302, 492
  Tunnel, 291
Distributivity, xii, 145, 157, 185, **477**
Dodecaphonsim, 358

Domain
  Quantitative, **175**
  Scott, xi, **151**, **157**, **163**, 172, 252, 292, 341, 438, 440, 450
    Embedding, **157**
    Redundancy, 166
Domestic quarrel, 303
Don Camillo, 7
Double
  Conditional, 268, 342
  Cut, 334
  Negation, *see* Negation (double)
  Stability, **297**
Duality $\cdot \perp \cdot$, **141**, 197, **292**, **373**, **376**, 404, **407**, **470**
  Bipolar, 181
  Polar, 194
    X^p, **141**
  Pole, **141**
Dungeon, 332
Dupond and Dupont, 332
Dynamics, 145

Egg vs. Hen, 266
Ehrhard, T., 254, **270**, **341**
Eigenvariable, **47**, 50, 56, 77, 242, 347
  Boolean, **246**, 443
Einstein, A., xii, 28, 369, 391
ELL, 5, 358, 359, **364**, **477**
  Bounds, **366**
  Expressiveness, **367**
L'entarteur, 29, 40
Epistemic logic, xi, **23**, 27, 28, 36, 309, 371, 495, 499
ε-substitution, **57**
Equality, **58**, 372
  vs. Isomorphism, 311
Ershov, Y., **151**
Essence vs. Existence, **3**, **5**, 7, 11–13, 69, 104, 107, **136**, 147, 171, 181, 208, 259, 266, 301, 313, 327, 328, 357, 468, 492, 501
Essenin-Volpin, A., 359, 448
η-expansion, **152**, 154, 211, 233, 276, 298, 351, **389**, 403, 409, 436, 505
Ethics **E**, **304**, 308, 321
  Complete, **304**, 311, 319
  Projection, **319**
Ethic lemma, **471**
Euclidian space, **375**
Excluded middle, *see Tertium non datur*
Execution, **408**, 417
Existence, *see* Property
Expansive vs. Recessive, **17**, 26, 275, 282, 283, 286
Explicit logic, 27, 74
Explicit mathematics, **499**
Exponentials !, ?, §
  Why not $?A$, **186**
Exponentials !, ?, §, 11, 14
  CBS, **341**
  Contraction, **208**
  Dereliction, **208**
  Heterodox, **364**
  Non-uniform, 328
  Of course $!A$, **186**, 208
  Orthodox, **331**
  Promotion, **208**
  Weakening, **208**
Extensional vs. Intensional, **371**
Extension cord, 276
Extraspective vs. Introspective, 253, **298**, 409, 421, 424
Extrication, **392**

Fact, **197**
Factor, **452**
Faggian, C., **328**
Faithfulness, **251**, 306, 324, 343, 389
  Category-theoretic, **251**
Faith 𝔉𝔦𝔡𝔢𝔰, Ω, **282**, 284, 285, 295, 327
Falsum **f**, **46**
Fax 𝔉𝔞𝔵, 274, **275**, **278**, 287

Feedback σ, **420**
Equation, 409, 418, **420**
Independance, 422
Lopsided, 422, **428**, **431**
Negative, 422
Positive, 422, **426**
Felscher, W., **258**, 266
Fiddling, *see* Intensional, 155, 371
Fifth generation, 90
Finitism, **7**, **448**
Generalised, 9
Hilbertian, 448
Fixed point
λ-calculus, 16, **104**
Programs, 16
Flatness, 449
Fleury, A., **236**
Focalisation, 43, **211**, 251, 262
Focus, **212**
Fock space, **459**, **505**
Følner sequence, **458**
Formalism, **5**, 8, 23, 128, 185, 219
Formula
Π^0_1, Σ^0_1, **17**, 67
Π^0_n, Σ^0_n, **29**
Π^1, 120, 251, 325
Π^1_n, Σ^1_n, **30**
Π^n, Σ^n, **30**
Closed, 41
Gödel, *see* Gödel formula
Negative, 262, 267
Positive, 262
Prenex, **29**
Recessive, **7**, 99
Rosser, **38**
Unknown, 256, 276
Fragment, **13**, 270
Frankenstein, V., 266
Frege, G., **140**, 369, 371, 500
French railways, 138
Freshness, 448
Freyd, P., 227
Friedman, H., **68**, **133**
Full completeness, *see* Faithfulness
Function
Ackermann, **132**
Ambiguous, 39
Analytic, 176, 186, 340
vs. Application, 100
Gustave, **254**, 271, **484**
Naive, 103
Partial, 104
Provably recursive, **20**, **67**
HA, 131
$\mathbf{PA}_2$, 133
PA, 133
Stable, **168**, **185**, **341**
Adjoint, **181**
Linear, **178**
Skeleton Sk, **170**
Strongly, **270**
Functional (recursive), 163
Functor, **147**
Forgetful, 39, **116**, **117**
Natural transformation, **147**
Cartesian, **167**
Fuzzy logic, *see* Paralogic, 373, 506

Galileo, 496
Game, 146, **257**, 504
`Me`, `You`, **257**
Dumb, 302
Dummy move, 263
Giving up, 255
Negation, **261**
Stalling, 255
Strategy, **261**
Copycat, **276**
Winning, 258, **261**
of Truth, 9
Gandy, R. O., 87, **135**, 151, 158, 163
Genesis, 13, 311, 439
Gentzen, G., 9, 34, **41**, 65, 75, 128, **257**, 320, 503

Geometry of interaction, *see* GoI
Gödel–Escher–Bach, *see* *L'entarteur*
Gödel, K., **8**, 12, 21, **30**, 36, **129**, **258**, 447, 501
  Formula, 19, 25, 26, **35**
  Number, 17, 272
  Translation A^g, 8, **71**, 73, 87, 268
    Polarised, **87**
GoI, 11, 14, **145**, 233, 259, 369, 403, 409, **417**, 505
  First, **432**
    Additives, **438**
    Exponentials, 436
    Identity, **433**
    Mistakes, 437
    Multiplicatives, **433**
  Second, *see* Conduct, Beahaviour, Project
Gonthier, G., **437**
Great Wall, **446**, 501
Gregariousness, 332
Ground, 405, 414
Group
  Amenable, 450, **458**, 479
  i.c.c., **455**
  Locally finite, 450, **459**
Guerrini, S., **232**
Guntánamo, 278
Gustave, *see* Berry, G., 484

Hamano, M., **325**
Handicraft vs. Industry, 137
Hauptsatz, xiii, 31, **45**, **51**, 87, 185, 207, 503
  LJ, **73**
  LL, 193, **201**
  Commutations, 52
  Key cases, 51
  Semantic proof, 51
  Structural case, 52
    Cross-cuts, 53, 268
    Polarisation, 53
Hemlock, 203
Herbrand, J., **50**, 243
Herodotus, 371
Heyting, A., **33**, **97**, 258
Hilbert, D., 7, 8, **57**
  Hotel, 310, 435
  Program, **5**, **20**, 22, 27, 439
  Space, 296, 370, **393**
    Dimension, **394**, 453
    Dual, **395**
    ℓ^2, 340, **393**, 432, 454
    Pre-, **394**
    Real, 375
    Separable, **394**, 432
    Sum, **394**
    Tensor product, **394**
Honesty, 12, **23**, 37, 90, 201
Horn clause, **89**
  Head, 90
  Tail, 90
Horror vacui, 252, 254
Howard, W. A., **109**, **158**
HS, **262, 263**, 272, 284
  Daimon, 264
  Empire eA, 264
  Sequents, 264
Huet, G., 28, **114**
Hughes, D., **247**
Huysmans, J.K., xii
Hyland, M., **259**
Hypercoherence, 254, **270**
Hyperfinite factor $\mathcal{R}$, 432, 447
Hypersequentialisation, **264**
Hyperspace, 358
Hypocrisy, 128

Iconoclasts vs. Iconodules, 357, 432
Idempotency, 410
Identity
  Axiom, **44**, **70**, 75
  Native, **389**
  Police-style, **389**

Idiom, 438, **443**, **466**
Imbrication, 370
vs. Par, 391
Imperfect, **13**, **208**
Linear connective, **185**
Logic, 328, 357
Implication
Intuitionistic $A \Rightarrow B$, **170**, 186
Linear $A \multimap B$, **179**, 186
Material, 46, 185, 203, 328
Implicit vs. Explicit, 31, 75, 140, 202, 206, 214, 499
Impunity, 8, 331, 357
Incarnated, *see* Design (material)
Incarnation *see* Mystery, Subtyping
Incarnation $|\mathfrak{G}|$, $|\mathbf{G}|$, 277, **297**, **300, 301**, 313, 326
Intersection, 308
Strategy (induced), 303
Incest, 107, 116, 224
Incidence graph, **404**
Incompleteness, 332, 502
First theorem, 16, **18**
Second theorem, 18, **19**
Induction, 31, 498, **501**
Infinity, 3–6, 11, 14, 35, 46, 69, 104, 109, 208, 331
vs. Complexity, 360
Infinite, 14, 69, 209, 331, 357
Qualitative, **447**
Quantitative, **447**
Inseparability, **39**
Integer
Church, **104**
Cro-Magnon, 119
Dedekind, **30**, 61, **118**, 119, 361
Natural, 359
Non-standard, 359
System **F**, **119**
Integral
Lebesgue, 429
Riemann, 429
Intensional, 298, 500
Intensional vs. Extensional, **298**, 328, 389
Interaction net, **237**
Internal vs. External, 304, 306, 327, **448**
Interpretation
Asymmetric, **175**
Hexagon, **176**
CCC
NJ, **152**
Coherent
F, **171**
LL, **189**, **192**
Functional, **98**
NJ, **100**
Application, **101**
Conditional, **101**
Graph, **101**
Injection, **101**
Pairing, **100**
Projection, **100**
Intersubjectivity, 391, **495, 496**
Intuitionism, **7**, 10
Intuitionistic logic, 11
LJ, **70**
NJ, **75**
Invariant mean, **458**
Inversion, **210**, 232, 262, 267
Isomorphism, 334
COH, **182, 183**, **187**
Canonical, 263, 310
CCC, **154**
de Curry–Howard, 348
Curry–Howard, 10, **107**, 115, 127, 266, 299
Iteration of theories, **10**, 26

James Bond 007, 14
Jarry, A., xii
Jarry, A., 499
Jivaro, *see* AI, 385
Joshua, 332

Jurassic Park, **68**, 360, 439, 448
Jus primæ noctis, 257

Kafka, F., xi
Kanovitch, M., **205**
Keaton, B., 137
Kelly, M., **195**
Kepler, J., xii, 9, 27, 497
Kleene, S. C., 50, 104, **131**, 151, 163
Kolmogorov, A., **97**
König's Lemma, 60, 109
Kreisel, G., xiii, **10**, 13, 20, **64**, 67, 97, 98, **109**, 151, 159, 163, 189, 259, 502
Kripke, S., *see* Model, **91**
Krivine, J. L., **105**, 138
Kronecker, L., 20, 35, 359
Kubrick, S., 3, 26

ℓ^1, ℓ^∞, **340**, **452**
La Palice, J., 213, 491
Labeled deductive system, 90, 299, 369
Lafont, Y., 142, 209, **237**, 335
λ-calculus
  Syntax, **105**
λ-calculus, 157, 316, 348, 432, 448
  β-conversion, **105**
  Confluence, **106**
  Contractum, **105**
  Differential, **341**
  η-conversion, **152**
  η-expansion, **153**
  λ-term
    Underlying, **116**
    $x, \lambda xt, (t)u$, **105**
  Pure, **103**
  Redex, **105**
    Family, **106**
  Translation
    Booleans, **104**
    Exponential, **108**
    Integers, **104**
    Pairs, **104**
    Product, **108**
    Sum, **108**, **117**
  Typed, **107**
    Normalisation (strong), **109**
λμ-calculus, **352**
Lambek, J., 196, **217**
  Calculus, 196, **217**
    Natural deduction, **237**
Lamping, J., **437**
Lateralisation, 469, **482**
Lattice
  Continuous, 151
  Orthomodular, 370, 393
Laurent, O., **344**, **352**
Lavoisier, A. L. de, 266
Layer, **468**
  −1, **42**, **140**, 197, 251, 259, 301, 306, 324, 325, 370, 410, 492, 496
  −2, **142**, **196**, 251, 306, 313, 328, 372, 392, 422
  −3, **145**, 204, 257, 259, 306, 320, 328, 361, 420, 492, 496
Lewis Carroll, 55
Liar, *see* Blair
Light logics, 5, 14, 35, 358, 360, 361, **364**
Limit
  Inductive $\varinjlim$, **154**, 256
    Direct, **165**
  Projective $\varprojlim$, **148**, 256
Lincoln, P., **236**
Linear logic, xii, 8, 11, 14, 58, **178**, 267, 269, 330, 448
  LL, **188**
  Compact, **226**
  Cyclic, 196, **218**
  Intuitionistic **ILL**, **216**
  Polarised, **343**
    LLP, **343**

Teleologic, 218, **237**
Linguistic turn, xi
Link, 257, 365
Axiom, **221**, 264, 267, 344, 348, 349, 403
Crossing, 223, 238, 239
Cut, **221**, 264, 265, 283, 333, 344
$\exists$, 347
⅋, **221**, 264, 344, 403
$\oplus$, 347
$\forall$, 347
$\otimes$, **221**, 344, 403
Splitting, **231**
?, **241**, 333, 344, 349
&, 347
Literal, 256, 325, **326**, 346
LLL, 5, 358, 359, **364**, **485**
Bounds, **365**
Expressiveness, **366**
Lloyola, I. de, 201, 332
Localisation, 257
Local vs. Global, 229
Location, 102
Locative
Product $\lfloor\times\rfloor$, **311**, 405
Projection, 315
vs. Spiritual, 50, 102, 108, **137**, 263, 308, **309**, 311, 315, 320, 322, 324, 352, 376, 445
Loch Ness, 155
Locus ξ, **272**
Incomparable, **272**
Parity, **272**
Logic
+ Control, 90, 369
du Gendarme, 495
Police style, 27, 371
Programming, 49, 57, **89**
PROLOG, **90**, 138, 333
CWA, **91**
Linear, **210**, 211
Resolution, **89**
of Rules, **139**, 144
Lorenzen, P., **258**, 266
School, **258**, 329
Loser, 492
Ludics, 28, 67, 137, 247, 251, 256, 259, 263, 266, **272**, **306**, 338, 343, 344, 349, 389
Exponential, **342**
Łukasiewicz, J., 144, 238

Mac Lane, S., **195**
Madoff, B., 3, 142, 371
Mafia, 309
Majorisability, **158**
Modulus of continuity, **158**
Mallarmé, S., 97, 496
Map, *see* Application
Martin-Löf, P., **113**, **128**, 138, 357
Matrioshka-turtles, 3, 10, 140, 181
Matthew, 259, 342
Maurel, F., **342**, 350
Meander, 259
Mechanism, 26
Median, **395**
Mémé Octavie, 329
Mendelsohn, E., 502
Menù del Cavaliere, **209**
Meowing dog, 128, 495
Meta-, xi, 35, 98, 99, **500**
Circle, 332
Mathematics, **20**
Method, 21
Pasta, *see Zia* Ermenegilda, 329
Semantics, 97
Spectacles, 29
System, xii, 3, 10, 449
Metaphor, 24, 97, 146, **207**, 259, 438
Microcosm, *see* Viewpoint
Milner, R., **447**
Mixture, 381
ML, **135**
Modality, **209**, 360

Linear, **209**
Necessity, 4
$\Box$, **92**, **209**
Possibility, 4
$\Diamond$, **93**, **209**
Modal logic, xi, xiii, 4, 11, 12
S4, 92, **208**
S5, xiii, 92, 185, 215
Model, 141
Herbrand, 57, 91
Kripke, xiii, 4, **91**, 142, 202, 215, 332, 381, 392, 459, **500**
Phase, 142, **197**
Completeness, **199**
Fact, **197**
Soundness, **198**
Tautological, **199**
Schizophrenic, 50
Scott, **172**
Theory, xi
Three-valued, **51**, 128, 175
Topological, **92**, 142
Modus Ponens, 6, 8, 41, 44, 76, 408, 418
Molière, J. B., 492
Monet, C., 15
Money vs. Barter, 499
Monism, 260
Monkey and nut, 332, 449
Monster, *see* Function (Ackermann), 357
Morning star, 140, 371
Morphism vs. Object, 144, 501
Morphology, 12
Morphy, P., 263, 276
Mostowski, A., 214
μ-calculus, **351**
Cotype, **350**, 352
Covariable α, **350**, 352
μ-abstraction $\mu\alpha t$, **350**
Naming $[\alpha]t$, **350**
Multi-level marketing, 3, 26
Multi-valued logic, 51, 492
Multiplicatives $\otimes$, ⅋, $\multimap$, **182**
Tensor product $\mathfrak{A} \otimes \mathfrak{B}$, **316**
Multiplicatives $\otimes$, ⅋, $\multimap$
Adjunction $(\mathfrak{F})\mathfrak{A}$, **316**, 374
CBS, **340**
PCS, **374**
QCS, **387**
Munich school, **67**, 334
Mustard pot, 443
Mystery of incarnation, 137, 301, 304, **311**, 313

Namibia, xi
Natural deduction, **75**
NJ, **75**
NK, **87**
Conclusion, **75**
Contractum, **78**
Cut, **78**
Commutative, **84**
Degree, **80**
Elimination, **76**
Hypothesis, **75**
Discharged, **75**
Main, **81**, 120, 220
Introduction, **76**
Lambek calculus, **237**
Linear, **219**
Normalisation, 34, 51, **78**
Strong, **81**, **87**, 109, **135**
Weak, **81**, **86**
Normal form, **80**
Bounds, **88**
Premise
Main, **77**
Minor, **77**
Proof, **75**
Redex, **78**
Reduction, **80**
$\Rightarrow$, **79**
$\exists$, **84**
$\forall$, **79**
$\vee$, **83**

$\wedge$, **78**
Commutative, **84**
Immediate, **78**
Signature, **81**
Negation
$\neg A$, 6
as Failure, **91**, 139, 494
Double, **72**, **156**
Linear $\sim A$, **180**
Polar, **141**
Procedural, **91**
Nestorianism, 26
Neutrals
Additive $\top$, **0**, 279, 300, 308, 344
Multiplicative **1**, $\bot$, 256, 263, **317**, **319**, 330
Multiplicative **1**, $\boldsymbol{\bot}$, 346
Newspeak, **58**, 97, 298, 499
Nickau, H., **259**
Nietzsche, F., 28, 139
Nilpotency, 409, **419**, 422, 425
Strong, 437, 490, 494
Non-associative logic, **219**
Non-commutative
Geometry, 392, 410
Logic, 45, 196, 407, **410**
Cotensors $\parr$, $\ltimes$, $\rtimes$, **412**
Tensors $\otimes$, $\olessthan$, $\ogreaterthan$, **412**
Non-locality, 220
Non-monotonic logic, **24**, 34, 253, 332, 441, 498
Nonsense, 55, 185
Normal, **452**
Normalisation, 34, **51**, **78**, **80**, **81**, **86–88**, **109**, **121**, 124, 126, **135**, 224, 252, **264**, **265**, **284**, **288**, 334, 346, 352, 419
Normal form $[\![\mathfrak{R}]\!]$, *see* Normalisation, **264**, **284**, **291**, **333**, **346**
$\ll \mathfrak{D} \mid \mathfrak{E} \gg$, 292
Commutation, **286**
Conversion, **285**
$\mathfrak{D}$ai, **285**
Failure, **285**
$\mathfrak{F}$ides, **285**
Normal form $\sigma[\![h]\!]$, **421**, **429**, **431**, 462
Nostradamus, M., 181
Notation, xi, **184**
Polish, 103
Numerisation, 16

Objective vs. Subjective, 391, 493, **494**
Object vs. Subject, 140, 370, **371**
Occurrence, 50, **102**, 108, 223, 309, 324, 326, 348, 449
Ockham, G. d', 329
Okada, M., **200**
ω-Rule, **66**, 320
Ong, L., **259**
Onion (logical), 31, 504
Opacity, 332
Operator
Adjoint, **396**
Bounded, **396**
Closable, 426, 428, **441**
Diagonalisable, **400**
Hermitian, **379**, **398**
Positive, **379**, **398**
Square root $\sqrt{u}$, **398**
Normal, **398**
Partial isometry, 398, **400**, 408, 435
Domain, Image, **400**
Partial symmetry, **420**, 493
Kernel, carrier, **420**
Projection, **398**
Self-adjoint, *see* Hermitian
Spectral radius $\varrho(u)$, **397**, 402
Spectrum $\mathrm{Sp}(u)$, **397**
Symmetry, **398**
Trace-class, 392, 401, **460**
Unbounded, 397, 423, 426, **440**
Unitary, **398**, 406
Opportunism, 294
Oracle, 163

Order variety, 411, **412**
Ordinal, 332, 334
  ϵ_0, 9, 34, **66**, 68
  Γ_0, 68, 159
  Notation, 67
  Panzerdivisionen, *see* Munich school
  Provably recursive, **68**
  vs. Theory, **68**
Orthonormal basis, 384, **394**, 395, 432
Orwell, G., 58, 97, 147, 298, 320, 353, 495

Pandora's box, 35
Pandora box, 209, 342
Para-proof θ, 403
Paraconsistent logic, **23**, 38, 90, 185, 299
Paradise, *see* Saddam
Paradox, *see* Antinomy, 5, 15, **98**, 448
  Cantor, **15**, 157, 159, 321
  EPR, **391**
  Gödel, 321
  Liar, *see* Blair
  Richard, **16**, 33
  Subjective, **491**
Paragraph §, **367**
Parallel or, **166**, 247
Paralogic, 27, 186, 227, 495
Parametricity, 325
Paraproof θ, **261**
Parigot, M., **352**
Parity `Even, Odd`, 273
  Relative `Me, You`, 273
Parsimony, **328**
Partial vs. Total, **252**, 282, 284
Pascal, B., 11
Paternity, *see* Logic (relevant)
Pauli matrices, 380, **382**
PCS, 373
  Carrier, **373**
Peano, G., 5, **31**
PER, **326**
Perennialisation, 446, **478**
Perenniality, 14, 46, 358
  of Perenniality, 209, **331, 332**
Perfect, **13**, **202**
  Causality, **202**, 214
    Action, **203**
    Linearity, **203**
    Reaction, **204**
  vs. Imperfect, 11, 13, **202**, 330, 357, 362, 503
  Implication, **202**
  Linear connective, **182**
  Logic, 252, 264, 324, 332, 357
  Mode and time, 207
  Resource, **204**
    Miracle, 204
Petri net, **205**, 226
Petrol (twopence), 329
π-calculus, **448**
Pitchfork $\Xi \vdash \Lambda$, **273**
  Atomic, **273**
  Comb, **273**
  Handle, **273**
  Handle ξ, 338
  Main, **284**
  Paritary, **273**
  Polarity, **273**
  Tine, **273**
Platonism vs. Formalism, 12, 107
Plausibility, *see* Covenant
Pleonasm, *see* Truth (tarskian)
Plotkin, G., **166**, 389
Poincaré, H., 7, 159, 298, 369
Polar, *see* Duality
Polarisation, 53, 87, **156**, 247, **251**, 364, 468
  Classical, **267**
  Intuitionistic, **269**
  Objections, **256**
  Systeme **F**, **270**
Polarity, 31, 75, 83, 130, 148, 204, **213**
  Exponentials, **215**
  Negative, **210**

Positive, **211**
Polar decomposition, 398
Police of time, 214
Polymorphism, **135**, 434
Pons Asinorum, 186, 207, 216, 222, 260, 403
Popper, K., **28**
Popular democracy, 369
Positive vs. Negative, 214
Potentiality, 4, 17, 327, 389, 392
Prawitz, D., 34, **75**, 87, 116, 146
Precedence $\preceq$, 277, 283, **293**
Predicativity, **159**
Prenex form, **29**, 55, **321**, 325
Principia Mathematica, 21, 107, **127**
Procedurality, xiii, 47, 90, 98, 208, 219, 239, 255, 256, 261, 263, 274, 276, 279, 282, 329, 333, 367, 494
Procrastination, 213
Procustus bed, 381, 390
Programming
Declarative, 90
Logic, *see* Logic Programming, 49, **89**
Projection lemma, **319**
Project $a \cdot + \cdot \alpha + A$, **467**
Alien, **467**
Application, **473**
Carrier, **469**
Disjoint, **469**
Faithful, **472**
Connected, 478
Daimon, **475**
Idiom, **467**
Minimal, **467**
Injection, **472**
Partial, **471**
Plot, **467**
Projection, **472**
Pseudo-trace, **466**
Successful, 490
Variant, **467**, 475
Wager, 464, **467**
Proof, **98**
Automated, 49
Cut-free, **45**, 49, 200
vs. Model, 141, 495
Proof-net $\mathfrak{R}$, 82, **223**, 251, 361, 409
Additive, **245**, 299
Slice, **245**, 443
Box, **240**
Conclusion, **224**
Correctness, **228**, 470
Design, 279, **284**, **288**
Closed, **284**
Paritary, **284**
Partial, **284**
Protoslice, 291
Empire eA, **230**
Border, **230**
Imperialism, **230**, 244
Main gate, **230**
Principal choice, **231**
Simultaneous, **231**
η, **233**
Euler–Poincaré, **233**, 284, 345
Exponential, **241**, **333**
Intensional, 299
λ-calculus, **348**
Light, **364**
Neutrals, **235**
Polarised, 339, **344**, 352
$\lessdot$, **344**
vs. Ludics, **349**
Propagation, **288**
Protoslice, **290**
Quantifier, **242**
Sequentialisation, **227**
Structure, **223**
Normalisation, **224**
Switch, **228**, 403
Jump, **243**
Main, **243**

Switching, **288**
Proof-search, 210
Proof theory, xi
Property
Decidable, **17**
Disjunction, **74**, **129**, **312**, 322, **339**
Existence, **74**, **129**, 324, **339**
Expansive, **17**, 31, 172
Falsifiable, 28
Recessive, **17**, 31
Semi-decidable, **18**
Signature, **50**, 75
Subformula, 21, 31, **49**, 82
Proust, M., 28
Provability, 141, 253
Pseudo-
Design $\mathfrak{Fides}$, **282**
Fax $\mathfrak{Fax}$, **276**, **279**, 288
Psychology, 445
Ptolemy, 494
Pure vs. Typed, 137
Purity of methods, 21
Push vs. Pull, 436

QCS, 154, **369**, 492, 504
Canonical, **380**
Carrier, 380
Domain Fin_C, **377**
Preorder $\preccurlyeq_C$, **378**
Semi-norm $\| \cdot \|_C$, **378**
Quantifiers $\forall, \exists$, 57, **320**, 347, 348
First order, **127**
Multiplicative, 268
Numerical, 320, 324
Second order, **60**, **115**, **485**
Spiritual, **320**, 328
Quantum, **369**, 438
Coherent space, *see* QCS
Computation, 370
Computing, 372
Logic, xi, xii, 11, 369, 393
Physics, 7, 335, 358, 372, **380**
Measure, 203
Question, 497
Question vs. Answer, 328, 372, **409**

Ramification I, **272**
$\mathfrak{Ram}_{(\lambda,I)}$, **295**
Ravenna, 357
Reader's Digest, xi
Realisability θ ® A, **131**, 322
Typed, **131**
Record, 313
Field, 313
Reducibility, **113**, 255
Candidate, **124**, 136, 255, 361
Linear, 194
Parametric, **124**
Simplicity, **112**, **122**
Simply typed, **121**
Formalisation, **133**
vs. Essence, 138
Reduction
$\leadsto$, **105**
$\leadsto_1$, **106**
Left, *see* Loch Ness
Optimal, 437
Referee, 28, 301
Reflexion, 134
Regnier, L., **233**
Rehabilitation, 259
Reification, 359, 392, 439
Relevant logic, 46, 185, **203**
Renoir, J., 445
Rescaling, **475**
Reservoir $\mathbb{X}$, **318**, 319, 322, 324
§**G**, **318**
Resolvant $\mathrm{res}(h, \sigma)$, **429**, 431, **441**
Resource, 447
Rétoré, C., **236**
Retroaction σ
Equation, 146
Retrospection, 507
Reverse mathematics, *see Jurassic Park*

Revision, 333
Revisionism, *see* AI
Richard, J., **16**
Riemann, B., 140, 369
Right-handed-cup, 298
Rigor mortis, 189
Robinson, R., **31**
Roman bus, 282
Rosser, J. B., 104
  Variant, 19, 22, 30, 32, **37**, 142, 299
Ruet, P., 196, **410**
Rule
  Admissible, 141
  Dessin, **274**
    Focus, **274**
    Negative, **274**
    Positive, **274**
    Proper/improper, **275**
  of the Game, 259, **261**, 303
  Main, **284**
  Promotion, **364**, 437
  Structural, **45**, **71**
    Contraction, 35, **45**, **71**, 88, 186, 201, 267, 336, **362**, 410, 437, 446, **478**
    Dereliction, **362**, 437
    Digging, **362**, 437
    Exchange, **45**, **71**
    Mix, **236**, 339
    Relative, **267**
    Reverse contraction, 202
    Weakening, **45**, **71**, 185, 201, 255, 267, 274, 279, 282, 285, 286, **319**, 328, 336, **362**, 437, **478**
Ruminants, 393, 422
Russell, B., 21, 116, **127**

Saddam Hussein, 26, 203
Schellinx, H., **217**, 344
Schema
  Comprehension, **61**, 103, 125, 139
    Naive, **5**, 87
  Induction, **33**, **65**, 99
    Transfinite, 66
  Reflexion, 4, 10, 57, **64**
Schizophrenia, 44
Scholastics, 238
  Ancient, **239**
  New, 238
Schönfinkel, M., 502
Schütte, K., 9, 51, **61**, **66**, 128, 320
Scientism, xi, xii, 3, 27, 439
Scott, D., *see* Domain, 149, **151**, **163**, 502
Scott, P. J., **236**
Sea, 331
Second chance, 60
Second order logic, **60**
Sectarianism, 34, 78, 90, 110, 159, 253, 332
Self-reference, *see L'entarteur*
Selinger, P., **352**, **372**
Semantics, **97**, 260, **499**
  Algebraic, 92, 97, **201**
  Denotational, 97, 145
  Game, 97
  Operational, 97, 145
  Phase, 92, 142, **197**
  of Proofs, **97**
  vs. Syntax, 97, 303, 370
Semi-calculability, 441
Semi-commutation, **213**
Semi-direct product, **456**
Sense, 496
Sense vs. Denotation, **140**, 206, 329, **371**
Sequent
  Classical $\Gamma \vdash \Delta$, **42**
  Intuitionistic $\Gamma \vdash A$, **70**
Sequentialisation, 251, 254
Sequent calculus
  HS, 264
  LJ, 41, 55, 60, **70**
    Cut, **70**
    Decision, **73**

Identity, **70**
Logical rules, **71**
LK, 41, **43**
Cut, **44**
Decision, **49**
Extension cord, 44, 146
Identity, **44**
Plugging, 44, 146
Right formulation, **48**
Signification, **43**
Subformula, **48**
Symmetry, **47**
LL, **188**
Cut, **192**
Dereliction, **189**
Identity, **188**
Mixed, **191**, 267
HS, 343
LC, **338**
LLP, **343**
Second order, **61**
Subformula, **62**
Sesquilinear form, **394**
Set
Π^0_n, Σ^0_n, **29**
Complete, **29**
Hyperarithmetic, **30**, 37
vs. Property, 100, 125
Recursive, **29**
Saturated, **91**, 143, **151**
Semi-recursive, r.e., **29**
Theory, xi, 3, **5**, 332, 335, 358, 371, 393, 495
Naive, 5, 146
ZF, 5, 21, 39, 58, 359
Seventh day, 439
Sex of angels, 448
Sherlock Holmes, 498
Shift $\uparrow, \downarrow, \updownarrow$, 262, **304**, 324
Shinjuku, 23
Short cut, **424**
σ-equivalence, **349**
Signature, **50**, 274
Simon, H., xii, 9, 27
Size, 361, **365**, **424**
Skunk $\mathfrak{Skunk}$, T, 286, **294**, **300**, 308, 310, 326
Positive $\mathfrak{Skunk}^+$, T^+, **294**, **300**, 301, 304
Slice, **289**
$<_{\mathfrak{S}}$, **289**
$\ll_{\mathfrak{S}}$, **289**
Balanced, 289
Maul, 289
sN, *see* Strongly normalisable
Socrates, 203
So far, so good, 17, 31
Solipsism, 7, 12
Solovay, R., **40**
Sophism, xi, 10, 143, 181, 329, 358, 371, 440, 449, 496
Sorbonne, *see* Scolastics, 12
Soundness, 97, 140
Space-time, **382**
Specification, 137
Spectral calculus, **396**, **399**, 420
Spin, 104, 370, **383**, 391, 492
Bool, 136, **385**
Bool$_{\vec{A}}$, 136, **386**
Spirit vs. Matter, 447
Stability, **168**, **185**, **297**, 387, 503
$*$-isomorphism, **402**
Stars vs. Galaxies, 292, 439
State ρ, **401**
Faithful, **451**
Stone–Čech compactification, 451
Stoup $\vdash \Gamma; A$, **337**, 339, 343, 348
Stream, 163
Strongly normalisable, **109**
Sub-
Chronicle, **280**
Dessin, **277**
Formula, *see* Property, **49**, 62
Locus, **272**

Strict, **272**
Typing, 103, **135**, 137, **313**, 326, 343, 363, 376, 406, 413
Coercion, **313**
Inheritance, 313
Subjectivism, 7, 298, 494, 501, 505
Subjectivity, **371**, 493, 495
Subject vs. Object, 505
Substitution, 346
Explicit, **349**
Lemma, **125**
Substraction, *see* Loch Ness
Subtyping, 434
Sudden death, 283
Superego, *see* Typing, 266, 301
Superposition, 245, 348, 372, 443
Supreme Court, 20
Surjective pairing, **153**
Syllogism, 144, **238**, 417
Acronym, **239**
Syntax, 98
Synthetic connective, *see* **HS**, **214**
System
Hilbert-style, **41**, 110, 418
Martin-Löf, **113**
Dependant product, **113**
Dependant sum, **113**
First version, 114, **128**
Semi-formal, 66
System **F**, 109, **115**, 361
Connective
Conjunction, **116**, 121
Disjunction, **116**
Existence, **117**
Free structure, **118**
Binary integers, **119**
Binary trees, **119**
Booleans, **118**
Constructor, **119**
Destructor, **119**
Integers, **119**
Lists, **119**
Reduction, **115**
Term, **115**
Extraction, $\{t\}B$, **115**, 125, 488
Generalisation, ΛXt, **115**
Typability, **136**
Type, **115**
$X, A \Rightarrow B, \forall XA$, **115**
System **T**, 115, **130**, 158
Iterator, **130**
Recursor, **130**, 135
System **F**, 432, 434

Tableaux, 41
Tait, W. W., **61**, 106, **121**
Takeuti, G., **61**, 117
Conjecture, **61**, **126**
Tarski, A., 13, **37**, 139, 213, 491
Teleologisation, 350
Temperament, 358
Temporal logic, xi, 214
Temporary logic, 332
Term
Abstraction $\{x; A\}$, **61**, **128**, 159
Termination, 107
Tertium non datur, 7, 8, **74**, 142
Terui, K., **201**
Test $\theta \in A^t$, 28, 260, 261, 303, 403
vs. Refutation, 260
Theorem
Adjunction, **464**, **473**
Analytical, **292**
Associativity, 274, 284, **295**, 422, **423**, **462**
Bipolar, **373**, **376**, **404**
Blair, 26
Böhm, 292, **298**, 349
Cauchy–Schwarz, **394**
Chinese remainder, **36**, 312
Church–Rosser, **80**, **105**, 116, 225, 264, 292, 296, 334, 346, 352
Closed graph, 441
Closure, **300**

Commutation, **321**
Completeness, 30, **59**
Compositionality of truth, **490**, 496
Deduction, **42**, 54, 76, 110
Dini, 164, 430, 440
Distributivity, **474**
Elimination, *see Hauptsatz*
Faithfulness, **326, 327**
GNS, 397, **402**, 451, 457
Hahn–Banach, 374, 377, **395**
Herbrand, 50, **55**, 57, 74, 391
Incarnation, **301**
Incompleteness, 8, **15**, 30, 285, 498, 501
Krein–Milman, **395**
Kruskal, 68
Linearisation, **471**
Monotonicity, **297**
Murry–von Neumann, **457**
Normalisation
Faulty, **124**, 225
Strong, **135**, 252, **265**, 334, 346, 352
System **F**, **121**, 126
Weak, **80**
Normal form, **422**, **431**, 432
Representability, **121**
Representation, **170**
Scott, **173**
Separation, 274, 277, **293**, 300, 342
Sequentialisation, **228**, **244**, **345**, 405
Soundness, **59**
Stability, **297**
Subjective consistency, **491**
Tarski, 16, 25, **36**, 63
Turing, 30
Thom, R., 26
Thomas Aquinus, **3**, 357
Thomism, 3, 7, 13, 209, 331, 502
Thought, 28
Three-valued logic, **51**
Tickle, 445
Time, 202
Logical, 214
Topology
T_0, **164**, 292
Bonification, 164
Extremely disconnected, **450**
Norm, **439**
Separable, 394
Strong, **439**
Weak, **396**, **439**
Weakened, **395**
Torino school, **135**
Tortoise Principle, **55**, 62, 230, 408, **419**, 427
Totalitarian, 27
Totalitarism, xi, 499
Totality, **252**
Tower of exponentials, 49, 54, 67, **89**, 108, 132, 209, 357, 360–362, 366, 368
Trace, 227, 382, 447, 451, **454**
Central, **465**
Cyclicity, **400**, 465
Norm, 380, **401**
Translation
Gödel A^g, 8, **71**, 73, 87
Polarised, **156**
LJ $\mapsto$ **NJ**, **83**
NJ $\mapsto$ **LJ**, **83**
NJ $\mapsto$ **S4**, **209**
Transparency, 442, **497**
Tree
Infinite branch, **60**
Positive, **345**
Search, **59**
Trip $\uparrow A$, $\downarrow A$, $\updownarrow A$, **406**, 409
Short, **407**, **424**
Switch, **406**
Non-commutative, **413**
Triviality, 329
Troelstra, A. S., 110
Truism, 9, 10, 47, 64, 213

Truth, 128, 141, 258, 299, **325**
  La Palice, 213, 491
  Predicate, **36**
    Bounded, **64**
  Tarskian, 4, 13, **25**, 47, 63
Turing, A., 16, 25
Turnstile $\vdash$, **43**
Twist, **388**, 390, 392, 401
Twist, 417
Type theory, **126**
  Martin-Löf, **113**
  Russell, **127**
Typing, 107, 116
  vs. Essence, 107, **137**

Ultrafinitism, 359
Undecidability, 16, 19, 25, 32, **38**, 498
Underground, *see* Layer
Underlining, 267, 333, 345, 348, 365
Unification, 50, **56**, 89
Unified logic, 270
Uniformity, 342
  External, **326**
  Internal, **327**

Valéry, P., 371
Van Glabbeek, R. D., **247**
Van Tonder, A., **372**
Variable, **56**, 445
  Bound, 105
    α-conversion, **105**, 445, **466**
  Head, 81, **120**, 348, 353
  Hidden, 370
Vauzeilles, J., **332**
Verum **v**, **46**
Vicious circle, **224**, 226, 361, 408
Viewpoint, **488**
vN algebra, *see* von Neumann (algebra)
von Neumann, J., **370**, 381
  Algebra, **451**
    Finite, **453**
    Predual, **452**
    Trace, **454**
    Type $\mathbf{I}_\infty$, 432, **453**
    $\mathbf{I}_n$, **453**
    $\mathbf{II}_\infty$, **453**
    $\mathbf{II}_1$, **453**
    Type $\mathbf{II}_1$, 392, 432
    III, **453**
  Algebra $\mathcal{A}[\mathfrak{G}]$
    Hyperfinite, **456**
  Group algebra $\mathcal{A}[\mathfrak{G}]$, **454**

W^*-algèbre, **451**
Wave
  Function, **380**
    Preselection, 381, 384, **386**, 388–390
    Reduction, **381**, **386**
  vs. Particle, 386
Wave vs. Particle, **494**
Weak
  Reflexivity, 379
Weapons of mass destruction, *see* WMD
Weight ϖ, **265**, 299, 347
Weil, A., 9, 34
Wells, J., **136**
Western, 4
Whitehead, A. N., 21, **127**
Winning, 253, **299**
Witness ¥$\boldsymbol{p}$, **480**
Wittgenstein, L., 181
WMD, 26, 143

Xerxes, xii, 371

Yes man, 294
Yetter, D., 217

Zeno of Elea, 55, 500
Zermelo, E., 5
Zia Ermenegilda, 329